DISORDERS OF THE AUTONOMIC NERVOUS SYSTEM

The Autonomic Nervous System

A series of books discussing all aspects of the autonomic nervous system. Edited by Geoffrey Burnstock, Department of Anatomy and Developmental Biology, University College London, UK

Volume 1
Autonomic Neuroeffector Mechanisms
edited by *G. Burnstock* and *C.H.V. Hoyle*

Volume 2
Development, Regeneration and Plasticity of the Autonomic Nervous System
edited by *I.A. Hendry* and *C.E. Hill*

Volume 3
Nervous Control of the Urogenital System
edited by *C.A. Maggi*

Volume 4
Comparative Physiology and Evolution of the Autonomic Nervous System
edited by *S. Nilsson* and *S. Holmgren*

Volume 5
Disorders of the Autonomic Nervous System
edited by *D. Robertson* and *I. Biaggioni*

Additional volumes in preparation

Autonomic Ganglia
E.M. McLachlan

Central Nervous Control of
Autonomic Function
D. Jordan

Autonomic Control of the Respiratory
System
P. Barnes

Autonomic Innervation of the Skin
I.L. Gibbins and *J.L. Morris*

Nervous Control of Blood Vessels
T. Bennett and *S. Gardiner*

Nervous Control of the Eye
G. Burnstock and *A.M. Sillito*

Autonomic–Endocrine Interactions
K. Unsicker

Nervous Control of the
Gastrointestinal Tract
M. Costa

Nervous Control of the Heart
J.T. Shepherd and *S.F. Vatner*

This book is part of a series. The publisher will accept continuation orders which may be cancelled at any time and which provide for automatic billing and shipping of each title in the series upon publication. Please write for details.

DISORDERS OF THE AUTONOMIC NERVOUS SYSTEM

Edited by

D. Robertson and I. Biaggioni

Clinical Research Center, Vanderbilt University
Nashville, Tennessee, USA

CRC Press
Taylor & Francis Group
Boca Raton London New York

CRC Press is an imprint of the
Taylor & Francis Group, an **informa** business

First published 1995 by Harwood Academic Publishers

Published 2019 by CRC Press
Taylor & Francis Group
6000 Broken Sound Parkway NW, Suite 300
Boca Raton, FL 33487-2742

© 1995 by Taylor & Francis Group, LLC
CRC Press is an imprint of Taylor & Francis Group, an Informa business

First issued in paperback 2019

No claim to original U.S. Government works

ISBN 13: 978-0-367-45598-9 (pbk)
ISBN 13: 978-3-7186-5146-7 (hbk)

Visit the Taylor & Francis Web site at
http://www.taylorandfrancis.com

and the CRC Press Web site at
http://www.crcpress.com

British Library Cataloguing in Publication Data

Robertson, David
    Disorders of the Autonomic Nervous
    System. – (Autonomic Nervous System
    Series, ISSN 1047-5125; Vol. 5)
    I. Title II. Biaggioni, Italo III. Series
    616.88

Contents

Preface to the Series — Historical and Conceptual Perspective of The Autonomic Nervous System Book Series

The pioneering studies of Gaskell (1886), Bayliss and Starling (1899), and Langley and Anderson (see Langley, 1921) formed the basis of the earlier and, to a large extent, current concepts of the structure and function of the autonomic nervous system; the major division of the autonomic nervous system into sympathetic, parasympathetic and enteric subdivisions still holds. The pharmacology of autonomic neuroeffector transmission was dominated by the brilliant studies of Elliott (1905), Loewi (1921), von Euler and Gaddum (1931), and Dale (1935), and for over 50 years the idea of antagonistic parasympathetic cholinergic and sympathetic adrenergic control of most organs in visceral and cardiovascular systems formed the working basis of all studies. However, major advances have been made since the early 1960s that make it necessary to revise our thinking about the mechanisms of autonomic transmission, and that have significant implications for our understanding of diseases involving the autonomic nervous system and their treatment. These advances include:

(1) Recognition that the autonomic neuromuscular junction is not a 'synapse' in the usual sense of the term where there is a fixed junction with both pre- and postjunctional specialization, but rather that transmitter is released from mobile varicosities in extensive terminal branching fibres at variable distances from effector cells or bundles of smooth muscle cells which are in electrical contact with each other and which have a diffuse distribution of receptors (see Hillarp, 1959; Burnstock, 1986a).

(2) The discovery of non-adrenergic, non-cholinergic nerves and the later recognition of a multiplicity of neurotransmitter substances in autonomic nerves, including monoamines, purines, amino acids, a variety of different peptides and nitric oxide (Burnstock *et al.*, 1964; Burnstock, 1986b; 1993b; Burnstock and Milner, 1992; Rand, 1992; Snyder, 1992).

(3) The concept of neuromodulation, where locally released agents can alter neurotransmission either by prejunctional modulation of the amount of transmitter released or by postjunctional modulation of the time course or intensity

of action of the transmitter (Marrazzi, 1939; Brown and Gillespie, 1957; Vizi, 1979; Kaczmarek and Leviton, 1987).

(4) The concept of cotransmission that proposes that most, if not all, nerves release more than one transmitter (Burnstock, 1976; Hökfelt, Fuxe and Pernow, 1986; Burnstock, 1990a) and the important follow-up of this concept, termed 'chemical coding', in which the combinations of neurotransmitters contained in individual neurones are established, and whose projections and central connections are identified (Furness and Costa, 1987).

(5) Recognition of the importance of 'sensory-motor' nerve regulation of activity in many organs, including gut, lungs, heart and ganglia, as well as in many blood vessels (Maggi and Meli, 1988; Burnstock, 1990a; 1993a), although the concept of antidromic impulses in sensory nerve collaterals forming part of 'axon reflex' vasodilatation of skin vessels was described many years ago (Lewis, 1927).

(6) Recognition that many intrinsic ganglia (e.g., those in the heart, airways and bladder) contain integrative circuits that are capable of sustaining and modulating sophisticated local activities (Burnstock et al., 1987). Although the ability of the enteric nervous system to sustain local reflex activity independent of the central nervous system has been recognized for many years (Kosterlitz, 1968), it has been generally assumed that the intrinsic ganglia in peripheral organs consisted of parasympathetic neurones that provided simple nicotinic relay stations.

(7) The major subclasses of receptors to acetylcholine and noradrenaline have been recognized for many years (Dale, 1914; Ahlquist, 1948), but in recent years it has become evident that there is an astonishing variety of receptor subtypes for autonomic transmitters (see Br. J. Pharmacol. [1991], 102, 560–561). Their molecular properties and transduction mechanisms are being characterized. These advances offer the possibility of more selective drug therapy.

(8) Recognition of the plasticity of the autonomic nervous system, not only in the changes that occur during development and aging, but also in the changes in expression of transmitter and receptors that occur in fully mature adults under the influence of hormones and growth factors following trauma and surgery, and in a variety of disease situations (Burnstock, 1990b).

(9) Advances in the understanding of 'vasomotor' centres in the central nervous system. For example, the traditional concept of control being exerted by discrete centres such as the vasomotor centre (Bayliss, 1923) has been supplanted by the belief that control involves the action of longitudinally arranged parallel pathways involving the forebrain, brain stem and spinal cord (Loewy and Spyer, 1990).

In addition to these major new concepts concerning autonomic function, the discovery by Furchgott that substances released from endothelial cells play an important role, in addition to autonomic nerves, in local control of blood flow, has made a significant impact on our analysis and understanding of cardiovascular function (Furchgott and Zawadski, 1980; Ralevic and Burnstock, 1993). The later identification of nitric oxide as the major endothelium-derived relaxing factor (Palmer *et al.*, 1988) (confirming the independent suggestion by Ignarro and by Furchgott) and endothelin as an endothelium-derived constricting factor (Yanagisawa *et al.*, 1988) have also had a major impact in this area.

In broad terms, these new concepts shift the earlier emphasis on central control mechanisms towards greater consideration of the sophisticated local peripheral control mechanisms.

Although these new concepts should have a profound influence on our considerations of the autonomic control of cardiovascular, urogenital, gastrointestinal and reproductive systems and other organs like the skin and eye in both normal and disease situations, few of the current textbooks take them into account. This is largely because revision of our understanding of all these specialist areas in one volume by one author is a near impossibility. Thus, this book series of 14 volumes is designed to overcome this dilemma by dealing in depth with each major area in separate volumes and by calling upon the knowledge and expertise of leading figures in the field. Volume 1, published in early 1992, dealt with the basic mechanisms of *Autonomic Neuroeffector Mechanisms* which set the stage for later volumes devoted to autonomic nervous control of particular organ systems, including *Heart; Blood Vessels; Respiratory System; Eye; Skin; Gastrointestinal Tract* and *Urogenital System* (volume 3, published in early 1993). Another group of volumes will deal with *Central Nervous Control of Autonomic Function; Autonomic Ganglia; Autonomic–Endocrine Interactions; Development, Regeneration and Plasticity of the Autonomic Nervous System* (volume 2, published in 1992) and *Comparative Physiology and Evolution of the Autonomic Nervous System* (volume 4, published in 1994).

Abnormal as well as normal mechanisms will be covered to a variable extent in all these volumes, depending on the topic and the particular wishes of the Volume Editors, but this volume is specifically devoted to *Disorders of the Autonomic Nervous System*.

A general philosophy followed in the design of this book series has been to encourage individual expression by Volume Editors and Chapter Contributors in the presentation of the separate topics within the general framework of the series. This was demanded by the different ways that the various fields have developed historically and the differing styles of the individuals who have made the most impact in each area. Hopefully, this deliberate lack of uniformity will add to, rather than detract from, the appeal of these books.

G. Burnstock
Series Editor

REFERENCES

Ahlquist, R.P. (1948) A study of the adrenotropic receptors. *Am. J. Physiol.*, **153**, 586–600.

Bayliss, W.B. (1923) *The Vasomotor System*. London: Longman.

Bayliss, W.M. and Starling, E.H. (1899) The movements and innervation of the small intestine. *J. Physiol. (Lond.)*, **24**, 99–143.

Brown, G.L. and Gillespie, J.S. (1957) The output of sympathetic transmitter from the spleen of a cat. *J. Physiol. (Lond.)*, **138**, 81–102.

Burnstock, G. (1976) Do some nerve cells release more than one transmitter? *Neuroscience*, **1**, 239–248.

Burnstock, G. (1986a) Autonomic neuromuscular junctions: Current developments and future directions. *J. Anat.*, **146**, 1–30.

Burnstock, G. (1986b) The non-adrenergic non-cholinergic nervous system. *Arch. Int. Pharmacodyn. Ther.*, **280** (suppl.), 1–15.

Burnstock, G. (1990a) Co-Transmission. The Fifth Heymans Lecture - Ghent, February 17, 1990. *Arch. Int. Pharmacodyn. Ther.*, **304**, 7–33.

Burnstock, G. (1990b) Changes in expression of autonomic nerves in aging and disease. *J. Auton. Nerv. Syst.*, **30**, 525–534.

Burnstock, G. (1993a) Introduction: Changing face of autonomic and sensory nerves in the circulation. In: *Vascular Innervation and Receptor Mechanisms: New Perspectives*. Edited by L. Edvinsson and R. Uddman, pp. 1–22. Academic Press Inc., San Diego, USA.

Burnstock, G. (1993b) Physiological and pathological roles of purines: an update. *Drug Dev. Res.*, **28**, 195–206.

Burnstock, G., Campbell, G., Bennett, M. and Holman, M.E. (1964) Innervation of the guinea-pig taenia coli: Are there intrinsic inhibitory nerves which are distinct from sympathetic nerves? *Int. J. Neuropharmacol.*, **3**, 163–166.

Burnstock, G., Allen, T.G.J., Hassall, C.J.S. and Pittam, B.S. (1987) Properties of intramural neurones cultured from the heart and bladder. In *Histochemistry and Cell Biology of Autonomic Neurons and Paraganglia. Exp. Brain Res. Ser. 16*, edited by C. Heym, pp. 323–328. Heidelberg: Springer Verlag.

Burnstock, G. and Milner, P. (1992) Structural and chemical organisation of the autonomic nervous system with special reference to nonadrenergic, noncholinergic transmission. In *Autonomic Failure*, 3rd edn, edited by R. Bannister, pp. 107–125. Oxford: Oxford University Press.

Dale, H. (1914) The action of certain esters and ethers of choline and their reaction to muscarine. *J. Pharmacol. Exp. Ther.*, **6**, 147–190.

Dale, H. (1935) Pharmacology and nerve endings. *Proc. Roy. Soc. Med.*, **28**, 319–332.

Elliott, T.R. (1905) The action of adrenalin. *J. Physiol. (Lond.)*, **32**, 401–467.

Furchgott, R.F. and Zawadski, J.V. (1980) The obligatory role of endothelial cells in the relaxation of arterial smooth muscle by acetylcholine. *Nature*, **288**, 373–376.

Furness, J.B. and Costa, M. (1987) *The Enteric Nervous System*. Edinburgh: Churchill Livingstone.

Gaskell, W.H. (1886) On the structure, distribution and function of the nerves which innervate the visceral and vascular systems. *J. Physiol. (Lond.)*, **7**, 1–80.

Hillarp, N.-Å. (1959) The construction and functional organisation of the autonomic innervation apparatus. *Acta Physiol. Scand.*, **46** (suppl. 157), 1–38.

Hökfelt, T., Fuxe, K. and Pernow, B. (Eds.) (1986) Coexistence of neuronal messengers: A new principle in chemical transmission. In *Progress in Brain Research*, vol. 68, Amsterdam; Elsevier.

Kaczmarek, L.K. and Leviton, I.B. (1987) *Neuromodulation. The Biochemical Control of Neuronal Excitability*, pp. 1–286. Oxford: Oxford University Press.

Kosterlitz, H.W. (1968) The alimentary canal. In *Handbook of Physiology*, vol. IV, edited by C.F. Code, pp. 2147–2172. Washington, DC: American Physiological Society.

Langley, J.N. (1921) *The Autonomic Nervous System*, part 1. Cambridge: W. Heffer.

Lewis, J. (1927) *The Blood Vessels of the Human Skin and Their Responses*. London: Shaw & Sons.

Loewi, O. (1921) Über humorale Übertrangbarkeit der Herznervenwirkung. XI. Mitteilung. *Pflugers Arch. Gesamte Physiol.*, **189**, 239–242.

Loewy, A.D. and Spyer, K.M. (1990) *Central Regulations of Autonomic Functions*. New York: Oxford University Press.

Maggi, C.A. and Meli, A. (1988) The sensory-efferent function of capsaicin-sensitive sensory nerves. *Gen. Pharmacol.*, **19**, 1–43.

Marrazzi, A.S. (1939) Electrical studies on the pharmacology of autonomic synapses. II. The action of a sympathomimetic drug (epinephrine) on sympathetic ganglia. *J. Pharmacol. Exp. Ther.*, **65**, 395–404.

Palmer, R.M.J., Rees, D.D., Ashton, D.S. and Moncada, S. (1988) Arginine is the physiological precursor for the formation of nitric oxide in endothelium-dependent relaxation. *Biochem. Biophys. Res. Commun.*, **153**, 1251–1256.

Ralevic, V. and Burnstock, G. (1993) *Neural-Endothelial Interactions in the Control of Local Vascular Tone*, pp.1–117. Medical Intelligence Unit, R.G. Landes Company, Medical Publishers, Austin, Texas.

Rand, M.I. (1992) Nitrergic transmission: nitric oxide as a mediator of non-adrenergic, non-cholinergic neuro-effector transmission. *Clin. Exp. Pharm. Physiol.*, **19**, 147–169.

Snyder, S.H. (1992) Nitric oxide: first in a new class of neurotransmitter? *Science*, **257**, 494–496.

Vizi, E.S. (1979) Prejunctional modulation of neurochemical transmission. *Prog. Neurobiol.*, **12**, 181–290.

von Euler, U.S. and Gaddum, J.H. (1931) An unidentified depressor substance in certain tissue extracts. *J. Physiol.*, **72**, 74–87.

Yanagisawa, M., Kurihara, H., Kimura, S., Tomobe, Y., Kobayashi, M., Mitsui, Y., Yazaki, Y., Goto, K. and Masaki, T. (1988) A novel potent vasoconstrictor peptide produced by vascular endothelial cells. *Nature*, **332**, 411–415.

David Robertson

Italo Biaggioni

Contributors

Appenzeller, Otto
New Mexico Health Enhancement
  and Marathon Clinics
Research Foundation
1559 Eagle Ridge
NE Albuquerque
New Mexico 87122, USA

Axelrod, Felicia B.
New York University
School of Medicine
530 First Avenue
New York, NY 10016, USA

Benowitz, Neal
San Francisco General Hospital
  Medical Center
University of California
San Francisco, CA 94110, USA

Biaggioni, Italo
Autonomic Dysfunction Center
Vanderbilt University
Nashville, TN 37232-2195, USA

Converse Jr., R.L.
Department of Internal Medicine
University of Texas
  Southwestern Medical Center
Dallas, Texas 75235-9034, USA

Convertino, Victor A.
Armstrong Laboratory
Brooks Air Force Base
Texas 78235-5117, USA

Frankel, Hans L.
National Spine Injuries Centre
Stoke Mandeville Hospital
Bucks, UK

Freeman, Roy
Division of Neurology
New England Deaconess Hospital
Suite 4A, 110 Francis Street
Boston, MA 02110, USA

Gifford Jr., R.W.
Institute of Rehabilitation & Medicine
400 E 34th Street
New York, NY 10016, USA

Jacobsen, Tage N.
Suite 278A, St Paul's Hospital
1081 Burrard Street
Vancouver, BC V6Z 1Y6
Canada

Jost, C.M.T.
Department of Internal Medicine
University of Texas
  Southwestern Medical Center
Dallas, Texas 75235-9034, USA

Lipsitz, Lewis A.
Hebrew Rehabilitation
  Center for Aged
1200 Center Street
Roslindale, MA 02131, USA

Manger, William M.
National Hypertension Association,
  Inc. (High Blood Pressure)
324 East 30th Street, New York
NY 10016, USA

Mathias, Christopher J.
Department of Medicine
St Mary's Hospital
Praed Street
London W2 1NY, UK

Mosqueda-Garcia, Rogelio
Autonomic Dysfunction Unit
Vanderbilt University
Nashville, TN 37232-2195, USA

Onrot, Jack
Suite 278A
St Paul's Hospital
1081 Burrard Street
Vancouver, BC V6Z 1Y6
Canada

Polinsky, Ronald J.
Sandoz Pharmaceuticals Corporation
East Hanover
NJ 07936, USA

Robertson, David
Clinical Research Center
Vanderbilt University
Nashville, TN 37232-2195, USA

Robertson, Rose Marie
Department of Medicine
Division of Cardiology
Vanderbilt University
The Medical Center
Nashville, TN 37232-2195, USA

Ryan, Sheila M.
Dallas, TX 75235-9034, USA
Department of Anesthesiology
Beth Israel Hospital
Boston, MA, USA

Schatz, Irwin J.
Department of Medicine
John A. Burns School of Medicine
University of Hawaii
Honolulu, Hawaii 96813, USA

Stainback, Raymond
Adult Echocardiography Laboratory
505 Parnassus Avenue
Moffitt Hospital Room M-342A
San Francisco, CA 94143-0214
USA

Tseng, Ching-Jiunn
Department of Medical Education
   and Research
Kaohsiung Veterans General Hospital
Kaohsiung, Taiwan

Tung, Che-Se
Departments of Pharmacology
   and Physiology
National Defense Medical Center
Taipei, Taiwan

Victor, Ronald G.
Department of Internal Medicine
University of Texas
Southwestern Medical Center

Geoffrey Burnstock – Editor of The Autonomic Nervous System Book Series

1 Introduction

David Robertson

Departments of Medicine, Pharmacology and Neurology,
Clinical Research Center, Vanderbilt University,
*Nashville, Tennessee 37232-2195, USA**

INTRODUCTION

In spite of the interest of 19th century physiologists in the autonomic nervous system, it was not until 1925 that a real understanding of the clinical presentation of autonomic failure emerged. This work came from careful clinical investigations carried out by Samuel Bradbury and Cary Eggleston (Bradbury and Eggleston, 1925). While there had been previous reports of people with hypotension or syncope (Beales, 1856; Hill and Barnard, 1897), it had not been clearly related to the autonomic nervous system prior to the pivotal studies of 1925. This report appeared in the initial volume of the *American Heart Journal* and two years later, it was supplemented by autopsy data in one subject, confirming that adrenocortical atrophy was not the explanation for the blood pressure abnormality. That autopsy excluded the nervous system from study at the request of the family, thus preventing the investigators from documenting the accuracy of their hypothesis that the pathologic lesion lay at the level of the autonomic nervous system.

Nevertheless, from our current vantage point, it is remarkable how thoroughly these investigators were able to define their patients' illness using a battery of practical bedside and laboratory investigations. They described the orthostatic hypotension with an unchanged pulse, reduced sweating, reduced basal metabolic rate, nocturia, impotence, constipation, supine hypertension, and anemia. They detected parasympathetic failure (since atropine was ineffectual), in the face of end-organ

*Supported in part by NIH grants HL44589 and RR00095 and NASA grants NAG 9-563 and NCC 2-696. Photograph of Dr. Drager courtesy of Dr. Thomas James.

1

sensitivity to pilocarpine. They also noted sympathetic denervation hypersensitivity. They achieved success with pressure garments on the legs and abdomen in improving symptoms. Although these investigators ultimately moved into other areas of study and practice (geographic medicine and clinical pharmacology), their 1925 contribution remains the single most important intellectual achievement in our understanding of clinical autonomic failure.

BRADBURY-EGGLESTON SYNDROME

This disorder is best defined by the characteristics of the first patient in a series of three individuals reported by Bradbury and Eggleston (1925). It appears to be due to peripheral autonomic impairment, and its features are discussed in Chapter 5. It appears that there are about 25,000 Americans who have this disorder. Many names have been applied to this. Bradbury and Eggleston called the disorder postural hypotension. Later investigators have used such terms as idiopathic orthostatic hypotension, progressive autonomic failure, pure autonomic failure, peripheral autonomic failure, and peripheral dysautonomia. Given our current level of understanding of this syndrome or group of disorders, it is probably unwise to be too rigid in regard to nomenclature. It is likely that in subsequent years great strides will be made in understanding the underlying pathophysiology. From this improved understanding of pathophysiology should emerge a more appropriate nomenclature.

After 1925 a number of reports began to appear in the clinical literature describing patients similar to those of Bradbury and Eggleston. Sometimes abnormalities in other neurological systems were encountered along with the autonomic failure. In 1946, Natalie Maria Briggs (Briggs, 1946) systematically reviewed the literature and described all cases of postural hypotension seen at the Mayo Clinic in Rochester, Minnesota from 1927 to 1945. Her careful analysis made clear that many neurological abnormalities could accompany autonomic failure and that a large portion of patients with autonomic failure had evidence of some other neurological problem outside the autonomic nervous system. This work appeared as a thesis, a copy of which is in the library of the University of Minnesota. Unfortunately, because the work did not appear in publication, it did not get the attention that it by all rights should have had.

SHY-DRAGER SYNDROME

Shy and Drager (1960) described the pathology of two patients with autonomic failure accompanied by significant central involvement. Their description built upon the observations of Dr. Briggs and has led to improved understanding of the distinction between autonomic failure *sui generis*, and autonomic failure with evidence of other neurological involvement. This disorder, or group of disorders has been referred to as the Shy-Drager syndrome. About a decade after the observations of Shy and Drager, Oppenheimer used the term multiple system atrophy to describe these patients. A limitation of this term is its nonspecificity; but it remains widely used in the United Kingdom.

CONTEMPORARY AUTONOMIC RESEARCH

Since the pivotal studies of these five investigators, an increasing number of physicians and scientists worldwide are beginning to address the problems of pathophysiology and therapy that beset patients with autonomic disorders. Our understanding is growing, but unfortunately, it is still at a very primitive level, in part because of the limited support available for the conduct of investigations in these patients, but also because contemporary methodology has not readily been able to define etiology. With improved tools, including especially molecular biology, we may expect rapid improvements during the next decade in our understanding of these disorders. The purpose of this book is to describe what is known about these disorders in a disease-oriented context. It is hoped that this presentation of autonomic disorders will be interesting to young investigators and will encourage them to undertake work in this important and exciting field.

BIOGRAPHIES OF KEY EARLY INVESTIGATORS IN CLINICAL AUTONOMIC RESEARCH

Samuel Bradbury, M.D.

Samuel Bradbury, M.D., (Figure 1.1) was born in the Germantown section of Philadelphia, Pennsylvania on 30 April 1883, the son of Samuel and Martha Washington (Chapman). He obtained his medical degree at the University of Pennsylvania in 1905. On 26 September 1914, he married Althea Norris Johnson and they had four children: 1) Samuel, 2) Emily Carey (Mrs. Alfred Vail), 3) Althea Norris (Mrs. David Lowshak), and 4) Wilmer Johnson. He worked in Philadelphia hospitals until 1911. From 1912 through 1917, he was assistant visiting physician at the City Hospital in New York. He was visiting physician there from 1917 through 1924. Between 1924 and 1927, he was assistant visiting physician at Bellevue Hospital, then from 1924 through 1927, Chief of Medicine at the Cornell Clinic, and Assistant Professor of Clinical Medicine at Cornell University Medical College during the period 1921 through 1927. He served as consulting physician at New York Infirmary for Women and Children in the years 1924 through 1927. In 1927, he took the post of Director of the Outpatient Department of the Philadelphia Hospital in Philadelphia, where he was Director of Service and Coordinator of the Department of Medicine in the Germantown Hospital.

Dr. Bradbury was a Diplomate of the American Board of Internal Medicine, a member of the American Medical Association, the Medical Society of the State of Pennsylvania, the Philadelphia County Medical Society, the College of

FIGURE 1.1   Samuel Bradbury, M.D. (1883–1947)

Physicians of Philadelphia, the American Clinical and Climatological Association, the Academy of Medicine of New York, the American Association for the Advancement of Science, and the Harvey Society. He was author of the monograph *Internal Medicine – Treatment* (1923) and also *What Constitutes Adequate Medical Service?* (published in 1927 by the Committee of the United Hospital Fund of New York). He was also a contributor to the *Encyclopedia of Medicine* and the *Textbook of Medicine* by American authors on the subject of adequate medical care in 1937. Dr. Bradbury died 30 August 1947.

Cary Eggleston, M.D.

Cary Eggleston (Figure 1.2) was born in Brooklyn, New York, 18 August 1884. He obtained his medical degree at Cornell in 1907. From 1908 through 1910, he served as an intern at the New York Hospital and also in 1910 served briefly at the Sloan Hospital for Women. He became an Instructor in Pharmacology and Medicine at Cornell Medical College in 1911, was promoted to Assistant Professor in 1918, and Associate Professor in 1939. He served in this capacity until 1953 when he assumed the role of consultant physician until 1954 at which time he retired.

His activities included being visiting physician in the Outpatient Department of the New York Hospital from 1910 through 1912, attending physician in the Department of Contagious Diseases in 1911, associate attending physician from

FIGURE 1.2   Cary Eggleston, M.D.

1932 until 1950, and work in the tuberculosis clinic of Bellevue Hospital from 1910 until 1912, assistant and associate visiting professor at those institutions from 1921 until 1934, and a visiting physician from 1934 until 1954. He was Chief of the Clinic in the Children's Department of the de Milt Dispensary from 1911 until 1912, consulting cardiologist at the Willard Parker Hospital 1932 through 1954 and consulting physician in the New York Infirmary for Women and Children 1935 through 1954. He served as consulting cardiologist at the Ruptured and Crippled Hospital 1940 through 1954.

He was associate editor of the *New York Medical Journal* 1912 through to 1918. He edited Progress and Therapeutics in the *American Journal of Medical Sciences* and became vice president of the United States Pharmacopeial Convention in 1940, before subsequently serving 9 years, 1941 through 1950, as President of this organization. He was a fellow of the American Medical Association, the Society for Clinical Investigation, a fellow of the American College of Physicians, a fellow of the American Heart Association, a fellow of the American Academy of Compensation Medicine, and a fellow of the New York Academy of Medicine. His specialities were clinical pharmacology and therapeutics, diseases of the heart and circulation, and digitalis use. During much of his career Dr. Eggleston resided at 215 E. 72nd Street, New York 21, New York.

FIGURE 1.3   Natalie Maria Briggs. M.D. (1915–1988)

Natalie Maria Briggs, M.D.

In October 1946, Natalie Maria Briggs, M.D., (Figure 1.3) published her analysis of postural hypotension cases in the literature and reviewed all cases seen at the Mayo Clinic between 1927 and 1945. She clearly recognized the important neurological concomitant of autonomic failure and was awarded a Master of Science degree for this work by the University of Minnesota. Dr. Briggs died on 27 April 1988 in Washington.

George Milton Shy, M.D.

George Milton Shy (Figure 1.4) was born in Trinidad, Colorado on 30 September 1919, the son of James C. and Zella May (Henderson) Shy. He was educated at Oregon State University where he received his B.S. degree in 1940, and at the University of Oregon, where he received his medical degree in 1943. He interned at Coffee Memorial Hospital, Portland, Oregon (1942–43), and served as assistant resident in Medicine at the Royal Victoria Hospital in Montreal (1943–44). After serving in the Army Medical Corps from 1944 to 1946, he spent two years in Neurology at Queen's Square Hospital in London, returning to Montreal to finish his training in neurology at the Montreal Neurology Institute. He received his master's degree from McGill University in 1949, and accepted the position of Assistant Professor of Neurology at the University of Colorado in 1951. He left in 1953 to

FIGURE 1.4   George Milton Shy, M.D. (1919–1967)

become Clinical Director of Neurology at the National Institutes of Health and Associate Professor of Neurology at Georgetown University. He became Scientific Director of the Intramural Program in 1960, but two years later he became Chairman of the Department of Neurology at the University of Pennsylvania. In 1967, he left to assume duties as Chairman of the Department of Neurology and Director of the New York Neurology Institute at Columbia University. Unfortunately, he died of myocardial infarction soon after assuming the duties.

Glenn Albert Drager, M.D., Ph.D.

Glenn Albert Drager (Figure 1.5) was born in Irving, Kansas on 15 September 1917. He received his bachelor's degree from the University of Wyoming in 1940 and his Ph.D. in Anatomy from the University of Minnesota in 1943. While in Minnesota, he became acquainted with Dr. Natalie Maria Briggs, who ultimately came to obtain her master's degree at the University of Minnesota. Dr. Drager later became Instructor in Neuroanatomy in Galveston, progressing through the ranks to Associate Professor in 1946. He obtained his medical degree in 1954. He then went to serve as a clinical associate at the National Institutes of Health from 1955 through 1959, where he worked with Dr. Milton Shy. In 1959, he became Associate Professor of Neurology at Baylor University (1959–1964), moving back to Galveston as Associate Professor of Neurology and Anatomy in 1964. His major

FIGURE 1.5   Glenn Albert Drager, M.D. Ph.D.
(1917-1968)

interests were the relationship of the hypophysis and the hypothalamus to endocrine and autonomic function. He died near Galveston, Texas of an accidental gunshot wound in 1968.

REFERENCES

Beales, R. (1856) On syncope senilis. *Lancet.*, **2**, 102.

Bradbury, S. and Eggleston, C. (1925) Postural hypotension: A report of three cases. *Am. Heart J.*, **1**, 73–86.

Briggs, N.M. (1946) Neurological Manifestations Associated with Postural Hypotension. *Thesis* B-MnU-M-46-70, University of Minnesota.

Hill, L. and Barnard, H. (1897) The influence of the force of gravity on the circulation. *J. Physiol.* (Lond.), **21**, 323–352.

Shy, G.M. and Drager, G.A. (1960) A neurological syndrome associated with orthostatic hypotension. *Arch. Neurol.*, **2**, 511–527.

2 Brainstem and Cardiovascular Regulation

Ching-Jiunn Tseng[1,3] and Che-Se Tung[2]

Departments of Pharmacology[1] and Physiology[2],
National Defense Medical Center, Taipei, Taiwan, R.O.C.

Department of Medical Education and Research[3],
Kaohsiung Veterans General Hospital, Kaohsiung, Taiwan, R.O.C.

The neural control of the cardiovascular system is very complex. By using traditional and newly developed methods for tracing neural pathways, important sites in the control of cardiovascular activity have been located in the pontomedullary and hypothalamic regions. A brief introduction to the afferent and efferent neural pathway and the neuroanatomy and functional role of some of the nuclei (for example, nucleus tractus solitarii, area postrema, dorsal and ventrolateral medulla, parabrachial nucleus and locus coeruleus) known to be important to central cardiovascular regulation are reported in this chapter.

KEY WORDS: Neural pathways, area postrema, hypothalamus, medulla, locus coeruleus

INTRODUCTION

The cardiovascular system is regulated by different mechanisms including local, humoral and neural factors. Since arterial pressure is the product of cardiac output and total peripheral resistance, the mechanisms responsible for the control of arterial pressure operate by affecting either the cardiac output or total peripheral resistance (Guyton *et al.*, 1974).

Central control of the cardiovascular system is complex and still not well understood. The central nervous system (CNS) receives sensory information from many sources and sends out appropriate regulatory signals to several effector organs. It has been suggested that disturbances of these regulatory mechanisms may be involved in chronic elevation in systemic arterial pressure (Reis, 1981). For instance, abnormalities in central autonomic control may occur early in the course of borderline hypertension, and then over a period of years, through a sequence of secondary changes, lead to essential hypertension (Reis, 1981). Whether central

9

autonomic control plays a role in borderline or essential hypertension is still unknown. Indeed, the neuronal circuitry in the brain is so complex that we still do not know the precise mechanisms of the depressor effect of some of the most commonly employed antihypertensive drugs, such as α-methyldopa and clonidine. But the more we understand about central cardiovascular regulatory mechanisms, the more we can assess its role in cardiovascular disease, and the more rationally we can approach new drug development. The aim of this chapter is to introduce key features of the brainstem neural cardiovascular control system.

In 1870, Dittmar and Owsjannikow introduced the concept that the CNS regulates the circulation (Rothschuh, 1973). Until early 1940, it was still considered that the lower brainstem, the so-called vital center, was almost solely responsible for the integration of cardiovascular reflexes, and hence the activity of the autonomic nervous supply to the heart and the blood vessels. However, after 1960, it became increasingly evident that there were many other centers involved in neural cardiovascular regulatory reactions.

Brain structures that participate in the control of cardiovascular parameters have been identified by electrical or chemical stimulation, by ablation of the structure using either mechanical, electrical or chemical procedures, or by anatomical and electrophysiological techniques (Stock et al., 1983). More recently, new methods for tracing neural pathways have been developed, for example, immunohistochemical methods (Ross, Ruggiero and Reis, 1985), autoradiographic techniques (Young and Kuhar, 1979), and the retrograde axonal transport of horseradish peroxidase and fluorescent dyes from axon terminals to cell bodies (Dahlstrom and Fuxe, 1964) for identification of transmitter-specific pathways. By using these methods, important sites in the control of central cardiovascular activity have been located in the pontomedullary and hypothalamic regions (Elliot, 1979). The diencephalon also functions to integrate autonomic responses, and some nuclei in this region process information from the brainstem as well as from limbic, cerebral cortex, cerebellar and other brain regions (Korner, 1971).

AFFERENT NEURAL PATHWAYS

The central cardiovascular regulatory system provides not only a tonic control of autonomic outflow to the peripheral circulation, but also integrates primary cardiovascular reflexes following environmental disturbance. Central cardiovascular control is achieved in the medulla oblongata through integration of input from receptors in the chest and neck regions (Reis et al., 1984). The carotid sinus and aortic depressor nerves convey cardiovascular information from baroreceptors and chemoreceptors to the CNS. Stimulation and transection of these nerves alter the resting level of arterial pressure (Fink et al., 1980; Ciriello and Calaresu, 1978). The carotid sinus nerve afferent fibers in the cat terminate throughout the rostrocaudal extent of the nucleus of the solitary tract (NTS), with the densest labelling in the caudal part of the nucleus (Ciriello and Calaresu, 1981). The aortic depressor nerve has similar afferent fibers in the solitary tract. The densest projections have been found ipsilaterally in the interstitial nucleus and dorsolateral aspect of the solitary complex (Wallach and Loewy, 1980).

Many direct anatomical projections have been demonstrated from the NTS to other important cardiovascular control areas in the brainstem and higher centers. For example, the area postrema, appears to receive information from peripheral baroreceptors and from the NTS. This circumventricular organ, seems to contain a cardiovascular pathway conveying information to other vasomotor neurons located elsewhere in the brainstem including the rostral ventrolateral medulla and possibly the spinal cord (Ross et al., 1984). The rostral ventrolateral medulla, which also has connections with the NTS, has been shown to play an important role in the regulation of vasomotor tone. It is useful to review in some detail the important nuclei in the brainstem since these are sites where active investigations are done.

NUCLEUS TRACTUS SOLITARII

The nucleus of the solitary tract (NTS) is located in the dorsal medulla. In the rat, the rostral limit of the NTS is located at the level of the caudal pole of the facial motor nucleus. At this level, the NTS is a bilateral structure, that when extending caudally, come medial and close to the walls of the fourth ventricle. At the level of the posterior tip of the area postrema, the neuronal groups within the NTS fuse medially and form the commisural nucleus of the NTS. This part of the NTS lies dorsal to the central canal and the dorsal motor nucleus of the vagus. The caudal limit of the NTS is located at the level of the pyramidal decussation.

The NTS has long been identified as a site where the first synapse of the baroreceptor reflex is located (Seller and Illert, 1969; Spyer, Mifflin and Withington-Wray, 1987). Therefore, it is possible that within this region the afferent information and the descending influences are integrated. For instance, termination of afferent input from arterial or cardiopulmonary baroreceptor in the great vessels of the thorax and neck reaches the NTS through the glossopharyngeal (IX) nerve and the vagus (X) nerve (Seller and Illert, 1969; Spyer, Mifflin and Withington-Wray, 1987; Cottle, 1964). The NTS provides important inhibitory control over sympathetic outflow, since its destruction (bilaterally) produces fulminating hypertension (Doba and Reis, 1974). The mechanism of hypertension is likely to be due to central interruption of baroreflexes in the NTS since electrolytic lesions resulted in disappearance of baroreflexes and the pattern of redistribution of blood flow was similar to that produced by sinoaortic denervation (Doba and Reis, 1974; Palkovits, 1980).

It has been suggested that glutamate or a closely related excitatory substance is an important neurotransmitter of baroreceptor afferents terminating in the NTS (Talman, Perrone and Reis, 1980). Microinjection of glutamate or its analogs into the NTS evokes a dose-dependent fall in arterial blood pressure, bradycardia, and apnea (Talman, Perrone and Reis, 1980). Administration within the NTS of glutamate receptor antagonists, such as glutamate diethylester (GDEE) or kynurenic acid, increased arterial pressure and blunted baroreflex activation (Talman *et al.*, 1980; Leone and Gordon, 1989). The nature of the glutamate receptor involved in the baroreflex arc and in the responses to glutamate is still under discussion. Both NMDA and non-NMDA receptors (including the metabotropic receptor) have been implicated.

Inhibitory amino acids have also been proposed to modulate cardiovascular activity in the NTS. Stimulation of $GABA_B$ receptors within the NTS can result in pressor responses. For instance, intra-NTS microinjection of baclofen, a gama-aminobutyric acid receptor agonist results in a dose-dependent rise of arterial blood pressure and heart rate. Baclofen also evokes a depression of baroreceptor-mediated responses (Florentino, Varga and Kunos, 1990; Sved and Sved, 1990). In contrast, very high doses of $GABA_A$ receptor agonists (such as muscimol) need to be administered to elicit changes in blood pressure (Bousquet *et al.*, 1982). Furthermore, bicuculline, a $GABA_A$ receptor antagonist, did not affect heart rate after its administration into the NTS (Dimicco *et al.*, 1979). This suggests that in the NTS the actions of GABA on blood pressure and heart rate are mediated by activation of $GABA_B$ receptors.

In addition to the excitatory and inhibitory neurotransmitters, the NTS is richly innervated by neurons containing a number of potential neurotransmitters or neuromodulators, such as noradrenaline, adrenaline, acetylcholine, serotonin, angiotensin II (AII), vasopressin, thyrotropin-releasing hormone, corticotropin-releasing factor, somatostatin, oxytocin, β-endorphin, enkephalins, neuropeptide Y (NPY), substance P, neurotensin, atrial natriuretic peptide, and adenosine. In the NTS, noradrenaline seems to have an inhibitory cardiovascular role. Microinjection of noradrenaline results in a dose-dependent decrease in arterial blood pressure and heart rate (DeJong, 1974). These effects seem to be mediated by activation of both α_2-adrenergic and β-adrenergic receptors (Kubo and Misu, 1981; Tung *et al.*, 1983).

Cholinergic systems have been identified within the NTS (Palkovits and Jacobowitz, 1974; Helke, Sohl and Jacobowitz, 1980) and functional studies have demonstrated a potential involvement in cardiovascular regulation. Microinjection of nicotine into the NTS results in a hypotensive and bradycardic effect (Kubo and Misu, 1981; Robertson, Tseng and Appalsamy, 1988; Tseng *et al.*, 1993). Nicotine effects were inhibited by prior administration of hexamethonium in the NTS (Kubo and Misu, 1981). Like nicotine, microadministration of acetylcholine and several cholinergic agonists into the NTS reduced blood pressure and heart rate (Criscione, Reis and Talman; 1983; Sapru, 1989). However, these cardiovascular effects have been shown to be mediated by muscarinic receptors. Taken together, these results may indicate the presence of both muscarinic and nicotinic receptors in the NTS.

Although arginine vasopressin (AVP) is synthesized in the paraventricular and supraoptic nuclei of the hypothalamus, AVP-containing fibers have been shown to terminate in the NTS (Sofroniew and Schrell, 1981). After microinjection into the

NTS, AVP evoked an increase in blood pressure (Brattström, DeJong and DeWeid, 1988). Local administration of a V_1-receptor subtype antagonist has been shown to eliminate the hemodynamic responses of AVP (Vallejo, Carter and Lightman, 1984; King and Pang, 1987).

NPY is a 36-amino acid peptide for which binding sites have been identified in different brain sites including the NTS (Harfstrand et al., 1986; Nakajima, Yashima and Nakamura, 1986). Originally it was reported that microinjection of NPY into the NTS resulted in either a pressor response (with 470 fmol) or a depressor effect (4.7 pmol, Carter, Vallejo and Lightman, 1985). However, in other studies, dose-dependent hypotension and bradycardia have been observed (Tseng et al., 1988). These cardiovascular effects have been shown to be inhibited by the microinjection of specific antiserum (Tseng et al., 1988). The physiological relevance of NPY in the NTS remains unknown.

AII may influence autonomic function by either gaining access within the brain through circumventricular organs or by being locally produced in several brain nuclei. Among other sites, high densities of AII-binding sites have been identified in the NTS and in the area postrema. The cardiovascular effects of exogenously administered AII vary according to the dose. Pressor responses have been reported with doses between 50 to 500 ng (Casto and Phillips, 1984). Others have reported a biphasic depressor-pressor effect with doses ranging between 100 to 250 ng (Campagnole-Santos et al., 1989; Rettig, Healy and Printz, 1986). In a study evaluating the effects of different doses of AII, it was clear that a depressor effect is predominant at concentrations below 20 ng, whereas doses above 200 ng resulted in a pressor effect in normotensive animals (Mosqueda-Garcia et al., 1990). Interestingly, only a depressor effect was observed in renal hypertensive rats, irrespective of the dose of AII which was microinjected in the NTS (Mosqueda-Garcia et al., 1990).

Endogenous opioid peptides have been implicated in a number of autonomic functions including cardiovascular regulation. Receptors for enkephalins, β-endorphin, and dynorphin have been identified in the NTS (Khachaturian, Lewis and Schafer, 1985) and pharmacological studies have suggested a cardiovascular regulatory role. Studies using specific antagonists and selective antiserum directed either against enkaphalins or to β-endorphin indicate that in the NTS there are enkephalinergic (pressor) or endorphinergic (depressor) mechanisms (Petty and DeJong, 1982, 1983). The NTS is the only other site (outside the hypothalamus) where β-endorphin is synthesized (Finley, Maderdrut and Petruz, 1981; Khachaturian, Lewis and Schafer, 1985). Microinjection of β-endorphin into the NTS results in hypotension and bradycardia (Petty and DeJong, 1982; Mosqueda-Garcia and Kunos, 1987). In normotensive animals these effects seem to be mediated by μ-opiate receptors (Mosqueda-Garcia and Kunos, 1987). On the other hand, in hypertensive rats, δ-receptors seem to mediate the β-endorphin cardiovascular response. Although these results suggest that the presence of hypertension regulates the type of opioid receptor involved in the cardiovascular response of β-endorphin, the relevance of opioid peptides in the development of hypertension is unknown.

The endogenous nucleoside adenosine has been studied for its potential role as neuromodulator in a number of autonomic functions (Fredholm *et al.*, 1987). The highest density of adenosine uptake sites within the CNS has been observed in the NTS (Marangos *et al.*, 1982). Several groups have characterized the potent cardiovascular effects of purinergic activation in this nucleus (Barraco *et al.*, 1988; Tseng *et al.*, 1988; Mosqueda-Garcia *et al.*, 1989). These effects were dose-dependent and inhibited by specific adenosine-receptor antagonists. Although adenosine has been considered an inhibitory neuromodulator, in the NTS the effects of adenosine are more compatible with neuronal activation. Further work has demonstrated that the cardiovascular effects of adenosine in the NTS are compatible with a specific interaction with the excitatory amino acid glutamate (Mosqueda-Garcia *et al.*, 1991). The cardiovascular effects of adenosine can be inhibited by either adenosine receptor antagonists or by glutamate receptor antagonists. In addition, using microdialysis techniques, these studies demonstrated that the excitatory effects of adenosine on cardiovascular activity may be mediated by increased levels of glutamate in the NTS (Mosqueda-Garcia *et al.*, 1991).

Neurons with cell bodies in the NTS project to many rostral sites (such as paraventricular hypothalamic nucleus, parabrachial nucleus, ventrolateral medulla, etc.), some of which themselves send efferents directly to the sympathetic preganglionic nuclei in the intermediolateral column of the spinal cord (Palkovits and Zaborszky, 1977).

AREA POSTREMA

The area postrema (AP) is a circumventricular organ lying in the midline dorsal surface of the caudal medulla oblongata. The AP has been observed in all mammalian species and is characterized by an extremely rich capillary plexus (Roth and Yamamoto, 1968). The capillaries have an unusually weak blood-brain barrier which allows neural elements in the AP to be directly exposed to substances born in the peripheral circulation (Dempsey, 1973). Furthermore, evidence exists that portions of the AP are bathed in cerebrospinal fluid from the overlying fourth ventricle (Krisch and Leonhardt, 1987). The AP is closely adjacent to other structures with known cardiovascular function, including the NTS, dorsal motor nucleus of the vagus, and the A2 catecholaminergic cell group.

It was reported that the AP might be partly responsible for the elaboration of cerebrospinal fluid (Wislocki and Putnam, 1924), but emetic, cardiovascular and gustatory functions were also recognized. For instance, electrical stimulation of the AP in cats resulted in a sudden drop of arterial pressure (Ransom and Billingsley, 1916). Borison and Wang (1953) discovered that ablation of a lateral portion of the AP can abolish the emetic response to cardiac glycosides in cats, and they termed this area a "chemoreceptive trigger zone" for vomiting. Chronic ablation of the AP in dogs revealed a lasting mild hypotension associated with persistent tachycardia and a small reduction in cardiac output (Barnes, Ferrario and Conomy, 1979).

The relevance of the AP for the effects of AII has been documented in dogs by chemical (Ferrario, Dickinson and McCubbin, 1970), electrical (Barnes, Ferrario and Conomy, 1979) and lesion studies (Ferrario, Gildenberg and McCubbin, 1972). On the other hand, in the rat there is conflicting evidence regarding the importance of the AP for the effects of AII. Intravertebral injection of AII was less effective in increasing BP than intracarotid infusion (which does not provide blood supply to the AP region, Haywood *et al.*, 1980). Lesion studies have indicated that ablation of the AP does not eliminate the pressor response to acute i.v. (Zanderberg, Palkovits and DeJong, 1977, Haywood *et al.*, 1980) or i.c.v. (Haywood *et al.*, 1980) administration of AII and does not interfere with the development of hypertension in one kidney of renal hypertensive rats (Haywood *et al.*, 1980). In direct contrast with these findings, some evidence has indicated that AP ablation in the rats reduces basal blood pressure (BP) and heart rate (HR) (Skoog and Mangiapane, 1988), it may eliminate the hypertensive response to chronic AII infusion (Fink, Bruner and Mangiapane, 1987) and it may prevent not only chronic hypertension in high renin hypertensive models (Fink *et al.*, 1986) but also in other models in which plasma renin activity is known not to be elevated (Fink, Bruner and Mangiapane, 1987; Mangiapane *et al.*, 1989). Controversy also exists regarding the effects of activation of AP neurons on BP and HR. Although one study suggested that electrical stimulation of the AP increases BP with high frequency of stimulation (Mangiapane, Bruner and Fink, 1985), another study using low frequency of stimulation demonstrated a decrease in these hemodynamic variables (Fergurson and Marcus, 1988). Although "*in vivo*" electrophysiological studies show that a number of AP neurons respond to iontophoretic application of AII (Carpenter, Briggs and Strominger, 1984), "*in vitro*" experiments in AP neuronal cells have been unsuccessful in demonstrating effects on AP neurons (Brooks, Hubbard and Sirett, 1983). Interestingly, one study reported a modest pressor and bradycardiac effect after a large dose of AII in the AP (Casto and Phillips, 1984). This effect was suggested to be mediated by an increase in sympathetic vasoconstrictor outflow. A more recent study clarified some of this controversy (Mosqueda-Garcia *et al.*, 1990). When studying the dose-related effects of microinjection of AII into the AP, it is now evident that microinjection of low doses of AII (2–20 ng) into the AP results in hypotensive effects. In contrast, higher doses seem to produce a pressor effect by, presumably, activation of other cardiovascular centers outside the AP (Mosqueda-Garcia *et al.*, 1990). The results from this and other studies (Rettig, Healy and Printz, 1986) indicate that stimulation of AII receptors in the AP is depressor in nature in normotensive animals. Furthermore, these cardiovascular effects, irrespective of the depressor or pressor actions, are not associated with activation of central sympathetic outflow (Mosqueda-Garcia *et al.*, 1990).

Neuronal cell groups within the AP are also quite responsive to microinjection of adenosine. Microinjection of adenosine and ATP into the AP produced hypotension and bradycardia which was blocked by a specific adenosine antagonist (Tseng *et al.*, 1988). The physiological role of adenosine and other peptides, such as AII, in the AP remain to be elucidated.

VENTROLATERAL MEDULLA (VLM)

It is now widely accepted that the rostral VLM (RVLM) plays an important role in the regulation of the cardiovascular system. Early evidence first indicated that tonic arterial pressure maintenance requires the integrity of this area (Reis and Cuenod, 1965). Electrical stimulation within a region of the RVLM (known to have a dense population of adrenaline-containing neurons) caused hypertension, whereas electrolytic destruction of neurons within this region resulted in hypotension (Dampney *et al.*, 1982). Microinjection of glutamate into the same area elicited a sustained increase in arterial pressure (Dampney *et al.*, 1982).

Two important vasomotor control areas within the VLM have been reported: the rostral (C1 area, RVLM) and caudal (A1 area, CVLM) vasomotor areas (Reis, 1981). Immunohistochemical studies show that the RVLM is closely associated with the lateral reticular nucleus, nucleus ambiguus and the nucleus reticularis paragigantocellularis, while the CVLM is located near the caudal lateral reticular nucleus, nucleus ambiguus, and the ventrolateral nucleus reticularis ventralis. Stimulation of neurons within the RVLM causes an increase in arterial pressure through projections to preganglionic sympathetic neurons in the intermediolateral column of the thoracic spinal cord (Hökfelt *et al.*, 1974; Ross *et al.*, 1983). On the other hand, stimulation of neurons within the CVLM leads to a reduction of pressure, (Blessing and Reis, 1982). Pharmacological studies suggest that the rostral area is an important determinant of arterial pressure and that its activity is in some way modulated by the caudal area. Activation of the caudal area may inhibit cells within the rostral vasomotor area and cause hypotension (Willette *et al.*, 1984).

The RVLM had long been recognized as the source of tonic control of the spinal cord as an important site of action of antihypertensive drugs (for both the α_2-adrenergic receptor agonist, clonidine (Bousquet *et al.*, 1975), the β-blocker, propranolol (Sun and Guyenet, 1990)). Bilateral electrolytic lesions of RVLM reduced arterial pressure and heart rate to values comparable to spinal cord transection within two minutes and abolished the reflex hypotension and bradycardia evoked by baroreceptor stimulation (either electrical vagal stimulation or carotid sinus stretch) (Ross *et al.*, 1984). It appears, therefore, that the RVLM is not only important for providing tonic activity to the sympathetic cardiovascular neurons but is also important in mediating the baroreceptor reflex (Dembowsky and McAllen, 1985; Granata *et al.*, 1985). Evidence has demonstrated that baroreceptor-evoked inhibition of the bulbospinal neurons located in the subretrofacial nucleus of the RVLM was the result of a $GABA_A$ receptor mediated process (Sun and Guyenet, 1985).

In addition to glutamate, adrenaline and GABA, many other neurotransmitters have been reported to be involved in cardiovascular regulation of the RVLM. Recently, by using an acetylcholine synthesis inhibitor and a muscarinic receptor antagonist, evidence suggests that the cholinergic innervation in the RVLM exerts an excitatory influence on vasopressor neurons (Arneric *et al.*, 1990). In addition, microinjection of nicotine increases blood pressure and heart rate in a dose-dependent manner (Tseng *et al.*, 1993). These effects may mediate part of the pressor effects of smoking.

The involvement of peptides in the central neural regulation of the circulation has been discussed in detail elsewhere (Reid and Rubin, 1987). For instance, microinjection of NPY into the RVLM produced pressor and bradycardic effects (Tseng *et al.*, 1988), and immunohistochemical evidence suggests that most of the NPY terminals in the intermediolateral column of the thoracic spinal cord come from the neurons in the RVLM (Tseng *et al.*, 1993).

There are other nuclei in the brainstem that play an important role in central cardiovascular control. The dorsal medulla, parabrachial nucleus and locus coeruleus are all involved in the complex central cardiovascular control. There are interconnections and projections among these nuclei, and usually the interconnections and projections are bilateral and reciprocal.

DORSAL MEDULLA

Pressor responses can be elicited from both RVLM and dorsal medulla by either glutamate or by electrical stimulation in rats, rabbits, cats, and pigs (Chai *et al.*, 1988). This study demonstrates that both the dorsal medulla and the RVLM contain neural perikarya that mediate pressor effects. Bilateral lesioning of the dorsal medulla, particularly in the vicinity of the nucleus reticularis parvocellularis, observed a profound hypotension and abolition of the vasomotor component of the cerebral ischemic response (Kumada, Dampney and Reis, 1979).

However, a recent study suggests that the dorsal medulla does not appear to be essential for the genesis of basal vasomotor tone (Yardley, Andrade and Weaver, 1989). Indeed, the increase in arterial pressure and heart rate elicited by stimulation of the dorsal medulla could be associated with two different cardiovascular response patterns. At the majority of effective stimulation sites, generalized vasoconstriction with decreases in both renal and femoral conductance occurred. At the other effective sites, renal vascular constriction and hind-limb vascular dilatation occur.

A recent study by Chai *et al.* (1991) demonstrated stimulation of the paramedial reticular nucleus (PRN) could inhibit or abolish pressor responses during stimulation of dorsal medulla and RVLM, and thus suggests that PRN may be a part of an inhibitory modulatory system affecting responses to activation of the sympathetic pressor regions of the medulla.

PARABRACHIAL NUCLEUS (PB)

The PB received projections from many nuclei which are important in central cardiovascular regulation, including the amygdala, lateral paraventricular and preoptic areas of hypothalamus, midline cerebellum, NTS, dorsal medullary reticular formation and the nucleus ambiguus (Marovitch, Kumada and Reis, 1982). The ascending pathway of the PB includes reciprocal projections from the PB to the paraventricular hypothalamus, nucleus amygdala. Enkephalin and neurotensin have been suggested to be present in the cells of this nucleus. The descending pathway includes projections to the NTS, and a direct pathway from neurons in the ventrolateral aspect of the PB to the region of the intermediolateral column (IML) of thoracic spinal cord has been shown.

Stimulation of the PB produced a pressor response in the cat and rat (Marovitch, Kumada and Reis, 1982; Hade *et al.*, 1988), and bradycardia in the rabbit (Hamilton and Reid, 1981). The pressor effect caused by stimulation of PB is a result of an increase of both total peripheral resistance and cardiac output.

LOCUS COERULEUS (LC)

The LC/subcoeruleus (SC) contains the densest aggregate of NA-containing neurons in the brainstem, with widespread efferent projections throughout the brain and spinal cord. This group plays a major role in several neurobiological regulations including control of blood pressure, neuroendocrine secretions and modulation of behavioral mechanisms (Svensson, 1987). Ascending projections from the LC provides the sole NA innervation of the cerebral, hippocampal, and cerebellar cortices. Also, a part of this ascending projection ends in parvicellular division of the paraventricular nucleus of the hypothalamus, which is believed to be involved in both modes of hypothalamic control in neuroendocrine and autonomic responses. In addition, descending projections from the LC to the spinal cord are known to terminate at the central part of the dorsal horn, with a very sparse innervation of the superficial dorsal horn.

Many different types of receptors have been identified in the LC: opioid, muscarinic, serotonergic, peptidergic and alpha and beta-adrenergic receptors. Stimulation of these receptors produced different responses. This implied that the LC may be divided into functional subdivisions dependent on the region of the LC, the neurotransmitter/neuropeptide(s) contained with the neurons and their efferent projections (Holets *et al.*, 1988; Illes and Regnold, 1990). Physiological studies demonstrated that the LC impulse activity is often correlated with peripheral autonomic activity, so that stimuli which activate the LC also activate the autonomic system. One of the best characterized responses is the so-called defense reactions which involve autonomic responses. In the long run, such stress-elicited autonomic actions are thought to contribute to a more permanent elevated arterial blood pressure (Folkow, 1984).

The LC sends projections rostrally from the pons to several areas, including those in which NA participates in blood pressure and fluid volume control (Bhaskaran and Freed, 1988). The LC receives both adrenergic input and excitatory amino acid input from the nucleus paragigantocellularis. The interactions with baroreceptors is suggested by the presence of fibers from the LC that terminate in the A1 and A2 areas which in turn are linked to efferent baroreflex axes such as C1, C2, dorsal vagal nucleus and nucleus ambiguus. The LC may also be involved in the cardiovascular response that occurs after changes in blood volume such as moderate hemorrhage or changes in posture (Elam, Svensson and Thoren, 1985). Furthermore, a sustained natriuresis can be induced by cholinergic stimulation of the LC which is attenuated by lesions of the anterioventral third ventricle (De Luca *et al.*, 1991). In short, this evidence indicates that the LC is involved in cardiovascular regulation.

EFFERENT NEURAL PATHWAY

The sympathetic preganglionic fibers originate from the intermediolateral column of the spinal cord between C8 and L4 (Henry and Calaresu, 1972; Oldfield and McLachlan, 1981), but a small number are located in the nueleus intercalatus of the spinal thoracic and lumbar segment.   Their efferent fibers synapse with the postganglionic neurons in sympathetic ganglia, which innervate the heart and smooth muscles of vessels.

The precise origin of each descending vasomotor pathway from supraspinal sites is still not clear.   Mono or polysynaptic connections may be involved.   A monosynaptic pathway between cells in the caudal raphe nuclei and sympathetic preganglionic neurons in the rat spinal cord have been reported recently (Bacon, Zagon and Smith, 1990).   The descending pathways arise from several bulbar excitatory and inhibitory nuclei, over which both tonic and reflex control of the activity in sympathetic preganglionic neurons occurs. The excitatory pathways are in the dorsal lateral part of the lateral funiculus, which normally mediates the tonic activity of the intermediolateral motor neurons in intact animals. The descending inhibitory pathways have been located in the dorsal lateral funiculus and in the ventrolateral ventral funiculus. Both noradrenergic and serotonergic neurons are involved in this descending inhibition (Lewis and Coote, 1991; Zagon and Bacon, 1991). There is evidence that the descending inhibitory pathways are involved in the modulation of baroreceptor reflexes (Coote and MacLeod, 1974).

REFERENCES

Americ, S.P., Guiljano, R., Ernsberger, P., Underwood, M.D. and Reis, D.J. (1990) Synthesis, release and receptor binding of acetylcholine in the C1 area of the rostral ventrolateral medulla: contributions in regulating arterial pressure. *Brain Res.*, **511**, 98–112.

Bacon, S.J., Zagon, A. and Smith, A.D. (1990) Electron microscopic evidence of a monosynaptic pathway between cells in the caudal raphe nuclei and sympathetic preganglionic neurons in the rat spinal cord. *Exp. Brain Res.*, **79**, 589–602.

Barnes, K.L., Ferrario, C.M. and Conomy, J.P. (1979) Comparison of the hemodynamic changes produced by electrical stimulation of the area postrema and NTS in the dog. *Circ. Res.*, **45**, 136–143.

Barraco, R.A., Janusz, C.J., Polasek, P.M., Parizon, M. and Roberts, P.A. (1988) Cardiovascular effects of microinjection of adenosine into the nucleus of the tractus solitarius. *Brain Res. Bull.*, **20**, 129–132.

Bhaskaran, D. and Freed, C.R. (1988) Changes in neurotransmitter turnover in locus coeruleus produced by changes in arterial blood pressure. *Brain Res. Bull.*, **21**, 191–199.

Blessing, W.W. and Reis, D.J. (1982) Inhibitory cardiovascular function of neurons in the caudal ventrolateral medulla of the rabbit: relationship to the area containing A1 noradrenergic cells. *Brain Res.*, **253**, 161–171.

Borison, H.L. and Wang, S.C. (1953) Physiology and pharmacology of vomiting. *Pharmacol. Rev.*, **5**, 193–230.

Bousquet, P., Feldman, J., Velly, J. and Block, R. (1975) Role of ventral surface of the brainstem in the hypotensive effect of clonidine. *Eur. J. Pharmacol.*, **34**, 151–156.

Bousquet, P., Feldman, J., Bloch, R. and Schwartz, J. (1982) Evidence for a neuromodulatory role of GABA at the first synapse of the baroreceptor reflex pathway. Effects of GABA derivatives injected into the NTS. *Naunyn-Schmiedebergs Arch. Pharmacol.*, **319**, 168–171.

Brattström, A., DeJong, W. and DeWeid, D. (1988) Vasopressin microinjections in to the nucleus tractus solitarii decrease rate and blood pressure in anaesthetized rats. *J. Hypertens. (Suppl.)*, **6**, S521–S524.

Brooks, M.J., Hubbard, J.I. and Sirett, N. (1983) Extracellular recording in rat area postrema *in vitro*, and the effects of cholinergic drugs, serotonin and angiotensin II. *Brain Res.*, **261**, 85–90.

Campagnole-Santos, M.J., Diz, D.I., Santos, R.A.S., Khosla, M.C., Brosnihan, K.B. and Ferrario, C.M. (1989) Cardiovascular effects of angiotensin-(1–7) injected into the dorsal medulla of rats. *Am. J. Physiol.*, **257**, H324–H329.

Carpenter, C.O., Briggs, D.B. and Strominger, N. (1984) Behavioral and electrophysiological studies of peptide-induced emesis in dogs. *Fed. Proc.*, **43**, 2952–2954.

Carter, D.A., Vallejo, M. and Lightman, S.L. (1985) Cardiovascular effects of neuropeptide Y in the nucleus tractus solitarius of rats: relationship with noradrenaline and vasopressin. *Peptides*, **6**, 421–425.

Casto, R. and Phillips, M.I. (1984) Cardiovascular actions of microinjections of angiotensin II in the brainstem of rats. *Am. J. Physiol.*, **246**, R811–R816.

Chai, C.Y., Lin, R.H., Lin, A.M.Y., Pan, C.M., Lee, E.H.Y. and Kuo, J.S. (1988) Pressor responses from electrical or glutamate stimulations of the dorsal or ventrolateral medulla. *Am. J. Physiol.*, **255**, R709–R717.

Chai, C.Y., Lin, A.M.Y., Su, C.K., Hu, S.R., Yuan, C., Kao, L.S., Kuo, J.S. and Goldstein, D.S. (1991) Sympathoadrenal excitation and inhibition by lower brainstem stimulation in cats. *J. Auton. Nerv. Syst.*, **33**, 35–46.

Ciriello, J. and Calaresu, F.R. (1978) Separate medullary pathways mediating reflex vagal bradycardia to stimulation of buffer nerves in the cat. *J. Auton. Nerv. Syst.*, **1**, 13–22.

Ciriello, J. and Calaresu, F.R. (1981) Projections from buffer nerves to the nucleus of the solitary tract: an anatomical and electrophysiological study in the cat. *J. Auton. Nerv. Syst.*, **3**, 299–310.

Coote, J.H. and MacLeod, V.H. (1974) Evidence of the involvement in the baroreceptor reflex of a descending inhibitory pathway. *J. Physiol. (Lond.)*, **241**, 477–496.

Cottle, M.K (1964) Degeneration studies of primary afferents of IXth and Xth cranial nerves in the cat. *J. Comp. Neurol.*, **122**, 329–345.

Criscione, L., Reis, D.J. and Talman, W.T. (1983) Cholinergic mechanisms in the nucleus tractus solitarii and cardiovascular regulation in the rat. *Eur. J. Pharmacol.*, **88**, 47–55.

Dahlstrom, A. and Fuxe, K. (1964) Evidence for the existence of monoamine containing neurons in the central nervous system. I. Demonstration of monoamines in cell bodies of brainstem neurons. *Acta. Physiol. Scand.*, **62**, 1–55.

Dampney, R.A., Godschild, A.K., Robertson, L.G. and Montgomery, W. (1982) Role of ventrolateral medulla in vasomotor regulation: a correlative anatomical and physiological study. *Brain Res.*, **249**, 223–235.

DeJong, W. (1974) Noradrenaline, central inhibitory control of blood pressure and heart rate. *Eur. J. Pharmacol.*, **88**, 47–55.

De Luca, L.A., Franci, C.R., Saad, W.A., Camargo, L.A.A. and Antunes-Rodrigues, J. (1991) Natriuresis, not seizures, induced by cholinergic stimulation of the locus coeruleus is affected by forebrain lesions and water deprivation. *Brain Res. Bull.*, **26**, 203–210.

Dembowsky, K. and McAllen, R.M. (1985) Baroreceptor inhibition of subretrofacial neurons, evidence from intracellular recordings in the cat. *Neurosci. Lett.*, **111**, 139–143.

Dempsey, E.W. (1973) Neural and vascular ultrastructure of the area postrema in the rat. *J. Comp. Neurol.*, **150**, 177–200.

Dimicco, J.A., Gale, K., Hamilton, B. and Gillis, R.A. (1979) GABA receptor control of parasympathetic outflow to heart; characterization and brainstem localization. *Science*, 204, 1106–1108.

Doba, N. and Reis, D.J. (1974) Role of central and peripheral adrenergic mechanisms in neurogenic hypertension produced by brainstem lesions in the rat. *Circ. Res.*, 34, 293–301.

Elam, M., Svensson, T.H. and Thoren, P. (1985) Differentiated cardiovascular afferent regulation of locus coeruleus neurons and sympathetic nerves. *Brain Res.*, 358, 77–84.

Elliot, J.M. (1979) The central noradrenergic control of blood pressure and heart rate. *Clin. Exp. Pharmacol. Physiol.*, 6, 569–579.

Fergurson, A.V. and Marcus, P. (1988) Area postrema stimulation induced cardiovascular changes in the rat. *Am. J. Physiol.*, 255, R855–R860.

Ferrario, C.M., Dickinson C.J. and McCubbin, J.W. (1970) Central vasomotor stimulation by angiotensin. *Clin. Sci.*, 39, 239–245.

Ferrario, C.M., Gildenberg, P.L. and McCubbin, J.W. (1972) Cardiovascular effects of angiotensin mediated by the central nervous system. *Circ. Res.*, 30, 257–262.

Fink, G.D., Kennedy, F., Bryan, W.J. and Werber, A. (1980) Pathogenesis of hypertension in rats with chronic aortic baroreceptor deafferentation. *Hypertension*, 2, 319–325.

Fink, G.D., Bruner, C.A., Pawloski, C.M., Blair, M.L., Skoog, K.M. and Mangiapane, M.L. (1986) Role of the area postrema in hypertension after unilateral artery constriction in the rat. *Fed. Proc.*, 45, 875.

Fink, G.D., Bruner, C.A. and Mangiapane, M.L. (1987) Area postrema is critical for angiotensin-induced hypertension in rats. *Hypertension*, 9, 355–361.

Finley, J.C.W., Maderdrut, J.L. and Petruz, P. (1981) The immunocytochemical localization of enkephalin in the central nervous system. *J. Comp. Neurol.*, 198, 541–565.

Florentino, A., Varga, K. and Kunos, G. (1990) Mechanism of the cardiovascular effects of GABA receptor activation in the nucleus tractus solitarii of the rat. *Brain Res.*, 535, 264–270.

Folkow, B. (1984) Stress and blood pressure. In: *Adrenergic Blood Pressure Regulation*. Symp Excerpta Medica: Amsterdam, pp. 87–93.

Fredholm, B.B., Duner-Engstrom, M., Fastbom, J., Jonzon, B., Lindgren, D., Norstedt, C., Pedta, F. and Van den Ploeg, I. (1987) Interactions between the neuromodulatory adenosine and the classic transmitters. In: *Topics and Perspectives in Adenosine Research*, Gerlach, E. and Becker, B.F. (eds.) Springer-Verlag: Berlin, pp. 499–508.

Granata, A.R., Ruggiero, D.A., Park, D.H., Joh, T.H. and Reis, D.J. (1985) Brainstem area with C1 epinephrine neurons mediates baroreflex vasodepressor responses. *Am. J. Physiol.*, 248, H547–H567.

Guyton, A.C., Coleman, T.G., Cowley, A.W. and Manning, R.D. (1974) A system analysis approach to understanding long range arterial blood pressure control and hypertension. *Circ. Res.*, 35, 159–176.

Hade, J.S., Mifflin, S.W. Donta, T.S. and Felder, R.B. (1988) Stimulation of parabrachial neurons elicits a sympathetically mediated pressor response in cats. *Am. J. Physiol.*, 255, H1349–H1358.

Hamilton, C.A. and Reid, J.L. (1981) Changes in alpha-adrenoreceptors during long-term treatment of rabbits with prazosin. *J. Cardiovasc. Pharmacol.*, 3, 977–985.

Harfstrand, A., Fuxe, K., Agnati, L.F., Benfenati, F. and Goldstein, M. (1986) Receptor autoradiographic evidence for high densities of ^{125}I-neuropeptide Y binding sites in the nucleus tractus of the solitarius of the normal male rat. *Acta. Physiol. Scand.*, 128, 195–200.

Haywood, J.R., Fink, G.D., Buggy, J., Phillips, M.I. and Brody, M.J. (1980) The area postrema plays no role in the pressor action of angiotensin in the rat. *Am. J. Physiol.*, 239, H108–H113.

Helke, C.J., Sohl, B.D. and Jacobowitz, D.M. (1980) Choline acetyltransferase activity in discrete brain nuclei of DOCA-salt hypertensive rats. *Brain Res.*, 193, 293–298.

Henry, J.L. and Calaresu, F.R. (1972) Topography and numerical distribution of neurons of the thoraco-lumbar intermedioarterial nucleus in the cat. *J. Comp. Neurol.*, 144, 205–214.

Hökfelt, T., Fuxe, K., Goldstein, M. and Johansson, D. (1974) Immunohistochemical evidence for the existence of adrenaline neurons in rat brain. *Brain Res.*, 66, 235–251.

Holets, V.R., Hökfelt, T., Rökaeus A., Terenius, L. and Goldstein, M. (1988) Locus coeruleus neurons in the rat containing neuropeptide Y, tyrosine hydroxylase of galanin and their efferent projections to the spinal cord, cerebral cortex and hypothalamus. *Neuroscience*, 24, 893–906.

Illes, P. and Regnold, J.T. (1990) Interaction between neuropeptide Y and noradrenaline on central catecholamine neurons. *Nature*, 344, 62–63.

Khachaturian, H., Lewis, M.E. and Schafer, M.K.H. (1985) Anatomy of the CNS opioid systems. *Trends Neurosci.*, **8**, 111–119.

King, K.A. and Pang, C.C. (1987) Cardiovascular effects of injections of vasopressin into the nucleus tractus solitarius in conscious rats. *Br. J. Pharmacol.*, **90**, 531–586.

Korner, P.I. (1971) Integrative neural cardiovascular control. *Physiol. Rev.*, **51**, 312–367.

Krisch, B. and Leonhardt, H. (1987) The functional and structural border between the CSF and blood milieu in the circumventricular organs (organum vasculosum laminae terminals, subfornical organ, area postrema) of the rat. *Cell. Tissue Res.*, **195**, 485–497.

Kubo, T. and Misu, Y. (1981) Pharmacological characterization of the α-adrenoreceptors responsible for a decrease of blood pressure in the nucleus tractus solitarii of rats. *Naunyn-Schiedeberg's Arch. Pharmacol.*, **317**, 120–125.

Kumada, M., Dampney, R.A.L. and Reis, D.J. (1979) Profound hypotension and abolition of the vasomotor component of the cerebral ischemic response produced by restricted lesions of medulla oblongata in rabbit. *Circ. Res.*, **44**, 63–70.

Leone, C. and Gordon, F.J. (1989) Is L-glutamate a neurotransmitter of baroreceptor information in the nucleus of the tractus solitarius? *J. Pharmacol. Exp. Ther.*, **250**, 953–962.

Lewis, D.I. and Coote, J.H. (1991) Excitation and inhibition of rat sympathetic preganglionic neurones by catecholamines. *Brain Res.*, **530**, 229–234.

Mangiapane, M.L., Bruner, C.A. and Fink, G.D. (1985) Area postrema role in the control of arterial pressure and regional vascular resistance in the rat. *Fed. Proc.*, **44**, 1346.

Mangiapane, M.D., Skoog, K.M., Rittenhouse, P., Blair, M.L. and Sladek, C.D. (1989) Lesion of the area postrema region attenuates hypertension in spontaneously hypertensive rats. *Circ. Res.*, **64**, 129–135.

Marangos, P.J., Patel, J., Clark-Rosenberg, R. and Martino, A.M. (1982) [^{3}H]-nitrobenzylthionosine binding as a probe for the study of adenosine uptake sites in the brain. *J. Neurochem.*, **39**, 184–191.

Marovitch, S., Kumada, M. and Reis, D.J. (1982) Role of the nucleus parabrachialis in cardiovascular regulation in cat. *Brain Res.*, **232**, 57–75.

Mosqueda-Garcia, R. and Kunos, G. (1987) Opiate receptors and the endorphin-mediated cardiovascular effects of clonidine in rats: evidence for hypertension-induced μ-subtype to δ-subtype changes. *Proc. Natl. Acad. Sci. (USA)*, **84**, 8637–8641.

Mosqueda-Garcia, R., Tseng, C.J., Appalsamy, M. and Robertson, D. (1989) Modulatory effects of adenosine on baroreflex activation in the brainstem of normotensive rats. *Eur. J. Pharmacol.*, **174**, 119–122.

Mosqueda-Garcia, R., Tseng, C.J., Appalsamy, M. and Robertson, D. (1990) Cardiovascular effects of microinjection of angiotensin II in the brainstem of renal hypertensive rats. *J. Pharmacol. Exp. Ther.*, **255**, 374–381.

Mosqueda-Garcia, R., Tseng, C.J., Appalsamy, M., Beck, C. and Robertson, D. (1991) Cardiovascular excitatory effects of adenosine in the nucleus of the solitary tract. *Hypertension*, **18**, 494–502.

Nakajima, T., Yashima Y. and Nakamura, K. (1986) Quantitative autoradiographic localization and neuropeptide Y receptors in the rat lower brainstem. *Brain Res.*, **380**, 144–150.

Oldfield, B.J. and McLachlan, E.M. (1981) An analysis of the sympathetic preganglionic neurons projecting from the upper thoracic spinal roots of the cat. *J. Comp. Neurol.*, **196**, 329–345.

Palkovits, M. (1980) The anatomy of central cardiovascular nerves. In *Central Adrenaline Neurons, Basic Aspects and Their Role in Cardiovascular Functions*, pp. 3–17. Oxford, Pergamon Press.

Palkovits, M. and Jacobowitz, D.M. (1974) Topographical atlas of catecholamine and acetylcholinesterase containing neurons in the rat brain. II. Hind brain (mesencephalon, rhombencephalon). *J. Comp. Neurol.*, **157**, 29–42.

Palkovits, M. and Zaborszky, L. (1977) Neuroanatomy of central cardiovascular control nucleus tractus solitarii, afferent and efferent neuronal connections in relation to the baroreceptor reflex arc. *Prog. Brain Res.*, **47**, 9–34.

Petty, M.A. and DeJong, W. (1982) Cardiovascular effects of β-endorphin after microinjection into the nucleus tractus solitarii of the anesthetised rat. *Eur. J. Pharmacol.*, **81**, 449.

Petty, M.A. and DeJong, W. (1983) Enkephalins induce a centrally mediated rise in blood pressure in rats. *Brain Res.*, **260**, 322.

Ransom, S.W. and Billingsley, P.R. (1916) Vasomotor reactions from stimulation of the floor of the fourth ventricle. Studies in vasomotor reflex arcs III. *Am. J. Physiol.*, **41**, 85–89.

Reid, J.L. and Rubin P.C. (1987) Peptides and central neural regulation of the circulation. *Physiol. Rev.*, **67**, 725–749.

Reis, D.J. (1981) The brain and arterial hypertension, evidence for a neural imbalance hypothesis. In *Disturbances in Neurogenic Control of Circulation*, edited by F.A. Abboud, H.A. Fozzard, J.P. Gilmore and D.J. Reis, pp. 87–104. Bethesda, American Physiological Society.

Reis, D.J. and Cuenod, M. (1965) Central neural regulation of carotid baroreceptor reflexes in the cat. *Am. J. Physiol.*, **209**, 1267–1269.

Reis, D.J., Granata, A.R., Joh, T.H., Ross, C.A., Ruggiero, D.A. and Park, D.H. (1984) Brainstem catecholamine mechanisms in tonic and reflex control of blood pressure. *Hypertension*, **6**, 7–15.

Rettig, R., Healy, D.P. and Printz, M.P. (1986) Cardiovascular effects of microinjections of angiotensin II into the nucleus tractus solitarii. *Brain Res.*, **364**, 233–240.

Robertson, D., Tseng, C.J. and Appalsamy, M. (1988) Smoking and mechanisms of cardiovascular control. *Am. Heart J.*, **115**, 258–263.

Ross, C.A., Ruggiero, D.A., Joh, T.H., Park, D.M. and Reis, D.J. (1983) Adrenaline synthesizing neurons in the rostral ventrolateral medulla: a possible role in tonic vasomotor control. *Brain Res.*, **273**, 356–361.

Ross, C.A., Ruggiero, D.A., Joh, T.H., Park, D.H. and Reis, D.J. (1984) Rostral ventrolateral medulla: selective projections to the thoracic autonomic cell column from the region containing C1 adrenaline neurons. *J. Comp. Neurol.*, **228**, 168–184.

Ross, C.A., Ruggiero, D.A. and Reis, D.J. (1985) Projections from the nucleus tractus solitarii to the rostral ventrolateral medulla. *J. Comp. Neurol.*, **242**, 511–534.

Roth, G.I. and Yamamoto, W.S. (1968) The microcirculation of the area postrema in the rat. *J. Comp. Neurol.*, **133**, 329–340.

Rothschuh, K.E. (1973) History of Physiology, New York, Krieger Publishing Company.

Sapru, H.N. (1989) Cholinergic mechanisms subserving cardiovascular function in the medulla and spinal cord. In: Cirillo, J., Caverson, M.M. and Polosa, C. (eds) *Progress in Brain Research*, **81**, 171–179.

Seller, H. and Illert, M. (1969) The localization of the first synapse in the carotid sinus baroreceptor reflex pathway and its alteration of afferent input. *Pflugers Arch.*, **306**, 1–19.

Skoog, K.M. and Mangiapane M.L. (1988) Area postrema and cardiovascular regulation in rats. *Am. J. Physiol.*, **254**, H963–H969.

Sofroniew, M.V. and Schrell, U. (1981) Evidence for a direct projection from oxytocin and vasopressin neurons in the hypothalamic paraventricular nucleus to the medulla oblongata: immunohistochemical visualization of both the horseradish peroxidase transported and the peptide produced by the same neurons. *Neurosci. Lett.*, **22**, 211–217.

Spyer, K.M., Mifflin, S.W. and Withington-Wray, D.J. (1987) Diencephalic control of the baroreceptor reflex at the level of the nucleus of the tractus solitarius. *Organization of the Autonomic Nervous System, Control and Peripheral Mechanisms*, pp. 307–314. New York, Alan R. Liss.

Stock, G., Schmelz, M., Knuepfer, M.M. and Forssman, W.G. (1983) Functional and anatomical aspects of central nervous cardiovascular regulation. In *Current Topics in Neuroendocrinology, Central Cardiovascular Control, Basic and Clinical Aspects*, edited by D. Ganten and D. Paff, pp. 1–31. Berlin: Springer-Verlag.

Sun, M.K. and Guyenet, P.G. (1985) GABA-mediated baroreceptor inhibition of reticulospinal neurons. *Am. J. Physiol.*, **249**, R672–R680.

Sun, M.K. and Guyenet, P.G. (1990) Excitation of rostral medullary pacemaker neurons with putative sympathoexcitatory function by cyclic AMP and β-adrenoceptor agonists *in vitro*. *Brain Res.*, **511**, 30–40.

Sved, A.F. and Sved, J.C. (1990) Endogenous GABA acts on GABA-B receptors in nucleus tractus solitarius to increase blood pressure. *Brain Res.*, **526**, 235–240.

Svensson, T.H. (1987) Peripheral, autonomic regulation of locus coeruleus noradrenergic neurons in brain: putative implications for psychiatry and psychopharmacology. *Psychopharmacology*, **92**, 1–7.

Talman, W.T., Perrone, M.H. and Reis, D.J. (1980) Evidence for L-glutamate as the neurotransmitter of primary baroreceptor afferent nerve fibers. *Science*, **709**, 803–814.

Talman, W.T., Perrone, M.H., Scher, P., Kwo, S. and Reis, D. (1981) Antagonism of the baroreceptor reflex by glutamate diethylester, an antagonist of L-glutamate. *Brain Res.*, **217**, 186–191.

Tseng, C.J., Biaggioni, I., Appalsamy, M. and Robertson, D. (1988) Purinergic receptors in the brainstem mediate hypertension and bradycardia. *Hypertension*, **11**, 191–197.

Tseng, C.J., Lin, H.C., Wang, S.D. and Tung, C.S. (1993) Immunohistochemical study of catecholamine enzymes and neuropeptide Y (NPY) in the rostral ventrolateral medulla and bulbospinal projection. *J. Comp. Neurol.*, **334**, 294–303.

Tseng, C.J., Mosqueda-Garcia, R., Appalsamy, M. and Robertson, D. (1988) Cardiovascular effects of neuropeptide Y in rat brainstem nuclei. *Circ. Res.*, **64**, 55–61.

Tseng, C.J., Appalsamy, M., Robertson, D. and Mosqueda-Garcia, R. (1993) Effects of nicotine on brainstem mechanisms of cardiovascular control. *J. Pharmacol. Exp. Ther.*, **265**, 1511–1518.

Tung, C.S., Onuora, C.O., Robertson, D. and Goldberg, M. (1983) Hypertensive effect of yohimbine following selective injection into the nucleus fractus solitarii of normotensive rats. *Brain Res.*, **277**, 193–195.

Vallejo, M., Carter, D.A. and Lightman, S.L. (1984) Hemodynamic effects of arginine-vasopressin microinjections into the nucleus tractus solitarius: a comparative study of vasopressin, a selective vasopressin receptor agonist and antagonist, and oxytocin, *Neurosci. Lett.*, **52**, 247–252.

Wallach, J.H. and Loewy, A.D. (1980) Projections of the aortic nerves to the nucleus tractus solitarius in the rabbit. *Brain Res.*, **188**, 247–251.

Willette, R.N., Punnen, S., Krieger, A.J. and Sapru, H.N. (1984) Interdependence of rostral and caudal ventrolateral medullary areas in the control of blood pressure. *Brain Res.*, **321**, 169–174.

Wislocki, G.B. and Putnam, T.J. (1924) Further observations on the anatomy and physiology of the area postrema. *Anat. Rec.*, **27**, 151–156.

Yardley, C.P., Andrade, J.M. and Weaver, L.C. (1989) Evaluation of cardiovascular control by neurons in the dorsal medulla of rats. *J. Auton. Nerv. Syst.*, **29**, 1–12.

Young, W.S. III and Kuhar, M.J. (1979) A new method for receptor autoradiography, [^{3}H] opioid receptors in rat brain. *Brain Res.*, **179**, 255–270.

Zagon, A. and Bacon, S.J. (1991) Evidence of a monosynaptic pathway between cells of the ventromedial medulla and the motoneuron pool of the thoracic spinal cord in rat, electron microscopic analysis of synaptic contacts. *Eur. J. Neurosci.*, **3**, 55–65.

Zanderberg, P., Palkovits, M. and DeJong, W. (1977) The area postrema and control of arterial blood pressure: absence of hypertension after excision of the area postrema in rats. *Pflugers Arch.*, **372**, 169–173.

3 Evaluation of Autonomic Failure

Rogelio Mosqueda-Garcia

Autonomic Dysfunction Unit, Vanderbilt University,
Nashville, TN 37232-2195, USA

Autonomic dysfunction can manifest in a myriad of symptoms. Through the years a number of different tests have been advocated to document the impact that many diseases have on the autonomic nervous system or to evaluate the extent or presence of autonomic dysfunction. Information obtained through autonomic testing is valuable in the diagnosis of many disorders including orthostatic hypotension due to primary autonomic failure or multiple system atrophy (Shy–Drager syndrome), diabetes mellitus, amyloidosis, porphyria, and neurocardiogenic syncope, among others. Autonomic testing can also be helpful to assess the adequacy and impact of therapeutic measurements of many of these diseases.

This chapter presents information on the methods available to evaluate autonomic dysfunction. The tests have been subdivided according to the end organ response that they primarily evaluate and they include: 1) Integrated evaluation of blood pressure and heart rate. 2) Predominant evaluation of blood pressure responses. 3) Predominant evaluation of heart rate responses. 4) Evaluation of sweating and skin autonomic responses. 5) Test of pupillary innervation. 6) Determination of sympathetic function.

A review of the methods, including the description of each method, the theoretical aspect of each test, how the test is performed as well as the interpretation of the results with advantages and disadvantages of each test, is presented.

KEY WORDS: Hypotension, diabetes, Valsalva maneuver, Shy-Drager, porphyria, amyloidosis, mental stress, blood pressure

INTRODUCTION

The autonomic nervous system regulates a number of different vegetative functions that are essential for homeostasis. This system provides innervation to the heart, blood vessels, glands, visceral organs, and smooth muscles. Autonomic regulation of visceral function involves, at the most simple level, reflex arcs which include sensors, afferent fibers, central processing, efferent branches, and neuroeffector junctions in target organs or tissues (Shepherd and Mancia, 1986). Because of the complexity and extent of autonomic innervation in the organism, it is not surprising that disruption of one or more parts of the reflex arc results in dysautonomia which is manifested in a myriad of symptoms.

As in any other field of medicine, the most valuable information for evaluating autonomic dysfunction comes from the clinical history. A complete and detailed examination of relevant clinical symptoms helps to establish the probable cause and extent of the disease. Consequently, every effort should be made to obtain a detailed autonomic clinical history. This is complemented by a selection of appropriate autonomic tests. Selective application of simple physiological and pharmacological examination often yields valuable information of autonomic nervous system function. Many of these tests measure a change in end-organ function (for instance, blood pressure or heart rate) in response to a stimulus. Some others evaluate the integrity of autonomic structures such as the skin or the pupil.

Information about the functional state of the autonomic nervous system has proven to be valuable in the diagnosis of many disorders including, orthostatic hypotension (Robertson and Robertson, 1985), diabetes mellitus (Rothschild *et al.*, 1987; Fisher and Frier, 1989), amyloidosis and porphyria (Fouad, Tarazi and Bravo, 1985; Robertson and Robertson, 1985). It should be emphasized, however, that no patient will require the performance of each of the tests that will be described in this chapter. The final selection will depend on the individual case. It is usually expedient to start with simple and noninvasive tests. Depending on the clinical symptoms and the initial results, specific tests can be applied that with reveal partial or selective dysfunction.

In addition, autonomic tests are helpful to evaluate the adequacy and impact of therapeutic measurements. For instance, some antihypertensive drugs may precipitate orthostatic hypotension. A simple orthostasis test will indicate the degree of hypotension caused by the antihypertensive agent and it may help in the selection of final dosage. Unfortunately, assessment of autonomic reflex arc function is often neglected in routine medical visits. In many instances, valuable information is not obtained even though many of these tests do not require special equipment or additional time (i.e., determination of standing blood pressure).

The purpose of this chapter is to provide information on the methods available to evaluate autonomic dysfunction. The description of each method contains a discussion of **A**) theoretical aspects of each test, **B**) how the test is administered, **C**) interpretation of the results, and **D**) potential disadvantages of each method. Previous reviews have divided autonomic testing in noninvasive and invasive procedures. In this chapter, on the other hand, we have divided tests of autonomic function according to the final organ response which is evaluated (Table 1).

I.   INTEGRATED EVALUATION OF BLOOD PRESSURE AND HEART RATE

I.1.   THE ORTHOSTASIS TEST

A) Compensatory alterations in autonomic tone are necessary to maintain adequate tissue perfusion to vital organs during and after assumption of the upright position. Upon standing, the increase of gravitational forces results in regional hemodynamic changes that include a pooling of blood in the lower extremities. If the subject remains standing, approximately 500 ml of blood is trapped in the distensible

TABLE 3.1

I. INTEGRATED EVALUATION OF BLOOD PRESSURE AND HEART RATE.

   1. The Orthostasis Test.

   2. Tilt Test.

   3. Valsalva Maneuver.

   4. Cold Pressor Test.

   5. Mental Stress.

   6. Handgrip Test.

   7. The Phenylephrine Test.

   8. The Clonidine Test.

II. PREDOMINANT EVALUATION OF BLOOD PRESSURE RESPONSES.

   1. The Tyramine Test.

   2. The Phentolamine Test.

   3. The Glucagon Test.

   4. The Yohimbine Test.

   5. The Cuff Occlusion Test.

   6. The Saline Infusion Test.

III. PREDOMINANT EVALUATION OF HEART RATE RESPONSES.

   1. Sinus Arrhythmia.

   2. The Atropine Test.

   3. The Isoproterenol Test.

   4. The Propranolol Test.

IV. EVALUATION OF SWEATING AND SKIN AUTONOMIC RESPONSES.

   1. Skin Wrinkling Test.

   2. Skin Vasomotor Control Tests.

   3. Sweating Tests.

   4. The Axonal Reflex Sweating Test.

V. TESTS OF PUPILLARY INNERVATION.

   1. The Ciliospinal Reflex Test.

   2. The Pupillary Ephedrine and Methacholine Tests.

VI. DETERMINATION OF SYMPATHETIC FUNCTION.

   1. Plasma Catecholamines.

   2. Microneurography.

---

veins below the level of the heart, plasma moves to the interstitial fluid and a decrease in venous return occurs (Rushmer, 1976). These changes are detected by stretch- or baro-receptors which relay information to the central nervous system which in turn decrease parasympathetic tone (importantly, cardiac vagal tone) and increase sympathetic outflow (increase in noradrenaline secretion and in peripheral resistances) (Spyer, 1984). Although less well-understood, there are probably many other inputs into autonomic control of the circulation, including cerebellar, neurourstibular, visual, abdominal and skeletal muscle afferent fibers. Evaluation of the orthostasis test may include blood pressure measurements, heart rate recordings and/or plasma catecholamine determinations.

B) Blood pressure and heart rate are measured twice with the patient in the basal (supine) state. Heart rate can be recorded with an electrocardiogram (EKG). The patient is then instructed to stand for 5 minutes. Blood pressure and heart rates are again determined twice. If plasma catecholamines are measured, an intravenous line should be placed one hour before initial testing. A basal blood sample is usually obtained after determination of cardiovascular parameters and it is repeated after 5 minutes of standing. Some authors have advocated catecholamine determinations 5 min before standing, and 2, 5, and 10 min after standing (Dunlap and Pfeifer, 1989). Furthermore, occasional patients will demonstrate a decline in blood pressure only after standing for 30 minutes or more (Streeten and Anderson, 1992).

C) When the erect posture is assumed by a normal subject, the heart rate rises about 20–30 beats per minute (bpm). This transient tachycardia is maximal around the 15th beat after standing. This is followed by a decline continuing to around the 30th beat. Five minutes after standing, the heart rate is usually about 12 bpm over basal level (Robinson et al., 1966; Ewing, 1978). In young subjects, heart rate can be raised as much as 25 bpm, while in older subjects the heart rate increment may be as small as 8 bpm (Ewing, 1978). Measurement of the shortest R–R interval after starting to stand (around the 15th second) and the longest R–R interval (around the 30th second) enables a ratio to be calculated (Ewing et al., 1978, 1985). This was originally quantified as the 30:15 ratio. A value less than 1.0 is considered abnormal. The magnitude of this ratio, however, decreases with increasing age and some authors have characterized a ratio below 1.04 in young subjects as abnormal (McLeod and Tuck, 1987). The reflex tachycardia is thought to be maintained by a decrease in parasympathetic outflow, since the response can be abolished with atropine but not with beta blockers (Ewing, 1978).

With the assumption of upright posture, the systolic pressure usually declines about 5 mm Hg, (measured by a cuff) or in a minority of subjects does not change (Robertson, 1981). The initial response may be followed by a further decrease in pressure and at the end of 30 seconds by a recovery. Some groups have considered a decrease in systolic blood pressure below 10 mm Hg as normal, a fall between 11 and 29 mm Hg as borderline and decreases more than 30 mm Hg as abnormal (Robertson, 1981; Robinson et al., 1983; McLeod and Tuck, 1987). One to two minutes after standing, circulatory readjustments occur and systolic blood pressure is not different from supine pressure in normal subjects.

Diastolic blood pressure upon standing rises between 5 to 10 mm Hg and, as with systolic blood pressure, a subsequent fall is observed. Falls in diastolic blood pressure greater than 10 to 15 mm Hg have been considered abnormal (Robertson, 1981; McLeod and Tuck, 1987). Contrary to the heart rate ratio, the arterial blood pressure responses (systolic or diastolic) to a change in posture does not alter with age (Kaijser and Sachs, 1985; McLeod and Tuck, 1987). In normal subjects, plasma levels of noradrenaline are increased about twofold with assumption of the standing position (Vendsalu, 1960; Ziegler and Lake, 1985).

In order to evaluate autonomic dysfunction, the key point to consider is that the increase in heart rate is dependent on the prevailing blood pressure level. A fall in blood pressure greater than 15 mm Hg coupled with an increase in heart rate greater than 25 bpm suggest inadequate effective volume (Robertson, 1981; Fouad, Tarazi and Bravo, 1985). A fall in blood pressure greater than 15 mm Hg associated with a heart rate change less than 10 bpm is suggestive of a baroreceptor or sympathetic nervous system defect (Robertson, 1981; Fouad, Tarazi and Bravo, 1985). A fall in blood pressure more than 15 mm Hg associated with a fall in heart rate suggest a parasympathetic component to the hypotension such as that seen in vasovagal syncope (Goldstein et al., 1982).

In the most severe cases of pure autonomic failure, there may be no increase in heart rate on assumption of the upright posture, even though blood pressure may drop more than 50 mm Hg (Onrot et al., 1986). These findings suggest a complete loss of efferent sympathetic and parasympathetic control of the heart and vasculature. The normal increase in plasma noradrenaline produced by upright posture is lost in autonomic failure. It must be noted, however, that a modest increase in plasma noradrenaline can be observed even in the most severe patients, because of the decrease in liver blood flow and noradrenaline clearance produced by upright posture in these patients (Meredith et al., 1992).

In patients with the orthostatic intolerance syndrome, there is often a very large increase in heart rate on assumption of the upright posture even if the drop in blood pressure is quite modest (Fouad, Tarazi and Bravo, 1985). Increments in heart rate of 50 bpm or greater are often seen in this group of patients.

D) When using an inflatable cuff for blood pressure determination, it is important that the arm is extended horizontally when the subject is in the standing position beeause the hydrostatic effect of the column in the dependent arm may give falsely elevated blood pressure readings (McLeod and Tuck, 1987). A misleading 15:30 heart rate ratio will be produced either if the trace is counted from where the subject completes standing or if this ratio is measured too rigidly (exactly at the 15 and 30 beats) (Dunlap and Pfeifer, 1989). A tilt table should not be used for this test as the biphasic heart rate response to standing is not observed during passive tilting.

I.2. PASSIVE UPRIGHT TILT TEST

A) Passive tilting from the horizontal to the vertical position increases the venous pressure in the feet. In most individuals, this represents an increase from 5 to 10 mm Hg to a level approximately 90 mm Hg (Walker and Longland, 1950). This rise in hydrostatic pressure causes a distension of the vascular bed below the heart. With continued passive standing, the high hydrostatic pressure causes a further progressive loss of blood volume through transudation of blood from capillaries into surrounding tissues. After 30 minutes of quiet standing, 15% of the blood volume may be lost from the circulation (Hickler, Hoskins and Hamlin, 1960). This results in a fall in central venous pressure and decreased diastolic cardiac filling. Cardiac output is decreased by about 30% with an associated fall in systolic blood pressure (approximately 15 mm Hg within 10 seconds of upright position) (Lagerlof *et al.*, 1951). Activation of baroreceptor mechanisms reflexly increases heart rate and peripheral vascular resistances that attempt to compensate tissue perfusion within 30 seconds of upright positioning. Although the tilt study evaluates similar mechanisms than the orthostasis test, some differences are present.

B) The subject is placed on a tilt table. If catecholamines are going to be evaluated, an intravenous line should be placed at this time. A rest period (30 to 60 minutes) is allowed before basal determinations. Heart rate should be continuously monitored with an EKG and blood pressure with either an intraarterial line or with a blood pressure cuff. The subject can be tilted at 60 degrees upright for a 30 to 60 minute period. Frequent blood pressure and heart rate determinations are made during this period as well as close monitoring of any symptoms experienced by the subject (lightheadedness, dizziness, tremor, headaches, nausea, etc). If the test is tolerated, at the end of the period, a venous sample for catecholamines should be obtained. Alternatively, catecholamines could be drawn a few moments after the subject experiences a presyncopal event.

An alternate method is to tilt the subject at 15° intervals. After an initial supine resting period, the tilt is stopped at 15° for 3 minutes. At the end of this period, the tilt is increased to 30°, 45°; 60°, and finally to 75°, with intervals of 3 minutes between each increase. The subject will be maintained at 75° for the rest of the time until 30 minutes of tilt are completed (Mosqueda-Garcia, Ananthran and Robertson, 1993).

C) In young normal individuals, the systolic pressure on tilting frequently increases at first, and then it falls to or slightly below control levels (Hickler, Hoskins and Hamlin, 1960). Diastolic blood pressure shows a persistent rise and pulse pressure often narrows after several minutes of tilting. Overall, mean arterial pressure on upright tilting is maintained in the normal subject (Figure 3.1). Heart rate increases from the beginning of the tilt without the biphasic effects observed after standing. Plasma noradrenaline in normal subjects increases by 80 to 100% from basal levels (Hickler, Hoskins and Hamlin, 1960; Bannister, Sever and Gross, 1977).

In patients with vasovagal syncope, the mean arterial pressure is often well-maintained during the first minutes of upright tilt. This is usually associated with a normal rise in pulse rate. During tilt study, blood pressure becomes unstable and fluctuating, and in some instances, a marked increase in heart rate is observed (Hickler, Hoskins and Hamlin, 1960). In some cases, there is a narrowing of pulse pressure with a decrease in mean arterial pressure that can continue to pronounced vasodilation in spite of sustained heart rate. It is thought that a progressive increase in cholinergic mechanisms may contribute to some of the

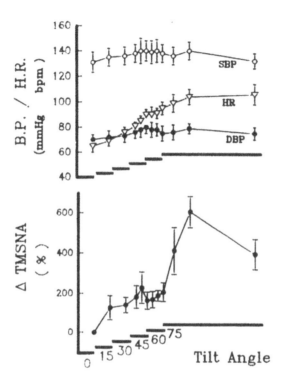

FIGURE 3.1 Hemodynamic responses during progressive upright tilt in normotensive health volunteers. Subjects were monitored continuously through an intraarterial line and after 30 minutes, they were tilted at 15° intervals every three minutes. A final tilt of 75° was maintained for 30 minutes. Values represent mean ± s.e.m. of 4 subjects.

symptoms (sweating, salivation, nausea) that the patient experiences before sudden appearance of bradycardia and hypotension (Hickler, Hoskins and Hamlin, 1960; Strasberg *et al.*, 1989). Interestingly, plasma noradrenaline has been reported to be maximal at the time of the vasovagal syncope (Hickler, Hoskin and Hamlin, 1960; Vingerhoets, 1984). On the other hand, others have failed to observe an increase in plasma noradrenaline before or during the vasodepressor syncope (Goldstein *et al.*, 1982; Ziegler *et al.*, 1986). In one report, a withdrawal in sympathetic nerve activity occurred during syncope (Wallin and Sundlof, 1982). In another report, evidence of a progressive decrease in microneurographic activity during upright tilt in patients with vasovagal syncope has been documented (Mosqueda-Garcia, Ananthran and Robertson, 1993). Some authors have advocated the addition of isoproterenol infusion to the upright tilt (Almquist *et al.*, 1989) to increase the sensitivity of the test to detect patients with vasovagal episodes. The addition of isoproterenol, however, is likely to decrease the specificity of the upright tilt (Kapoor and Brant, 1992).

In patients with oligemia or in whom volume is restricted, there is a large fall in systolic blood pressure, a large rise in pulse rate, and a normal rise on diastolic blood pressure on tilting (Fouad, Tarazi and Bravo, 1985). In patients with autonomic failure, the blood pressure consistently falls. The degree of blood pressure decrease depends on the severity of the autonomic dysfunction (Hickler, Hoskins and Hamlin, 1960; Fouad, Tarazi and Bravo, 1985).

D) A major limitation of the tilt study is the presence of false positive responses. It has been estimated that as many as 20% of healthy young adults will have syncope on passive upright tilting (Hickler, Hoskin and Hamlin, 1960). The incidence seems to be smaller among older subjects.

I.3. VALSALVA MANEUVER

A) Changes in systemic blood pressure are detected by baroreceptors which relay information to the central nervous system for subsequent efferent modulation of cardiovascular function (Spyer, 1984). The best characterized baroreflex arc involves activation of baroreceptors in the aortic arc and carotid sinus by a sudden increase in blood pressure (Spyer, 1984). The baroreceptors relay information through the IX and X cranial nerves to the nucleus of the solitary tract (NTS), where the first synapse of the baroreflex is located (Palkovits and Zaborszky, 1977). Information from this nucleus to other brainstem and higher centers modulate sympathetic and parasympathetic activity to blood vessels, heart and adrenal medulla (Palkovits and Zaborszky, 1977). In man, the Valsalva maneuver tests several of the components of the baroreflex arc and is a useful method to evaluate reflex mechanisms that are involved in circulatory control (Sharpey-Schafer 1956; Bannister, Sever and Gross, 1977).

B) The Valsalva maneuver is performed during monitoring of intra-arterial blood pressure and continuous EKG recording. In addition to a pressure transducer and a chart recording, the magnitude of the Valsalva maneuver should be monitored by measuring the pressure by which the subject exhales. A simple device to measure this is a mercury manometer adapted with a suitable mouthpiece. The subject should remain in the supine position. After basal blood pressure and heart rate determinations, the subject is instructed to exhale forcibly into the manometer to create a pressure of 40 mm Hg for 20 seconds. Then, following release of the Valsalva maneuver, blood pressure and heart rate are further followed for an additional 1 to 2 minutes.

C) The Valsalva maneuver has been subdivided into four phases that describe and relate the mechanical and cardiovascular changes (Robertson, 1981; McLeod and Tuck, 1987; Pfeifer and Peterson, 1987). During phase I, the sudden increase in intrathoracic pressure produced by strained expiration is transmitted to the aorta, causing a transient increase in stroke volume and blood pressure (Figure 3.2). In phase II, as continued strained expiration is maintained, venous return is reduced and cardiac output falls. This produces a decrease in blood pressure, which in turn causes reflex tachycardia. With the release of raised intrathoracic pressure, blood pressure abruptly falls during phase III. About 5 seconds later, during phase IV,

VALSALVA

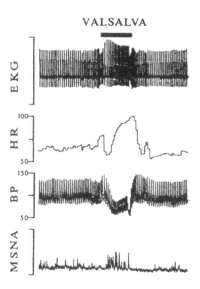

FIGURE 3.2   Representative tracing of a Valsalva maneuver. A 24 year old normotensive subject was asked to exhale forcibly into a mercury manometer to maintain 40 mmHg for 20 seconds. The two upper panels represent the changes in cardiac function (EKG and heart rate), whereas the third panel is a recording from the changes in intraarterial blood pressure. Changes in nerve sympathetic traffic are presented in the bottom panel (MSNA). Note the reflex changes in HR and MSNA evoked by blood pressure effects.

there is a marked rise in arterial blood pressure that exceeds basal levels. This blood pressure overshoot is the result of an increase in cardiac output which accompanies the rise in venous return to the heart, and the residual increase in peripheral vascular resistance due to the reflex increase in sympathoadrenal activity. The increase in stroke volume and blood pressure will in turn trigger a reflex bradycardia (Figure 3.2).

The normal tachycardia during phase II of the Valsalva maneuver is at least 20–25 bpm above baseline. This tachycardia is blunted or fails to occur in patients with sympathetic nervous system dysfunction (Robertson, 1981). During the blood pressure overshoot (phase IV), the mean blood pressure rises at least 10 mm Hg in normal subjects. A failure of blood pressure to rise in phase IV is also consistent with sympathetic failure (Ewing, 1978). On the other hand, absence or a blunted bradycardia during a blood pressure overshoot suggests selective parasympathetic dysfunction (Leon, Shaver and Leonard, 1970). The Valsalva maneuver is altered by many drugs and many disease states. In congestive heart failure, the blood pressure rises at the onset of the phase I and remains raised through the Valsalva maneuver. These patients also have no diminution in pulse pressure. At the end of the maneuver blood pressure and pulse pressure fall immediately to the baseline level without any overshoot phenomenon (Eckberg, 1980).

Methods for assessment of the Valsalva maneuver have included a relation of the increase in diastolic blood pressure during phase IV to the decrease in pulse pressure during the early part of the maneuver (Ewing *et al.*, 1985). Another method relates the decrease in heart rate after the Valsalva maneuver to the basal heart rate (Akselrod *et al.*, 1981). The Valsalva ratio is another method of evaluation frequently used. It has the particular advantage of monitoring only heart rate responses and obviates the placement of an intra-arterial line (Levin, 1966; Ewing, Campbell and Clarke, 1980). The ratio considers the longest R–R interval shortly after the maneuver to the shortest R–R interval during the maneuver. Generally, Valsalva ratios equal to or greater than 1.4 are considered normal (Levin, 1966). Some authors, however, have considered 1.2 as the lower limit in normal subjects (Ewing *et al.*, 1985; Van Lieshout, *et al.*, 1989). Ratios below 1.10 are clearly abnormal.

D) The Valsalva maneuver should be standardized to avoid increased variation in heart rate and blood pressure response. Often, if subjects are not instructed to strain properly, the cardiovascular changes may falsely indicate abnormality. Although the Valsalva ratio is a useful screening test, it is a less sensitive determination of the cardiovascular changes that occur during the Valsalva maneuver. For instance, a normal Valsalva ratio has been reported in patients who fail to show overshoot of blood pressure during the maneuver (Van Lieshout *et al.*, 1989).

Difficulties in interpreting Valsalva responses are present in patients with some cardio-respiratory diseases. Some individuals may have difficulty maintaining straining during expiration for the time necessary to initiate reflex responses. An adequate training period should always be attempted before making final determinations during the Valsalva maneuver. Furthermore, the Valsalva ratio has been reported to be affected by age (McLeod and Tuck, 1987).

I.4. COLD PRESSOR TEST

A) Different types of stress are known to increase the activity of the sympathetic nervous system in normal subjects, including mental stress, loud noise and pain (Bannister and Mathias, 1988). The application of ice to the neck or hands elicits a rapid reduction in forearm and skin blood flow with a concomitant increase in systemic blood pressure (Hines and Brown, 1936; Godden, Roth and Hines, 1965). The response is most likely initiated by a reflex mediated through afferent pain and temperature fibers from the skin (Jamieson, Ludbrook and Wilson, 1971). These somatic impulses are relayed through the spinothalamic tract to several brain areas. The cardiovascular response is mediated by efferent sympathetic vasoconstrictor fibers (that increase blood pressure), and efferent sympathetic impulses directed to the heart (that increase heart rate).

B) Several variations of the cold pressor test have been developed. Some of them include arm, feet, or face immersions into ice-cold water (Lloyd-Mostyn and Watkins, 1975; Hosking, Bennett and Hampton, 1978; Scriven *et al.*, 1984). Perhaps the most commonly used and easily performed is the cold pressor test in the arm. Blood pressure and heart rate are determined in the right arm with the subject in the supine position in a quiet room. The subject then places his left hand in a container of half ice and half water for 60 seconds. Blood pressure and heart rate are determined again in the right arm at the end of the 60 second period.

C) Normal responses to the cold pressor test average a 20 mm Hg increment in systolic blood pressure and about 10 bpm increase in heart rate (Robertson, 1981). As with other autonomic tests, considerable interindividual variation occurs. A normal cold pressor test indicates a functioning reflex arc with intact sensory nerves, spinothalamic tracts, suprapontine and intrathalamic relays, descending sympathetic pathways, peripheral sympathetic nerves, and vascular receptors. A negative cold pressor test is consistent with a lesion in one or more of these components but is not diagnostic of a lesion site.

D) Some authors have observed an increased pressor response in subjects with atherosclerosis, or, in contrast, a minimal response in normal individuals (Robertson, 1981). An increased response is typical of baroreflex failure. Another drawback of this test is that the cold stimulus is often considered very painful and a majority of the subjects are reluctant to have it repeated.

I.5. MENTAL STRESS

A) Mental activity, when performed with a time limit or under some psychological pressure, affects the autonomic nervous system (Falkner et al., 1981; Schatz, 1986). This test evaluates sympathetic efferent activity originating from the cortex and does not involve activation of the afferent limb of the reflex arc (Schatz, 1986).

B) Baseline blood pressure and heart rate are determined with the subject in the supine position. The subject is instructed to subtract serial sevens beginning at 200 and counting for 60 seconds as fast as he or she can. Blood pressure, and heart rate is recorded at the end of the behavioral task (Falkner, et al., 1981).

C) In normal subjects blood pressure will usually increase about 15 mm Hg and the heart rate will rise 10 bpm. As mentioned above, this test is particularly useful because it stimulates efferent sympathetics without depending upon afferent input from the arm or the thorax. In conjunction with other autonomic tests, the mental stress could provide important information of the lesion site. For instance, an abnormal Valsalva maneuver plus a negative mental arithmetic test suggests a lesion in brain areas that regulate cardiovascular function, or in the efferent sympathetic nerves, or a defect in vascular responsiveness.

D) Some individuals are less stressed by mental arithmetic than others (either because they perform calculations easily or because they are not concerned whether their answers are right or wrong) (Bennett et al., 1978). These subjects may have a diminished mental arithmetic response in spite of a normal sympathetic nervous system. Another limitation of this test is habituation to the procedure and/or learning of the behavioral task. These two factors will reduce the mental stress experienced by the subject and will be reflected as a diminished cardiovascular response (Dunlap and Pfeifer, 1989).

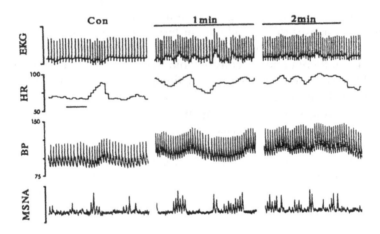

FIGURE 3.3   Changes in cardiovascular function evoked by a handgrip test in a healthy volunteer. A 28 year old white male was asked to maintain 30% of maximal voluntary contraction for 3 minutes. A one minute tracing is presented for control (Con), and for 1 and 2 minutes of sustained handgrip. Note the changes in heart rate (EKG and HR), blood pressure (BP) and muscle sympathetic nerve activity (MSNA).

I.6. HANDGRIP TEST

A) Reflex increases in heart rate and blood pressure have been observed after fatiguing muscle exercise (Lind, Taylor and Humphreys, 1964). This response is believed to be mediated by afferent fibers activated by metabolic products of anaerobic metabolism (Goodwin, McCloskey and Mitchell, 1972; Hultman and Sjoholm, 1982) and by activation of central mechanisms (Freyschuss, 1970; Goodwin, McCloskey and Mitchell, 1972). The changes in blood pressure produced by handgrip relate to activation of the sympathetic nervous system, whereas the early response in heart rate results primarily from parasympathetic withdrawal (Ewing *et al.*, 1974; Mark *et al.*, 1985).

B) The handgrip technique requires a dynamometer and sphygmomanometer. The subject is asked to grip the dynamometer as strong as he can on three successive trials. The final test will be done at 30% of the mean force of the maximal voluntary contraction. After basal blood pressure and heart rate are determined, the subject is then asked to maintain this level of pressure for 3 minutes. Blood pressure and heart rate are again determined. The response to the test is taken as the difference in blood pressure and heart rate just before releasing the grip and the resting value.

C) An average response is an increase of 20 and 16 mm Hg in systolic and diastolic blood pressure, respectively (Ewing, Irving and Kerr, 1974) (Figure 3.3). Values between 11 and 15 mm Hg for diastolic blood pressure are considered borderline and a value below 10 mm Hg is abnormal (Eckberg, 1980). An increase in 10 bpm is considered normal. The response is not affected by age (McLeod and Tuck, 1987) and negative or positive responses are interpreted like those of the cold pressor test.

Subjects with abnormally low blood pressure responses to this test are patients with damaged autonomic pathways including diabetic and uremic neuropathy (Eckberg, 1980). An abnormal handgrip test, however, has been observed in patients with mitral valve stenosis (Dunlap and Pfeifer, 1989).

D) There is considerable interindividual variation. The main source of error is not reaching maximal gripping that will produce muscle fatigue. Increases in blood pressure, however, have been reported with a grip 15% of the maximum (Dunlap and Pfeifer, 1989).

I.7. THE PHENYLEPHRINE TEST

A) Phenylephrine is an α_1-adrenoreceptor agonist with selective intrinsic activity in α-adrenergic receptors. Activation of these receptors in vascular beds evoke a vasoconstrictor effect that results in a sudden increase in blood pressure (Parks et al., 1961). In addition to the vascular response, baroreflex bradycardia can be evaluated with the administration of phenylephrine (Eckberg, 1980).

B) The patient is placed in a quiet environment with an intravenous infusion of 5% dextrose and constant monitoring of blood pressure (with an intra-arterial line) and heart rate. Phenylephrine is given as bolus injections (12.5, 25, 50, 100, 200, 400 μg) until the blood pressure increases as a mean of 30–35 mm Hg. An interval of 5 to 10 minutes is left between doses.

C) Blood pressure in normal subjects increases approximately 25 mm Hg with 200 μg of phenylephrine (Robertson et al., 1984). An increase in blood pressure greater than 25 mm Hg following 50 μg bolus injection suggests hypersensitivity to α-adrenergic agents. This occurs in conditions associated with reduced neuronal levels of noradrenaline, and with baroreflex failure (Robertson et al., 1984).

Heart rate falls with pressor doses of phenylephrine in normal subjects. The sensitivity of the baroreceptor can be assessed by relating increments in pressure to decrements in heart rate following phenylephrine administration (Mosqueda-Garcia et al., 1990). The systolic blood pressure of each beat after phenylephrine is related to the pulse interval of the next beat. This is done until the peak of systolic blood pressure. The relation between the pulse interval and systolic blood pressure should be linear, and the slope describing this line represents an index of baroreceptor sensitivity (Bristow, Brown and Cunninham, 1971).

Patients with hypertension are known to exhibit a resetting of the baroreflex slope (Bristow et al., 1969). Since baroreflex sensitivity decreases with age (Gribbin et al., 1971; Amorim et al., 1988), proper adjustments of predicted normal values should be done.

D) Standard stimulus is not easily achieved in pharmacological testings. A number of factors that affect drug distribution may also affect the final concentration of the drug at the relevant receptor site. For this reason, testing during the drug's distribution phase is preferable. Some groups have used infusion pumps that will deliver the same amount of the agent at identical speed, when comparing drug effects at different intervals. A common problem in calculating the final increment in blood pressure arises from determination of basal blood pressure. In response to breathing cyclic oscillations of blood pressure occur. Evaluation of the pressor effect of phenylephrine should take into consideration these changes in blood pressure. Failure to do this will produce significant changes in estimation of phenylephrine pressor response.

I.8. CLONIDINE TEST

A) Clonidine is an α_2-adrenergic agonist that affects blood pressure by a central effect (Laubie and Schmitt, 1977). Acting in the lower brainstem, clonidine decreases sympathetic outflow (Van Zwieten and Timmermans, 1979). Although this drug has been used in the treatment of hypertension, in autonomic testing, clonidine is used for two main purposes. In the most widely applied, it is used to suppress plasma noradrenaline levels in subjects with excessive sympathetic activation to distinguish them from patients with pheochromocytoma (Bravo and Gifford, 1984). In addition, clonidine has been used to evaluate patients with severe dysautonomia (Robertson *et al.*, 1983; Kooner *et al.*, 1991).

B) An intravenous line is placed in a patient's arm. Basal determinations of blood pressure, heart rate, and plasma catecholamines are done after 30 minutes rest. Clonidine (0.3 mg) is given orally and blood pressure and heart rate are monitored at 15 minutes intervals for the following 4 hours. At 2 hours and at 3 hours following administration, a plasma catecholamine is obtained and blood pressure and heart rate are carefully observed for the entire period.

C) In subjects with excessive sympathetic drive, plasma noradrenaline falls to normal levels. On the other hand, a failure of clonidine to decrease high baseline levels of noradrenaline (over 1000 pg/ml), even if hypotension develops, strongly suggests the presence of pheochromocytoma (Robertson, 1981; Bravo and Gifford, 1984).

In the evaluation of patients with severe dysautonomia, clonidine is used to assess how much sympathetic noradrenergic function remains in a given subject (Robertson *et al.*, 1983; Kooner *et al.*, 1991). In patients with dysautonomia, a fall in blood pressure and heart rate indicates that there is residual autonomic control of heart rate and vasculature. This, however, may be very attenuated and may be associated with very low baseline blood pressures in the upright posture. On the other hand, an increase in blood pressure or even a failure of blood pressure to fall in response to clonidine in a patient being evaluated for dysautonomia confirms that the sympathetic denervation of the vasculature is essentially complete (Robertson *et al.*, 1983). In patients with efferent parasympathetic lesions, the bradycardic effect of clonidine is attenuated or absent (Kooner *et al.*, 1991).

D) Clonidine has a sedative effect that can be significant in some patients. If hypotension has occurred in a patient being evaluated for dysautonomia, the patient should be monitored during the following 4–5 hours until the effect of clonidine has worn off. Depending on the degree of dysautonomia, hypotension or hypertension may occur. If treatment is needed, this can best be achieved by head-up or head-down tilt of the patient, or the administration of physiological saline solution.

II. TESTS THAT PREDOMINANTLY EVALUATE BLOOD PRESSURE RESPONSES

II.1. THE TYRAMINE TEST

A) The naturally occurring amine, tyramine, when given intravenously releases noradrenaline from sympathetic nerve endings which produces a transient rise in blood pressure (Bannister and Mathias, 1988). Tyramine is an indirectly acting sympathetic agent that requires intact re-uptake process and adequate noradrenaline in nerve endings for its action (Scriven et al., 1982). Tyramine probably acts by releasing noradrenaline from cytoplasmic pools without affecting granular storags (Bannister and Mathias, 1988). The administration of this amine can assess sympathetic neural stores of noradrenaline.

B) This test is carried out in the same manner as the phenylephrine test. It is necessary to have an intra-arterial recording of blood pressure since alterations in resistance artifactually raise sphygmomanometric blood pressures. When both the arterial line and the venous line have been placed, tyramine is administered in bolus doses beginning with 250 μg and then continuing with 500, 1000, 2000, 4000 and 6000 μg at 10–15 minutes intervals.

C) In situations where normal body stores of noradrenaline are present, a normal pressor response to tyramine is expected (Scriven et al., 1982; Scriven et al., 1984). Little change in blood pressure occurs below 15 μg/kg/min in normal subjects (approximately 2000 μg) (Scriven et al., 1984). A typical normal response is a 25 mm Hg increase in blood pressure after 2000 to 3000 μg of tyramine intravenously. The pressor response to tyramine is dependent on the amount of noradrenaline released.

In most of the patients with multiple system atrophy, the tyramine response is intact (Polinsky et al., 1981). Patients with peripheral autonomic failure, in contrast, exhibit significantly enhanced pressor responses (Schatz, 1986). In situations where neuronal stores of noradrenaline are low, the response to tyramine will be reduced. This has been observed when sympathetic neurons to vascular resistance vessels have been destroyed or when neural stores of noradrenaline are reduced by drugs or disease states. In situations associated with excessive neuronal stores of noradrenaline (i.e. pheochromocytoma), the pressor response to tyramine may be increased. Some authors indicated that an increase of 20 mm Hg or more in blood pressure after injection of 1000 μg or less of tyramine is diagnostic of pheochromocytoma (Robertson, 1981). Tyramine responses are also greatly enhanced in persons receiving monoamine oxidase inhibitors. On the other hand, failure of blood pressure to increase significantly after 6000 μg is characteristic of loss of neuronal stores of noradrenaline (Robertson, 1981; Bannister and Mathias, 1988).

II.2. THE PHENTOLAMINE TEST

A) Phentolamine is a 2-substituted imidazoline with α-adrenergic blocking activity. This drug produces competitive α-adrenergic blockade that is relatively transient. Phentolamine decreases blood pressure with an increase in venous capacity and decrease in peripheral resistance. The dilation is predominantly due to a direct action on vascular smooth muscle (Das and Parratt, 1971).

B) Blood pressure and heart rate are determined with the patient in resting state. An intravenous line is placed and after 30 minutes a bolus dose of 0.5 mg of phentolamine is given as a test dose. If the patient tolerates this without significant response in the first minute following administration, a 4.5 mg bolus injection is given. Blood pressure and heart rate are determined immediately afterwards and at 1 minute intervals for 5 minutes.

C) A positive test is a blood pressure fall of 35/25 mm Hg and is suggestive of pheochromocytoma (Robertson, 1981; Bravo and Gifford, 1984). The average fall in blood pressure in pheochromocytoma patients is 78/53 mm Hg. False positive responses may occur in the presence of stroke, uremia, and in patients receiving sedative, narcotic, antihypertensive, or vasoconstrictive agents (Robertson, 1981).

D) The pronounced fall in blood pressure in patients with pheochromocytoma after phentolamine makes this test somewhat dangerous and it should be employed with caution. This test should never be done with blood pressure lower than 170/110 mm Hg in the basal state. Phentolamine may also cause cardiac stimulation which seems to be more than just a reflex response to peripheral vasodilation. Due to this, cardiac arrhythmias have been reported after phentolamine administration (Das and Parratt, 1971).

II.3. THE GLUCAGON TEST

A) Glucagon, a single-chain polypeptide hormone, has actions which are considered antagonistic to those of insulin. In high concentrations, glucagon has a positive cardiac inotropic effect and can affect motor, vascular, and secretory functions in the gastrointestinal tract (Jaspan, 1981).

B) Blood pressure and heart rate are determined with the patient in the basal state. A cold pressor test is performed as described above. Following this, a 1 mg bolus of glucagon is administered through an intravenous dextrose line. Blood pressure is recorded every 30 seconds for 5 minutes.

C) Glucagon can be used to stimulate release of catecholamine from a pheochromocytoma patients (Bravo and Gifford, 1984). This test has been applied in patients with multiple endocrine adenoma syndromes to rule out the presence of this catecholamine secreting tumor. A positive test is considered when glucagon increases blood pressure by 20/15 mm Hg greater than the peak blood pressure response of the cold pressor test (Robertson, 1981).

D) As with other pharmacological tests, caution must be taken in the application of this test. Because significant increases in blood pressure occasionally occur, phentolamine must always be available to lower excessive blood pressure elevations.

This test should not be performed if the basal blood pressure is greater than 170/110 mm Hg nor if the patient has received narcotics, sedatives, or antihypertensive medications.

II.4. THE YOHIMBINE TEST

A) Yohimbine, an α_2-adrenoceptor antagonist, is an indole alkaloid found in a variety of botanical sources (Goldberg and Robertson, 1983). Yohimbine has been shown to produce a dose related rise in blood pressure with concomitant elevation of plasma catecholamines and enhancement of cardiovascular reflexes (Goldberg, Hollister and Robertson, 1983). The use of yohimbine provides a means of assessing the effect of enhancement in central sympathetic outflow on blood pressure and heart rate.

B) After a resting period and under intermittent monitoring of blood pressure and heart rate (every 5–10 minutes), yohimbine is administered. The usual dose is 5 mg p.o., and heart rate and blood pressure are followed for the next 3 hours.

C) The yohimbine test is particularly useful in detecting how much the residual autonomic nervous system can be activated in patients with dysautonomias. An increase in blood pressure of 20 mm Hg or greater indicates significant retention of sympathetic noradrenergic neurons to the vasculature. A failure of yohimbine to increase blood pressure suggests that loss of sympathetic noradrenergic fibers has proceeded very far in the subject being tested. Plasma noradrenaline levels before and after yohimbine will generally show an increment in noradrenaline of at least 25% in healthy subjects (Goldberg, Hollister and Robertson, 1983). This may be absent in patients with dysautonomia.

D) In addition to acting centrally to increase sympathetic flow, yohimbine may also block inhibitory presynaptic α_2-adrenoreceptors located in sympathetic noradrenergic nerve terminals. This will result in an increase in noradrenaline release and may contribute to the pressor effects of the drug (Goldberg and Robertson, 1983). The interpretation of this test is also compounded by the variable magnitude of postsynaptic adrenoreceptor hypersensitivity present in patients with autonomic failure (Biaggioni, Robertson and Robertson, 1993).

II.5. CUFF OCCLUSION TEST

A) The cardiovascular system uses several mechanisms for adapting to changes in volume. The cuff occlusion test provides a means of assessing the capability of the cardiovascular system to withstand an acute intravascular volume loss.

B) Blood pressure is determined in the resting subject. Immediately after, blood pressure cuffs are placed around the thighs and inflated to a level of 10 mm Hg below diastolic blood pressure to occlude venous return. If this pressure is maintained for 10 minutes, more than a liter of blood can be pooled in the legs following this maneuver (Robertson, 1981).

C) The normal response is no change or a small reduction in blood pressure (no more than 10 mm Hg). Reduction of blood pressure more than this amount strongly suggests a sympathetic nervous system lesion. Impairment of volume and pressure reflexes will also result in an abnormal response. Subjects whose blood volume is abnormally low at baseline may also have an excessive reduction in blood pressure. In this case, however, there is usually a compensatory tachycardia.

D) There is very limited information regarding the sensitivity of this test. Presently, there is no data correlating catecholamine values and the decrease of blood pressure evoked by the cuff occlusion test in dysautonomia patients. Similarly, no information is available concerning the effects of age and other factors.

II.6. THE SALINE INFUSION TEST

A) The saline infusion test allows to study the function of buffering cardiovascular reflexes. Under normal conditions, reflex adjustment mechanisms reduce the cardiovascular impact of an acute volume overload. The saline infusion test may be considered as the inverse of the cuff occlusion test with both procedures involving similar circulatory reflexes (such as baroreceptor and/or renal adaptative processes).

B) The subject remains supine and an intravenous line is placed for administration of saline. A Foley catheter is placed for determination of urine output. Blood pressure, heart rate, and urine output are monitored for at least one hour before volume load. A three thousand milliliters of physiological saline solution is infused over one hour. During and at the end of this period, blood pressure, urine output, and urinary sodium content are recorded.

C) Normal subjects will increase their blood pressure no more than 10 mm Hg during this maneuver. A subject with intact cardiovascular reflexes will excrete approximately 6 ml of urine per minute and 15 mEq of sodium per hour. Patients with severe autonomic impairment will have a blood pressure elevation greater than 20 mm Hg, excrete about 23 ml of urine per minute and excrete 80 mEq of sodium per hour.

D) This test must be conducted with extreme caution. Acute saline load may precipitate congestive heart failure in patients with compromised cardiovascular function. This limits the application of the saline infusion test in many patients with autonomic dysfunction and in elderly subjects.

III. TESTS EVALUATING HEART RATE RESPONSES

III.1. SINUS ARRHYTHMIA

A) Heart rate has a cyclic variation which is coupled to respiratory function. During inspiration heart rate increases whereas it slows during expiration (Kollai and Koizumi, 1981). This arrhythmia if measured during the R–R interval, is known as R–R variation or sinus arrhythmia. The variation in the heart rate is mainly under parasympathetic control (Wheeler and Watkins, 1973). Anticholinergic drugs, such as atropine, importantly reduce R–R variation. A small sympathetic influence is suggested by the effects of β-adrenergic drugs on R–R variation (Weinberg and Pfeifer, 1986).

B) Several techniques have been proposed to analyze sinus arrhythmia. The most routinely used method implies finding the difference between the maximum and minimum heart rate recorded during deep breathing at a rate of six breaths per minute (Ewing *et al.*, 1981). Recording of the R–R variation in the mornings after an overnight fast in a quiet atmosphere has been recommended. Continuous EKG and breathing recording is necessary. Minimal and maximal heart rate are calculated from the shortest and longest R–R interval on the electrocardiographic tracing.

Other methods of sinus arrhythmia evaluation include calculating the standard deviation of the mean R–R interval during a period of regular and quiet breathing (Ewing *et al.*, 1981), the expiration-inspiration ratio (Sundqvist, Almer and Lilja, 1979), and the circular mean resultant method (Weinberg and Pfeifer, 1984). The expiration-inspiration method is based on the calculated ratio of the longest R–R interval during expiration to the shortest R–R interval during inspiration. The circular mean resultant method is based on mathematical calculations involving vector analysis of R–R variation during each breathing cycle.

C) An increase in heart rate equal to or more than 15 bpm has been considered normal (Mackay *et al.*, 1980). Increases less than 5 bpm in elderly subjects or less than 10 bpm in young individuals are considered abnormal. Subjects under the age of 40, generally have 10–30 bpm difference between the shortest and fastest R–R interval (Persson and Solders, 1983). For subjects over 50 years old, a 5–20 bpm difference is normal (Wheeler and Watkins, 1973). A difference of less than 5 bpm suggests dysfunction of the vagus nerve, the thoracic afferent fibers, or the vasomotor center. The expiration-inspiration ratio also decreases with age (Smith, 1982). In general, ratios below 1.2 are considered abnormal up to age 40.

D) The major pitfalls in evaluating sinus arrhythmia come from ignoring factors that affect this test. For instance, patient position affects R–R variation (Dunlap and Pfeifer, 1989). Therefore, this test should be done always in the same supine position. Some training of the procedure is usually beneficial to standardize this test. The subject should be instructed and several trials should be done before final determinations.

III.2. THE ATROPINE TEST

A) Neural control of the heart involves parasympathetic modulation that is evident with cholinergic blocking agents (Wheeler and Watkins, 1973). Atropine is a widely used muscarinic antagonist which can be given intravenously. This allows one to assess the prevailing level of parasympathetic control of the heart rate.

B) This test requires constant EKG monitoring. Through an intravenous infusion of 5% dextrose, 0.04 mg/kg of atropine sulfate is given over 5 minutes. Heart rate is observed during the following 15 minutes. Complete efferent cardiac parasympathetic blockade is achieved with this dose. Blockade is achieved within five minutes and lasts for about 30 minutes.

C) In normal subjects, the heart rate will increase more than 30 bpm. When atropine is given slowly or at lower doses, a transient decrease in heart rate precedes the tachycardia. The increase in heart rate is probably the result of blocking vagal effects on the sino-atrial node pacemaker. Failure of the heart rate to increase after atropine suggests reduced parasympathetic function. A blunted response is also seen in congestive heart failure.

D) The response to atropine is more noticeable in young adults, in whom vagal tone is high. In infants or elderly subjects, atropine may fail to accelerate the heart rate. In addition to the usual muscarinic antagonism in the periphery, atropine can block α_1-adrenoceptors during the distribution phase of a large intravenous dose. Furthermore atropine may induce atrio-ventricular (AV) dissociation during the first minute of administration. This occurs before full parasympatholytic effect of the drug is manifest and may be of some concern for some patients. However, the transient AV dissociation evoked by atropine is asymptomatic and usually missed if the EKG is not being carefully monitored. This dose of atropine may produce significant side effects due to cholinergic blockade, such as dry mouth, urinary retention, constipation, and mental disturbances, especially in the elderly.

III.3. ISOPROTERENOL TEST

A) Isoproterenol is a β-adrenergic agonist that can stimulate both β_1 and β_2-adrenergic receptors. Because of positive chronotropic and inotropic effects of isoproterenol, cardiac output increases. Peripheral vascular resistance may also decrease with intravenous infusion of isoproterenol.

B) The subject is placed in a quiet environment with an intravenous infusion of 5% dextrose and constant EKG monitoring. Intra-arterial recording of blood pressure is optional, as this test addresses the effects on heart rate. Following a control period and some placebo injections, isoproterenol is administered in a bolus dose of 0.1 μg. If no response occurs, the dose is increased stepwise (0.25, 0.5, 1.0, 2.0, 4.0 μg, etc), until an increase in heart rate of 30–35 beats occurs. One or two placebo injections are given randomly during the test. Curves are constructed relating the increase in heart rate to the dose of isoproterenol given. From the dose-response curve, the dose which should give an increase in heart rate of 25 bpm is calculated (Vestal, Wood and Shand, 1979).

C) This test provides a measure of the functional state of the β-adrenergic receptor. Normal values for the isoproterenol test decline with age (Vestal, Wood and Shand, 1979). Young people (21 to 40 years old) will usually respond to 1 μg with 25 bpm increase in heart rate. Patients above 60 years old require approximately 5 μg to accomplish the same response. Patients with pure autonomic failure and other conditions associated with reduced circulating catecholamines may have hypersensitivity to isoproterenol and respond to smaller doses than would be expected for age-matched subjects (Robertson et al., 1984; Baser et al., 1991). The dose of isoproterenol that increases the heart rate 25 bpm usually moderately reduces blood pressure. If sympathetic dysfunction or α-blockade is present, the blood pressure reduction may be profound.

D) Sinus arrhythmia can significantly alter heart rate in some individuals. To compensate for this, baseline heart rate is determined as the mean of three shortest contiguous respiratory rate (RR) intervals during the 3 minutes prior to the injection; the isoproterenol heart rate is determined by measurement of the 3 shortest continuous RR intervals during 3 minutes following the isoproterenol injection.

III.4. PROPRANOLOL TEST

A) Propranolol is a β-adrenergic blocking agent. In autonomic testing, propranolol is used to determine the prevailing level of β_1 and β_2 receptor activation in a given subject (Weiner, 1980). In contrast to the phenylephrine test, the tyramine test, and the isoproterenol test, the propranolol test is meant to achieve a high level of β-adrenoceptor blockade in all subjects tested.

B) The patient is maintained supine with a slow intravenous infusion of 5% dextrose and constant EKG monitoring. After a resting period, propranolol is infused intravenously at 1.1 mg/min for 10 minutes. At the end of this period heart rate is evaluated. If a high level of β-adrenoceptor blockade is needed to be continued beyond this point, an infusion of 0.05 mg/min can be continued for the duration of the study.

C) Propranolol decreases heart rate and cardiac output, prolongs mechanical systole, and may decrease blood pressure (Dollery, Paterson and Conolly, 1969). A normal response is a fall in heart rate of at least 5 bpm. In patients with a high level of sympathetic activation or pheochromocytoma, the fall in heart rate may be much greater. A failure of the heart rate to fall by 5 bpm in response to propranolol suggests inadequate baseline sympathetic tone consistent with dysautonomia.

D) Propranolol should not be used in patients with asthma. This β-blocker can cause atrio-ventricular dissociation and cardiac arrest in patients treated with drugs that prolong A–V conduction. In addition, heart failure may develop after intravenous administration of propranolol. It should not be used in patients with pheochromocytoma because of the risk of precipitating a hypertensive episode. Other limitations with propranolol are similar to the restrictions described above for other pharmacological tests.

IV. TESTS THAT EVALUATE SWEATING AND SKIN AUTONOMIC RESPONSES

IV.1. SKIN WRINKLING TEST

A) Autonomic effector organs in human skin such as blood vessels and sweat glands are innervated by separate sets of sympathetic fibers (Dale and Feldberg, 1934; Fagius and Wallin, 1980). Although sympathetic skin activity is not baroreflex responsive, it does respond to thermoregulatory and emotional stimuli (Delius *et al.*, 1972a). The basis for the skin wrinkling test is the observation that intact innervation of the fingers is important in mediating skin wrinkling in response to application of water (Bull and Henry, 1977).

B) In a temperature controlled room, both hands are immersed in tap water at 40°

for 30 minutes. Under direct light, the palm and surface of the finger are observed for wrinkling.

C) The finger wrinkling response is abolished by upper thoracic sympathectomy. The test is also abnormal in some patients with diabetic autonomic dysfunction, the Guillain-Barre syndrome and other peripheral sympathetic dysfunction in limbs. The test is valuable because it is easy to apply and reflects peripheral rather than cardiac sympathetic innervation. Furthermore, it is of particular help when unilateral lesions are suspected since right-left asymmetry of finger wrinkling is particularly easy to detect.

D) As with other tests, there is high interindividual variation. Furthermore, there is not a clear way to grade the wrinkling response.

IV.2. TESTS OF SKIN VASOMOTOR CONTROL

A) Sympathetic skin efferents modulate vasoconstrictor and vasodilator fibers which in turn affect skin blood flow (Uvnas 1954; Vallbo *et al.*, 1979). The relevance of each fiber (vasoconstrictor vs vasodilator) is not uniform and marked regional differences are present. For instance, reflexes originating in the thorax act to increase sympathetic tone in the dorsal handvein through activation of vasoconstrictor fibers, without participation of vasodilator fibers (Roddie, Shepherd and Whelan, 1957). On the other hand, skin blood flow to the forearm and other regions, is regulated by both vasoconstrictor and vasodilator fibers (Roddie, Sherpherd and Whelan, 1957).

B) Peripheral vasomotor control to the skin can be evaluated using plethysmography, skin temperature, heat flow discs, laser doppler measurements, and xenon clearance. In venous occlusion plethysmography, the extremity or digit is enclosed in a plethysmograph and the venous return is occluded for several seconds (Aminoff, 1980). During the time that arterial inflow is unimpeded the volume increment can be recorded by a volume transducer. The slope of the volume:time trace is an index of blood flow. A related method includes the use of a strain-gauge plethysmograph which will record changes in circumference instead of volume (Robertson, 1981).

In the venoconstriction test, a #21 gauge intravenous needle is placed in a superficial dorsal handvein. This is connected to a pressure transducer, calibrated to accurately measure pressures over the range 2–40 mm Hg. A pediatric blood pressure cuff is wrapped around the wrist of the arm. Just prior to measurement of vasoconstriction, the forearm should be raised by bending the elbow so that the hand is directly above the elbow. At this time, the pediatric blood pressure cuff is inflated to 300 mm Hg and the hand is again allowed to rest horizontally (Robertson, 1981).

Laser Doppler flowmeter uses a laser light beam. A portion of the light strikes non-moving structures and is reflected back with no frequency shift (Doppler broadening). The spectral broadening is related to the red cells in the flow field and the instrument measures the speed of the red blood cells (Stern *et al.*, 1977).

C) Using one of the above procedures, changes in blood flow can be measured. Alterations in blood flow can be triggered in a normal person with immersion of one hand in hot water. This maneuver causes reflex vasodilation in the other hand as well as an increase in blood flow. This reflex is thought to be due to an elevation in central temperature (Pickering, 1932). Another procedure that measures changes in blood flow to the hand is the application of radiant heat to the trunk (Kerslake and Cooper, 1950). A rapid increase (within 2 minutes) in the blood flow normally results from vasodilation which is mediated thorough a reflex pathway above C5 (Appenzeller and Schnieden, 1963).

In the venoconstriction test, the pressure as recorded by the transducers becomes a function of venous tone while the cuff is elevated. After three minutes of quiet rest, venous tone will have stabilized and the introduction of a sympathetic stimulus, such as the Valsalva maneuver or a deep breath will provoke an immediate increase in venous tone unless sympathetic nervous system outflow is blocked. After a period of reperfusion, the process can be repeated. The response is defined as the mean of the two (from a total of four) maximal increases in venous tone. A normal response is increase in venous tone (5–35 mm Hg) following the Valsalva maneuver or deep breath. Subjects who do not have as much as 5 mm Hg response have lesions either in the afferent neurons from the thoracic receptors, lesions in the vasomotor center, or impairment in the efferent sympathetic outflow to the arm. Absent forearm venoconstriction responses are characteristic of autonomic failure and therapy with noradrenergic neuron-depleting drugs such as guanethidine and guanadrel.

D) Problems with these methods include assumptions that limit the interpretation of the results. For instance, laser Doppler velocimeter assumes static vascular geometry and laminar flow. However, vascular geometry changes, flow is nonlaminar, and red cells have a spining motion as they flow. Other disadvantages of this method include a difficult calibration for different tissues and the low penetration of the laser beam (1 mm).

In the venoconstriction test, failure to raise the hand prior to inflation of the cuff will lead to excessive blood remaining in the hand after cuff inflation which reduces the sensitivity of subsequent venous pressure measurements. The increase in hand blood flow seen after radiant heat to the trunk will be present if the central temperature is below 36.5°C. If central temperature is below that, vasodilation may be delayed over 15 minutes.

IV.3. SWEATING TESTS

A) There are two types of sweat glands, ecrine and apocrine (Low, 1984). Ecrine glands receive a rich supply of blood vessels and sympathetic fibers. They, however, are unusual in that sympathetic innervation is cholinergic (Dale and Feldberg, 1934; Fagius and Wallin, 1980). As a consequence, secretion of this type of gland is abolished by sympathectomy and atropine. The secretion of apocrine glands, in contrast, is not abolished by sympathectomy.

B) The majority of tests that quantify the sweat response are based on a color change induced by the reaction of sweat droplets with a chemical indicator. For instance, a powder is prepared by mixing 30 grams of quinizarin (1,4-

dihydroxyanthraquinone) with 30 grams of powdered sodium carbonate, and 60 grams of starch (Guttmann, 1947). It is then dusted over the skin with a pad of cotton wool with moderate pressure used to ensure that the orifices of all sweat ducts become filled. Even distribution of powder is important. Once the powder is applied, the patient is warmed either with a heat cradle or by placing one hand in hot water (43°) until the central body temperature has risen 1° above normal. Sweating is indicated by a color change of the quinizarin powder from blue-grey to dark purple. Individual sweat ducts initially appear as small dark dots whereas anhidrotic areas remain unchanged in color (Guttmann, 1947).

Alternatives to quinizarin powder include starch iodine paper, povidone iodine or a mixture of iodine, alcohol and castor oil (Low, 1984). Starch-containing paper should be kept in a closed vessel with iodine until use. It should then be applied to the skin with transparent adhesive tape. The use of povidone iodine soap (Betadine) is achieved by mixing soap and water as for a surgical scrub and placing it on the areas of skin to be tested. The soap film on the skin should then be allowed to dry before the test is undertaken. Sweating appears by a darkening in the color of the soap film in the areas where sweat glands are active. With the iodine-alcohol-castor oil mixture the sweat droplets that form are retained and appear violet-black. The number of droplets can be counted either directly or on photographs, and the diameter of the droplet provides an estimate of the volume of the droplet (Low, 1984).

C) Sweating abnormalities are common in autonomic disorders. Large areas of anhidrosis are found in patients with autonomic neuropathy due to central or peripheral causes (List and Peet, 1938). The pattern of anhidrosis may be of value in localizing lesions causing Horner's syndrome (Morris, Lee and Lim, 1984), and peripheral nerve abnormalities (Guttmann, 1940). Documentation of reduced sweating confirms that the neurological lesion is not confined to noradrenergic neurons.

D) The indicators have the disadvantage of staining clothes and may persist for several days in crevices. Addition of vinegar to the soap and water facilitates cleaning.

IV.4. THE AXONAL REFLEX SWEATING TEST

A) Postganglionic sympathetic lesions causing anhidrosis can be distinguished from preganglionic or central lesions by iontophoresis or injection of a cholinomimetic substance into the skin of the anhidrotic area (Hyndman and Wolkin, 1941). Sweating in response to administration of parasympathicomimetic drugs is thought to be dependent of an axon reflex (Chalmers and Keele, 1952). After denervation, the sweat gland does not respond to acetylcholine or related substances and it provides a sensitive index for intactness of its nerve supply (Low, 1984).

B) Intradermal acetylcholine (one-tenth milliliter of a 1:10,000 acetylcholine solution) is injected in a test area. Some authors (Coon and Rothman, 1941) have administered intradermal nicotine (usually 10^{-5} g/ml) or, alternatively, pilocarpine (List and Peet, 1938).

C) In response to acetylcholine, the sweating begins on the injection wheal within 5 to 10 seconds after administration and is thought to be due to a direct muscarinic action. The response is maximal 5 minutes after administration and may reach a diameter of 2 to 3 cm. There is in addition a second response. This develops within 2 minutes and is characterized by a sudden outbreak of sweat droplets in an area about 5 cm in diameter surrounding the wheal. This is probably a nicotinic action. A localized pilomotor effect is also present.

Intradermal nicotine causes a sudden outbreak of sweat droplets over an area of approximately 5 cm in diameter. This response lasts between 2 to 4 minutes and also occurs with piloerection. Both of these effects are the result of the axonal reflex. Acetylcholine is often chosen over nicotine, as the response to the former is more prominent. Sweating around the injection site indicates that the local postganglionic nerve supply to the gland is intact.

D) There is variation to the cholinomimetic response between women and men. Some studies have indicated that there is a stronger sweat response in men (Kahn and Rothman, 1942). Also, racial differences have been noted (Gibson and Shelley, 1948). The response to intradermal parasympathomimetic agents was stronger in black males, following by white males, black females, and white females (Gibson and Shelley, 1948). The amount of sweating has been reported to decrease with age (Silver, Montagna and Karacan, 1964), although more recently, others have been unable to confirm this (MacMillan and Spalding, 1969). Circadian changes in sweating also influences the results of these tests. In one study, the sweating threshold was lower in the morning than in the afternoon or evening (Crocford *et al.*, 1970).

V. TEST OF PUPILLARY INNERVATION

V.1. THE CILIOSPINAL REFLEX TEST

A) The size of the pupil is influenced by parasympathetic and sympathetic impulses (Smith, 1988). Sympathetic fibers affect pupillary reflexes acting on the iris, blood vessels and on the dilator muscle. Noradrenaline secreted by postganglionic sympathetic nerves acts on α-adrenergic receptors that contract the radial smooth muscle of the dilator pupillae. Under intact sympathetic innervation mydriasis develops in response to a painful stimuli. On the other hand, sympathetic dysfunction results in a small pupil which fails to dilate in darkness.

B) In a quiet room with dim lighting, the size of the pupil is observed before and during painful stimulation of the skin of the face, neck or upper trunk on the ipsilateral side.

C) A normal response is a bilateral pupillary dilation of 1–2 cm and indicates intact sympathetic innervation of the eye. It is a marker for pre- or postganglionic (as opposed to higher) sympathetic lesions.

D) Considerable variation and difficulties in obtaining adequate measurements of pupillary changes are the main drawbacks of this test.

V.2. THE PUPILLARY EPHEDRINE AND METHACHOLINE TESTS

A) In addition to sympathetic influence, the size of the pupil is also affected by parasympathetic tone (Smith, 1988). Parasympathetic innevation modulates the activity of the pupillary sphincter. Different pharmacological agents (with sympathetic or parasympathetic properties) affect pupillary size. Cholinomimetic drugs acting on muscarinic receptors block the responses of the pupillary sphincter and ciliary body (Bourgon, Pilley and Thompson, 1978). On the other hand, sympathomimetic substances have mydriatic action which is probably mediated by α-adrenergic receptors (Smith, 1988).

B) Sympathomimetic drugs are preferred over anticholinergics as the mydriatic actions of the former agents last only a few hours and do not increase intraocular pressure. In a dimly lit room, two drops of a 1:1000 solution of adrenaline are instilled in the conjunctival sac. In the methacholine test, 2 drops of a 2.5% freshly prepared solution is instilled into the conjunctival sac (Lepore, 1985). Other drugs which have been administered include cocaine (4%), hydroxyamphetamine (1%), and pilocarpine (0.125%) (Lepore, 1985).

C) In normal subjects, little or no change in the size of the pupil occurs in response to adrenaline. In patients with denervated postganglionic sympathetic nerves, denervation hypersensitivity leads to a widely dilated pupil.

Acute parasympathetic dysfunction results in a large pupil with poor reaction to light. In the methacholine test, the pupil becomes considerably smaller in patients with parasympathetic paralysis. In pupils with intact innervation, methacholine or pilocarpine cause no contraction.

Instillation of 4% cocaine into the conjunctival sac causes pupillary dilation when the sympathetic innervation is intact, whereas the response is reduced or absent when there is a lesion in the oculosympathetic pathways. In normal subjects, instillation of hydroxyamphetamine causes pupillary dilation by releasing noradrenaline from postganglionic sympathetic fibers. An absence of mydriasis after instillation of this drug suggests a lesion of postganglionic oculosympathetic fibers.

D) A mydriatic response to adrenaline does not necessarily indicate autonomic impairment. This response has also been reported to occur in hyperthyroidism, diabetes, acute pancreatitis, renal hypertension, and glaucoma (Robertson, 1981).

VI. DETERMINATION OF SYMPATHETIC FUNCTION

VI.1. PLASMA CATECHOLAMINES

A) One of the most useful tests in the evaluation of autonomic dysfunction and orthostatic intolerance is determination of plasma catecholamines (Robertson, 1981; Low, 1984; McLeod and Tuck, 1987; Dunlap and Pfeifer, 1989). After noradrenaline is released from the nerve terminals, this substance is removed from the synaptic cleft by several uptake processes with only a small fraction reaching the circulation (Bevan, Bevan and Duckles, 1980). The fraction of noradrenaline present in the bloodstream derives mainly from nerve endings of blood vessels, including arterioles, which are one of the main determinants of total peripheral resistance. In contrast, the contribution of the adrenal medulla for plasma noradrenaline is minimal under resting conditions (Brown et al., 1981).

Noradrenaline reaches the bloodstream mainly by diffusion into the capillary system at the adventitial surface of arteries and in minor proportion by diffusion directly into the lumen of the vessel (Bevan, Bevan and Duckles, 1980). In most vascular beds, excluding the liver and lungs, the rate of noradrenaline spillover exceeds the rate of noradrenaline removal. This results in an arteriovenous increment in plasma noradrenaline (Goldstein et al., 1985).

B) Determination of plasma catecholamines can be done as a single procedure or (more often) in combination with other autonomic tests. The conditions for taking blood samples must be standardized. To obtain an adequate baseline value, at least 30 minutes should be allowed between placement of the catheter and the drawing of the sample. Proper care should be taken during sample collection, storage and processing of the sample. For instance, hemolysis may artificially affect the values of noradrenaline in plasma. The blood should be collected in a previously chilled tube and processed as soon as possible in order to avoid possible changes in noradrenaline values. Once the plasma is obtained, storage at $-70°C$ is recommended. Specific details of catecholamine measurement are provided in several papers (Robertson et al., 1979; Ziegler, 1989).

C) In general, changes in plasma noradrenaline has been shown to correlate well with factors that affect sympathetic tone (Goldstein, 1987). For instance, interference with central sympathetic function and/or blockade of ganglionic neurotransmission decreases plasma noradrenaline. In normal subjects, plasma noradrenaline approximately doubles when assuming the standing position or in response to upright tilt (Hickler, Hoskins and Hamlin, 1960; Bannister, Sever and Gross, 1977). In contrast, patients with progressive autonomic dysfunction (including diabetics) have little or no increase in plasma noradrenaline and this correlates with their orthostatic intolerance (Bannister, Sever and Gross, 1977). In patient with pure autonomic failure, basal levels of noradrenaline are lower than in normal subjects (Polinsky, 1988). Similar low values are observed in patients with sympathectomy and in patients with tetraplegia. Plasma catecholamines and their metabolites are also of diagnostic value in conditions such as pheochromocytoma and dopamine beta hydroxylase deficiency (DBH). Plasma catecholamines are particularly important in pheochromocytoma if the sample is obtained during a paroxysmal hypertensive event (Bravo and Gifford, 1984). The absolute absence of noradrenaline in presence of abnormally high values of dopamine and its metabolites suggest DBH deficiency (Robertson et al., 1986).

D) Several factors should be considered when inferring sympathetic function from plasma catecholamine values. Besides considerable variation in plasma catecholamines that reflect inter- and intra-assay variability and sensitivity, the following kinetic and physiological factors determine the final plasma values of noradrenaline: 1) Values of venous blood noradrenaline is the ratio of the spillover rate into the blood stream and clearance (Bevan, Bevan and Duckles, 1980). Changes in noradrenaline clearance by pharmacological agents (such as propranolol, Cryer *et al.*, 1980) or by pathophysiological conditions (such as congestive heart failure, Esler *et al.*, 1985; Leimbach *et al.*, 1986) could increase noradrenaline levels without real increases in sympathetic neuronal function. **2)** During experimental manipulations, sympathetic outflow and changes in blood flow may vary from one vascular bed to another (Goldstein, 1987). **3)** Modulation of noradrenaline release by presynaptic systems and multiple removal processes for this catecholamine also determine the cleft-plasma concentration gradient of noradrenaline (Goldstein, 1987). **4)** The changes in synaptic sympathetic function are not reflected immediately (and sometimes entirely) on plasma. Similarly, when studying rapid opposite hemodynamic changes within one simple test (such as in the Valsalva maneuver), it is unlikely to detect changes on plasma catecholamines. In summary, if these factors are not carefully considered, peripheral venous blood catecholamine levels can be, at best, an insensitive index of sympathetic activity. Some, but not all, of these problems can be circumvented with measurement of noradrenaline clearance using a steady-state infusion of labelled noradrenaline to determine actual spillover rate (Esler *et al.*, 1980).

VI.2. MICRONEUROGRAPHY

A) The recording of sympathetic nerve activity through an electrode placed in a peripheral nerve is known as microneurography (Delius *et al.*, 1972b). This technique allows recordings of action potentials from sympathetic fibers in awake, unanesthetized subjects. The autonomic features of peripheral sympathetic nerves, with each fascicle encapsulated by connective tissue of high impedance, allows

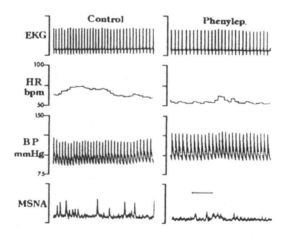

FIGURE 3.4   Cardiovascular changes produced by phenylephrine administration. These tracings represent the hemodynamic changes evoked by an infusion of saline (0.1 mg/kg/min, control) or phenylephrine (0.8 µg/kg/min) in a normotensive healthy volunteer. Note the increases in blood pressure and the lower levels of heart rate and muscle sympathetic nerve activity during the phenylephrine infusion.

recording selectively from that fascicle, without confounding activity from neighboring fascicles. With microneurographic recordings it is possible to evaluate two types of sympathetic fibers: skin and muscle sympathetics (Delius *et al.*, 1972a, 1972b; Vallbo *et al.*, 1979; Wallin and Sundlof, 1982; Mark *et al.*, 1985; Fagius and Wallin, 1980).

Muscle sympathetic nerve activity (MSNA) consists of bursts of impulses coupled to cardiac rhythm and blood pressure (Delius *et al.*, 1972b). Several studies have indicated that variations on muscle sympathetic nerve activity are determined by fluctuations of diastolic blood pressure (Delius *et al.*, 1972b). Accordingly, muscle sympathetic activity is modulated by baroreflex activation (Sanders, Fergurson and Mark, 1988) (Figure 3.4). For instance, activation of carotid sinus baroreceptors inhibits muscle sympathetic activity (Delius *et al.*, 1972b). In contrast, baroreceptor denervation increases muscle sympathetic activity and eliminates its cardiac rhythmicity (Delius *et al.*, 1972b).

Recordings from muscle sympathetic fibers has been shown to correspond to efferent activity since local anesthesia of the nerve proximal to the recording eliminates the activity, but anesthesia distal to the recording site does not (Delius *et al.*, 1972b). Furthermore, ganglionic blockade with trimethaphan reversibly eliminates the activity. Significant correlations between microneurography and maneuvers that affect central sympathetic tone have been observed (Delius *et al.*, 1972b). It has been suggested that impulse traffic in sympathetic muscle nerves is an important determinant of forearm venous plasma concentration of noradrenaline (Rea *et al.*, 1990a). Accordingly, several studies demonstrated a highly significant correlation between the changes on muscle sympathetic nerve activity and plasma noradrenaline concentration during stimulation of sympathetic drive (Rea *et al.*, 1990a). Peak levels of plasma noradrenaline, however, lag behind the increase in muscle sympathetic nerve activity (Rea *et al.*, 1990a). Furthermore, measurements of arterial noradrenaline spillover are felt to underestimate the degree of sympathetic activation induced by hypotension when compared with measurements of muscle sympathetic nerve activity (Rea *et al.*, 1990b).

B) Recordings of microneurographic activity can be obtained from the peroneal nerve. At the level of the fibular head, the approximate location of this nerve is done with transdermal electrical stimulation (10–60 V, 0.01 msec duration). This stimulation produces muscle contraction of the foot. Following this, a reference tungsten electrode, with a shaft diameter of 200 μm, is placed subcutaneously. A similar electrode, with an uninsulated tip (1–5 μm), is inserted into the nerve and used for recording of muscle sympathetic nerve activity. Placement of the recording electrode is guided by electrical stimulation (1–4 V, 0.01 msec duration). Electrical stimulation is performed with a stimulator connected to an isolation unit. When the electrode is placed within the nerve, the electrical stimulation produces muscle twitches but not paresthesia. The electrode is then switched to a recording mode, and fine adjustments of its position are made to obtain satisfactory multiunit recordings of sympathetic nerve activity.

Recorded signals are fed to a preamplifier (1,000 fold amplifications) and filtered using a band width between 700 and 2,000 Hz. The filtered signal is rectified, amplified 100-fold, and integrated in a resistance-capacitance network using a time constant of 0.1 second. The final signal is monitored using a storage oscilloscope and recorded on paper in a polygraph from which a mean voltage is used for measurements. Sympathetic nerve activity is expressed as burst/min × mean burst amplitude.

C) This test allows direct assessment of sympathetic nerve activity to the skeletal muscle vasculature (Delius *et al.*, 1972b). Microneurography is particularly informative about the autonomic effects of physiological maneuvers and pharmacological agents. In normal subjects sympathetic nerve activity increases during phase II of the Valsalva maneuver (hypotensive phase) (Figure 3.2) and in response to held expiration. Whereas arousal stimuli does not have an effect on sympathetic nerve activity, it is suppressed during phase IV (blood pressure overshoot), and by vasopressor agents (Figure 3.4).

Patients with DBH deficiency have normal to moderate increase MSNA (Rea *et al.*, 1990c). This indicates the integrity of neural sympathetic cardiovascular reflexes in these patients. At the present time, there is not adequate information concerning the therapeutic or diagnostic value of microneurography in patients with pure autonomic failure or multiple system atrophy.

D) The main limitation of this technique is the requirement of sophisticated equipment. Furthermore very limited information in normal subjects is available. Currently microneurography must be considered a research procedure.

REFERENCES

Akselrod, S., Gordon, D., Ubel, F.A., Shannon, D.C., Barger, A.C. and Cohen, R.J. (1981) Power spectrum analysis of heart rate fluctuation: A quantitative probe of beat-to-beat cardiovascular control. *Science*, **213**, 220–222.

Almquist, A., Goldenberg, I.F., Milstein, S., Chen, MY., Chen, X. and Hansen, R., *et al.* (1989) Provocation of bradycardia and hypotension by isoproterenol and upright posture in patients with unexplained syncope. *N. Engl. J. Med.*, **320**, 346–351.

Aminoff, M.J. (1980) Peripheral sympathetic function in patients with a polyneuropathy. *J. Neurol. Sci.*, **44**, 213.

Amorim, D.S., Neto, J.A.M., Maciel, B.C., Gallo, L., Fiho, J.T. and Manco, J.C. (1988) Cardiac autonomic function in healthy elderly people. *Geriatric Cardiovasc. Med.*, **1**, 65–71.

Appenzeller, O. and Schnieden, H. (1963) Neurogenic pathways concerned in reflex vasodilation in the hand with special reference to stimuli affecting pathway. *Clin. Sci.*, **25**, 413–421.

Bannister, R. and Mathias, C. (1988) Testing autonomic reflexes. In: *Autonomic Failure*, edited by R. Bannister, pp. 289–307, New York: Oxford University Press.

Bannister, R., Sever, P. and Gross, M. (1977) Cardiovascular reflexes and biochemical responses in progressive autonomic failure. *Brain*, **100**, 327–344.

Baser, S.M., Brown, R.T., Curras, M.T., Baucom, C.E., Hooper, D.R. and Polinsky, R.J. (1991) Beta-receptor sensitivity in autonomic failure. *Neurology*, **41**, 1107–1112.

Bennett, T., Farquhar, I.K., Hosking, D.J. and Hampton, J.R. (1978) Assessment of methods for estimating autonomic nervous control of the heart in patients with diabetes mellitus. *Diabetes*, **27**, 1167–1178.

Bevan, J.A., Bevan, R.N. and Duckles, S.P. (1980) Adrenergic regulation of vascular smooth muscle. In: *Handbook of Physiology 2. The Cardiovascular System*, edited by D.F. Bohr, A.P. Smylo and H.V. Sparks, pp. 515–566, Bethesda, Maryland: American Physiological Society.

Biaggioni, I., Robertson, R.M. and Robertson, D. (1993) Manipulation of norepinephrine metabolism with yohimbine in the treatment of autonomic failure. *J. Clin. Pharmacol.*, in press.

Bourgon, P., Pilley, S.F.J. and Thompson, H.S. (1978) Cholinergic supersensitivity of the iris sphincter in Adie's tonic pupil. *Am. J. Ophthalmol.*, **85**, 873–877.

Bravo, E.L. and Gifford R.W. (1984) Pheochromocytoma: diagnosis, location and management. *N. Engl. J. Med.*, **311**, 1298–1303.

Brown, M.J., Jenner, D.A., Allison, D.J. and Dollery, C.T. (1981) Variations in individual organ release of noradrenaline measured by an improved radioenzymatic technique: limitations of peripheral venous measurements in the assessment of sympathetic nervous activity. *Clin. Sci.*, **61**, 585–590.

Bristow, J.D., Brown, E.B. and Cunninham, D.J.C. (1971) Effect of bicycling on baroreflex regulation of pulse interval. *Circ. Res.*, **28**, 582–592.

Bristow, S.D., Honour, A.S., Pickering, G.W., Sleight, P. and Smyth, H.S. (1969) Diminished baroreflex sensitivity in high blood pressure. *Circ. Res.*, **39**, 48–54.

Bull, C. and Henry, J.A. (1977) Finger wrinkling as a test of autonomic function. *Br. Med. J.*, **1**, 1551.

Chalmers, T.M. and Keele, C.A. (1952) The nervous and chemical control of sweating. *Br. J. Dermatol.*, **64**, 43.

Coon, J.M. and Rothman, S. (1941) The sweat response to drugs with nicotine-like action. *Pharmacol.*, **73**, 1.

Crocford, C., Davies, C.T.M. and Weiner, J.S. (1970) Circadian changes in sweating threshold. *J. Physiol. (Lond.)*, **207**, 26P.

Cryer, P.E., Rizza, R.A., Haymond, M.W. and Gerich, J.E. (1980) Epinephrine and norepinephrine are cleared through beta-adrenergic, but not alpha-adrenergic, mechanisms in man. *Metabolism*, **29** (suppl. 1), 1114–1118.

Dale, H.H. and Feldberg, W. (1934) The chemical transmission of secretory impluses to the sweat glands of the cat. *J. Physiol. (Lond.)*, **82**, 121–128.

Das, P.K. and Parratt, J.R. (1971) Myocardial and haemodynamic effects of phentolamine. *Br. J. Pharmacol.*, **41**, 437–444.

Delius, W., Hagbarth, K-E., Hongell, A. and Wallin, B.G. (1972a) Manoeuvres affecting sympathetic outflow in human skin nerves. *Acta Physiol. Scand.*, **84**, 177–191.

Delius, W., Hagbarth, K.E., Hongell, A. and Wallin, B.G. (1972b) General characteristics of sympathetic activity in human muscle nerves. *Acta Physiol. Scand.*, **84**, 65–81.

Dollery, C.T., Paterson, J.W. and Conolly, M.E. (1969) Clinical pharmacology of beta-receptor-blocking drugs. *Clin. Pharmacol. Ther.*, **10**(6), 765–781.

Dunlap, E.D. and Pfeifer, M.A. (1989) Autonomic function testing. In: *Handbook of Research Methods in Cardiovascular Behavioral Medicine*, edited by N. Schneiderman, S.M. Weiss, and P.G. Kaufmann, pp. 91–106, New York: Plenum Press.

Eckberg, D. (1980) Parasympathetic cardiovascular control in human disease: A critical review of methods and results. *Am. J. Physiol.*, **239**, H581–H593.

Esler, M.D., Hasking, G.J., Willet, I.R., Leonard, P.W. and Jennings, G.L. (1985) Noradrenaline release and sympathetic nervous system activity. *J. Hypertens.*, **3**, 117–129.

Esler, M., Jackman, G., Kelleher, D., Skews, H., Jennings, G., Bobik, A. and Kooner, P. (1980) Norepinephrine kinetics in patients with idiopathic autonomic insufficiency. *Circ. Res.*, **46**, 1–47.

Ewing, D.J. (1978) Cardiovascular reflexes and autonomic neuropathy. *Clin. Sci. Mol. Med.*, **55**, 321–329.

Ewing, D.J., Irving, J.B. and Kerr, F. (1974) Cardiovascular responses to sustained handgrip in normal subjects and in patients with diabetes mellitus: a test of autonomic function. *Clin. Sci.*, **46**, 295–306.

Ewing, D.J., Campbell, I.W., Murray, A., Neilson, J.M.M. and Clarke, B.F. (1978) Immediate heart-rate response to standing: simple test for autonomic neuropathy in diabetics. *Br. Med. J.*, **1**, 145–147.

Ewing, D.J., Campbell, I.W. and Clarke, B.F. (1980) The natural history of diabetic autonomic neuropathy. *Q. J. Med.*, **49**, 95–108.

Ewing, D.J., Borsey, D.Q., Bellavere, F. and Clarke, B.F. (1981) Cardiac autonomic neuropathy in diabetes: comparison of measure of RR-interval variation. *Diabetologia*, **21**, 18–31.

Ewing, D.J., Martyn, C.N., Young, R.J. and Clarke, B.F. (1985) The value of cardiovascular autonomic tests: 10 years experience in diabetes. *Diabetes Care*, **85**, 491–498.

Fagius, J. and Wallin, B.G. (1980) Sympathetic reflex latencies and conduction velocities in normal man. *J. Neurol. Sci.*, **47**, 443–451.

Falkner, B., Kushner, H., Onesti, G. and Angelakos, S. (1981) Cardiovascular characteristics in adolescents who develop essential hypertension. *Hypertension*, **3**, 521–527.

Fisher, B.M. and Frier, B.M. (1989) Usefulness of cardiovascular tests of autonomic function in asymptomatic diabetic patients. *Diabetes Res. Clin. Pract.*, **6**, 157–160.

Fouad, F.M., Tarazi, R.C. and Bravo, E.L. (1985) Orthostatic hypotension. Clinical experience with diagnostic tests. *Cleveland Clin. Q.*, **52**, 561–568.

Freyschuss U. (1970) Cardiovascular adjustment to somatomotor innervation. *Acta Physiol. Scand.*, **342**(suppl), 1–63.

Gibson, T.E. and Shelley, W.B. (1948) Sexual and racial differences in the response of sweat glands to acetylcholine and pilocarpine. *J. Invest. Dermatol.*, **11**, 137.

Godden, J.O., Roth, G.M. and Hines, E.A. (1965) The changes in the intraarterial pressure during emersion of the hand in ice cold water. *Circulation*, **12**, 963–973.

Goldberg, J.R. and Robertson, D. (1983) Yohimbine: a pharmacological probe for study of the α_2-adrenoceptor. *Pharmacol. Rev.*, **35**, 143–179.

Goldberg, M.R., Hollister, A.S. and Robertson, D. (1983) Influence of yohimbine on blood pressure, autonomic reflexes and plasma catecholamines in man. *Hypertension*, **5**, 772–778.

Goldstein, D.S. (1987) Catecholamines in plasma and cerebrospinal fluid: sources and meanings. *Brain Peptides and Catecholamines in Cardiovascular Regulation*, edited by J.P. Buckley and C.M. Ferrario, pp. 15–26, New York: Raven Press.

Goldstein, D.S., Spanarkel, M., Pitterrnan, A., Toltzis, R., Gratz, E., Epstein, S. and Keiser, H.R. (1982) Circulatory control mechanisms in vasodepressor syncope. *Am. Heart J.*, **104**, 1071–1075.

Goldstein, D.S., Zimlichman, R., Stull, R., Folio, J., Levinson, P.D., Keiser, H.R. and Kopin I.J. (1985) Measurement of regional neuronal removal of norepinephrine in man. *J. Clin. Invest.*, **76**, 15–21.

Goodwin, G.M., McCloskey, D.I. and Mitchell, H.J. (1972) Cardiovascular and respiratory responses to changes in central command during isometric exercise at constant muscle tension. *J. Physiol. (Lond.)*, **226**, 173–190.

Gribbin, B., Pickering, T.G., Sleight, P. and Peto, R. (1971) Effect of age and high blood pressure on baroreflex sensitivity in man. *Circ. Res.*, **29**, 424–431.

Guttmann, L. (1940) Topographic studies of disturbances of sweat secretion after complete lesions of peripheral nerves. *J. Neurol. Neurosurg. Psychiatry*, **3**, 197–210.

Guttmann, L. (1947) Management of quinizarin sweat test. *Postgrad. Med. J.*, **23**, 353–366.

Hickler, R.B., Hoskins, R.G. and Hamlin, J.T. (1960) The clinical evaluation of faulty orthostatic mechanisms. *Med. Clin. of North America*, **44**, 1237–1250.

Hines, E.A. and Brown, G.E. (1936) The cold pressor test for measuring the reactibility of the blood pressure: data concerning 571 normal and hypertensive subjects. *Am. Heart J.*, **11**, 1–9.

Hosking, D.J., Bennett, T.J. and Hampton, J.R. (1978) Diabetic autonomic neuropathy. *Diabetes*, **7**, 1043–1049.

Hultman, E. and Sjoholm, H. (1982) Blood pressure and heart rate response to voluntary and non-voluntary static exercise in man. *Acta Physiol. Scand.*, **115**, 499–501.

Hyndman, O.R. and Wolkin, J. (1941) The pilocarpine sweating test. I. A valid indicator in differentiation of preganglionic and postganglionic sympathectomy. *Arch. Neurol. Psychiatry*, **45**, 992–1006.

Jamieson, G.G., Ludbrook, J. and Wilson, A. (1971) Cold hypersensitivity in Raynaud's phenomenon. *Circulation*, **44**, 254–264.

Jaspan, J.B. (1981) Glucagon: basic pathophysiological considerations. *Glucagon in Gastroenterology and Hepatology*, edited by J. Picazo, pp. 1–24, Lancaster, England: MTP Press.

Kahn, D. and Rothman, S. (1942) Sweat response to acetylcholine. *J. Invest. Dermatol.*, **5**, 431.

Kaijser, L. and Sachs, C. (1985) Autonomic cardiovascular responses in old age. *Clin. Physiol.*, **5**, 347–357.

Kapoor, W.N. and Brant, N. (1992) Evaluation of syncope by upright tilt testing with isoproterenol. A nonspecific test. *Ann. Int. Med.*, **116**, 358–363.

Kerslake, D.McK. and Cooper, K.E. (1950) Vasodilation in the hand in response to heating the skin elsewhere. *Clin. Sci.*, **9**, 31–47.

Kollai, N.M. and Koizumi, K. (1981) Cardiovascular reflexes and interrelationships between sympathetic and parasympathetic activity. *J. Auton. Nerv. Syst. Function*, **4**, 135–148.

Kooner, J.S., Birch, R., Frankel, H.L., Peart, W.S. and Mathias, C.J. (1991) Hemodynamic and neurohormonal effects of clonidine in patients with preganglionic and postganglionic sympathetic lesions. *Circulation*, **84**, 75–83.

Lagerlof, H., Eliasch, I.I., Werko, L. and Berglund, E. (1951) Orthostatic changes of pulmonary and peripheral circulation in man. *Scand. J. Clin. Lab. Invest.*, **3**, 85–91.

Laubie, M. and Scmitt, H. (1977) Sites of action of clonidine: centrally mediated increase in vagal tone, centrally mediated hypotensive and sympatho-inhibitory effects. *Prog. Brain Res.*, **47**, 337–356.

Leimbach, W.N., Wallin, B.G., Victor, R.G., Aylward, P.E., Sundlof, G. and Mark, A.L. (1986) Direct evidence from intraneural recordings for increased central sympathetic outflow in patients with heart failure. *Circulation*, **73**, 913–919.

Leon, D.F., Shaver, J.A. and Leonard, J.J. (1970) Reflex heart rate control in man. *Am. Heart J.*, **80**, 729–738.

Lepore, F.E. (1985) Diagnostic pharmacology of the pupil. *Clin. Neuropharmacol.*, **8**, 23–37.

Levin, A.B. (1966) A simple test of cardiac function based upon the heart rate changes induced by the Valsalva manoeuvre. *Am. J. Cardiol.*, **18**, 90–95.

Lind, A.R., Taylor, S.H. and Humphreys, P.W. (1964) The circulatory effects of sustained voluntary muscle contraction. *Clin. Sci.*, **27**, 229–244.

List, C.F. and Peet, M.M. (1938) Sweat secretion in man III. Clinical observations on sweating produced by pilocarpine and mecholyl. *Arch. Neurol. Psychiatry*, **40**, 269.

List, C.F. and Peet, M.M. (1938) Sweat secretion in man, II. Autonomic distribution of disturbances in sweating associated with lesions of the sympathetic nervous system. *Arch. Neurol. Psychiatry*, **39**, 27–43.

Lloyd-Mostyn, H. and Watkins, B.J. (1975) Defective innervation of the heart in diabetic autonomic neuropathy. *Brit. Med. J.*, **3**, 15–17.

Low, P.A. (1984) Quantitation of autonomic function. Peripheral Neuropathy, edited by P.J. Dyck, P.K. Thomas, E.H. Lambert and R. Bunge. pp. 1139–1165, Philadelphia: W.B. Saunders Co.

Mackay, J.D., Page M.Mc.B., Cambridge, J. and Watkins, B.J. (1980) Diabetic autonomic neuropathy — the diagnostic value of heart rate monitoring. *Diabetologia*, **18**, 471–478.

MacMillan, A.L. and Spalding, J.M.K. (1969) Human sweating response to electrophoresed acetylcholine: a test of postganglionic sympathetic function. *J. Neurol. Neurosurg. Psychiatry*, **32**, 155.

Mark, A., Victor, R.G., Nerhed, C. and Wallin, B.G. (1985) Microneurographic studies of the mechanisms of sympathetic nerve responses to static exercise in humans. *Circ. Res.*, **57**, 461–469.

McLeod, J.G. and Tuck, R.R. (1987) Disorders of the autonomic nervous system: part 2. Investigation and treatment. *Ann. Neurol.*, **21**, 519–529.

Meredith, I.T., Eisenhofer, G., Lambert, G.W., Jennings, G.L., Thompson, J. and Esler, M.D. (1992) Plasma norepinephrine responses to head-up tilt are misleading in autonomic failure. *Hypertension*, **19**, 628.

Morris, J.G.L., Lee, J. and Lim, C.L. (1984) Facial sweating in Horner's syndrome. *Brain*, **107**, 751–758.

Mosqueda-Garcia, R., Tseng, C-J., Biaggioni, I., Robertson, R.M. and Robertson, D. (1990) Effects of caffeine on baroreflex activity in humans. *Clin. Pharmacol. Ther.*, **48**, 568–574.

Mosqueda-Garcia, R., Ananthran, V. and Robertson, D. (1993) Progressive sympathetic withdrawal preceding syncope evoked by upright posture. *Circulation*, **88**, 1–84.

Onrot, J., Goldberg, M.R., Hollister, A.S., Biaggioni, I., Robertson, R.M. and Robertson, D. (1986) Management of chronic orthostatic hypotension. *Am. J. Med.*, **80**, 454–464.

Palkovits, M. and Zaborszky, L. (1977) Neuroanatomy of central cardiovascular control. Nucleus tractus solitarii: afferent and efferent neuronal connections in relation to the baroreceptor reflex arc. *Prog. Brain Res.*, **47**, 9–34.

Parks, V.J., Sandinson, A.G., Skinner, S.L. and Whelan, R.F. (1961) Sympathomimetic drugs in orthostatic hypotension. *Lancet*, **1**, 1133–1136.

Persson, A. and Solders, G. (1983) R–R variations in Guillain-Barre syndrome: a test of autonomic dysfunction. *Acta Neurol. Scand.*, **67**, 294–300.

Pfeifer, M.A. and Peterson, H. (1987) Cardiovascular autonomic neuropathy. In: *Diabetic Neuropathy*, edited by P.J. Dyck, P.K. Thomas, A.K. Asbury, A.I. Winegrad and D. Porte. pp. 48–79, Philadelphia: Saunders.

Pickering, G.W. (1932) The vasomotor regulation of heat loss from skin in relation to external temperature. *Heart*, **16**, 115–135.

Polinsky, R.J. (1988) Neurotransmitter and neuropeptide function in autonomic failure. In: *Autonomic Failure*, edited by R. Bannister, pp. 321–347, New York: Oxford University Press.

Polinsky, R.J., Kopin, I.J., Ebert, M.H. and Weise, V. (1981) Pharmacologic distinction of different orthostatic hypotension syndromes. *Neurology*, **31**, 1–7.

Rea, R.F., Eckberg, D.L., Fritsch, J.M. and Goldstein, D.S. (1990a) Relation of plasma norepinephrine and sympathetic traffic during hypotension in humans. *Am. J. Physiol.*, **58**, R982–R986.

Rea, R.F., Grossman, E. and Goldstein, D.S. (1990b) Relation between sympathetic nerve activity and norepinephrine spillover in humans. *Clin. Res.*, **38**, 828A.

Rea, R., Biaggioni, I., Robertson, R.M., Haile, V. and Robertson, D. (1990c) Reflex control of sympathetic nerve activity in dopamine-beta-hydroxylase deficiency. *Hypertension*, **15**, 107–112.

Robertson, D. (1981) Clinical pharmacology: assessment of autonomic function. In: *Clinical Diagnostic Manual for the House Officer*, edited by K.L. Baughman and B.M. Greene, pp. 86–101, Baltimore: Williams & Wilkins.

Robertson, D., Johnson, G.A., Robertson, R.M., Nies, A.S., Shand, D.G. and Oates, J.A. (1979) Comparative assessment of stimuli that release neuronal and adrenomedullary catecholamines in man. *Circulation*, **59**, 637–643.

Robertson, D., Goldberg, M.R., Hollister, A.S., Wade, D. and Robertson, R.M. (1983) Clonidine raises blood pressure in severe idiopathic orthostatic hypotension. *Am. J. Med.*, **74**, 193–200.

Robertson, D., Hollister, A.S., Carey, E.L., Tung, C.S., Goldberg, M.R. and Robertson, R.M. (1984) Increased vascular beta$_2$-adrenoceptor responsiveness in autonomic dysfunction. *J. Am. Col. Cardiol.*, **3**, 850–856.

Robertson, D. and Robertson, R.M. (1985) Orthostatic hypotension. Diagnosis and therapy. *Modern Concepts In Cardiovasc. Disease*, **54**, 7–12.

Robertson, D., Goldberg, M.R., Onrot, J., Hollister, A.S., Wiley, R., Thompson, J G. and Robertson, R.M. (1986) Isolated failure of autonomic noradrenergic neurotransmission: evidence for impaired β-hydroxylation of dopamine. *N. Engl. J. Med.*, **214**, 1494–1497.

Robinson, B.F., Epstein, S.E., Beiser, G.D. and Braunwald, E. (1966) Control of the heart rate by the autonomic nervous system. Studies in man on the inter-relation between baroreceptor mechanisms and exercise. *Circ. Res.*, **19**, 400–411.

Robinson, B.J., Johnson, R.H., Lambie, D.G. and Palmer, K.T. (1983) Do elderly patients with an excessive fall in blood pressure on standing have evidence of autonomic failure? *Clin. Sci.*, **64**, 587–591.

Roddie, I.C., Shepherd, J.T. and Whelan, R.F. (1957) The vasomotor nerve supply to the skin and muscle of the human forearm. *Clin. Sci.*, **16**, 67.

Rothschild, A.H., Weinberg, C.R., Halter, J.B., Porte, D. and Pfeifer, M.A. (1987) Sensitivity of R–R variation and Valsalva ratio in assessment of cardiovascular diabetic autonomic neuropathy. *Diabetes Care*, **10**, 735–741.

Rushmer, R.F. (1976) Effects of posture. Part 1. Circulatory response to arising. In: *Cardiovascular Dynamics*. pp. 363–384, Philadelphia: WB Saunders.

Sanders, J.S., Fergurson, D.W. and Mark, A.L. (1988) Arterial baroreflex control of sympathetic nerve activity during elevation of blood pressure in normal man: dominance of aortic baroreceptors. *Circulation*, **77**, 279–288.

Schatz, I.J. (1986) Clinical diagnosis. In: *Orthostatic Hypotension*, edited by I.J. Schatz, pp. 67–89, Philadelphia: Davis Co.

Scriven, A.J.I., Dollery, C.T., Murphy, M.B., Macquin, I. and Brown, M.J. (1982) Blood pressure and plasma norepinephrine concentrations after endogenous release by tyramine. *Clin. Pharmacol. Ther.*, **33**, 710–716.

Scriven, A.J.I., Brown, M.J., Murphy, M.B. and Dollery, C.T. (1984) Changes in blood pressure and plasma catecholamines caused by tyramine and cold exposure. *J. Cardiovasc. Pharmacol.*, **6**, 954–960.

Sharpey-Schafer, E.P. (1956) Circulatory reflexes and chronic disease of the afferent nervous system. *J. Physiol. (Lond.)*, **134**, 1–10.

Shepherd, J.T. and Mancia, G. (1986) Reflex control of the human cardiovascular system. *Rev. Physiol. Biochem. Pharmacol.*, **105**, 1–99.

Silver, A., Montagna, W. and Karacan, I. (1964) Age and sex differences in spontaneous adrenergic and cholinergic human sweating. *J. Invest. Dermatol.*, **43**, 255.

Smith, S.A. (1982) Reduced sinus arrhythmia in diabetic autonomic neuropathy: diagnostic value of an age-related normal range. *Br. Med. J.*, **285**, 1599–1601.

Smith, S.A. (1988) Pupillary function in autonomic failure. In: *Autonomic Failure*, edited by R. Bannister, pp. 393–412, New York: Oxford University Press.

Spyer, K.M. (1984) Central control of the cardiovascular system. *Recent Advances in Physiology*, edited by P.F. Baker, pp. 163–200, Edinburgh: Churchill Livingstone.

Stern, M.D., Lappe, D.L., Bowen, P.D., Chimosky, J.E., Holloway G.A., Keiser H.R and Bowman R.L. (1977) Continuous measurement of tissue blood flow by laser-Dopplers spectroscopy. *Am. J. Physiol.*, **232**, 441.

Strasberg, B., Rechavia, E., Sadie, A., Kusinec, J., Mager, A., Sclarovsky, S. and Agmon, J. (1989) The head-up tilt table test in patients with syncope of unknown origin. *Am. Heart J.*, **118**, 923–927.

Streeten, D.H.P. and Anderson, A.H. Jr. (1992) Delayed orthostatic hypotension intolerance. *Arch. Intern. Med.*, **152**, 1066–1072.

Sundqvist, G., Almer, L.-O. and Lilja, B. (1979) Respiratory influence on heart rate in diabetes mellitus. *Br. Med. J.*, **1**, 924–925.

Uvnas, B. (1954) Sympathetic vasodilator outflow. *Physiol. Rev.*, **34**, 608.

Vallbo, A.B., Hagbarth, K.E., Torebjork, H.E. and Wallin, B.G. (1979) Somatosensory, propioceptive and sympathetic activity in human peripheral nerves. *Physiol. Rev.*, **59**, 919.

Van Lieshout, J.J., Wieling, W., Wesseling, K.H. and Karemaker, J.M. (1989) Pitfalls in the assessment of cardiovascular reflexes in patients with sympathetic failure but intact vagal control. *Clin. Sci.*, **76**, 523–528.

Van Zwieten, P.A. and Timmermans, P.B.M.W.M. (1979) The role of central α-adrenoceptors in the mode of action of hypotensive drugs. *Trends Pharmacol. Sci.*, **1**, 39–42.

Vendsalu, A. (1960) Studies on adrenaline and noradrenaline in human plasma. *Acta Physiol. Scand.*, **173**, 49–70.

Vestal, R.E., Wood, A.J.J. and Shand, D.G. (1979) Reduced β-adrenoceptor sensitivity in the elderly. *Clin. Pharmacol. Ther.*, **26**, 181–186.

Vingerhoets, A.J.J.M. (1984) Biochemical changes in two subjects succumbing to syncope. *Psychosomatic Med.*, **46**, 95–103.

Walker, A.J. and Longland, C.J. (1950) Venous pressure measurement in foot as aid to investigate disease in leg. *Clin. Sci.*, **9**, 101.

Wallin, B.G. and Sundlof, G. (1982) Sympathetic outflow to muscles during vasovagal syncope. *J. Auton. Nerv. Syst.*, **6**, 287–291.

Wheeler, T. and Watkins, P.J. (1973) Cardiac denervation in diabetes. *Br. Med. J.*, **4**, 584–593.

Weinberg, C.R. and Pfeifer, M.A. (1984) An improved method for measuring heart rate variability: assessment of cardiac autonomic function. *Biometrics*, **40**, 855–865.

Weinberg, C.R. and Pfeifer, M.A. (1986) Development of a predictive model for symptomatic neuropathy in diabetes. *Diabetes*, **35**, 873–880.

Weiner, N. (1980) Drugs that inhibit adrenergic nerves and block adrenergic receptors. In: *The Pharmacological Basis of Therapeutics*, edited by A.G. Gilman, L.S. Goodman and A. Gilman, pp. 176–210, New York: MacMillan.

Ziegler, M.G. (1989) Catecholamine measurement in behavioral research. In: *Handbook of Research Methods in Cardiovascular Behavioral Medicine*, edited by N. Schneiderman, S.M. Weiss, P.G. Kaufmann, pp. 167–184, New York: Plenum Press.

Ziegler, M.G., Echon, C., Wilner, K.D., Specho, P., Lake, C.R. and McCutchen, J.A. (1986) Sympathetic nervous withdrawal in the vasodepressor (vasovagal) reaction. *J. Auton. Nerv. Syst.*, **17**, 273–278.

Ziegler, M.G. and Lake, C.R. (1985) Noradrenergic responses to postural hypotension: Implications for therapy. In: *The Catecholamines in Psychiatric and Neurologic Disorders*, edited by C.R. Lake and M.G. Ziegler, pp. 121–136, Woburn: Butterworths.

4 Age-Related Changes in the Autonomic Nervous System

Sheila M. Ryan and Lewis A. Lipsitz

Department of Anesthesiology, Beth Israel Hospital,
and Hebrew Rehabilitation Center for Aged, Boston, Massachusetts

Aging results in alterations in selected areas of the autonomic nervous system. Most of the evidence for age-related changes in autonomic function are derived from investigations of autonomic control of cardiovascular function. Results suggest that baroreflex-mediated cardioacceleration in response to hypotensive stress is impaired with aging. Hypertension further impairs baroreflex function. Basal plasma noradrenaline levels increase with aging and with healthy elderly people exhibit a heightened and prolonged noradrenaline response to orthostatic stress. The response of cardiac and vascular beta receptors to adrenergic stimulation is blunted, despite normal lymphocyte beta receptor density. Instead, aging is associated with functional uncoupling of the beta receptor activation unit. In contrast, alpha-mediated vascular responsiveness appears to be preserved during aging.

Impaired temperature regulation in elderly persons results from a combination of physiological age-related changes including decreased temperature perception and discrimination, delayed shivering and vasodilation in the heat. Thus, elderly people are at increased risk for the development of highly morbid clinical syndromes such as orthostatic hypotension, postprandial hypotension and hypothermia.

KEY WORDS: Aging, cardiovascular function, baroreflexes, hypotension, hypothermia, catecholamines

INTRODUCTION

Clinical observations suggest that the autonomic nervous system is impaired with aging. Impairment is evident in the increased prevalence of postural hypotension, thermoregulatory disorders, and abnormal cardiovascular responses of elderly persons to clinical tests of autonomic function such as the Valsalva maneuver, standing, and lowering body negative pressure. In addition, pharmacological studies using adrenergic agonists suggest that sympathetic responsiveness is blunted in the elderly. Although plasma noradrenaline levels increase with normal aging, adrenergic receptor affinity for adrenergic agonists decreases.

Several methodologic problems confound the interpretation of research data in this area. Difficulties include quantifying the degree to which research findings are

61

influenced by disease rather than aging. The fact that clinical tests of autonomic function rely on reflex responses makes specifying the actual level of abnormality very difficult. In addition, previous human studies have not been controlled for body composition, physical exercise and diet. This lack of control may cause further difficulty with interpretation of the results; e.g. obesity has been shown to alter noradrenaline kinetics in young individuals (Sowers *et al.*, 1982; Schwartz, Halter and Bierman, 1983).

Animal studies must be interpreted with caution because of considerable differences in autonomic responses among different animal species (Fleisch, Maling and Brodie, 1970). In addition, some studies do not compare mature and senescent animals but, rather look at the differences between young and mature animals. This approach may lead to confusion between true age-related changes and those that occur as a result of growth and development. Despite these complexities, overall evidence suggests that some deterioration in selected areas of autonomic nervous system function occurs with normal aging. This chapter outlines the evidence for autonomic function changes with aging, concentrating primarily on changes in the cardiovascular system.

PHYSIOLOGICAL ALTERATION IN AUTONOMIC NERVOUS SYSTEM FUNCTION WITH AGING

BAROREFLEX FUNCTION

The baroreflex is a complex neural feedback loop that rapidly restores the blood pressure to its normal set-point during transient physiologic perturbations. This neural feedback mechanism involves: tonic signals from baroreceptors located in the carotid arteries, aortic arch, right atrium and lung to inhibitory vasomotor centers of the brainstem; central nervous system integration of these signals; efferent sympathetic and parasympathetic nerve activity; and effector organ response. Most of the current information about baroreflex function and aging is derived from measurements of cardiovascular responses to stimulation of baroreceptors. A progressive age-related impairment in baroreflex function has been well demonstrated by a reduction in the cardioinhibitory response to hypertensive stimuli such as phenylephrine infusion and phase IV of the Valsalva maneuver, and by a blunted cardioacceleratory response to hypotensive stressors such as upright posture, lower body negative pressure or nitroprusside infusion.

Although many investigators use these measurements to infer decreases in baroreceptor sensitivity, such data do not localize the site of age-related changes in baroreflex function. The fact that plasma noradrenaline responses to hypotensive stresses are heightened and prolonged in healthy elderly individuals (see below) suggests that baroreflex activation is intact with aging. The data reviewed below suggest that aging is associated primarily with a decline in beta-adrenergic responsiveness at the site of the end-organ.

Age-associated elevations in blood pressure have been considered a possible cause and/or consequence of baroreflex impairment. Both normal aging and hypertension exert independent effects of baroreflex sensitivity (Gribbin *et al.*, 1971; Shimada *et al.*, 1985). A hypothesis has been suggested that the decrease in arterial distensibility that accompanies aging and hypertension results in diminished baroreceptor stretch, less tonic inhibition of the brainstem sympathetic outflow results in increased circulating noradrenaline which in turn may result in further vasoconstriction, blood pressure elevation, and baroreflex impairment (Rowe and Troen, 1980; Pfeifer *et al.*, 1983; Linares and Halter, 1987). This hypothesis is supported by elevated basal plasma noradrenaline levels and heightened plasma noradrenaline response to baroreceptor stimulation in aged subjects. However, it is not entirely consistent with the attenuated plasma noradrenaline response observed in hypertensive subjects (Jansen *et al.*, 1989). The relationship between aging, hypertension, and baroreflex function remains to be fully elucidated.

Heart rate and blood pressure variability

Sympathetic and parasympathetic influences on the heart result in considerable beat-to-beat heart rate variability which is commonly used as a measure of autonomic function. Healthy human aging is associated with attenuation of the normal respiratory sinus arrhythmia (Pfeifer *et al.*, 1982), the cardioacceleratory response to cough (Wei *et al.*, 1983), and the cardioinhibitory response to phase IV of the Valsalva maneuver (Shimada *et al.*, 1986). Recently, spectral analysis of heart rate time series data has been used to examine and quantify the sympathovagal influences that contribute to heart rate variability (Pomeranz, Macauley and Caudill, 1985). High frequency heart rate oscillations (0.15-0.40 Hz) represent parasympathetic control of heart rate and can be abolished by atropine. Low frequency oscillations (0.01-0.10 Hz) represent primarily sympathetic inputs, although there is evidence for some parasympathetic contribution as well. Low frequency oscillations are enhanced by maneuvers that activate sympathetic nervous system activity, particularly head-up tilt in healthy young subjects. Recent studies show a marked attenuation or loss of the low frequency response during head-up tilt in elderly subjects (Jarisch *et al.*, 1987; Simpson and Wicks, 1988; Lipsitz *et al.*, 1990). This finding is consistent with reduced baroreflex control of heart rate in elderly subjects exposed to postural stress.

In contrast to heart rate variability, blood pressure variability increases with age. This variability is evident in repeated measurements several minutes apart during activities of daily living (Jonsson *et al.*, 1990) and during beat-to-beat blood pressure monitoring (Mancia *et al.*, 1980; Furman *et al.*, 1990). Wider fluctuations in blood pressure may be due to loss of baroreflex-mediated heart rate responses that ordinarily buffer transient changes in blood pressure.

THE EFFECT OF AGING ON CATECHOLAMINES

BASAL PLASMA LEVELS

Significant evidence suggest that basal plasma noradrenaline levels increase with age (Ziegler, Lake and Kopin, 1976; Barnes et al., 1982; Veith et al., 1986). However, the prolongation of noradrenaline response to sympathetic nervous system stimuli may falsely elevate supine resting levels if subjects are not given a sufficient length of time to achieve truly basal conditions (Young et al., 1980). Noradrenaline is released at postganglionic sympathetic nerve terminals and undergoes extensive local metabolism and reuptake into the neuron. Circulating noradrenaline levels represent "spill-over" into the blood stream. Therefore, plasma noradrenaline levels are only indirect measures of noradrenaline release at the synapse. Despite this, venous plasma noradrenaline levels do appear to correlate with sympathetic nervous system activity, measured by microelectrode recordings of peroneal and median nerves (Goldstein et al., 1983; Wallin, 1988).

In contrast, adrenaline is released directly into the bloodstream in response to sympathetic stimulation of the adrenal medulla, and is then transported via the circulation to target organs. Adrenaline is removed through non-neuronal uptake and metabolism. Plasma adrenaline levels are probably not affected by aging (Veith et al., 1986; Morrow et al., 1987).

NORADRENALINE KINETICS

Age-related elevations in plasma noradrenaline may be due to many factors, including increased appearance at the synapse, increased spill-over into the systemic circulation, decreased reuptake by presynaptic neurons, decreased local metabolism, and decreased systemic clearance. To determine whether elevations in plasma noradrenaline levels reflect heightened sympathetic nervous system activity, or merely decreased clearance, recent investigations have used radiotracer methods to examine noradrenaline kinetics. Using tritiated noradrenaline infusion and a two compartment model to estimate noradrenaline disposition, Veith et al (1986) demonstrated a 32% increase in arterialized noradrenaline appearance, and a 19% decrease in clearance in healthy elderly subjects (mean age = 68 +/− 5 years) compared to young (age = 27 +/− 6 years). Other studies have shown similar age-related increases in appearance rates (Rubin et al., 1982; Hoeldtke and Cilmi, 1985; Morrow et al., 1987) and decreases in clearance (Esler et al., 1981; Morrow et al., 1987), still others show no difference in the clearance (Rubin et al., 1982; Hoeldtke and Cilmi, 1985). The apparent conflict in results may reflect different blood sampling methodologies — venous versus arterialized and fasting versus non-fasting conditions. Several authors have demonstrated that arterialized noradrenaline samples may provide more accurate estimations of catecholamine kinetics, as venous noradrenaline reflects that which has escaped metabolism in the capillary bed (Best and Halter, 1982; Chang et al., 1986).

TABLE 4.1

Catecholamines and aging

	Noradrenaline	Adrenaline
Basal Plasma Levels	Increase	No change
Stimulated Plasma Levels (tilt, isometric exercise)	Increase	No change
Metabolism	Largely local through reuptake	Non neuronal uptake
Measured	Circulation levels reflect "spill-over" into bloodstream	Circulating levels released directly into bloodstream
Kinetics		
Appearance	Increase	No change
Clearance	Increase	No change

In contrast to the increased basal noradrenaline levels and altered noradrenaline kinetics seen in aging, adrenaline appearance and clearance are unaffected by age (Morrow *et al.*, 1987). (Table 4.1)

Circadian rhythm and catecholamines

Catecholamines exhibit diurnal variation which is preserved during aging. Plasma noradrenaline levels are highest during the late morning to early afternoon and fall gradually during the night. Elderly subjects have higher noradrenaline levels during a 24 hour period than young subjects. This elevation is most pronounced during the night (Prinz *et al.*, 1979) and is associated with increased nocturnal wakefulness and less stage 4 sleep in elderly subjects.

Stimulated catecholamine levels in aging

Postural stress induced by active standing and head-up tilt result in an increase in plasma noradrenaline which is exaggerated in the old. Young *et al* (1980) demonstrated that not only were the levels higher in older subjects after upright posture, but also the time to return to baseline level was prolonged. Plasma noradrenaline responses to isometric exercise (Ziegler, Lake and Kopin, 1976; Palmer, Ziegler and Lake, 1978), the cold compressor test (Ziegler, Lake and Kopin, 1976; Palmer, Ziegler and Lake, 1978), psychological stimuli (Barnes *et al.*, 1982) and graded levels of cardiac work (Fleg, Tzankoff and Lakatta, 1985) all also increase more in old compared with young subjects.

The effect of aging on the sympathetic nervous system

Much of our current knowledge about age-related changes in sympathetic nervous system function is derived from studies of cardiovascular response to hypotensive stress. Heart rate and vascular responses provide quantifiable indices of sympathetic efferent response. However, little information is available on age-related alterations in afferent pathways and central integral of baroreceptor input. The fact that plasma noradrenaline levels are heightened and prolonged during hypotensive stress, but heart rate responses are blunted in elderly subjects, suggest that aging results in impaired beta-mediated adrenoreceptor responses to sympathetic activation (Feldman, 1986).

Alterations in beta and alpha adrenergic receptors will be discussed separately.

BETA RECEPTORS

β-1 Adrenergic receptors

In the normal individual stimulation of β-1 receptors using intravenous infusion of sympathetic agonists such as isoproterenol results in cardioacceleration. The dose of isoproterenol required to raise the heart rate is increased with aging in humans (Vestal, Wood and Shand, 1979) and aged animals (Lakatta *et al.*, 1975; Abrass, Davis and Scarpace, 1982). The normal increase in heart rate induced by isoproterenol may also be due to a reflex cardioacceleratory response to the vasodilatation produced by simultaneous activation of β-2 receptors. Evidence discussed below indicates that β-2 receptors function is impaired with senescence.

β-2 Adrenergic receptors

Van Brunnelen *et al* (1981) measured alterations in forearm blood flow in 15 healthy subjects (7 older than 50 years) using venous occlusion plethysmography. In response to increasing doses of isoproterenol, older individuals demonstrated a resistance to vasodilatation, suggesting that aging may be accompanied by a decline in β-2 adrenoreceptor function. Pan *et al* (1986) also investigated the β-2 response to isoproterenol but used the dorsal hand vein as an isolated system not dependent on reflex responses. Isoproterenol-mediated vascular relaxation was impaired in elderly individuals. In contrast, nitroglycerine-mediated relaxation was normal in both young and elderly subjects. As nitrate-induced vasodilatation was intact, the impairment in β-agonist response was not due to structural alterations in the vessel wall. This finding supports an age-related decline in β-2 mediated vasodilatation in the elderly. As nitroglycerine acts independently of cAMP, the defect may lie at the level of cAMP production.

In vitro experiments have utilized aortic rings isolated from rats and beagles. These experiments have also demonstrated an age-related decrease in vasodilatation during β-agonist stimulation with isoproterenol (Cohen and Berkowitz, 1974; Fleisch and Hooker, 1976; Simpkins, Field and Ress, 1983; Shimizu and Toda, 1986). This result further supports a decline in β-2 function during aging.

LYMPHOCYTE STUDIES

The human lymphocyte is rich in β-3 receptors. It has thus provided an accessible model for the study of β receptor structure and function. However, it is not fully known whether age-related changes in the receptors of human lymphocytes reflect changes in receptors in less accessible, but more relevant areas of the body such as the heart, lungs and blood vessels.

In order to activate a response the β receptor complex requires an intact membrane receptor, adenylate cyclase which catalyses cAMP production, and a guanine regulatory protein, required for the binding of the agonist to the receptor.

An early study by Schocken and Roth suggested that β-2 receptor density on human mononuclear cells decreased with age (Schocken and Roth, 1977). Since then, numerous studies using similar radioligand techniques have shown no alteration in the receptor density in human lymphocytes (Abrass and Scarpace, 1981; Feldman *et al.*, 1984). Therefore, despite the high levels of circulating noradrenaline found in elderly individuals, down regulation of beta receptors does not occur with aging.

More recently, beta receptor responsiveness has been investigated by monitoring the cAMP response to adrenergic stimulation. Lymphocyte cAMP production in response to isoproterenol stimulation is reduced in elderly subjects (Dillon *et al.*, 1980). A decrease in lymphocyte adenylate cyclase activity has also been demonstrated in elderly subjects (Krall *et al.*, 1981; Abrass and Scarpace, 1982). Receptors need to be in the high affinity state to produce adequate cAMP to effect activation of the beta response. Feldman demonstrated a reduction in the number of high affinity beta receptors in lymphocytes of elderly persons (Feldman *et al.*, 1984). These studies suggest that aging is associated with functional uncoupling of the beta receptor complex.

Alpha receptors

Although current evidence suggests that beta adrenergic responsiveness decreases with normal aging, age-related changes in the alpha-adrenergic system are not as well delineated. The data currently available suggest that alpha receptors show less functional alteration than beta receptors during normal aging.

VASCULAR RESPONSE TO ALPHA ADRENERGIC ACTIVATION

In vitro studies of isolated human arteries and veins suggest that the alpha-mediated response to noradrenaline is not altered with aging (Scott and Reid, 1982; Stevens, Kipe and Moulds, 1982). This result is supported by the *in vivo* observation of preserved alpha-2-mediated phenylephrine-induced vasoconstriction of the dorsal hand vein (Pan *et al.*, 1986). However, Elliot *et al* (1982) found a reduced pressor effect of phenylephrine in elderly individuals. Clinical data from studies of forearm vascular responses to posture change of lower body negative pressure have been inconsistent. In elderly subjects exposed to these hypotensive stresses, forearm vasoconstriction has been reported to be enhanced (Ebert *et al.*, 1982), blunted

(Shiraki *et al.*, 1987; Jungu *et al.*, 1989), or unchanged (Lipsitz *et al.*, 1991) compared to the young.

Lipsitz *et al* (1991) demonstrated a high degree of variability in forearm vascular response to posture change in healthy and multiply-impaired elderly subjects. The average forearm vasoconstrictor response to one minute of standing (measured by venous occlusion plethysmography) was attenuated in both groups of elderly subjects compared to the younger counterparts, although many elderly subjects, both with and without multiple chronic illnesses, demonstrated normal responses. Vascular response to orthostatic hypotension is a dynamic process not adequately reflected by isolated measurements that may be high or low at any given point in time until steady-state is achieved. Therefore, conflicting results of different clinical studies may be due to subject heterogeneity, different degrees of orthostatic stress, inconsistent timing of measurements, or different measurement techniques.

Altered vascular contractility in advanced age may also result from structural changes in the vasculature rather than functional changes in the receptors themselves. In old rats, reduced aortic contractility associated with a progressive coarsening of the medial layer and derangement of the elastic lamella of the vessels has been demonstrated (Tuttle, 1966). The baroreflex response in atherosclerotic animals is also blunted (Angell-James, 1974; Cox, Bagshaw and Detweiler, 1980). Wei *et al* (1986) measured the baroreflex response in Fischer rats, a strain that is free of atherosclerosis. They demonstrated that the baroreflex-mediated blood pressure response to carotid sinus stimulation was preserved. Preservation of baroreflex response in the absence of atherosclerosis supports the concept that the reduced baroreflex function seen with aging results in part from atherosclerotic changes in the vessel wall.

PLATELET STUDIES

The human platelet has been used as a model for the study of α-2 receptors just as lymphocytes have been used to study beta receptors. These studies, which employed radioligand agonist/antagonist techniques to investigate α-2 receptor density, yielded conflicting results. Brodde *et al* (1982) found that aging decreased the number of alpha receptors on platelet membranes using yohimbine (an α-2 antagonist) binding in healthy individuals aged 14 to 76 years. In contrast, Yokoyama *et al* (1984) demonstrated an increase in alpha receptors between the ages of 21 to 77 years using the nonselective alpha antagonist dihydroergocryptine. Others (Elliot *et al.*, 1982; Supiano *et al.*, 1987) have found no change in receptor density, also using dihydroergocryptine binding in whole platelets and membranes. Although it is not clear at this point whether receptor density in human platelets changes with age, the evidence so far does not uniformly support receptor down-regulation in response to increased circulating noradrenaline levels with aging. The influence of technical and gender differences on alpha receptor density in the populations studied have yet to be fully clarified. No control has been made for the actual age of the platelets themselves which may influence receptor levels, and potentially explain some of the conflicting results.

An age-related decline in adrenaline-mediated adenylate cyclase production by human platelets has been demonstrated (Supiano *et al.*, 1987). This finding suggests that despite little alteration in alpha adrenergic vascular function, aging may be associated with a decrease in the high affinity agonist binding sites on platelet alpha receptors. This finding is similar to the alterations observed in the beta receptors, but requires confirmation.

PARASYMPATHETIC NERVOUS SYSTEM

Age-related alterations in the parasympathetic nervous system are difficult to evaluate.   Most of the currently available clinical evidence for a decline in parasympathetic function is derived from studies of heart rate variability.

Normal human aging is associated with a decrease in heart rate variability (Gautschy, Weidmann and Gnadinger, 1986; O'Brien, O'Hare and Carrall, 1986) during rest and deep breathing. The ratio of R-R intervals during expiration and inspiration is a standard method of evaluating autonomic function and reflects the combination of parasympathetic and sympathetic influences on the heart rate (Pfeifer *et al.*, 1982). The expiration: inspiration R-R interval ratio is reduced with aging suggesting a reduction in parasympathetic control of the heart rate (Pfeifer *et al.*, 1983; Vita *et al.*, 1986). However, many physiological changes with aging may influence this finding. Impaired baroreceptor function decreased cardiac responsiveness to sympathetic and, perhaps, parasympathetic input and changes in lung and chest wall compliance which affect intrathoracic pressures and venous return to the heart during deep breathing may all influence heart rate variability. Therefore, R-R variability during respiration cannot be considered a pure test of parasympathetic function.

More recent studies have utilized spectral analysis of heart rate time series data to quantify sympathetic and parasympathetic influences on the heart. Using this technique, Lipsitz *et al* (1990) have shown reductions in overall heart rate variability with aging, as well as in sympathetic and parasympathetic components. They also demonstrated a greater attenuation of parasympathetic rather than sympathetic heart rate fluctuations in healthy elderly subjects. The existence of an age-related decline in cardiac parasympathetic activity is further supported by experiments in aged rats showing a reduction in heart rate response to vagal denervation and stimulation with methacholine (Kelliher and Conahan, 1980).

PATHOLOGICAL ABNORMALITIES IN AUTONOMIC NERVOUS SYSTEM FUNCTION WITH AGING

As a result of the age-related physiological changes discussed above, as well as the age-associated increase in the prevalence of diseases which further impair

TABLE 4.2

Conditions Associated with Orthostatic Hypotension

i. *Central nervous system diseases:*

Parkinson's disease

Shy Drager Syndrome

Progressive Supranuclear Palsy

Cerebrovascular disease

Wernicke's encephalopathy

ii. *Peripheral neuropathies:*

Diabetes mellitus

Alcoholism

Vitamin deficiency e.g. B12

Tabes dorsalis

Carcinoma (esp. bronchus/pancreas)

Amyloidosis

iii. *Primary:*

Idiopathic orthostatic hypotension (pure autonomic failure)

iv. *Physical deconditioning:*

Prolonged bedrest

v. *Volume depletion*

vi. *Varicose veins*

autonomic function, elderly people are prone to several clinical syndromes associated with autonomic dysfunction. These include orthostatic hypotension, postprandial hypotension and hypothermia. Each of these syndromes is discussed briefly below.

ORTHOSTATIC HYPOTENSION

Prevalence estimates of orthostatic hypotension in different groups of elderly people range from 4% to 33% (Mader, 1989). However, in normotensive healthy elderly individuals the prevalence is less that 7% (Mader, Josephson and Rubensiein, 1987). Although orthostatic hypotension is a cardinal feature of autonomic dysfunction in a young individual – often heralding the onset of autonomic failure – in the older person it is more likely to result from comorbidity (Table 4.2) and medication usage (Table 4.3) than from a syndrome of pure autonomic failure.

TABLE 4.3

Medications Associated with Orthostatic Hypotension

i.	Antihypertensives
ii.	Nitrates
iii.	Diuretics
iv.	Phenothiazines
v.	Tricyclic antidepressants
vi.	MAO inhibitors
vii.	L-Dopa

Orthostatic hypotension is generally defined as a decline in systolic blood pressure of 20 mmHg or diastolic blood pressure of 10 mmHg or greater upon assuming an upright posture, with or without symptoms. It should be regarded as a symptom rather than a disease entity, as well as an important risk factor for falls and syncope (Lipsitz and Fullerton, 1986; Tinetti, Williams and Mayewski, 1986).

PATHOPHYSIOLOGY OF ORTHOSTATIC HYPOTENSION

On assumption of the upright posture, approximately 500 ml of blood pools in the lower extremities and splanchnic circulation, thereby reducing venous return to the heart. The consequent unloading of cardiopulmonary and carotid baroreceptors reduces tonic inhibitory input to brainstem vasomotor centers in the nucleus tractus solitarius and results in efferent sympathetic activation and parasympathetic withdrawal. Within 10 seconds of standing, the healthy young subject demonstrates a brisk heart rate response due to vagal inhibition. The systolic blood pressure falls transiently for 10-20 seconds, but is rapidly restored by sympathetic-mediated cardioacceleration and vasoconstriction. Blood pressure may continue to fall if there is an excessive reduction in blood volume which is not counteracted by these normal physiological responses.

In the aged individual, baroreflex sensitivity is reduced. The early postural cardioacceleration observed in young people, is blunted in the elderly individual (Dambrink and Weiling, 1987). The early cardioacceleration is due to parasympathetic withdrawal. Despite the lack of heart rate acceleration on standing, only a small percent of normotensive elderly persons have orthostatic hypotension. Most are probably protected from orthostatic hypotension by alpha-mediated vasoconstriction, which is largely preserved with aging.

Age-related changes in the mechanical properties of the heart may predispose the elderly person to orthostasis, especially in volume depleted states. Increased cardiac stiffening can impair early ventricular diastolic filling (Miller et al., 1986; Bryg, Williams and Labovitz, 1987; Downes et al., 1989) and make elderly persons more dependent on preload to maintain stroke volume. In the absence of compensatory cardioacceleration, preload reduction during upright posture can then cause a marked reduction in cardiac output. The elderly person is also particularly vulnerable to intravascular volume losses through age-related reductions in renin and aldosterone (Epstein and Hollenberg, 1976), and elevation in atrial natriuretic peptide (Ohashi *et al.*, 1987) which cause renal salt and water wasting. This decline in renal salt and water conservation, as well as a decrease in the thirst response to plasma hypertonicity (Phillips *et al.*, 1984), predispose elderly people to dehydration. Therefore, changes in volume status that may occur during acute illness or a result of diuretic medications can significantly reduce cardiac output and result in orthostatic hypotension (Shannon *et al.*, 1986).

In Western societies, aging is associated with progressive elevations in systolic blood pressure. This elevation may be due to increased vascular stiffness and, possibly, increased activity of the sympathetic nervous system (see prior section). Hypertension is paradoxically associated with orthostatic hypotension (Lipsitz *et al.*, 1985) and explains a large proportion of the variance in orthostatic blood pressure responses among the elderly (Harris *et al.*, 1986). This association may be due simply to the phenomenon of regression to the mean, or to impairments in baroreflex function that are independently associated with hypertension. In a recent study of hemodynamic responses to head-up tilt in groups of young and old normotensive and hypertensive subjects, Jansen *et al* (1989) found that hypertension, rather than age, was associated with an early fall in systolic blood pressure. Therefore, in the absence of hypertension, aging may have little clinical impact on postural blood pressure homeostasis.

Most elderly persons are able to compensate for orthostatic stress despite the age-related changes outlined above that predispose them to postural hypotension. However, when exposed to additional hypotensive stresses such as diuretic-induced intravascular volume reduction, many elderly individuals lack the physiologic reserve to guard against hypotension (Shannon *et al.*, 1986). These individuals have age-related impairments in cardioacceleration and heightened plasma noradrenaline responses to postural stress (Table 4.4); they are often asymptomatic, with no other evidence of autonomic dysfunction, and may be described as having "physiologic orthostatic hypotension". In contrast, elderly persons with severe symptomatic orthostatic hypotension have a "pathological" condition due to specific diseases which impair autonomic function. These have symptoms of autonomic insufficiency and subnormal plasma noadrenaline responses to upright posture. They are chronically disabled by orthostatic symptoms, in contrast to individuals with physiologic orthostatic hypotension who become symptomatic only during periods of excessive hemodynamic stress.

POSTPRANDIAL HYPOTENSION

Like orthostatic hypotension, postprandial hypotension is a condition commonly seen in patients with autonomic failure (Robertson, Wade and Robertson, 1981; Mathias, 1990) as well as in multiply-impaired (Lipsitz *et al.*, 1983) and healthy

TABLE 4.4
Age-Related Changes Predisposing to Orthostatic Hypotension:

i.	Impaired heart rate response to hypotension
ii.	Decreased baroreflex sensitivity
iii.	Impaired renal salt and water conservation
iv.	Decreased thirst
v.	Impaired cardiac ventricular filling
vi.	Reduced renin, angiotensin and aldosterone levels

(Lipsitz and Fullerton, 1986) elderly people.   In ways similar to orthostatic hypotension, postprandial hypotension may also be viewed as a consequence of either age-related physiologic changes or pathologic abnormalities in autonomic function.   Although the mechanism of postprandial hypotension is unknown, asymptomatic elderly persons with the "physiologic" variant appear to have inadequate cardiovascular compensation for splanchnic blood pooling during food digestion. This is evident in the moderate decline in blood pressure after a meal and a blunted heart rate increase that is unable to compensate for reduced blood pressure (Lipsitz et al., 1986a). These individuals may become symptomatic if hypotensive medications are taken before a meal, or in the setting of volume depletion.

Elderly patients with pathologic postprandial hypotension have marked, symptomatic reductions in blood pressure that may result in syncope (Lipsitz et al., 1986b). These patients demonstrate an initial increase in plasma noradrenaline following a meal, but a subsequent inappropriate decline at the time that blood pressure is falling (Lipsitz et al., 1986b). They may also have clinical evidence of autonomic dysfunction (Ryan et al., 1990).

Previous studies that have examined the potential role of various gut peptides including insulin, in the pathophysiology of postprandial hypotension, have failed to find significant associations (Jansen et al., 1987). In autonomic failure, caffeine (Onrot et al., 1985) and somatostatin analogues (Hoeldtke et al., 1989) have proven beneficial in preventing postprandial splanchnic vasodilation, although the exact mechanisms are not fully understood.

Further research is needed. It is important that future investigations carefully define the populations under study in order to distinguish asymptomatic elderly people from those with autonomic failure and specific autonomic nervous system disease.

TEMPERATURE REGULATION

Temperature regulation is a complex function involving the integration of afferent signals from peripheral temperature receptors with central connections in the anterior hypothalamus and efferent pathways to the sweat glands and vasculature.

TABLE 4.5
Diseases Associated with Hypothermia

i.	Hypothyroidism
ii.	Diabetes mellitus
iii.	Wernicke encephalopathy
iv.	Renal failure
v.	Congestive heart failure
vi.	Stroke
vii.	Sepsis

TABLE 4.6
External Factors Predisposing the Elderly to Hypothermia:

i.	Poor heating and living conditions
ii.	Malnutrition
iii.	Sedentary lifestyle — immobility from arthritis, Parkinson's disease, cerebrovascular accidents, etc.
iv.	Medications: Phenothiazines Beta – blockers Antidepressants: tricyclic and monoamine oxidase inhibitors Anticholinergic Barbiturates Salicylate Alcohol

Efferent impulses are transmitted via the sympathetic nervous system to receptors in the sweat glands and vasculature which function to preserve or dissipate heat. Impairments in thermoregulation of elderly persons – either due to autonomic dysfunction or through complications of comorbidity and medication usage – increase their vulnerability to extremes of temperature (Collins and Exton-Smith, 1983; Wongsurawat, Davis and Morley, 1990) (Tables 4.5 and 4.6).

Hypothermia is defined as a decrease in body core temperature (esophageal, rectal or tympanic) below 35 degrees centigrade or 95 degrees fahrenheit. In UK surveys in 1975, 3.6% of individuals over 65 admitted to the hospital were hypothermic. Unfortunately, prevalence data on hypothermia can be difficult to interpret as much of it relies on death certificate information which may underreport the incidence of hypothermia (Collins, 1986). In addition, many hospital emergency rooms lack low-reading thermometers and therefore fail to detect hypothermia (Besdine, 1982).

Elderly hospitalized patients are also at risk of hypothermia. In Israel, despite a warm climate, 50% of patients identified as hypothermic developed it while in hospital (Kramer, Vandijk and Rosin, 1989). Medical conditions which impair thermoregulation in the elderly predispose them to the development of hypothermia even under relatively mild cold stress. Diseases such as Parkinson's disease and severe arthritis can immobilize the older person and thereby impair heat production. Malnutrition – by itself or in association with dementia, poor living conditions, cancer or other conditions – can result in a lowered basal metabolic rate and reduced heat production. Neuroleptic medications impair central heat regulation and can also immobilize an elderly person. Sepsis is frequently observed among elderly patients admitted with hypothermia (Kramer, Vandijk and Rosin, 1989). It appears that underlying medical conditions predisposing to hypothermia are more common that autonomic dysfunction *per se* (Besdine, 1979). Mortality in elderly persons with hypothermia is high and ranges from 30-80%. Mortality from hypothermia in association with myxoedema is particularly high.

Heat stroke also appears more prevalent in the elderly. In part, this may be attributed to physiologic changes that result in poor temperature perception, and thus lack of protective behavioral responses. Comorbidity, such as dementia and neuroleptic medications also impair the elderly persons capacity to detect and respond appropriately to elevated ambient temperatures.

Information on temperature regulation in the aged person is derived primarily from clinical studies. These depend on reflex responses of isolated limbs to temperature changes. Again, the use of varying methodologies make data interpretation and comparisons difficult. Nevertheless, aging appears to result in the following alterations in temperature control (Table 4.7).

CHANGES IN TEMPERATURE DISCRIMINATION IN OLD AGE

Young individuals are able to discriminate temperature differences of 1–2 degrees centigrade. In contrast, many elderly people are unable to detect differences in temperature closer than 2–4 degrees (Collins, Exton-Smith and Dore, 1981). Elderly individuals with poor temperature discrimination also demonstrate less ability to regulate their ambient temperature. This was demonstrated in a study by (Collins, Exton-Smith and Dore, 1981) in which subjects were asked to regulate room temperature by adjusting a thermostat. Elderly individuals with poor temperature discrimination lacked precision in adjusting the temperature, possibly due to impaired perception of ambient room temperature. This notion is supported by the report that elderly persons complain less than the young when exposed to a cold environment (Horvath *et al.*, 1955).

Alteration in temperature perception with aging may result in part from changes in the peripheral temperature receptors. These receptors are highly dependent on oxygen and therefore may be affected by diminished peripheral blood flow. Also, age-related alterations in skin collagen and elastic tissue may influence

TABLE 4.7

Physiological Changes Impairing Temperature Regulation:

i.	Decreased basal metabolic rate
ii.	Impaired perception of temperature changes
iii.	Impaired vasoconstriction in the cold
iv.	Reduced shivering response in the cold
v.	Delayed and reduced sweating in the heat
vi.	Impaired capacity to increase cardiac output
viii.	Impaired fluid conservation by the kidney

receptor function. The potential role of age-related changes in the hypothalamus on temperature perception and behavior is not well elucidated (Wongsurawat, Davis and Morley, 1990).

IMPAIRED THERMOREGULATION IN AGING

Basal metabolic rate/heat production

Aging is accompanied by a gradual decrease in basal metabolic rate (BMR), due in large part to decreased skeletal muscle mass. The reduction in BMR is evident at thermoneutral temperatures and in response to cold environments (Wagner, Robinson and Marino, 1974; Collins, Easton and Exton-Smith, 1981). Sedentary elderly individuals who lack physical conditioning and malnourished patients may not be able to generate sufficient heat to protect them from hypothermia when exposed to the cold.

Vasoconstriction/shivering

Peripheral vasoconstriction and shivering in response to cold exposure are important features of effective thermoregulation. Elderly individuals demonstrate considerable variability in their capacities to respond to cold exposure. However, in general, older people exhibit delayed and reduced vasoconstriction after cold exposure (Collins et al., 1981b; Collins, 1987; Wagner, Robinson and Marino, 1974). Vasoconstriction in the toe in response to reduced ambient air temperature is blunted in elderly individuals with atherosclerosis (Collins, 1987), suggesting that structural changes in aged vessels may contribute to alterations in temperature control. However, many of the studies of vascular responsiveness to cold exposure (Wagner, Robinson and Marino, 1974) utilize measures of forearm blood flow that do not distinguish muscle and skin blood flow. Again, reduced muscle mass may influence the results in elderly subjects.

Sudomotor and vasomotor neural activity can be recorded directly through peripheral cutaneous nerves of awake people (Bini et al., 1980). In thermoneutral conditions a constant volley of sudomotor and vasoconstrictor nerve activity occurs. Exposure to cold environments results in increased vasoconstrictor nerve activity and decreased sudomotor nerve activity; the opposite is true on exposure to heat. Collins (1983) reported that the vasoconstrictor activity, measured using photoelectric pulsimetry, in response to cold is attenuated in elderly subjects. If this finding proves reproducible in larger elderly populations it will provide more direct evidence of decreased vasoconstriction in elderly subjects. Further studies are required before the results can be considered conclusive.

Shivering is mediated through central hypothalamic pathways and can significantly increase muscular heat production. Healthy elderly individuals exhibit later and less intense shivering on exposure to cold (Collins et al., 1981b). The mechanism of this alteration in shivering response is not known.

CENTRAL TEMPERATURE REGULATION

Temperature exhibits a strong circadian rhythm with a nadir in the early hours of the morning. Basal temperatures in elderly subjects are not different from those of young subjects but the circadian rhythm of temperature may change with age. Vitiello et al (1986) studied the rectal temperature from 10 young and 8 elderly men during 24 hour periods. The elderly demonstrated a decrease in variation and amplitude of temperature over the 24 hours. The nadir during the early morning was higher in aged men at 36.7 degrees versus 36.4 degrees in younger individuals. The time of the nadir was similar in young and old subjects (between 2:00 - 6:00 a.m.).

VASODILATATION AND SWEATING

Sweating and vasodilatation normally occur in response to elevations in environmental temperature to prevent an excessive rise in the core temperature. Epidemiological studies suggest that elderly individuals are more vulnerable to heat stroke. Skin atrophy which accompanies normal aging results in a loss of sweat glands. As a result, elderly individuals exhibit reduced sweating in response to heat and neurochemical stimulation (Foster et al., 1976). In addition, the elderly have a higher core temperature threshold for the onset of sweating and vasodilatation. Vasodilatory responses to radiant heat on the forearm have been investigated using doppler skin blood flow measurements (Richardson, 1989). Young subjects demonstrated an increased forearm cutaneous blood flow in response to heat, measured as the increase in velocity and volume of blood cells flowing through the vessels. This response may result from the combination of increased blood flow through dilated veins, and from the opening of capillary vessels that increase the cutaneous vascular bed. Elderly subjects showed attenuated cutaneous blood flow responses to local heat, although it is not clear if this is secondary to reduced vasodilatation or less recruitment of capillary vessels. When the effect of age, cholesterol and plasma glucose on cutaneous blood flow response to ambient heat was investigated, age was the most important variable (Richardson, 1989).

Abnormalities in temperature regulation are frequently seen in individuals with other symptoms of autonomic dysfunction. A study by Collins *et al* (1977) suggests that orthostasis is more common in elderly people with a history of hypothermia. This requires confirmation. Temperature dysregulation may be part of a spectrum of autonomic impairment in the elderly person.

CONCLUSION

Distinguishing the physiologic changes that influence autonomic function from specific diseases that are more common in advanced age presents particular complexities. Normal human aging is probably not associated with autonomic failure *per se* – since healthy people adapt well to physiological stress under usual circumstances. However selected areas of the autonomic nervous system do appear to undergo age-related changes. These changes result in a decline in autonomic reserve capacity during exposure to extreme levels of stress. The development of diseases and taking of various drugs in advanced age may also independently influence autonomic function, or may interact with age-related changes to seriously impair homeostatic capacity. This can result in symptomatic autonomic dysfunction. The increased prevalence of orthostatic and hypothermia in association with medications or comorbidity provide examples of this kind of interaction. Specific diseases of the autonomic nervous system, such as the Shy-Drager syndrome, are rare but occur more frequently in elderly people and require careful diagnosis and individualized treatment. These syndromes are discussed in other chapters.

REFERENCES

Abrass, I.B. and Scarpace, P.J. (1981) Human lymphocyte beta-adrenergic receptors are unaltered with age. *J. Geront.*, **36**(3), 298–301.

Abrass, I.B. and Scarpace, P.J. (1982) Catalytic unit of adenylate cyclase: reduced activity in aged-human lymphocytes. *J. Clin. Endocrinol. Metab.*, **55**, 1026–1028.

Abrass, I.B., Davis, J.L. and Scarpace, P.J. (1982) Isoproterenol responsiveness and myocardial β-adrenergic receptor in young and old rats. *J. Geront.*, **37**(2), 156–160.

Angell-James, J.E. (1974) Arterial baroreceptor activity in rabbits with experimental atherosclerosis. *Circ. Res.*, **34**, 27–39.

Barnes, R.F., Raskind, M., Gumbrecht, G. and Halter, J.B. (1982) The effects of age on the plasma catecholamine response to mental stress in man. *J. Clin. Endocrinol. Metab.*, **54**, 64–69.

Besdine, R.W. (1979) Accidental hypothermia: The body's energy crisis. *Geriatrics*, **34**, 51–59.

Besdine, R.W. (1982) Accidental hypothermia in the elderly. *Medical Grand Rounds*, **1**(1), 36–47.

Best, J. and Halter, J.B. (1982) Release and clearance rates of epinephrine in man: importance of arterial measurements. *J. Clin. Endocrinol. Metab.*, **55**, 263–267.

Bini, G., Hagbarth, K.E., Hynninen, P. and Wallin, B.G. (1980) Thermoregulatory and rhythm-generating mechanisms governing the sudomotor and vasoconstrictor outflow in human cutaneous nerves. *J. Physiol.*, **306**, 537–552.

Brodde, O.E., Anlauf, M., Graben, N. and Bock, K.D. (1982) Age-dependent decreases of α_2-adrenergic receptor number in human platelets. *Eur. J. Pharmacol.*, **81**, 345–347.

Bryg, R.J., Williams, G.A. and Labovitz, A.J. (1987) Effects of aging on left ventricular diastolic filling in normal subjects. *Am. J. Cardiol.*, **59**, 971–974.

Chang, P.C., van der Krogt, J.A., Vermeij, P. and van Brummelen, P. (1986) Norepinephrine removal and release in the forearm of healthy subjects. *Hypertension*, **8**, 801–809.

Cohen, M.L. and Berkowitz, B.A. (1974) Age-related changes in vascular responsiveness to cyclic nucleotide and contractile agonists. *J. Pharmacol. Exp. Ther.*, **191**(1), 147–155.

Collins, K.J. (1983) Autonomic failure and the elderly. In: *Autonomic Failure*, edited by R. Bannister, pp. 487–587. Oxford: Oxford University Press.

Collins, K.J. (1986) Low indoor temperatures and morbidity in the elderly. *Age and Aging*, **15**, 212–220.

Collins, K.J. (1987) Increased morbidity and mortality from cardiovascular and respiratory disease in the elderly is strongly associated with cold winters in Britain. *Br. J. Hosp. Med.*, **38**(6), 506–513.

Collins, K.J. and Exton-Smith, A.N. (1983) Thermal homeostasis in old age. *J. Am. Ger. Soc.*, **31**(9), 519–524.

Collins, K.J., Dore, C., Exton-Smith A.N., Fox, R.H., MacDonald, I.C. and Woodward, P.M. (1977) Accidental hypothermia and impaired temperature homeostasis in the elderly. *Br. Med. J.*, **1**, 353–356.

Collins, K.J., Exton-Smith, A.N. and Dore, C. (1981) Urban hypothermia: preferred temperature and thermal perception in old age. *Br. Med. J.*, **282**, 175–177.

Collins, K.J., Easton, J.C. and Exton-Smith, A.N. (1981) Shivering thermogenesis and vasomotor responses with convective cooling. *J. Physiol. (Lond.)*, **382**, 76p.

Collins, K.J., Durnin, C.J.A. and Pluck, R.A. (1986) Peripheral venous compliance and vasomotor responses to cooling and warming in young and elderly subjects. *J. Physiol. (Lond.)*, **382**, 27p.

Cox, R.H., Bagshaw, R.J. and Detweiler, D.K. (1980) Alterations in carotid sinus reflex control of arterial hemodynamics associated with experimental hyperlipemia in the racing greyhound. *Circ. Res.*, **46**, 237–244.

Dambrink, J.H.A. and Weiling, W. (1987) Circulatory response to postural change in healthy male subjects in relation to age. *Clin. Sci.*, **72**, 335–341.

Dillon, N., Chung, S., Kelly, J. and O'Malley, K. (1980) Age and beta adrenoceptor-mediated function. *Clin. Pharmacol. Ther.*, **27**(6), 769–772.

Downes, T.R., Nomeir, A.M., Smith, K.M., Stewart, K.P. and Little, W.C. (1989) Mechanisms of altered pattern of left ventricular filling with aging in subject without cardiac diseases. *Am. J. Cardiol.*, **64**, 523–537.

Ebert, T.J., Huges, C.V., Tristani, F.R., Barney, J.A. and Smith, J.J. (1982) Effect of age and coronary heart disease on the circulatory responses to graded lower body negative pressure. *Cardiovasc. Res.*, **16**, 663–669.

Elliott, H.L., Sumner, D.J., McLean, K. and Reid, J.L. (1982) Effect of age on the responsiveness of vascular α-adrenoceptors in man. *J. Cardiovasc. Pharmacol.*, **4**(3), 388–392.

Epstein, M. and Hollenberg, N.K. (1976) Age as a determinant of renal sodium conservation in normal man. *J. Lab. Clin. Med.*, **87**(3), 41–417.

Esler, M., Skews, H., Leonard, P., Jackman, G., Bobik, A. and Kroner, P. (1981) Age-dependence of noradrenaline kinetics in normal subjects. *Clin. Sci.*, **60**, 217–219.

Feldman, R.D. (1986) Physiological and molecular correlates of age-related changes in the human β-adrenergic receptor system. *Fed. Proc.*, **45**, 48–50.

Feldman, R.D., Limbird, L.E., Nadeau, J. Robertson, D. and Wood, A.J. (1984) Alterations in leukocyte β-receptor affinity with aging. *N. Engl. J. Med.*, **310**(13), 815–819.

Fleg, J.L., Tzankoff, S.P. and Lakatta, E.G. (1985) Age-related augmentation of plasma catecholamines during dynamic exercise in healthy males. *J. Appl. Physiol.*, **59**(4), 1033–1039.

Fleisch, J.H. and Hooker, C.S. (1976) The relationship between age and relaxation of vascular smooth muscle in the rabbit and rat. *Circ. Res.*, **38**(4), 243–249.

Feisch, J.H., Maling, H.M. and Brodie, B.B. (1970) Beta-receptor activity in aorta. *Circ. Res.*, **26**, 151–162.

Foster, K.G., Ellis, F.P., Dore, C., *et al.*, (1976) Sweat responses in the aged. *Aging*, **5**, 91.

Furman. M.I., Kaplan, D.T., Ryan, S.M., Lipsitz, L.A. and Goldberg, A.L. (1990) A nonlinear parameter measures loss of complexity in heart rate time series data with aging. *Clin. Res.*, **38**(2), 491A.

Gautschy, B., Weidmann, P. and Gnadinger, M.P. (1986) Autonomic function tests as related to age and gender in normal man. *Klin. Wochenschr.*, **64**, 499–505.

Goldstein, D.S., McCarty, R., Polinsky, R.J. and Kopin, I.J. (1983) Relationship between plasma norepinephrine and sympathetic neural activity. *Hypertension*, **5**, 552–559.

Gribbin, B., Pickering, T.G., Sleight, P. and Peto, R. (1971) Effect of age and high blood pressure on baroreflex sensitivity in man. *Circ. Res.*, **29**, 424–431.

Harris, T., Kleinman, J. and Lipsitz, L.A. (1986) Is age or level of systolic blood pressure related to positional blood pressure change? *Gerontologist*, **26**, 59A Suppl.

Hoeldtke, R.D. and Cilmi, K.M. (1985) Effects of aging on catecholamine metabolism. *J. Clin. Endocrinol. Metab.*, **60**(3), 479–484.

Hoeldtke, R.D., Dworkin, G.E., Gasper, S.R., Israel, B.C. and Boden, G. (1989) Effect of the somatostatin analogue SMS-201-995 on the adrenergic response to glucose ingestion in patients with postprandial hypotension. *Am. J. Med.*, **86**, 673–677.

Horvath, S.M., Radcliffe, C.E., Hutt, B.K., *et al.*, (1955) Blood pressure reduction after oral glucose loading and its relation to age, blood pressure and insulin. *Am. J. Cardiol.*, **60**, 1087–1091.

Jansen, R.W., Lenders, J.W., Thien, T. and Hoefnagels, W.H. (1989) The influence of age and blood pressure on the hemodynamic and humoral response to head up tilt. *J. Am. Geriatr. Soc.*, **37**, 528–532.

Jarisch, W.R., Ferguson, J.J., Shannon, R.P. and Goldberger, A.L. (1987) Age-related disappearance of mayer-loke heart rate waves. *Experientia*, **43**, 1207–1209.

Jonsson, P.V., Lipsitz, L.A., Kelly, M.M. and Koestner, J. (1990) Hypotensive responses to common daily activities in institutionalized elderly. A potential risk for recurrent fall. *Arch. Intern. Med.*, **150**, 1518–1524.

Jungu, S., Takeshita, A., Imaizumi, T., Sakai, K. and Nakamuru, M. (1989) Age-related decreases in cardiac receptor control of forearm vascular resistance in humans. *Clin. Exp. Hypertens.*, **A11**(Suppl 1), 211-215.

Kelliher, G.J. and Conahan, S.T. (1980) Changes in vagal activity and response to muscarinic receptor agonists with age. *J. Gerontol.*, **35**(6), 842–849.

Krall, J.F., Connelly, M., Weisbart, R. and Tuck, M.L. (1981) Age-related elevation of plasma catecholamine concentration and reduced responsiveness of lymphocyte adenylate cyclase. *J. Clin. Endocrinol. Metab.*, **52**, 863–867.

Kramer, M.R., Vandijk, J. and Rosin, A.J. (1989) Mortality in elderly patients with thermoregulatory failure. *Arch. Intern. Med.*, **149**, 1521–1523.

Lakatta, E.G., Gerstenblith, G., Angell, C.S., Shock, N.W. and Weisfeldt, M.L. (1975) Diminished inotropic response to aged myocardium to catecholamines. *Circ. Res.*, **36**, 262–267.

Linares, O.A. and Halter, J.B. (1987) Sympathochromaffin system activity in the elderly. *J. Am. Geriatr. Soc.*, **35**, 448–453.

Lipsitz, L.A. and Fullerton, K.J. (1986) Postprandial blood pressure reduction in healthy elderly. . *J. Am. Geriatr. Soc.*, **34**, 267–270.

Lipsitz, L.A., Nyquist, R.P., Wei, J.Y. and Rowe, J.W. (1983) Postprandial reduction in blood pressure in the elderly. *N. Engl. J. Med.*, **309**, 81–83.

Lipsitz, L.A., Storch, H.A., Minaker, K.L. and Rowe, J.W. (1985) Intra-individual variability in postural blood pressure in the elderly. *Clin. Sci.*, **69**, 337–341.

Lipsitz, L.A., Plucino, F.C., Wei, J.Y. and Rowe, J.W. (1986a) Syncope in institutionalized elderly: the impact of multiple pathological conditions and situational stress. *J. Chronic. Dis.*, **39**(8), 619–630.

Lipsitz, L.A., Plucino, F.C., Wei, J.Y., Minaker, K.L. and Rowe, J.W. (1986b) Cardiovascular and norepinephrine responses after meal consumption in elderly (older than 75 years) persons with postprandial hypotension and syncope. *Am. J. Cardiol.*, **58**, 810–815.

Lipsitz, L.A., Meitus, J., Moody, G.B. and Goldberger, A.L. (1990) Spectral characteristics of heart rate variability before and during postural tilt. Relations to aging and risk of syncope. *Circulation*, **81**, 1803–1810.

Lipsitz, L.A., Bui, M., Steibeling, M. and McArdle, C. (1991) Forearm blood flow response to posture change in the very old: non-invasive measurement by venous occlusion plethysmography. *J. Am. Geriatr. Soc.*, **39**(1), 53–9.

Mader, S.L. (1989) Aging and postural hypotension. An update. *J. Am. Geriatr. Soc.*, **37**, 129–137.

Mader, S.L., Josephson, K.R. and Rubensiein, L.Z. (1987) Low prevalence of postural hypotension among healthy elderly. *JAMA*, **258**, 1511–1514.

Mancia, G. Ferrari, A., Gregorini, L., Parati, G. Pomidossi, G., Bertinieri, G. *et al.*, (1980) Blood pressure variability in man: its relation to high blood pressure, age and baroreflex sensitivity. *Clin. Sci.*, **59**, 401s–404s.

Mathias, C.J. (1990) Effect of food intake on cardiovascular control in patients with impaired autonomic function. *J. Neurol. Sci.*, **34**, 193–200.

Miller, T.R., Grossman, S.J., Schectman, K.B., Biello, D.R., Ludbrook, P.A. and Ehsani, A.A. (1986) Left ventricular diastolic filling and its association with age. *Am. J. Cardiol.*, **58**, 531–535.

Morrow, L.A., Linares, O.A., Hill, T.J., Sanfield, J.A., Supiano, M.A., Rosen, S.G., *et al.* (1987) Age differences in the plasma clearance mechanisms for epinephrine and norepinephrine in humans. *J. Clin. Endocrinol. Metab.*, **65**, 508–511.

O'Brien, I.A.D., O'Hare, P. and Carrall, R.J.M. (1986) Heart rate variability in healthy subjects: effects of age and derivation of normal ranges for tests of autonomic function. *Br. Heart J.*, **55**, 348–354.

Ohashi, M., Fujio, N., Nawata, H., Kato, K., Ibayashi, H., Kangawa, K., *et al.*, (1987) High plasma concentrations of human atrial natriuretic polypeptide in aged men. *J. Clin. Endocrinol. Metab.*, **64**, 81.

Onrot, J., Goldberg, M.R., Biaggioni, I., Hollister, A.S., Kingaid, D. and Robertson, D. (1985) Hemodynamic and humoral effects of caffeine in autonomic failure: Therapeutic implications for postprandial hypotension. *N. Engl. J. Med.*, **313**, 549–554.

Palmer, G.J., Ziegler, M.G. and Lake, C.R. (1978) Response of norepinephrine and blood pressure to stress increases with age. *J. Geront.*, **313**, 549–554.

Pan, H.Y.M., Hoffman, B.B., Pershe, R.A. and Blaschk, T.F. (1986) Decline in beta adrenergic receptor-mediated vascular relaxation with aging man. *J. Pharmacol. Exp. Ther.*, **239**(3), 802–807.

Pfeifer, M.A., Cook, D., Brodsky, J., Tice, D., Reenan, A., Swedine, S. *et al.*, (1982) Quantitative assessment of cardiac parasympathetic activity in normal diabetic man. *Diabetes*, **31**, 339–345.

Pfeifer, M.A., Weinberg, C.R., Cook, D., Best, J.D., Reenan, A. and Halter, J.B. (1983) Differential changes of autonomic nervous system function with age in man. *Amer. J. Med.*, **75**, 249–257.

Phillips, P.A., Rolls, B.J., Ledingham, J.G., Forslins, M.L., Morton, J.J., Crowe, M.J., *et al.*, (1984) Reduced thirst after water deprivation in healthy elderly men. *N. Engl. J. Med.*, **311**, 753–759.

Pomeranz, B., Macauley, R.J.B. and Caudill, M.A. (1985) Assessment of autonomic function in humans by heart rate spectral analysis. *Am. J. Physiol.*, **248** (Heart Circ Physiol 17), H151–H153.

Prinz, P.H., Halter, J., Benedetti, C. and Raskind, M. (1979) Circadian variation of plasma catecholamines in young and old men: relation to rapid eye movement and slow wake sleep. *J. Clin. Endocrinol. Metab.*, **49**, 300–304.

Richardson, D. (1989) Effects of age on cutaneous circulatory response to direct heat on the forearm. *J. Gerontol.*, **44**(6), M189–M194.

Robertson, D., Wade, D. and Robertson, R.M. (1981) Postprandial alterations in cardiovascular hemodynamics in autonomic dysfunctional states. *Am. J. Cardiol.*, **48**, 1048–1052.

Rowe, J.W. and Troen, B.R. (1980) Sympathetic nervous system and aging in man. *Endocrinol. Reviews*, **1**(2), 167–179.

Rubin, P.C., Scott, P.J.W., McLean, K. and Reid, J.L. (1982) Noradrenaline release and clearance in relation of age in blood pressure in man. *Eur. J. Clin. Invest.*, **12**, 121–125.

Ryan, S.M., Kelly, M.M., Striebeling, M. and Lipsitz, L.A. (1990) Impaired forearm vascular response to a meal in elderly syncope patients with postprandial hypotension. (PPH). *The Gerontologist*, **30**, 45A.

Schocken, D.D. and Roth, G.S. (1977) Reduced β-adrenergic receptor concentrations in aging man. *Nature*, **267**, 856–858.

Schwartz, R.S., Halter, J.B. and Bierman, E.L. (1983) Reduced thermic effect of feeding in obesity: role of norepinephrine. *Metabolism*, **32**, 114–117.

Scott, P.J.W. and Reid, J.L. (1982) The effect of age on the responses of human isolated arteries to noradrenaline. *Br. J. Clin. Pharmaco.*, **13**, 237–239.

Shannon, R.P., Wei, J.Y., Rosa, R.M., Epstein, F.H. and Rowe, J.W. (1986) The effect of age and sodium depletion on cardiovascular response to orthostasis. *Hypertension*, **8**(5), 438–443.

Shimada, K., Kitazumi, T., Ogura, H., Sadakane, N. and Ozawa, T. (1986) Effects of age and blood pressure on the cardiovascular responses to the Valsalva maneuver. *J. Am. Geriatr. Soc.*, **34**, 431–434.

Shimada, K., Kitazumi, T., Sadakane, N., Ogura, H. and Ozawa, T. (1985) Age-related changes of baroreflex function, plasma norepinephrine and blood pressure. *Hypertension*, **7**, 113–117.

Shimizu, I. and Toda, N. (1986) Alteration with age of the response to vasolidator agents in isolated mesenteric arteries of the beagle. *Br. J. Pharmacol.*, **89**, 769–778.

Shiraki, K., Sagawa, S., Yousef, M.K., Konda, N. and Miki, K. (1987) Physiological responses of aged men to head-up tilt during heat exposure. *J. Appl. Physiol.*, **63**(2), 576.

Simpkins, J.W., Field, F.B. and Ress, R.J. (1983) Age-related decline in adrenergic responsiveness of the kidney, heart and aorta of male rats. *Neurobiol. Aging*, **4**, 233–238.

Simpsons, D.M. and Wicks, R. (1988) Spectral analysis of heart rate indications reduced baroreceptor-related heart rate variability in elderly persons. *J. Gerontol. Med. Sci.*, **43**(1), M21–M24.

Sowers, J.R., Whitfield, L.A. Catania, R., Stern, N., Tuck, M.L., Domfeld, L., *et al.*, (1982) Role of sympathetic nervous system in blood pressure maintenance in obesity. *J. Clin. Endocrinol. Metab.*, **54**, 1181–1186.

Stevens, M.J., Kipe, S. and Moulds, R.F.W. (1982) The effect of age on the responses of human isolated arteries and veins to noradrenaline. *Br. J. Clin. Pharm.*, **14**, 750–752.

Supiano, M.A., Linares, O.A., Halter, J.B., Reno, K.M. and Rosen, S.G. (1987) Functional uncoupling of the platelet α_2-adrenergic receptor-adenylate cyclase complex in the elderly. *J. Clin. Endocrinol. Metab.*, **64**, 1160–1164.

Tinetti, M.E., William, T.F. and Mayewski, R. (1986) Fall risk index for elderly patients based on number of chronic disabilities. *Am. J. Med.*, **80**, 429–434.

Tuttle, R.S. (1966) Age-related changes in the sensitivity of rat aorta strips to norepinephrine and associated chemical and structural alterations. *J. Gerontol.*, 510–516.

van Brunnelen, P., Buhler, F.R., Kiowski, W. and Amann, F.W. (1981) Age-related decrease in cardiac and peripheral vascular responsiveness to isoprenaline: studies in normal subjects. *Clin. Sci.*, **60**, 571–577.

Veith, R.C., Featherstone, J.A., Linares, O.P.A. and Halter, J.B. (1986) Age differences in plasma norepinephrine kinetics in human. *J. Gerontol.*, **41**(3), 319–324.

Vestal, R.E., Wood, A.J.J. and Shand, D.G. (1979) Reduced β-adrenoceptor sensitivity in the elderly. *Clin. Pharmacol. Ther.*, **26**(2), 181–186.

Vita, G., Princi, P., Calabro, R., Toscano, A., Manna, L. and Messina, C. (1986) Cardiovascular reflex tests assessment of age-adjusted normal range. *J. Neurol. Sci.*, **75**, 263–274.

Vitiello, M.V., Smallwood, R.G., Avery, D.H., Pascually, R.A., Martin, D.C. and Prinz, P.N. (1986) Circadian temperature rhythms in young adult and aged men. *Neurobiol. Aging*, **7**(2), 97–100.

Wagner, J.A., Robinson, S. and Marino, R.P. (1974) Age and temperature regulation of humans in neutral and cold environments. *J. Appl. Physiol.*, **37**(4), 562–565.

Wallin, B.G. (1988) Relationship between sympathetic nerve traffic and plasma concentration of noradrenaline man. *Pharmacol. Toxicol.*, **63**(Suppl I), 9–11.

Wei, J.Y., Rowe, J.W., Kestenbaum, A.D. and Ben-Haim, S. (1983) Post cough heart rate response: influence of age, sex and basal blood pressure. *Am. J. Physiol.*, **245**, R18–R24.

Wei, J.Y., Mendelowitz, D., Anastasi, N. and Rowe, J.W. (1986) Maintenance of carotid baroflex function in advanced age in the rat. *Am. J. Physiol.*, **250** (Regulatory Integrative Comp. Physiol. 19) R1047–R1051.

Wongsurawat, N., Davis, B.B. and Morley, J.E. (1990) Thermoregulatory failure in the elderly. *J. Am. Gertiatr. Soc.*, **38**, 899–906.

Yokoyama, M., Kusui, A., Sakamoto, S. and Fukuzaki, H. (1984) Age associated increments in human platelet α_2 adrenoreceptor capacity. Possible machanism for platelet hyperactivity to epinephrine in aging man. *Thrombosis Res.*, **34**, 287–295.

Young, J.B., Roew, J.W., Pallota, J.V., Sparrow, D. and Handsberg, L. (1980) Enhanced plasma norepinephrine response to upright posture and oral and glucose administration in elderly human subjects. *Metabolism*, **2**, 532–539.

Ziegler, M.G., Lake, C.R. and Kopin I.J. (1976) Plasma noradrenaline increases with age. *Nature*, **261**, 333–335.

5 Pure Autonomic Failure

Roy Freeman

New England Deaconess Hospital, Division of Neurology,
110 Francis Street, Suite 4A Boston, MA 02110, USA

Pure autonomic failure is a rare degenerative disease of the nervous system characterized by progressive autonomic dysfunction without signs of central or extra-autonomic peripheral nervous system disease. Despite the absence of a satisfactory pathological examination, converging lines of evidence suggest that this disorder is predominantly due to peripheral postganglionic autonomic neuron degeneration. Prominent clinical symptoms include orthostatic hypotension, impotence, bowel dysfunction, bladder dysfunction, and disorders of sweating and thermoregulation. The primary biochemical abnormality is a low resting level of noradrenaline that does not increase with postural change or exertion. This disease progresses gradually and most patients, with some therapeutic intervention, are able to lead productive lives.

KEY WORDS: Autonomic dysfunction, postganglionic autonomic neuron degeneration, hypotension, impotence, bowel, bladder, thermoregulation, Shy-Drager syndrome, diabetes, Chagas, Guillain-Barr

INTRODUCTION

Bradbury and Eggleston first delineated the syndrome of progressive autonomic dysfunction in their landmark paper published in 1925. They comprehensively described three patients with an "extensive and peculiar disturbance in the functional activity of the vegetative nervous system". These patients all had incapacitating postural hypotension with cardiovascular, gastrointestinal, urogenital, thermoregulatory, sudomotor and pupillomotor dysfunction with "signs of slight and indefinite changes in the nervous system". Their insightful description remains the paradigm for the disorder characterized by autonomic dysfunction occurring without signs of central or peripheral nervous system disease.

Since these early descriptions the classification of those disorders producing autonomic failure has been laden with confusion (see Table 5.1). Early workers tended to emphasize orthostatic hypotension, the most disabling feature of autonomic failure, in their classification schemes. Primary orthostatic hypotension and idiopathic

TABLE 5.1
Terminology of autonomic failure disorders

Autonomic failure with central nervous system signs

  Primary autonomic failure

  Neurogenic orthostatic hypotension

  Shy-Drager syndrome

  Autonomic failure with multiple system atrophy

Autonomic failure without other nervous system signs

  Primary autonomic failure

  Neurogenic orthostatic hypotension

  Idiopathic orthostatic hypotension

  Bradbury-Eggleston syndrome

  Peripheral autonomic failure

  Progressive autonomic failure

  Pure autonomic failure

---

orthostatic hypotension thus were terms initially used to describe that group of disorders where the cause of autonomic failure, specifically orthostatic hypotension, was unknown (Bannister, Ardill and Fentem, 1967; Chokroverty *et al.*, 1969) This classification encompassed diseases with central nervous system dysfunction and diseases without any signs of neural disease other than autonomic dysfunction.

As early as 1933 the association of postural hypotension with the signs of central nervous system dysfunction was recognized (Barker, 1933) and even Bradbury and Eggleston's third case had brisk reflexes in the lower extremities with bilateral Babinski sign's (Bradbury and Eggleston, 1925). These abnormalities, however, were initially viewed as "incidental" (Barker, 1933). Shy and Drager first emphasized the role of the central nervous system in this disorder. In 1960 they described two patients with autonomic failure associated with iris atrophy, extra-ocular palsies, rigidity, tremor, loss of associated movements fasciculations, distal wasting and evidence of neuropathic or anterior horn cell involvement on electromyography and muscle biopsy (Shy and Drager, 1960). The relationship of the degeneration of preganglionic intermediolateral column sympathetic neurons to the autonomic dysfunction was recognized only several years later (Johnson *et al.*, 1966). In the ensuing years, the Shy-Drager syndrome referred to the combination of autonomic failure with signs of central nervous system disease, although idiopathic orthostatic hypotension was sometimes used interchangeably with the Shy-Drager syndrome to describe similar patients (Bannister, Ardill and Fentem, 1967; Thomas and Schirger, 1970). Adding further confusion, the disorder selectively affecting the autonomic nervous system, as initially outlined by Bradbury and Eggleston (1925), was often also called idiopathic orthostatic hypotension (Kontos, Richardson and Narvell, 1975; Low, Thomas and Dyck, 1978). Graham and Oppenheimer (1969), in an attempt to simplify the classification of the central nervous system degenerative diseases associated with autonomic failure, introduced the term multiple system atrophy to describe striatonigral degeneration and olivopontocerebellar atrophy that is sometimes associated with autonomic failure.

Converging lines of evidence suggest that autonomic failure in isolation, is predominantly due to degeneration of peripheral postganglionic autonomic neurons, however unequivocal pathological confirmation is still lacking. In addition to idiopathic orthostatic hypotension, this disorder has also been called progressive autonomic failure (Bannister, 1983; Cohen et al., 1987), the Bradbury-Eggleston syndrome (Robertson et al., 1984) and more recently pure autonomic failure (PAF) (Bannister, Mathias and Polinski, 1988). As our understanding of the pathology of this disorder or disorders matures, all these names will probably yield to a more appropriate designation based on etiologic grounds. In this manuscript PAF will be used to designate this disorder and the Shy-Drager syndrome (SDS) will be used to indicate autonomic failure characterized by central pathology.

CLINICAL FEATURES

GENERAL

The rarity of PAF has precluded community based epidemiological studies. The true incidence and prevalence of this disorder is not known, nor have there been any population-based prospective studies of the natural history. Several centres in the United States and the United Kingdom have accumulated series of patients but these case collections inevitably reflect the referral bias that is inherent to tertiary medical centre-based studies. Bannister, Mathias and Polinsky (1988) retrospectively compared the United Kingdom patients with autonomic failure to those of the United States. There were in total 46 PAF patients in comparison to 118 SDS patients, providing some indication of the comparative rarity of PAF. There was a 2 to 1 male to female sex ratio in the United Kingdom series with an opposite sex ratio in the United States series. In comparison there was a consistent 2 to 1 male to female sex ratio in those patients with SDS in both series. The age of onset in PAF was typically in the fourth or fifth decade, altough there was a wide age span, extending from 25 to 78 years (Bannister, Mathias and Polinsky, 1988).

PAF progresses gradually and some patients even report a plateau period. Many patients in the early stages of this disease lead productive lives despite their potentially disabling symptoms. Some evidence of the benign nature of PAF is given by the United States series in which less than 10% of the patients with PAF died during the period of follow-up, in contrast to a mortality of over 50% of those patients with SDS (Bannister, Mathias and Polinsky, 1988). The precise period of follow-up for his study, however, is not available.

ORTHOSTATIC HYPOTENSION

The evolution of man from a quadripedal to a bipedal animal, with the accompanying move from a horizontal to an erect posture placed considerable demands on the ability of the cardiovascular system to maintain adequate cerebral blood flow. The change from a supine to an upright position sets in motion a complex sequence of physiological reactions in response to the pooling of 500 to 1000 ml of blood in the lower extremities and splanchnic circulation (Hill, 1895). There is a reduction in the tonic inhibition exerted by the cardiopulmonary and arterial baroreceptors producing an increase in sympathetic autonomic outflow. This results in constriction of the capacitance and resistance vessels and a simultaneous increase in heart rate and myocardial contractility.   These compensatory mechanisms increase the peripheral resistance, venous return and cardiac output and thus limit the fall in blood pressure. The normal response to the assumption of the erect posture is a fall in systolic blood pressure, an increase in diastolic blood pressure (approximately 10 mmHg) and an increase in the pulse rate (approximately 10 beats per min.) (Crampton, 1920; Ellis and Haynes, 1936; Lutterloh, 1937; Borst *et al.*, 1982).

Early workers recognized that postural or perhaps more appropriately orthostatic hypotension (Laubry and Doumer, 1932) was due to a failure of the compensatory mechanisms that increase the peripheral resistance (Ellis and Haynes, 1936; Stead and Ebert, 1941; Hickham and Prior, 1951; Bickelman, Lippschutz and Brunges, 1961), venous return and cardiac output (Hickham and Prior, 1951; Bickelman, Lippschutz and Brunges, 1961). More recently the potential contribution of plasma volume contraction has been recognized (Ibrahim *et al.*, 1974; Wilcox *et al.*, 1984).

Orthostatic hypotension is the most incapacitating symptom of autonomic failure and, although it is not necessarily the first symptom, it is the symptom that usually leads patients to seek medical attention. Patients typically present with lightheadedness and presyncopal complaints that occur in response to sudden postural change, meals, exertion or prolonged standing. In contrast to neurally mediated syncope (vasovagal syncope), patients with fully-developed autonomic failure do not complain of accompanying nausea, vomiting, diaphoresis or pallor (Stead and Ebert, 1941). Complaints less easily recognized as hypotensive in origin, such as generalized weakness, fatigue, leg buckling, visual blurring, headache and neck pain also may be present. The visual complaints most likely represent retinal or occipital lobe ischemia. Neck pain, which may be the only symptom of orthostatic hypotension, is most likely a consequence of neck muscle ischemia. Loss of consciousness, when it occurs, may be of gradual onset or can occur suddenly, raising the possibility of a seizure or cardiac cause. Some patients may display coarse jerking movements and rarely focal neurological findings. The appearance of focal findings may suggest underlying cerebrovascular disease (Allen and Magee, 1934). Patients with autonomic failure accommodate to their orthostatic hypotension and are frequently able to tolerate significant falls in blood pressure (even to 40–50 mmHg) without symptoms (Stead and Ebert, 1941). This may in part be due to a shift to the left of the lower limit of cerebral autoregulation, the capacity to maintain constant cerebral blood flow despite changes in the perfusion pressure. Thomas and Bannister (1980) showed that cerebral blood flow was normally maintained in patients with autonomic failure until the mean arterial pressure fell to 40 mmHg in comparison to 60–70 mgHg in normal subjects.

Patients are typically most symptomatic during the morning and are more able to tolerate the upright posture as the day progresses. This symptomatic improvement reflects the increase in blood pressure that occurs throughout the day in autonomic failure. Supine hypertension, of uncertain etiology, also increases as the day progresses. Blood pressure may reach dangerously high levels in the supine position during the night. This reversal of the normal diurnal blood pressure cycle (Mann et al., 1983) should be borne in mind when conducting pharmacological studies.

THERMOREGULATION AND SWEATING

Disorders of sweating and thermoregulation are early symptoms of autonomic failure, although they usually do not lead patients to seek medical attention. Patients typically present with distal hypohidrosis or anhidrosis (Bradbury and Eggleston, 1925; Barker, 1933) that may gradually progress to total anhidrosis. Such patients can develop heat intolerance particularly in warmer climatic regions. Some patients report episodic unprovoked diaphoresis involving the whole body surface early in the course of their illness. This may be a consequence of denervation supersensitivity, although eccrine sweat glands are not customarily thought to obey Cannon's law of denervation (Cannon, 1939). Some patients will sweat excessively in the proximal body regions as a compensatory response to the loss of distal sweating sites. Focal areas of increased sweating such as the face, a limb, or the hemibody have also been reported (Allen and Magee, 1934).

UROGENITAL FUNCTION

Bladder symptoms are common in patients with PAF. Symptoms include frequency, nocturia, difficulty initiating urination, failure to maintain an adequate stream of urine, urinary retention and overflow incontinence. Patients are often subjected to unnecessary prostate surgery, which usually exacerbates the clinical condition by producing further incontinence.

Impotence in males is often the first symptom of autonomic failure and can precede the other autonomic symptoms by several years. Erectile function is predominantly mediated by the parasympathetic nervous system. An erection occurs when the arterial blood from the internal pudendal artery is directed to the corpora cavernosa by nitric oxide-induced relaxation of the smooth muscle surrounding the arterioles and the cavernosal sinusoids. Compression of the pene-trating cavernosal veins inhibits the venous outflow which, with contraction of the somatic nervous system innervated ischiocavernosus muscle, results in maximum distention of the tunica albuginea and the production of an erection. Involvement of the parasympathetic pelvic nerves (nervi erigentes) from spinal segments S2–4 is presumably responsible for erectile failure in PAF. Sympathetically mediated ejaculatory failure may precede the appearance of impotence, although impotence can occur with retained ability to ejaculate and experience orgasm. Retrograde ejaculation results when bladder neck closure, also controlled by the sympathetic nervous system, fails.

GASTROINTESTINAL SYMPTOMS

A variety of gastrointestinal symptoms accompany PAF although their presence is usually overshadowed by the cardiovascular, sudomotor and urogenital symptoms. Characteristic symptoms include anorexia, early satiety, heartburn, nausea and vomiting, abdominal pain, diarrhoea, constipation, weight loss and faecal incontinence (Bradbury and Eggleston, 1925; Barker, 1933; Camilleri *et al.*, 1985; Thatcher *et al.*, 1987). The anatomic and pathophysiological basis of these symptoms remain uncertain.

OTHER FEATURES

Patients may complain of visual blurring, poor adaptation to changes in light intensity, loss of libido, reduced lacrimation, dry mouth and nasal stuffiness. Bradbury and Eggleston drew attention to their patients youthful appearance, pallor, anaemia and elevation of the blood urea and nitrogen (Bradbury and Eggleston, 1925). Although forgotten for many years, the anaemia (and associated pallor) had been recently recognized as a common manifestation of PAF, although the anaemia is usually mild in degree, and responsive to erythropoietin (Biaggioni *et al.*, 1992).

INVESTIGATIONS

The anatomic location of the preganglionic autonomic nervous system renders it inaccessible to direct physiological testing. Furthermore, since most autonomic fibres are unmyelinated or lightly myelinated and therefore conduct nerve impulses slowly, conventional neurophysiologic techniques are an inadequate means to study autonomic function (see chapter 2). A group of tests assessing autonomic function and dysfunction has been developed to circumvent these problems by measuring the end-organ responses to various physiological and pharmacological perturbations; by determining the levels of autonomic neurotransmitters and neuromodulators in the plasma, cerebrospinal fluid and urine, and by quantifying autonomic receptor density and affinity. Some of these investigations permit the separation of preganglionic from postganglionic autonomic dysfunction and may thus differentiate PAF from SDS. The contrasting natural history and prognosis of SDS and PAF has provided impetus to the development of such techniques.

PHYSIOLOGICAL TESTS

CARDIOVASCULAR FUNCTION

The laboratory evaluation of the cardiovascular system in patients with autonomic failure includes measures of heart rate variation at rest and in response to deep respiration, the Valsalva manoeuvre and to postural change. These tests primarily provide an index of vagal cardiac function.   The blood pressure response to postural change (active standing or passive tilting), cold water immersion, mental stress tests and isometric exercise is also determined.   These tests primarily provide an assessment of sympathetic function.   Combinations of such tests provide a sensitive measure of autonomic function and have been utilized by many investigators evaluating patients with autonomic failure (Bannister, Sever and Gross, 1977; McLeod and Tuck, 1987b). Unfortunately, at present, these tests do not permit the separation of preganglionic from postganglionic neuron disease. More sophisticated tests assessing baroreceptor function (Eckberg et al., 1986) and spectral analysis of heart rate and blood pressure (Freeman et al., 1990) show some promise in this direction.

SUDOMOTOR FUNCTION

Testing of the eccrine sweat glands has provided a useful means of assessing and localizing sympathetic nervous system function in patients with autonomic failure. Thermoregulatory sweating can be tested by raising the body temperature with an external heating source.   The sweat response is assessed by measuring colour change of an indicator such as iodine with starch, quinizarin or alizarin-red (Guttman, 1947). Hypohidrosis and anhidrosis in patients with autonomic failure have been documented using this technique by many investigators (Bannister, Ardill and Fentem, 1967). Skin bioelectric recordings measuring skin conductance, skin resistance or the sympathetic skin potential provide an alternate measure of sudomotor function (Shahani et al., 1982; Low, 1984). These tests measure both central and peripheral aspects of the afferent sympathetic nervous system. Postganglionic sudomotor function can be determined by measuring sweat output after iontophoresis or intradermal injection of cholinergic agonists such as pilocarpine, nicotine or acetylcholine. These agents either stimulate sweat glands directly or effect a neighbouring population of sweat glands via an axonal reflex (Low, 1984).

Cohen et al. (1987) performed thermoregulatory sweat testing and the quantitative sudomotor axonal reflex test (Q-SART) — a measure of postganglionic sympathetic sudomotor function — on patients with autonomic failure.   Most patients with PAF had evidence of postganglionic sudomotor dysfunction determined by abnormalities on the Q-SART. However, some PAF patients with anhidrosis to radiant heating had normal postganglionic sudomotor function, suggesting a preganglionic lesion in such patients. This work is consistent with earlier anatomic work by this group demonstrating preganglionic and postanglionic loss in patients with autonomic failure (Low, Thomas and Dyck, 1978) (see below).

UROGENITAL FUNCTION

There has been no detailed report of urological function in patients with PAF. The urological evaluation of patients with autonomic failure includes the measurement of residual urinary volume by transurethral catheterization after micturition the excretory urogram (intravenous pyelogram); the cystometrogram, the voiding cystourethrogram, uroflometry and electromyography of the periurethral striated muscle. Urological evaluation may assist in determining the appropriate therapeutic strategy for patients with bladder dysfunction and perhaps more importantly will help avoid needless prostatic surgery (Kirby, 1988).

Clinical and laboratory testing of erectile function include the evaluation of nocturnal penile tumescence and rigidity, circumferential penile expansion measures, penile blood pressure measures, penile vascular ultrasonography, nerve conduction studies of the dorsal nerve of the penis and measurement of the latency of the bulbocavernosus reflex (Kirby, 1988; Shabsigh, Fishman and Scott, 1988).

GASTROINTESTINAL FUNCTION

Physiological studies assessing gastrointestinal function in PAF include use of standard X-rays, barium studies, endoscopy and colonoscopy followed by more specialized techniques such as radionuclide emptying studies, oesophageal manometry, gastrointestinal motility studies and anorectal manometry. Measurements of gastrointestinal peptides such as pancreatic polypeptide (PP), gastrin, cholecystokinin and motilin may provide an indirect index of gastrointestinal autonomic function.

Thatcher et al. (1987) demonstrated subclinical abnormalities in oesophageal manometry and liquid and solid gastric emptying in patients with PAF who lacked gastrointestinal symptomatology. In their series, the patients with gastrointestinal symptoms paradoxically had normal upper gastrointestinal motility studies. Camilleri et al. (1985) performed intestinal motility studies on a group of patients with neurogenic autonomic failure. They demonstrated postprandial antral hypomotility, absent gastric myoelectric complexes and abnormal non-propagating intestinal pressure bursts in patients with PAF.

Measurements of PP in response to a meal or insulin-induced hypoglycaemia provides an indirect measure of gastrointestinal vagal innervation, specifically vagal cholinergic pancreatic innervation. Polinsky et al. (1982) showed impaired PP release to insulin-induced hypoglycaemia in patients with autonomic failure. The PP levels did not correlate with plasma catecholamine levels, suggesting that the adrenergic and cholinergic systems may be affected to a different extent in individual patients. Polinsky et al. (1988) also demonstrated elevated basal gastrin levels and an enhanced gastrin response to insulin-induced hypoglycaemia in patients with PAF. They proposed that the high basal level was due to loss of vagal inhibition and the response to hypoglycemia was due to denervation supersensitivity. In contrast, patients with SDS had a low basal gastrin level and a reduced response to hypoglycaemia.

PUPILLOMOTOR FUNCTION

Pharmacological agents can be used to determine the integrity of the autonomic innervation of the pupil. The sympathetic innervation of the pupil is tested using cocaine (2–10%) and hydroxyamphetamine (1%). Topical installation of cocaine eyedrops, which prevents noradrenaline reuptake by the presynaptic neuron, will not cause mydriasis if there is dysfunction in the pre- or postganglionic sympathetic pathway to the eye. Hydroxyamphetamine, which releases intraneuronal noradrenaline from the postganglionic neuron, will dilate a normal pupil and a miotic pupil, constricted due to a preganglionic lesion (e.g. due to SDS). Hydroxyamphetamine will not dilate a miotic pupil due to a post-ganglionic sympathetic lesion (e.g. in PAF). Dilute pilocarpine (0.125%) will constrict a dilated denervated pupil due to a post-ganglionic parasympathetic lesion (Lepore, 1985). The augmented miotic response to methacholine (2.5%) occurs with both pre- and postganglionic parasympathetic lesions (Ponsford, Bannister and Paul, 1982).

MICRONEUROGRAPHY

The development of intraneural microneurography has permitted the direct measurement of sympathetic function. A tungsten microelectrode, several microns in size, is inserted into a fascicle of a distal sympathetic nerve to the skin or muscles. Using this technique sympathetic outflow to skin and muscle can be measured at rest and in response to various physiological perturbations (Delius *et al.*, 1972; Valbo *et al.*, 1979).

Dotson *et al.* (1990) performed cutaneous microneurographic studies on two patients with autonomic failure. They demonstrated a marked reduction in the sympathetic efferent nerve impulse activity in the patient with PAF, while the patient with SDS had spontaneous bursts of sympathetic efferent neural discharges, similar to those seen in normal subjects. Neither patient, however, exhibited the characteristic microneurographic response to those manoeuvres (such as deep inspiration, startle, mental stress and cold water immersion) that generate a reflex increase in efferent sympathetic activity. These studies support the evidence suggesting that PAF is predominantly due to degeneration of the peripheral postganglionic autonomic fibres, while the SDS is primarily due to a preganglionic autonomic defect with intact postganglionic neurons that are released from central control. However, the applicability of these findings to all patients with autonomic failure is uncertain, since most efforts to record nerve traffic in patients with severe disease have been unsuccessful.

PHARMACOLOGICAL TESTS

Pharmacological studies have enhanced our understanding of the pathophysiology of autonomic failure and may differentiate preganglionic from postganglionic autonomic dysfunction. Numerous investigators have demonstrated augmented blood-pressure responsiveness to the alpha adrenoreceptor agonists noradrenaline and phenylephrine (Chokroverty *et al.*, 1969; Kontos, Richardson and Narvell, 1975; Wilcox and Aminoff, 1976; Bannister *et al.*, 1979; Polinsky *et al.*, 1981; Robertson *et al.*, 1984), and exaggerated heart rate and blood-pressure responses to the beta adrenoreceptor agonists isoproterenol and terbutaline (Bannister *et al.*, 1981a; Robertson *et al.*, 1984) in patients with autonomic failure.

Polinsky *et al.* (1981) suggested that the slope or gain of the dose–response curve to pressor agents could provide a measure of baroreceptor modulation of blood pressure increases, while a shift to the left of the dose–response curve implied true denervation supersensitivity. The blood pressure response to pressors in both PAF and SDS patients showed increased gain suggesting impaired baroreceptor function. Only PAF patients had a shift to the left of the dose–response curve consistent with denervation supersensitivity.

The indirect sympathomimetic agonist tyramine, which releases noradrenaline from the adrenergic terminal, may separate preganglionic from postganglionic neuron dysfunction. Patients with SDS characteristically release normal amounts of noradrenaline from presumed intact but decentralized postganglionic neurons in response to a tyramine bolus, and show a normal or slightly enhanced blood pressure response to the noradrenaline released by that agent. In contrast, the amount of noradrenaline released by patients with PAF is reduced. The blood pressure response to tyramine in PAF depends not only on the amount of noradrenaline released but also on the alpha receptor sensitivity, the noradrenaline clearance rate and the baroreceptor function. The response is therefore unpredictable, particularly early in the illness (Kontos, Richardson and Narvell, 1975; Demanet, 1976; Polinsky *et al.*, 1981).

The mechanism of the enhanced response to sympathomimetic agents in autonomic failure is multifactorial, reflecting upregulation or an increase in the number and affinity of receptors due to reduced agonist exposure (see below), higher concentrations of the agonists due to lower plasma clearance rates and loss of baroreceptor and other homeostatic inhibitory mechanisms (Bannister *et al.*, 1979; Polinsky *et al.*, 1981)

RECEPTOR FUNCTION

Several radioligand binding studies have assessed receptor number and function in patients with autonomic dysfunction. Alpha adrenoreceptor number, quantified by measuring ^{3}H-dihydroergocryptine specific binding to platelet membrane alpha$_2$ receptors, is significantly increased in patients with PAF and SDS (Davies *et al.*, 1982; Kafka *et al.*, 1984). There is no change in the platelet alpha receptor affinity, determined by measuring the dissociation constant (K_d) for ^{3}H-dihydroergocryptine. Leukocyte beta$_2$ adrenoreceptors in patients with autonomic failure and SDS, assessed by ^{3}H-dihydroalprenolol binding, showed similar changes (Bannister *et al.*, 1981a). Since receptor number is modulated by the amount of agonist to which the receptor population is exposed, the increase in receptor number is most likely in response to the reduced plasma noradrenaline that accompanies these disorders. If there are similar vascular and cardiac receptor changes, they may be responsible in part for the exaggerated responses to sympathomimetic agents that accompany these disorders (Bannister *et al.*, 1978; Davies *et al.*, 1982; Kafka *et al.*, 1984).

BIOCHEMISTRY

Since the identification of noradrenaline as the primary postganglionic sympathetic nervous system neurotransmitter (von Euler, 1948) and the development of biochemical methods to measure this substance in the urine (Luft and von Euler, 1953; von Euler, Hellner and Purkhold, 1954) and plasma (Engelman, Portnoy and Lovenberg, 1968), catecholamines have been used to provide an indirect index of sympathetic nervous system activity. Luft and von Euler (1953) first documented biochemical abnormalities in autonomic failure. In two patients with orthostatic hypotension, they noted low noradrenaline levels in a 24 h urine collection. These patients were also unable to increase adrenaline excretion or manifest the characteristic adrenergic signs and symptoms in response to insulin-induced hypoglycemia.

In the ensuing years, numerous catecholamine measurement techniques have been used to provide an index of autonomic function. These include measurements of plasma noradrenaline concentration at rest and in response to physiological provocations, noradrenaline release rate from the sympathetic nerve varicosities and the "spill-over" rate to plasma, noradrenaline clearance and reuptake, urinary excretion of noradrenaline and noradrenaline metabolites and measurements of cerebrospinal fluid noradrenaline and noradrenaline metabolites.

PLASMA

Recumbent plasma noradrenaline levels are 200–300 pg/ml in normal subjects with a 2–3 fold increase on standing (Lake, Ziegler and Kopin, 1976). Hickler *et al.* (1959) first documented that patients with orthostatic hypotension were unable to increase their plasma noradrenaline concentration in response to postural change. Ziegler, Lake and Kopin, (1977) systematically compared plasma noradrenaline measurements of PAF patients with SDS patients. They noted low recumbent levels of plasma noradrenaline in PAF patients, that did not increase in response to standing or exercise. In contrast, the SDS patients had normal recumbent noradrenaline levels, but no increase in response to standing or exertion. This study confirmed the hypothesis that PAF was due to a peripheral sympathetic nervous system deficit, whereas the SDS was due to the inability to activate appropriately an intact peripheral sympathetic nervous system. These findings have been confirmed by other investigators (Polinsky *et al.*, 1981; Kopin *et al.*, 1983b; Robertson *et al.*, 1984). However, in these later studies, considerable overlap existed between the two patient groups, limiting the ability to make a definitive diagnosis based on plasma catecholamine measurements in individual patients.

Measurements of the plasma concentration of catecholamine metabolites such as 3-methoxy-4-hydroxyphenylglycol (MHPG), dihydroxyphenylethylene glycol and dihydroxyphenylacetic acid have also been used as an index of autonomic function. These are reduced in most patients with PAF (Polinsky, Jimerson and Kopin, 1984; Goldstein *et al.*, 1989).

CEREBROSPINAL FLUID

Although MHPG is the major brain metabolite of noradrenaline MHPG levels in the cerebrospinal fluid (CSF) cannot be used as a direct index of central noradrenaline metabolism because free plasma MHPG readily crosses the blood–brain barrier (Kopin *et al.*, 1983a). Polinsky, Jimerson and Kopin, (1984) measured CSF and plasma noradrenaline in patients with PAF and SDS. They noted that both patients with PAF and SDS had lower CSF MHPG levels than the controls. The low level in patients with PAF, however, is most likely a reflection of the low plasma MHPG level (see above), while in the SDS the low MHPG level is probably due to decreased central noradrenaline metabolism and thus decreased MHPG production.

URINE

Measurements of the urinary excretion of noradrenaline and noradrenaline metabolites can be used as a measure of sympathetic nervous system function. Since *o*-methylation is the predominant metabolic route of released noradrenaline and deamination the major metabolic route of intraneuronal metabolism, comparisons between the *o*-methylated metabolite normetanephrine (NMN), the deaminated metabolites DHPG and dihydroxymandelic acid, and the combined *o*-methylated and deaminated metabolites vanillylmandelic acid (3-methoxy-4-hydroxymandelic acid [VMA]) and MHPG may provide some insight into autonomic pathophysiology.

Kopin *et al.* (1983b) evaluated the urinary excretion of the major noradrenaline metabolites (VMA, MHPG and NMN) in patients with autonomic failure. Patients with PAF showed a significant decrease in total catecholamine excretion and individual catecholamine metabolite excretion in comparison to normal controls and patients with SDS. Patients with SDS showed a disproportionate reduction in urinary normetanephrine excretion. Normetanephrine represents the *o*-methylated metabolite of noradrenaline, and is thus the predominant metabolite of released (extra-neuronal) noradrenaline. The authors therefore suggested that their results in patients with SDS were consistent with the inability to release noradrenaline appropriately from an intact peripheral nervous system. Again there was overlap in measurements between patients groups and single measurements are not of value in the diagnosis of individual patients.

NORADRENALINE KINETICS

Plasma noradrenaline concentration is dependent not only on the release and "spillover" of noradrenaline from the sympathetic nerve terminals into the plasma but also on the noradrenaline clearance from the circulation. Static measurements of plasma noradrenaline concentration therefore may not provide an accurate index of sympathetic function. Investigators have performed noradrenaline kinetic studies on patients with autonomic failure using the isotope dilution method (Esler et al., 1980; Polinsky et al., 1985). Both studies showed a decreased noradrenaline release rate and decreased plasma clearance in patients with PAF. The plasma noradrenaline levels of the patients with PAF in Esler's study were only slightly reduced — possibly implying decreased neuronal uptake relative to the noradrenaline synthesis and release rate. These patients may have been at an earlier stage of their illness. In Polinsky's report the plasma levels of noradrenaline in patients with PAF were significantly reduced and correlated with the noradrenaline secretion rate, suggesting that the reduced plasma level is due to reduced noradrenaline release.

Polinsky et al., (1985) determined the proportion of noradrenaline clearance that is due to neuronal uptake (uptake$_{-1}$), by comparing the clearance of noradrenaline to that of isoproterenol, which is not taken up by sympathetic neurons. The neuronal noradrenaline uptake was significantly reduced in patients with PAF. In patients with SDS, the noradrenaline release rate neuronal uptake and total noradrenaline clearance were in the normal range.

PATHOLOGY

While the physiological, biochemical and pharmacological evidence is consistent with a predominantly (postganglionic) peripheral process, there are no satisfactory complete pathological studies in patients with PAF. Bradbury and Eggleston hypothesized a "localized lesion" in the "brain or spinal cord" as the cause of the "paralysis of the sympathetic vasoconstrictor endings" (Bradbury and Eggleston, 1925). The autopsy performed on their second patient did not include the brain or spinal cord. The patient did, however, have normal appearing adrenal glands and thyroid glands — the organs implicated in two prevailing hypothetical etiologies of the disorder (Bradbury and Eggleston, 1927). No comment was made on the peripheral nervous system.

Johnson et al. (1966) first recognized the importance of intermediolateral column degeneration in the causation of autonomic failure. They described the post mortem on a 66 year old male who had orthostatic hypotension, impotence, anhidrosis and abnormalities on autonomic function testing, without any signs of central nervous system disease. Pathological findings included severe intermediolateral column cell loss with scattered Lewy bodies in the brain and spinal cord without cell loss or gliosis of the pigmented nuclei. The peripheral nervous system pathological examination was unremarkable, showing only reduced nerve fibres with occasional degenerating fibres in some nerve bundles in the sympathetic ganglia. This patient died only 4 years after the onset of his symptoms and, given the limited period of follow-up, cannot be regarded as a definite case of PAF, despite the absence of central nervous system clinical signs on physical examination. Some cases may initially present clinically as PAF but during follow-up central nervous system involvement typical of the SDS may become apparent.

Several investigators have performed microscopic histochemical studies on the sympathetic vasomotor nerves of patients with autonomic failure. Kontos, Richardson and Narvell (1975) compared the walls of the deltoid muscle blood vessels in four patients with PAF to four control subjects. They showed the absence of catecholamine-specific histofluorescence in the adrenergic vasomotor terminals of the blood vessels, suggesting a peripheral pathology. The presence of denervation supersensitivity to noradrenaline and the absence of vasoconstriction to intra-arterial tyramine provided confirmation of a peripheral process.

Bannister *et al.* (1981b) examined the sympathetic perivascular nerves of patients with autonomic failure using catecholamine histofluorescence and electromicroscopy. They also showed an absence or a reduction in the perivascular adrenergic nerve catecholamine fluorescence at the adventitia-medial border of arteries of patients with autonomic failure due to both SDS and PAF and a reduction in the number of small granular noradrenergic vesicles. The depletion of noradrenergic vesicles was most marked in the patients with PAF.

Nanda *et al.* (1976, 1977) studied the adrenergic fibres of blood vessels of the quadriceps and palmaris longus muscles. They, in contrast to the above studies, showed that normal catecholamine histofluorescence was present in two patients with probable PAF and one patient with SDS. One patient with PAF had an unusual disorder characterized by normal parasympathetic and sympathetic cholinergic function, but the inability to release noradrenaline in response to postural change and indirect sympathomimetic agents. Furthermore, three patients with SDS in this report had an absent perivascular catecholamine histofluorescence.

Klein *et al.* (1980) performed electron microscopy on the saphenous veins of patients who had orthostatic hypotension without neurological signs (PAF). They showed loss of noradrenergic axons, terminals and vesicles in the wall of the saphenous veins of three patients with PAF who had the characteristic low resting plasma noradrenaline levels that did not increase with postural change (the hypoadrenergic response). They described normal noradrenergic terminals with normal small and large dense-core vesicles in two patients with PAF who had a significant increase in plasma noradrenaline in response to postural change (the hyperadrenergic response). They proposed a blunting of the effector response of the smooth muscle cell in these patients. An "intermediate" patient was also described with a reduced noradrenaline response to postural change but a normal number of axons and terminals on biopsy. Many of these terminals however, contained a dense amorphous material and resembled the appearance of degeneration induced by chemical sympathectomy.

Acknowledging the importance of the splanchnic autonomic outflow in the maintenance of postural normotension, Low, Thomas and Dyck (1978) analyzed the preganglionic autonomic nerve cell bodies and axons of the seventh thoracic spinal cord segment in an 85-year-old patient with presumed PAF and two patients with SDS. The intermediolateral column autonomic neuron cell bodies and axons in the patient with PAF were 52% and 41% of the respective mean control values. Although the depletion was not as severe as the patients with SDS, this study suggests some preganglionic cell loss in patients with PAF. There was no comparable reduction in the number of intermediate myelinated fibres (presumed gamma motor neuron axons) and large myelinated fibres (presumed alpha motor neuron axons) in patients with PAF. Despite the substantial difference between the intermediolateral column neuron counts of this patient and three control subjects aged 76 to 79, the advanced age of this single patient (85 years), may preclude drawing definitive conclusions for the present.

Dotson *et al.* (1990) demonstrated a population of empty Schwann cell bands and an abnormal increase in the number and distribution of small diameter unmyelinated axons in the sural nerve of a 61-year-old patient with PAF. In comparison, a patient with SDS showed normal structure and distribution of the unmyelinated sural nerve axons.

Since pathological evidence is the definitive standard for diagnosis, the anatomical substrate of PAF at this time remains uncertain. Furthermore some question exists as to whether PAF represents a single disorder or includes several heterogeneous disorders. Finally, the all too frequent patient, who presents with PAF and after an extended period of follow-up develops central nervous system signs, persists as a warning against drawing premature conclusions based on early physiological, biochemical and even pathological studies.

DIFFERENTIAL DIAGNOSIS

A variety of disorders is associated with autonomic failure (see Table 5.2). Careful history taking and physical examination with appropriate use of special investigations will usually permit separation of these disorders from PAF. A useful practical approach is to group the diseases associated with autonomic failure into those accompanied by predominantly central nervous system signs, those accompanied by predominantly peripheral nervous system signs, and those without neurological signs.

Of the disorders associated with central nervous system signs, SDS and Parkinson's disease are the most likely to produce clinically relevant autonomic dysfunction. Some extrapyramidal, cerebellar or pyramidal signs typically accompany the autonomic symptoms of SDS. Rarely, the autonomic features may precede the other neurological signs by several years, producing diagnostic confusion (see above). The autonomic symptoms accompanying Parkinson's disease are usually not as severe as those seen in SDS and PAF. They characteristically occur late in the course of the illness and are often associated with dopamine agonist therapy. There are many other disorders with autonomic dysfunction and central nervous system signs, however, the autonomic symptoms are usually of secondary importance and these disorders are unlikely to be confused with PAF. Many of these entities are considered elsewhere in this book.

TABLE 5.2

Classification of autonomic failure

Autonomic failure with central nervous system signs

    Shy-Drager syndrome (multiple system atrophy)

        Olivopontocerebellar atrophy

        Striatonigral degeneration

    Parkinson's disease

    Brain tumours (brainstem, cerebellum, diencephalon)

Wernicke's disease

Multiple infarcts

Syringomyelia and syringobulbia

Hydrocephalus

Multiple sclerosis

Myelopathies

        Traumatic

        Inflammatory

        Pernicious anaemia

        Degenerative

        Tabes dorsalis

Autonomic Failure with peripheral nervous system signs

    Peripheral neuropathies

        diabetes

        Amyloid

        Guillain-Barré syndrome

        Chronic inflammatory polyneuropathy

        Fabry's disease

        Pernicious anaemia

        Chagas disease

        Porphyria

        Uremia

        Alcoholic neuropathy

        Paraneoplastic neuropathy

TABLE 5.2
Continued

Toxic neuropathies (vacor, vincristine, perhexitene)

HIV neuropathy

Connective tissue diseases

Tangier disease

Hereditary sensory neuropathies

Botulism

Eaton-Lambert syndrome

Dopamine beta hydroxylase deficiency

Familial dysautonomia

Adie's syndrome

Menke's syndrome

Autonomic failure without neurological signs

Pure autonomic failure (idiopathic orthostatic hypotension)

Acute autonomic neuropathy

Medications

Antihypertensive agents

Tricyclic agents

Monoamine oxidase inhibitors

Antipsychotic agents

Aging

Hyperbradykinism

Endocrine

Adrenocortical deficiency

Pheochromocytoma

Hyperaldosteronism

Surgical sympathectomy

The peripheral neuropathies accompanying diabetes (Ewing, Campbell and Clarke, 1980), amyloid (Kelly *et al.* 1979) and Guillian-Barre syndrome (Tuck and McLeod, 1981) are the most frequent cause of clinically significant autonomic dysfunction accompanied by peripheral nervous system signs. The autonomic symptoms of diabetic neuropathy, amyloid neuropathy and paraneoplastic neu-ropathies (Chiappa and Young, 1973; Mamdani *et al.*, 1985; Van Lieshout *et al.*, 1986) may rarely precede the sensorimotor symptoms and signs. Under such circumstances these peripheral neuropathies may be confused with PAF. Autonomic dysfunction accompanying other peripheral neuropathies is usually mild and only rarely of clinical significance (McLeod and Tuck, 1987a). Dysautonomia with peripheral nervous system signs also accompanies Adie's syndrome, familial dysau-tonomia, congenital dopamine beta hydroxylase difficiency, Menke's syndrome and the Eaton-Lambert syndrome. These disorders are covered in detail elsewhere in this book.

Although their clinical features may be similar, the time course and natural history of the acute autonomic neuropathies should prevent confusion with PAF. These neuropathies may be post-viral (Yahr and Frontera, 1975), idiopathic (Appenzeller and Kornfeld, 1973; Hopkins, Neville and Bannister, 1974; Young *et al.*, 1975) or associated with diseases such as porphyria (Yeung Laiwah *et al.*, 1985) and botulism (Jenzer *et al.*, 1975). Other causes of autonomic dysfunction without neurological signs include medications, endocrine disease, surgical sympathectomy and possibly normal aging.

TREATMENT

Most therapeutic interventions in patients with autonomic failure are directed at the most disabling symptom, orthostatic hypotension. Treatment endeavors include the non-pharmacological such as external binders of the lower extremities and abdomen (Bradbury and Eggleston, 1925), elevating the head of the bed during sleep (MacLean and Allen, 1940) and a high sodium diet — all of which are generally of help primarily to the mildly afflicted.

Initial pharmacological interventions included the use of pressors such as ephedrine (Ghrist and Brown, 1928), ergotamine tartrate (Barker, 1933), phenyle-phrine (Capaccio and Donald, 1938) and amphetamine (Jeffers *et al.*, 1941). The use of the mineralocorticoids deoxycorticosterone via the parenteral route (Gregory, 1945) and some years later 9-alpha-fluorohydrocortisone orally (Hickler *et al.*, 1959) met with some therapeutic success.

More recent approaches have included other sympathomimetic agents such as phenylpropanolamine, methylphenidate, midodrine, clonidine, tyramine with a monoamine oxidase inhibitor, DL-threo-dihydroxyphenylserine and yohimbine; beta blockers with and without intrinsic sympathomimetic activity; other pressor agents such as caffeine, vasopressin, and somatostatin; prostaglandin synthesis inhibitors (indomethacin, ibuprofen, and naproxen); antihistamines and antisero-tonergic agents (Bannister, Ardill and Fentem, 1969; Thomas *et al.*, 1981, Onrot *et al.*, 1986, anonymous editorial, 1987). The sheer number of pharmacological agents for orthostatic hypotension underscores the inadequacy of the available treatment. Medications may benefit individual patients, usually the mildly afflicted, however the effects are often inconsistent, unsustained and complicated by such side effects as supine hypertension and fluid retention.

Constipation is the most troublesome gastrointestinal symptom of PAF but can be controlled by increasing dietary roughage and the judicious use of laxatives. Nausea, which is particularly common after meals may respond to metoclopramide. Diarrhoea is an infrequent complaint and can usually be treated with loperamide or diphenoxylate.

The treatment goal for the urological symptoms of autonomic failure is to improve bladder emptying to avoid the risk of infection due to stasis and to prevent overt incontinence. This is achieved by the institution of regular voiding patterns, intermittent catheterization, the use of cholinergic agonists (carbachol and bethanechol) and, only under rare circumstances, bladder neck surgery. Therapy for erectile failure entails the use of mechanical devices (Witherington, 1988), penile prosthetic implants or the auto-injection of papaverine or other vasoactive substances into the corpus cavernosum (Virag, 1982; Zorgniotti and Le Fleur, 1985).

Bradbury and Eggleston wrote in 1925 that "efforts to cure these patients, or to control their disorders, have been unavailing". Their efforts included "the administration of thyroxine, adrenaline, dried suprarenal substance, mixed glands, strychnin and digitalis, and the enforced consumption of sugar and of water". Sixty-six years later, although our therapeutic armamentarium is larger and our endeavours have met with some success, a degree of pessimism is undeniably warranted.

REFERENCES

Allen, E.V. and Magee, H.R. (1934) Orthostatic (postural) hypotention with syncope. *Med. Clin. N. Am.*, **18**, 585–595.

Anonymous Editorial (1987) Management of orthostatic hypotension. *Lancet*, **20**, 197–198.

Appenzeller, O. and Kornfeld, M. (1973) Acute pan-dysautonomia. *Arch. Neurol.*, **29**, 334–339.

Bannister, R., Ardill, L. and Fentem, P. (1967) Defective autonomic control of blood vessels in idiopathic orthostatic hypotension. *Brain*, **90**, 725–746.

Bannister, R., Ardill, L. and Fentem, P. (1969) An assessment of various methods of treatment of idiopathic orthostatic hypotension. *Q. J. Med.*, **38**, 377–395.

Bannister, R., Boylston, A.W., Davies, I.B., Mathias, C.J., Sever, P.S. and Sudera, D. (1981a) Beta-Receptor numbers and thermodynamics in denervation supersensitivity. *J. Physiol. (Lond.)*, **319**, 369–377.

Bannister, R., Crowe, R., Eames, R. and Burnstock, G. (1981b) Adrenergic innervation in autonomic failure. *Neurology*, **31**, 1501–1506.

Bannister, R., Davies, B., Holly, E., Rosenthal, T. and Sever, P. (1979) Defective cardiovascular reflexes and supersensitivity to sympathomimetic drugs in autonomic failure. *Brain*, **102**, 163–176.

Bannister, R., Mathias, C. and Polinski, R. (1988) Autonomic failure: a comparison between U.K. and U.S. experience. In *Autonomic Failure: A Textbook of Clinical Disorders of the Autonomic Nervous System*, edited by R. Bannister, 2nd edn, pp. 281–288.

Bannister, R., Sever, P. and Gross, M. (1977) Cardiovascular reflexes and biochemical responses in progressive autonomic failure. *Brain*, **100**, 327–344.

Bannister, R. (1983) Clinical features of progressive autonomic failure. In *Autonomic Failure*, edited by R. Bannister, pp. 67–73.

Barker, N.W. (1933) Postural hypotension, report of a case in review of the literature. *Med. Clin. N. Amer.*, **16**, 1301–1312.

Bickelman, N., Lippschutz, E.J. and Brunges, C.F. (1961) Hemodynamics of idiopathic orthostatic hypotension. *Am. J. Med.*, **30**, 26–38.

Biaggioni, I., Robertson, D., Haile, V. and Davis, R. (1993) Erythropoietin reverses the anemia of autonomic failure. *Clin. Auton. Res.*, **3**, 207.

Borst, C., Weiling, W., van Brederode, J.F.M., Hond, A., DeRijk, L.G. and Dunning, A.J. (1982) Mechanisms of initial heart rate response to postural change. *Am. J. Physiol.*, **243**, H676–H681.

Bradbury, S. and Eggleston, C. (1925) Postural hypotension: a report of three cases. *Am. Heart J.*, **1**, 73–86.

Bradbury, S. and Eggleston, C. (1927) Postural hypotension. An autopsy upon a case. *Am. Heart J.*, **iii**, 105–106.

Camilleri, M., Malagelada, J.R., Stanghellini, V., Fealey, R.D. and Sheps, S.G. (1985) Gastrointestinal motility disturbances in patients with orthostatic hypotension. *Gastroenterology*, **88**, 1852–1859.

Cannon, W.B. (1939) A law of denervation. *Am. J. Med. Sci.*, **198**, 737–750.

Capaccio, G.D. and Donald, C.J. (1938) Orthostatic hypotension: Report of a case treated with neosynephrin hydrochloride. *JAMA*, **110**, 1180–1182.

Chiappa, K.H. and Young, R.R. (1973) A case of paracarcinomatous pan-dysautonomia. *Neurology*, **23**, 423.

Chokroverty, S., Barron, K.D., Katz, F.H., Del Greco, F. and Shape, T. (1969) The syndrome of primary orthostatic hypotension. *Brain*, **92**, 743–769.

Cohen, J., Low, P., Fealey, R., Sheps, S. and Jiang, N.S. (1987) Somatic and autonomic function in progressive autonomic failure and multiple system atrophy. *Ann. Neurol.*, **22**, 692–699.

Crampton, C.W. (1920) The gravity resistent ability of the circulation; its measurement and significance (blood ptosis). *Am. J. Med. Sci.*, **160**, 721–737.

Davies, B., Sudera, D., Sagnella, G., Marchesi-Saviotti, E., Mathias, C., Bannister, R. *et al.* (1982) Increased numbers of alpha receptors in sympathetic denervation supersensitivity in man. *J. Clin. Invest.*, **69**, 779–784.

Delius, W., Hagbarth, K.E., Hongell, A. and Wallin, B.G. (1972) General characteristics of sympathetic activity in human muscle nerves. *Acta. Physiol. Scand.*, **84**, 65–81.

Demanet, J.C. (1976) Usefulness of noradrenaline and tyramine infusion tests in the diagnosis of orthostatic hypotension. *Cardiology*, **61**, 213–224.

Dotson, R., Ochoa, J., Marcettini, P. and Cline, M. (1990) Sympathetic neural outflow directly recorded in patients with primary autonomic failure: Clinical observations, microneurography and histopathology. *Neurology*, **40**, 1079–1085.

Eckberg, D.L., Harkins, S.W., Fritsch, J.M., Musgrave, G.E. and Gardner, D.F. (1986) Baroreflex control of plasma norepinephrine and heart period of healthy subjects and diabetic patients. *J. Clin. Invest.*, **78**, 366–374.

Ellis, L.B. and Haynes, F.W. (1936) Postural hypotension with particular reference to its occurence in disease of the central nervous system. *Arch. Intern. Med.*, **58**, 773–798.

Engelman, K., Portnoy, B. and Lovenberg, W. (1968) A sensitive and specific double isotope derivative method for the determination of catecholamines in biological specimens. *Am. J. Med. Sci.*, **255**, 259–268.

Esler, M., Jackman, G., Kelleher, D., Skews, H., Jennings, G., Bobik, A. *et al.* (1980) Norepinephrine kinetics in patients with idiopathic autonomic insufficiency. *Circulation Res.*, **46**, 47–48.

Ewing, D.J., Campbell, I.W. and Clarke, B.F. (1980) The natural history of diabetic autonomic neuropathy. *Q. J. Med.*, **49**, 95–108.

Freeman, R.L., Saul, J.P., Roberts, M.S., Berger, R.D., Broadbridge, C. and Cohen, R.J. (1990) Spectral analysis of heart rate in diabetic autonomic neuropathy. A comparison with standard tests of autonomic function. *Arch. Neurol.*, **48**, 185–190.

Ghrist, D.G. and Brown, G.E. (1928) Postural hypertension with syncope: its successful treatment with ephedrine. *Am. J. Med. Sci.*, **175**, 336–349.

Goldstein, D.S., Polinsky, R.J., Garty, M., Robertson, D., Brown, R.T., Biaggioni, I. *et al.* (1989) Patterns of plasma levels of catechols in neurogenic orthostatic hypotension. *Ann. Neurol.*, **26**, 558–563.

Graham, J.C. and Oppenheimer, D.R. (1969) Orthostatic hypotension and nicotine sensitivity in a case of multiple system atrophy. *J. Neural Neurosurg. Psych.*, **32**, 28–34.

Gregory, R. (1945) Treatment of orthostatic hypotension with particular reference to use of desoxycorticocosterone. *Am. H. J.*, **29**, 246–252.

Guttman, L. (1947) Management of the quinizarin sweat test. *Post Grad. Med. J.*, **23**, 353–355.

Hickam, J.B. and Prior, W.W. (1951) Cardiac output in postural hypotension. *J. Clin. Invest.*, **30**, 401–405.

Hickler, R.B., Thompson, G.R., Fox, L.M. and Hamlin, J.T. (1959) Successful treatment of orthostatic hypotension with 9-alpha-flurohydrocortisone. *NEJM*, **261**, 788–791.

Hill, L. (1895) The influence of the force of gravity on the circulation. *Lancet*, **1**, 338–339.

Hopkins, A., Neville, B. and Bannister, R. (1974) Autonomic neuropathy of acute onset. *Lancet*, **27**, 769–771.

Ibrahim, M.M., Tarazi, R.C., Dustan, H.P. and Bravo, E.L. (1974) Idiopathic orthostatic hypotension: circulatory dynamics in chronic autonomic insufficiency. *Am. J. Cardiol.*, **34**, 288–294.

Jeffers, W.A, Montgomery, H. and Burton, A.C. (1941) Types of orthostatic hypotension and their treatment. *Am. J. Med. Sci.*, **202**, 1–14.

Jenzer, G., Mumenthaler, M., Ludin, H.P. and Robert, F. (1975) Autonomic dysfunction in botulism B. A clinical report. *Neurology*, **25**, 150–153.

Johnson, R.H., Lee de, J., Oppenheimer, D.R. and Spalding, J.M.K. (1966) Autonomic failure with orthostatic hypotension due to intermediolateral column degeneration. A report of two cases with autopsies. *Q. J. Med.*, **35**, 276–292.

Kafka, M.S., Polinsky, R.J., Williams, A., Kopin, I.J., Lake, C.R., Ebert, M.H. *et al.* (1984) Alpha-adrenergic receptors in orthostatic hypotension syndromes. *Neurology*, **34**, 1121–1125.

Kelly, J.J., Kyle, R.A., O'Brien, P.C. and Dyck, P.J. (1979) The natural history of peripheral neuropathy in primary systemic amyloidosis. *Ann. Neurol.*, **6**, 1–7.

Kirby, R.S. (1988) Studies of the neurogenic bladder. *Ann. Royal Coll. of Surgeons*, **70**, 285–288.

Klein, R.L., McC. Baggett, J., Thureson-Klein, Å. and Langford, H.G. (1980) Idiopathic orthostatic hypotension: circulating noradrenaline and ultrastructural of saphenous vein. *J. Auton. Nerv. Syst.*, **2**, 205–222.

Kontos, H.A., Richardson, D.W. and Narvell, J.E. (1975) Norepinephrine depletion in idiopathic orthostatic hypotension. *Ann. Int. Med.*, **82**, 336–341.

Kopin, I.J. Gordon, E.K., Jimerson, D.C. and Polinsky, R.J. (1983a) Relation between plasma and cerebrospinal fluid levels of 3-methoxy-4-hydroxyphenylglycol. *Science*, **219**, 73–75.

Kopin, I.J., Polinsky, R.J., Oliver, J.A., Oddershede, I.R. and Ebert, M.H. (1983b) Urinary catecholamine metabolites distinguish different types of sympathetic neuronal dysfunction in patients with orthostatic hypotension. *J. Clin. Endocrinol. Metab.*, **57**, 632–637.

Lake, C.R., Ziegler, M.G. and Kopin I.J. (1976) Use of plasma norepinephrine for evaluation of sympathetic neuronal function in man. *Life Sci.*, **18**, 1315–1326.

Laubry, C. and Doumer, E. (1932) L'hypotension orthostatique. *Presse Med.*, **1**, 17–20.

Lepore, F.E. (1985) Diagnostic pharmacology of the pupil. *Clinical Neuropharmacology*, **8**, 27–37.

Low, P.A., Thomas, J.E. and Dyck, P.J. (1978) The splanchnic autonomic outflow in Shy-Drager syndrome and idiopathic orthostatic hypotension. *Ann. Neurol.*, **4**, 511–514.

Low, P.A. (1984) Quantitation of autonomic responses. In *Peripheral Neuropathy*, edited by P.J. Dyck, P.K. Thomas, E.H. Lambert and R. Bunge. pp. 1139–1165. W.B. Saunders, Philadelphia.

Luft, R. and von Euler, U.S. (1953) Two cases of postural hypotension showing a deficiency in release of nor-epinephrine and epinephrine. *J. Clinical Invest.*, **32**, 1065–1069.

Lutterloh, C.H. (1937) The clinical significance of the effects of posture on blood pressure. The postural test as a means of classifying hypotension. *Am. J. Med. Sci.*, **193**, 87–96.

MacLean, A.R. and Allen, E.V. (1940) *JAMA*, **115**, 2162–2167.

Mamdani, M.B., Walsh, R.L., Rubino, F.A., Brannegan, R.T. and Hwang, M.H. (1985) Autonomic dysfunction and Eaton Lambert syndrome. *J. Auton. Nerv. Syst.*, **12**, 315–320.

Mann, S., Altman D.G., Raftery, E.B. and Bannister, R. (1983) Circadian variation of blood pressure in autonomic failure. *Circulation*, **68**, 477–483.

McLeod, J.G. and Tuck, R.R. (1987a) Disorders of the autonomic nervous system: Part I. Pathophysiology and clinical features. *Ann. Neurol.*, **21**, 419–430.

McLeod, J.G. and Tuck, R.R. (1987b) Disorders of the autonomic nervous system: Part II. Investigation and Treatment. *Ann. Neurol.*, **21**, 519–529.

Nanda, R.N., Boyle, R.C., Gillespie, J.S., Johnson, R.N. and Keogh, H.J. (1977) Idiopathic orthostatic hypotension from failure of nordrenaline release in a patient with vasomotor innervation. *J. Neurol. Neurosurg. Psychiat.*, **40**, 11–19.

Nanda, R.N., Boyle, R.C., Gillespie, J.S., Johnson, R.N. and Keogh, H.J. (1976) Adrenergic innervation of blood vessels in patients with neurogenic orthostatic hypotension. *J. Neuropathol. Appl. Neurobiol.*, **2**, 49.

Onrot, J., Goldberg, M.R., Hollister, A.S. *et al.* (1986) Management of chronic orthostatic hypotension. *Am. J. Med.*, **80**, 454–464.

Polinsky, R.J., Goldstein, D.S., Brown, R.T., Keiser, H.R. and Kopin, I.J. (1985) Decreased sympathetic neuronal uptake in idiopathic orthostatic hypotension. *Ann. Neurol.*, **18**, 48–53.

Polinsky, R.J., Jimerson, D.C. and Kopin, I.J. (1984) Chronic autonomic failure: CSF and plasma 3-methoxy-4-hydroxyphenylglycol. *Neurology*, **34**, 979–983.

Polinsky, R.J., Kopin, L.T., Ebert, M.H. and Weise, V. (1981) Pharmacologic distinction of different orthostatic hypotension syndromes. *Neurology*, **31**, 1–7.

Polinsky, R.J., Taylor, I.L., Chew, P., Weise, V. and Kopin, I.J. (1982) Pancreatic polypeptide responses to hypoglycemia in chronic autonomic failure. *J. Clin. Endocrinol. Metab.*, **54**, 48–52.

Polinsky, R.J., Taylor, I.L., Weise, V. and Kopin, I.J. (1988) Gastrin responses in patients with adrenergic insufficiency. *J. Neurol. Neurosurg. Psych.*, **51**, 67–71.

Ponsford, R., Bannister, R. and Paul, E.A. (1982) Methacholine pupillary response in third nerve palsy in Adie's syndrome. *Brain*, **105**, 583–597.

Robertson, D., Hollister, A.S., Carey, E.L., Tung, C., Goldberg, M.R. and Robertson, R.M. (1984) Increased vascular β_2 adrenoreceptor responsiveness in autonomic dysfunction. *JACC*, **3**, 850–856.

Shabsigh, R., Fishman, I.J. and Scott, F.B. (1988) Evaluation of erectile impotence. *Urology*, **32**, 83–90.

Shahani, B.T., Halperin, J.J., Boulu, P. and Cohen, J. (1982) Sympathetic skin response — a method of assessing unmyelinated axon dysfunction in peripheral neuropathies. *J. Neurol. Neurosurg. Psych.*, **47**, 536–542.

Shy, G.M. and Drager, G.A. (1960) A neurological syndrome associated with orthostatic hypotension. *Arch. Neurol.*, **2**, 511–527.

Stead, E.A. Jr and Ebert, R.V. (1941) Postural hypotension: a disease of the sympathetic nervous system. *Arch. Int. Med.*, **67**, 546–562.

Thatcher, B.S., Achkar, E., Fouad, F.N., O'Donnell, J.K. and Revta, R. (1987) Altered gastroesophageal motility in patients with idiopathic orthostatic hypotension. *Cleve. Clin. J. Med.*, **54**, 77–82.

Thomas, D.J. and Bannister, R. (1980) Preservation of autoregulation of cerebral blood flow in autonomic failure. *J. Neurol. Sci.*, **44**, 205–212.

Thomas, J.E. and Schirger, A. (1970) Idiopathic orthostatic hypotension. A study of its natural history in 57 neurologically effected persons. *Arch. Neurol.*, **22**, 289–293.

Thomas, J.E., Schirger, A., Fealy, R.D. and Sheps S.G. (1981) Orthostatic hypotension. *Mayo Clin. Proc.*, **56**, 117–125.

Tuck, R.R. and McLeod, J.G. (1981) Autonomic dysfunction in Guillian-Barre syndrome. *J. Neurol. Neurosurg. Psych.*, **44**, 983–990.

Van Lieshout, J.J., Wieling, W., Montfrans, G.V., Settles, J.J., Speelman, J.D. Endert, E. *et al.* (1986) Acute dysautonomia associate with Hodgkin's disease. *J. Neurol. Neurosurg. Psych.*, **49**, 830–832.

Vallbo, A.B., Hagbarth, K.E., Torebjork, H.E. and Wallin, B.G. (1979) Somatosensory, proprioceptive and sympathetic activity in human peripheral nerves. *Physiol. Rev.*, **59**, 919–957.

Virag R. (1982) Intracavernous injection papaverine for erectile failure. *Lancet*, **2**, 938.

von Euler, U.S., Hellner, S. and Purkhold, A. (1954) Excretion of noradrenaline in the urine in hypertension. *Scand. J. Clin. Lab. Invest.*, **6**, 54–59.

von Euler, U.S. (1948) Identification of the sympathomimetic ergone in adrenergic nerves of cattle (sympathin N) with laevo-noradrenaline. *Acta. Physiol. Scand.*, **16**, 63–74.

Wilcox, C.S. and Aminoff, M.G. (1976) Blood pressure responses to noradrenaline and dopamine infusions in Parkinson's disease in the Shy–Drager syndrome. *Br. J. Clin. Pharmacol.*, **3**, 207–214.

Wilcox, C.S., Puritz, R., Lightman, S.L., Bannister, R. and Aminoff, M.J. (1984) Plasma volume regulation in patients with progressive autonomic failure during changes in salt intake or posture. *J. Lab. Clin. Med.*, **104**, 331–339.

Witherington, R. (1988) Suction device therapy in the management of erectile impotence. *Urologic Clinics of North America*, **15**, 123–128.

Yahr, M.D. and Frontera, A.T. (1975) Acute autonomic neuropathy. *Arch. Neurol.*, **32**, 132–133.

Yeung Laiwah, A.C., MacPhee, G.H.A., Boyle, P., Moore, M.R. and Goldberg A. (1985) Autonomic neuropathy in acute intermittent porphyria. *J. Neurol. Neurosurg. Psych.*, **48**, 1025–1030.

Young, R.R., Asbury, A.K., Corbett, J.L. and Adams, R.D. (1975) Pure pan-dysautonomia with recovery. *Brain*, **98**, 613–636.

Ziegler, M.G., Lake, C.R. and Kopin, I.J. (1977) The sympathetic nervous system defect in primary orthostatic hypotension. *N. Engl. J. Med.*, **296**, 293–297.

Zorgniotti, A.W. and LeFleur, R.S. (1985) Autoinjection of the corpus cavernosum with a vasoactive drug combination for vasculogenic impotence. *J. Urol.*, **133**, 39–41.

6 Shy-Drager Syndrome and Multiple System Atrophy

Ronald J. Polinsky

Sandoz Pharmaceuticals Corporation
East Hanover, New Jersey 07936, USA

Multiple system atrophy is a degenerative neurological disorder that may be attended by autonomic failure. Autonomic insufficiency commonly precedes neurological involvement which produces Parkinsonian features, cerebellar dysfunction and other signs of impaired central nervous system function. The disease generally begins in the sixth decade and leads to premature death about 8 years after the onset. Pharmacological and physiological investigations reveal defective baroreflex modulation of blood pressure. Peripheral sympathetic noradrenergic neurons are not appropriately activated. Several central nervous system neurotransmitter pathways are affected by the degenerative process. Neuronal loss and gliosis characterize the primary neuronal degeneration in numerous brain regions. Although the etiology remains obscure, considerable progress has been achieved in symptomatic treatment.

KEY WORDS: Autonomic failure, striatonigral degeneration, Shy-Drager syndrome, olivopontocerebellar atrophy

INTRODUCTION

Thirty years ago it would have been a relatively simple task to summarize the knowledge of multiple system atrophy with autonomic failure (MSA). Shy and Drager (1960) had only recently drawn attention to the clinical and pathological features of the disorder that bears their names. In their review of the literature they identified 40 cases with "idiopathic hypotension"; many of those patients would not be labelled as MSA according to more recent refinements in the clinical classification and criteria for diagnosis. Numerous articles have been published on various aspects of the disorder since the appearance of that seminal paper. In addition, the last three decades have witnessed substantial progress in biomedical research technology that has fostered major advances in the understanding and treatment of neurological disorders.

Concomitant with the progress in clinical investigation there has been an attempt to unify the terminology applied to disorders of the autonomic nervous system. A multitude of terms has been utilized, often interchangeably, in publications dealing with MSA. This indiscriminate, imprecise use of nomenclature hampers comparison among reports in the medical literature. Critical review of the clinical details is required to maintain the proper perspective. Nosology is particularly relevant to the separation between pure autonomic failure (PAF) and MSA since these distinct disorders are characterized by clear differences in clinical features, pathophysiology and prognosis. Pure autonomic failure is a syndrome of chronic autonomic dysfunction occurring in the absence of any identifiable etiology, peripheral neuropathy or central neurological signs (Bradbury and Eggleston, 1925). In contrast, MSA includes several degenerative neurological disorders that may be attended by autonomic failure (Graham and Oppenheimer, 1969). Implicit in the descriptive term multiple system atrophy is the neuronal loss affecting a variety of central nervous system regions. The primary involvement of several autonomic centers justifies inclusion of this involuntary movement disorder within the scope of a textbook focused on autonomic dysfunction. This chapter will review many of the accomplishments achieved in elucidating various aspects of this complex disorder. Since treatment of autonomic dysfunction is discussed in a separate section of this book, only those therapeutic aspects specifically relevant to MSA will be mentioned at the end of this chapter.

THE CLINICAL SYNDROME

Despite significant advances in our understanding of the pathophysiology involved in producing the symptoms of MSA, no diagnostic test has been developed. An accurate diagnosis requires careful assessment of the clinical features and natural history. Several neurodiagnostic procedures provide ancillary information to support the clinical impression. Distinct differences in prognosis between MSA and other autonomic nervous system disorders highlight the importance of a detailed clinical evaluation.

PRESENTATION

Although the disorder has been reported to begin as early as age 19 in rare patients (Yoshida et al., 1987), the average age of onset in large series' of patients falls in the sixth decade (Bannister et al., 1988). There does not appear to be a significant difference between the sexes in that regard; however, men are affected about twice as frequently as women (Table 6.1). This disease does not manifest any special predilection for a specific race, ethnic background, occupation or socioeconomic status.

TABLE 6.1

Clinical characteristics of patients with multiple system atrophy (Polinsky, unpublished observations).

Sex	Number	Age at onset (years)	
		Range	Mean ± SEM
Male	29	37–66	51.7 ± 1.1
Female	18	25–67	50.2 ± 2.9

Involvement of several brain systems in MSA results in a wide spectrum of clinical manifestations. A variety of specific symptoms may herald the onset of MSA, but two patterns emerge based on the designation of symptoms as autonomic or neurologic. Autonomic symptoms comprise the initial presentation in approximately 75% of patients with MSA (Polinsky and Nee, 1989). Among the initial complaints (Table 6.2), genitourinary dysfunction (impotence in men, symptoms referable to neurogenic bladder) occurs most frequently. These presenting symptoms may occur in isolation, which makes their clinical evaluation difficult except in retrospect. For example, impotence may precede other signs of autonomic failure by more than 10 years. Many patients undergo extensive evaluation and may even become frustrated with psychiatric approaches for management only to later learn that their symptoms were neurogenic in origin.

TABLE 6.2

Presenting symptoms in multiple system atrophy (Polinsky, unpublished observations).

Symptom	Number of patients
Genitourinary dysfunction	22
Lightheaded	9
Imbalance	7
Fatigue	2
Rigidity	2
Clumsy	1
Change in handwriting	1
Hypertension	1
Gait disorder	1
Syncope	1

In the early stage of the illness, symptoms reflecting abnormal cardiovascular control may be vague. Syncope, a severe consequence of the hypotension, is relatively uncommon as a presenting feature. However, a number of other symptoms (e.g. headache, neck pain, dimming of vision, yawning) may occur as early manifestations of impaired control of blood pressure. It should be emphasized that low blood pressure *per se* is not a definitive sign of autonomic dysfunction. A decrease in blood pressure upon standing indicates that there may be an inadequate compensatory response to cardiovascular changes induced by the postural stimulus. Whether a particular drop in blood pressure will cause symptoms is determined by the absolute blood pressure, the rate of change during the fall and the ability of cerebral autoregulation to maintain perfusion in the face of systemic hypotension. Another important consideration regarding blood pressure is that orthostatic hypotension reflects only one aspect of abnormal cardiovascular control. As will be discussed later in this chapter, patients with autonomic insufficiency respond in an exaggerated fashion to any physiological or pharmacological stimulus capable of raising or lowering blood pressure.

Imbalance is the most common presenting complaint in those patients whose illness begins with neurological symptoms. The gait disorder related to cerebellar or Parkinsonian features occurs in the absence of any detectable change in blood pressure. Other patients may notice stiffness, clumsiness or a change in handwriting at the onset of MSA. Acute respiratory stridor is an unusual but potentially catastrophic presentation that may necessitate tracheostomy (Williams, Hanson and Calne, 1979a).

CLINICAL COURSE

Progression is the hallmark of the natural history in patients with MSA. Clinical deterioration of autonomic and neurological function can occur in several patterns. The spectrum of symptoms may increase; most patients eventually develop sympathetic and parasympathetic involvement. This pandysautonomia affects virtually every automatic function throughout the body. Similarly, non-autonomic involvement can disrupt operation of the cerebellar, extrapyramidal and pyramidal systems to produce a complex array of abnormalities manifested primarily as a movement disorder. From a temporal standpoint, some patients exhibit a steady downhill progression while others appear to deteriorate in a step-wise fashion characterized by a series of plateaus lasting for variable periods of time. The changes in autonomic and neurological function can occur quite independently. Neurological symptoms appear approximately 5 years after onset in those patients who present with autonomic symptoms (Polinsky and Nee, 1989). Conversely, patients who first develop neurological involvement manifest autonomic dysfunction about 2 years following the initial presentation. These patterns facilitate clinical distinction among patients with MSA, PAF and Parkinson's disease. It is necessary to wait at least 5 years from the onset of autonomic symptoms before a diagnosis of PAF can be made with confidence. Early, severe involvement of the autonomic nervous system rarely occurs in Parkinson's disease.

Symptoms reflecting involvement of specific organs and neurological systems can also change with disease progression. Although postural hypotension is the most disabling feature of autonomic insufficiency, abnormal cardiovascular control may also be attended by a fixed heart rate and supine hypertension. Both extremes of blood pressure contribute to functional disability and compromise, particularly later in the course of the illness. Despite normal coronary arteries, angina may develop during periods of hypotension (Silverberg *et al.*, 1979). Symptoms can also have a positive or negative character. A period of spontaneous, painful erections may precede total impotence in men with MSA. Excessive sweating can last for several months before patients lose this important mechanism for dissipating excess body heat. Some individuals develop diarrhoea in contrast to the severe constipation experienced by the majority of patients. In general, there is a tendency towards loss of function as the illness advances.

Several patterns of neurological signs justify a clinical subclassification of patients with MSA. This is not surprising since the term is used to encompass two overlapping degenerative disorders, viz. olivopontocerebellar atrophy (OPCA) and striatonigral degeneration (SND). Imbalance, incoordination and dysarthria are the main clinical features observed in the OPCA variant. In those patients with the SND form of MSA, the rigidity and bradykinesia are more prominent than tremor in comparison with the usual balance of this clinical triad observed in Parkinson's disease. Each of these subgroups comprises about 25% of patients with MSA. The remaining 50% have features common to OPCA and SND; the original patients reported by Shy and Drager (1960) would be classified in this category. Neurological symptoms and signs referable to other systems are observed in all MSA variants (see Table 6.3). Dementia is particularly uncommon except in the end stage of the disease. As mentioned in the previous section, hoarseness and stridor may develop as a consequence of vocal cord paralysis (Williams, Hanson and Calne, 1979a; Teräväinen and Udd, 1982; Gilmartin *et al.*, 1984; Kenyon, Apps and Traub, 1984). Unilateral or bilateral involvement occurs in most patients. Although Shy and Drager (1960) included fasciculations in their description of "the full syndrome", they have not been frequently reported in the literature. Sleep apnea has also been observed in MSA (Guilleminault *et al.*, 1977; Lehrman *et al.*, 1978; Munschauer *et al.*, 1990); the mechanism involves a combination of obstructive and neurogenic components. The latter are particularly relevant since abnormal respiratory control may contribute to the premature death of patients with MSA.

DIAGNOSTIC STUDIES

Despite significant technological advances a diagnostic test for MSA is lacking. Similar to many other degenerative disorders, the diagnosis of MSA is based on the clinical assessment, i.e. history and neurological examination. Consequently, the primary non-research purpose for performing these investigative procedures is to identify other, more potentially treatable, etiologies.

TABLE 6.3

Miscellaneous neurological signs in 47 patients with multiple system atrophy (Polinsky, unpublished observations).

Signs	Number of patients
Snout reflex	38
Hyperreflexia	33
Babinski response (bilateral in 24)	33
Dysphagia	24
Hoarseness	19
Weakness	16
Nystagmus	12
Stridor	8
Dementia	4
Spasticity	1
Muscle atrophy	1
Hyporeflexia	1
Fasciculations	0

Cerebrospinal fluid (CSF) examination is generally unrevealing. Occasional patients may have an elevated protein level in CSF but this has no diagnostic significance. Similarly, the electroencephalogram (EEG) can show non-specific changes. Consistent with the clinical examination, standard electrophysiological assessment of sensory and motor nerves reveals normal latency and conduction times, except in those few patients who manifest objective clinical signs of neuropathy. Cohen et al. (1987) found peripheral somatic neuropathy in 7 of 36 patients with MSA. Although Shy and Drager (1960) noted denervation changes in muscle, we have not observed this pattern in our patients. However, electromyographic evidence of denervation in laryngeal muscles (Guindi et al. 1981) and the rectal sphincter (Singh and Fahn, 1980) has been reported.

Application of urodynamic and electromyographic strategies for investigating bladder function in patients with MSA has yielded several important observations. Voluntary and involuntary aspects of control appear to be affected (Kirby et al., 1986). Patients are unable to initiate a micturition reflex. Bladder filling causes involuntary detrusor contractions. In addition, MSA patients also develop severe urethral dysfunction, including a loss of urethral tone. Abnormal motor units recorded from the striated part of the urethral sphincter suggest reinnervation. The pattern of involvement may vary, e.g. some patients have detrusor areflexia while others are hyperreflexic (Salinas *et al.*, 1986). These abnormalities reflect the pandysautonomia in MSA which in this case results primarily from selective involvement of sacral anterior horn cells (Onuf's nucleus).

Although the EEG is not particularly valuable in evaluating patients with MSA, brainstem auditory evoked potentials may help to differentiate the disorder from PAF and Parkinson's disease. There is abnormal latency and amplitude (ratio of wave V/I) in most patients with MSA; this was not present in any PAF patients and occurred in only one patient with Parkinson's disease (Prasher and Bannister, 1986). The auditory pathway disruption was felt likely to occur in the superior olivary complex. In another study (Uematsu, Hamada and Gotoh, 1987), prolonged interpeak (I–III) latency correlated with the degree of pontine atrophy determined by computerized tomography (CT).

Neuroimaging techniques have proven to be more useful than the above methods in terms of supporting the clinical diagnosis. As expected, CT scanning is most helpful in those cases with cerebellar involvement. However, cerebellar and brainstem atrophy can be seen even in cases without significant clinical signs. Thus, an abnormal CT scan can facilitate an accurate diagnosis in patients with autonomic insufficiency. The CT scan is generally normal in those MSA patients who have only Parkinsonian features. Magnetic resonance imaging (MRI) has added a new dimension to neuroradiology through the opportunity to adjust scanning parameters in order to visualize structures not readily apparent with CT or other brain imaging modalities. Pontine and cerebellar atrophy can also be clearly visualized using MRI (Figure 6.1).

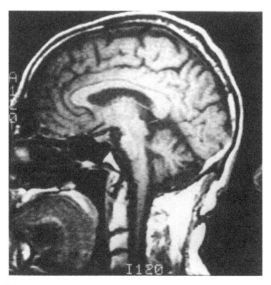

FIGURE 6.1   Sagittal MRI scan, T1-weighted (ME, TR 400, TE 20), shows severe atrophy of pons, cerebellar hemispheres and vermis in a patient with MSA.

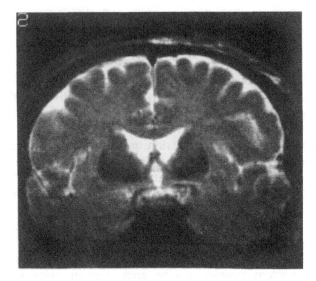

FIGURE 6.2a

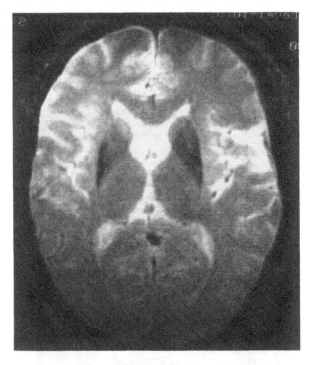

FIGURE 6.2b

FIGURE 6.2   (a) Coronal and (b) axial MRI scans, T2-weighted (SE, TR 2000, TE 100), in an MSA patient reveal severe hypointensity in the posterolateral putamina.

Using a strong (1.5 Tesla) magnetic source in combination with a T2-weighted spin-echo pulse sequence, it was possible to detect an abnormal decrease in putaminal signal intensity (Pastakia et al., 1985; Savoiardo et al., 1989). Although some changes may be seen in older normal subjects, the distinguishing feature in MSA is the particular involvement of the posterior and lateral portions of the putamina (Figure 6.2). Furthermore, there is a correlation in patients with MSA between rigidity and putaminal signal dropout (Brown et al., 1987). Iron deposition may contribute to the attenuation of signal intensity in the basal ganglia (Rutledge et al. 1987). Magnetic resonance imaging and CT scanning in PAF have only revealed changes compatible with age.

PROGNOSIS

Unfortunately, the long-term outlook for patients has not changed appreciably over the last 30 years, despite substantial advances in our understanding of the pathophysiology which have contributed to rational therapeutic approaches for managing various aspects of the disorder. Our inability to alter the prognosis highlights our difficulty in identifying the underlying cause of this devastating neurological disorder. Patients with MSA die approximately 8.5 years after the onset of their illness (Polinsky and Nee, 1989). There is a wide range in the duration of the illness, with a few cases who have died within several years of the onset and rare patients who have survived for more than 20 years. There is no difference in the duration between men and women, although patients with the OPCA variant may survive about 2–3 years longer on the average compared to the other subtypes. Age at onset, presenting syndrome (autonomic vs. neurological), and supine hypertension do not influence the duration of illness in MSA.

Several consequences of the degenerative disorder directly contribute to the premature demise of patients with MSA. Aspiration, apnea and presumably arrhythmias are the primary causes of death in most patients. The ability to handle secretions is clearly compromised in many patients; although swallowing may also be impaired it is rarely necessary to employ a feeding tube or gastrostomy. Restriction of the airway can result from a combination of abductor paralysis and rigidity. The latter may improve to some extent following administration of anti-parkinsonian medications (Gilmartin et al., 1984). The changes in blood pressure and arrhythmias during sleep may well be secondary to apneic episodes. Abnormal respiratory control and laryngeal obstruction are important factors that promote episodes of sleep apnea and respiratory stridor. Some investigators have recommended tracheostomy for patients who have even mild obstruction during sleep (Munschauer et al., 1990). From a prognostic standpoint a more conservative approach should be followed, since there is no evidence suggesting that tracheostomy increases longevity in MSA patients. Removal of the obstructive component does not improve the neurogenic element of this problem.

PATHOPHYSIOLOGY

Automatic control of vital functions is compromised early in the course of the disease in most patients. Hence, much of the research on MSA has focused on autonomic dysfunction. This section will deal primarily with neurotransmitter and neuropeptide function in MSA.

BLOOD PRESSURE CONTROL IN MSA

As discussed previously, orthostatic hypotension often becomes the most disabling feature of autonomic insufficiency. A variety of strategies have been applied to investigate circulatory control in autonomic failure. One particularly useful approach contrasts the results between MSA and PAF. Before discussing the abnormalities observed in these disorders, it is important to review some basic aspects of neurocirculatory control.

NEUROLOGICAL CONTROL OF THE CIRCULATION

Although several homeostatic mechanisms participate in the regulation of blood pressure, two specific nervous system elements play a major role in providing the vasomotor and cardiac responses required to maintain adequate blood pressure. The baroreflex arc responds rapidly to cardiovascular insults with sufficient magnitude so that it may be regarded as the initial defence against circulatory compromise (Figure 6.3). The postganglionic sympathetic neuron is the "final common pathway" through which efferent sympathetic outflow is transmitted to various organs and their regional circulation.

Blood pressure is monitored at strategic points within the cardiovascular system. The major baroreceptors are located in the carotid sinus and aortic arch, while low pressure mechanoreceptors are located in the atria; this geographic arrangement facilitates continuous assessment of arterial pressure to the brain and venous return to the heart respectively. Increased pressure sensed by these receptors is translated into an increased nerve firing frequency. Afferent baroreceptor activity is transmitted to the brainstem through the vagal and glossopharyngeal nerves. Although there is a convergence of afferent input to the nucleus of the tractus solitarii, the information is widely distributed to lower and higher nervous system centres. A thorough treatment of detailed connections involved in the central processing and control of blood pressure is beyond the scope of this chapter. Suffice it to say that central cardiovascular regulation involves input from many brain regions including the cerebral cortex, limbic system, basal ganglia and hypothalamus. Several brainstem centres and spinal cord reflexes also modulate blood pressure control. The efferent arm of the baroreflex is derived through coordination of sympathetic and parasympathetic activity originating from the intermediolateral column of the spinal cord (ILC) and dorsal motor nucleus of

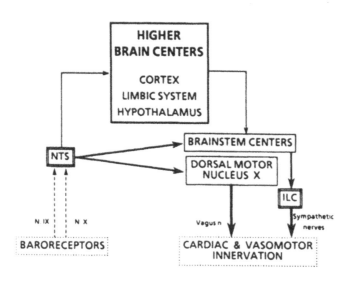

FIGURE 6.3   Schematic diagram of cardiovascular control by the nervous system. NTS – nucleus tractus solitarii.

the vagus (DMN) respectively. While vagal nerve fibres arising directly from the DMN innervate ganglia near peripheral organs, the ILC gives rise to preganglionic sympathetic fibers that supply the paravertebral chain of sympathetic ganglia. The postganglionic noradrenergic neuron provides sympathetic innervation to the heart and blood vessels throughout the body. From a functional standpoint, increased afferent baroreceptor activity (high blood pressure) causes a decrease in sympathetic efferent tone; vagal activity is correspondingly increased in response to the elevation in pressure. Thus, cardiac output and vasomotor tone are adjusted to maintain blood pressure within a reasonably narrow normal range.

A wide variety of neurotransmitters and neuropeptides have been associated with the neurological control of the circulation; noradrenaline (NA) has been most intensively studied in clinical investigations for two reasons. Firstly, the nordrenergic neuron provides the link that mediates central nervous system activity and the end-organ responses (Figure 6.4). Secondly, sensitive and specific analytical methods permit accurate measurement of NA and its metabolites in various biological fluids. A brief review of the functional characteristics of the sympathetic noradrenergic neuron will serve as a foundation to facilitate discussion of the biochemical and pharmacological approach used in studying patients with MSA.

Synthesis of NA occurs through a series of enzymatic reactions carried out within neurons. Although this pathway begins with tyrosine, dihydroxyphenylalanine (DOPA) is the catechol amino acid precursor for catecholamines. The rate-limiting step in catecholamine synthesis involves tyrosine hydroxylase. Dopamine

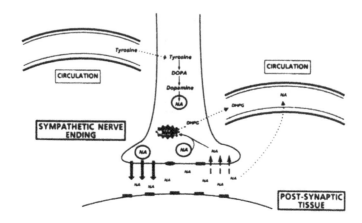

FIGURE 6.4   Functional characteristics of the postganglionic sympathetic noradrenergic neuron.

is converted to NA by dopamine-β-hydroxylase in the nerve ending where it is stored in granules. Nerve impulses release this stored NA into the synaptic cleft. Stimulation of postjunctional receptors by released NA results in vasoconstriction at neurovascular synapses except in large muscle beds where vasodilation is required during exercise. Presynaptic NA receptors provide feedback control by inhibiting further neurotransmitter release. An active neuronal uptake process is primarily responsible for terminating the actions of released NA. Local characteristics, e.g. innervation density and synaptic cleft width, may affect the relative importance of this mechanism. Neuronal uptake is extremely efficient; only a small amount (less than 10%) of the synaptic NA "escapes" into plasma. The metabolic fate of NA differs according to whether the transmitter is taken up into sympathetic neurons. Dihydroxyphenylglycol (DHPG) is formed after deamination of intraneuronal NA; cytoplasmic NA may be derived from neuronal uptake or leakage from storage vesicles. Extraneuronal NA metabolism involves O-methylation; thus, released NA that escapes the uptake process may be converted into normetanephrine, some of which is conjugated and excreted in the urine before deamination by monoamine oxidase. The importance of these functional characteristics will become more evident in consideration of the development of various strategies for evaluating sympathetic neuronal function.

PHYSIOLOGICAL ABNORMALITIES

As pointed out by Shy and Drager (1960), orthostatic hypotension is a prominent symptom of the disturbance in cardiovascular control in patients with MSA. The term does not convey the complexity and full consequences of the disruption in pathways mediating cardiac and vasomotor innervation. A variety of techniques have been employed to evaluate the baroreflex in man. These include carotid-sinus massage, carotid occlusion, sinus-nerve stimulation (or block), and neck-collar devices. These methods have not been systematically applied to investigate MSA. Administration of vasoactive drugs also yields a functional assessment of baroreflex integrity. This approach will be discussed in conjunction with the biochemical and pharmacological investigations. The present section will review some of the physiological aspects of abnormal blood pressure control in MSA.

The Valsalva manoeuvre in patients with MSA illustrates the basic characteristics of the problem with blood pressure control (Bannister and Mathias, 1988). During phase II, as hypotension develops, there is very little cardioaccelerator response. There is also no recovery of arterial pressure while intrathoracic pressure is increased; blood pressure continues to gradually fall as venous return is compromised. Absence of the normal overshoot in phase IV reflects impaired vasoconstriction in peripheral vasculature. The primary dysfunction in MSA prevents reflex activation of sympathetic outflow to compensate for changes in blood pressure caused by various stimuli. Thus, when patients with MSA assume the upright posture, blood pressure decreases because they are unable to counter the effects of blood pooling in the lower extremities induced by gravity. Although other factors likely contribute to the post-prandial hypotension observed in autonomic failure, a similar effect in the splanchnic circulation may play a role in this manifestation of impaired circulatory homeostasis (Robertson, Wade and Robertson, 1981). Using the technique of lower-body negative pressure, Bannister, Ardill and Fentem (1967) have shown that forearm blood flow does not increase in autonomic failure as observed in normal subjects. Furthermore, the calf volume increases much more rapidly than normal in patients with autonomic failure. From the results of physiological studies, it appears that sympathetic efferent outflow is primarily affected in MSA. Direct recording of sympathetic nerve activity more specifically localizes the abnormality to a preganglionic level. In a single patient with MSA, Dotson et al. (1990) observed episodes of sympathetic bursts that were not under reflex control. Thus, it appears that the sympathetic ganglion cells are decentralized.

In striking contrast to the deficit in baroreflex control is the preservation of cerebral autoregulation in MSA. Measurement of cerebral blood flow in patients with autonomic failure revealed that autoregulation functioned well over a wide range of systolic pressures between 40 and 170 mmHg (Thomas and Bannister, 1980). Failure occurred at a systolic pressure approximately 20 mmHg lower than observed in normal subjects. Furthermore, middle cerebral artery blood flow decreases to a lesser extent than expected on the basis of a reduction in systemic arterial pressure during tilting in patients with MSA (Briebach, Laubenberger and Fischer, 1989). This may explain why many patients appear to tolerate low blood pressures without experiencing symptoms of dizziness or syncope. The mechanism of this fortunate abnormality is unclear.

Another characteristic of impaired cardiovascular control in MSA is manifested in the diurnal pattern of blood pressure. Circadian variation in pressure is apparent, but the pattern is reversed from that obseved in normal subjects (Mann et al., 1983). The lowest pressures in the morning correlate with the clinical observation that this is the most difficult functional period for patients with autonomic insufficiency. High pressures at night probably contribute to the nocturnal diuresis in many of these patients. A reverse Trendelenberg position for sleeping has been recommended following the demonstration that head-up tilt at night produced an increase in weight attended by a reduction in sodium and water excretion (Bannister, Ardill and Fentem, 1969). This approach also minimizes the risks from supine hypertension.

Blood pressure during sleep also differs from normal in patients with MSA. A number of studies have confirmed the abnormal sleep pattern: reduction in duration of stages 3 and 4 [non-rapid eye movement (non-REM)] and REM sleep (Guilleminault et al., 1977; Lehrman et al., 1978; Martinelli et al., 1981; Coccagna et al., 1985). In normals, arterial pressure deceases in the successive sleep stages; sudden changes in blood pressure are associated with movement or REM sleep. In MSA patients with sleep apnea, there is generally little change in blood pressure during apneic episodes. However, patients without sleep apnea manifest sudden swings in pressure unassociated with movement during all stages of sleep; systemic pressure also increases in REM sleep (Martinelli et al., 1981). These changes in blood pressure may become apparent due to inadequate baroreflex modulation.

PHARMACOLOGICAL INVESTIGATIONS

Baroreflex and sympathetic neuronal function have been the primary targets of biochemical and pharmacological approaches used to investigate blood pressure control in MSA. As mentioned earlier, several other hormonal mechanisms play a more critical role in the chronic regulation of blood pressure through their effects on vasomotor tone, fluid balance and mineral homeostasis. Although the various mechanisms are discussed separately, it must be remembered that these systems act in concert with the autonomic nervous system to maintain circulatory homeostasis.

Despite limitations imposed by the sensitivity of NA to influence by a host of experimental and technical factors, plasma levels of the neurotransmitter serve as a meaningful index of sympathetic neuronal activity (Lake, Ziegler and Kopin, 1976). Wallin et al. (1981) found a good correlation between plasma NA levels and electrophysiological measurements of sympathetic nerve activity in man. Assumption of the upright posture is attended by a doubling of the plasma NA; this increment reflects the increased sympathetic nerve activity that counteracts the pooling of 500–700 ml of blood in the lower extremities due to gravity. In patients with MSA the supine plasma NA level is generally normal, but fails to increase appropriately upon standing (Ziegler, Lake and Kopin, 1977). The normal supine level suggests that postganglionic sympathetic neurons are relatively intact. Inability to adequately increase plasma NA in response to a postural stimulus is consistent with a baroreflex lesion that prevents activation of the postganglionic neurons. Although plasma NA levels are useful in characterizing the pathophysiology in MSA, they have limited diagnostic utility in individual patients. Since only 4% of patients with MSA manifest a plasma NA level less than 100 pg/ml (Polinsky, unpublished observations), observation of a low level supports the diagnosis of PAF in a patient without signs of central neurological involvement. Occasional patients with MSA manifest low levels of plasma NA suggesting peripheral sympathetic dysfunction. In a study of patients with PAF and MSA, Cohen *et al.*, (1987) observed that adrenergic denervation was uncommon in MSA, although the frequency of postganglionic sudomotor failure was similar to PAF. As discussed earlier, electromyographic studies in laryngeal (Guindi *et al.*, 1981) and rectal muscles (Singh and Fahn, 1980) have also revealed denervation changes.

Several other neurochemical approaches provide further evidence of the functional integrity of postganglionic sympathetic neurons in MSA. In addition to measuring plasma NA, it is also possible to measure levels of DOPA and DHPG. Since the rate-limiting step in catecholamine synthesis involves the conversion of tyrosine to DOPA, plasma levels of DOPA may serve as an index of NA synthesis (Goldstein *et al.*, 1987). Following uptake of released NA into presynaptic neurons, intraneuronal deamination by monoamine oxidase results in the formation of DHPG. Axoplasmic NA from leakage out of vesicles also contributes to DHPG. Normal catechol levels in MSA support a predominantly preganglionic dysfunction (Goldstein *et al.*, 1989). However, a small subgroup of patients have catechol patterns consistent with normal synthesis and decreased postganglionic sympathetic nerve activity or exocytotic release of NA.

Urinary excretion rates and apportionment of NA metabolites have also been used to evaluate sympathetic neuronal function in MSA. This strategy is based on differences in the metabolic fate between NA taken up into sympathetic neurons and that which escapes into plasma following release (Figure 6.5). As mentioned previously, intraneuronal NA metabolism yields deaminated metabolites; extra-neuronal O-methylation converts NA to normetanephrine. In patients with MSA, there is a disproportionate decrease in the urinary excretion of normetanephrine, although the total amount of NA metabolites is normal or slightly decreased (Kopin *et al.*, 1983b). Since normetanephrine is derived from released NA, this pattern of urinary metabolites is consistent with diminished activation of relatively intact postganglionic noradrenergic neurons.

The plasma level of NA is determined by the balance between spillover of released NA and neuronal uptake and subsequent metabolism. Thus, it is important to evaluate the uptake process in patients with autonomic failure. Substantial overlap in plasma NA levels between normals and various patient groups underscores the need for assessing neuronal uptake. Delayed clearance could elevate plasma NA levels and prolong the pressor effect of injected sympathomimetic amines (Bannister *et al.*, 1979). Despite the importance of this functional characteristic, few studies have been performed in patients with MSA. Two approaches provide the means for investigating neuronal uptake in man: (1) the plasma NA response to intravenously administered tyramine, an indirectly acting sympathomimetic drug; and (2) analysis of the kinetic disposition of infused radiolabelled NA and isoproterenol which is not taken up by sympathetic neurons. The increment in plasma NA following bolus injections of tyramine is normal in MSA, consistent with normal neuronal

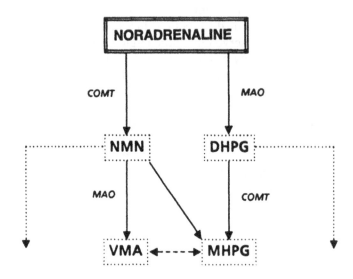

FIGURE 6.5   Noradrenaline metabolism.   COMT = catechol-ortho-methyltransferase;   MAO = monoamine oxidase; NMN = normetanephrine; VMA = vanillylmandelic acid.

uptake and stores of the catecholamine (Polinsky *et al.*, 1981a). Furthermore, the rates of disappearance of radiolabelled catecholamines, following infusion to steady-state levels, is similar in normal subjects and patients with MSA (Polinsky *et al.*, 1985). These findings further establish the functional integrity of peripheral sympathetic noradrenergic neurons in MSA. In contrast, patients with PAF manifest a reduced plasma NA increment to tyramine (Polinsky *et al.*, 1981a) and delayed removal of infused NA (Esler *et al.*, 1980; Polinsky *et al.*, 1985), reflecting impaired sympathetic neuronal function.

Pressor responsivity is the most clinically relevant aspect of sympathetic neuronal function. Exaggerated responses to drugs following nervous system lesions were described in the medical literature more than a century ago. Although the phenomenon of supersensitivity is still not completely understood, studies in experimental animals have clarified the distinction between denervation and decentralization (for review see Trendelenberg, 1963). These characteristic pharmacologic abnormalities form the basis for distinguishing pre- and postganglionic noradrenergic involvement. However, it is important to bear in mind that the lesion(s) in degenerative neurological disorders are rarely complete so that the pharmacological responses may not faithfully mirror the experimental situation. Chronic postganglionic denervation increases the pressor response to NA, while the effects of indirect sympathomimetics are reduced. Decentralization causes more modest changes in the blood pressure response and is not associated with a loss of neuronal NA stores; the increase in pressor sensitivity is non-specific.

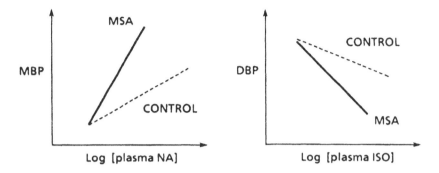

FIGURE 6.6   Exaggerated pressor and depressor responses in MSA. MBP = mean blood pressure, DBP = diastolic blood pressure, ISO = isoproterenol, MSA = multiple system atrophy.

Comparison of the blood pressure responses to pharmacological agents in patients with PAF and MSA illustrates the value of this approach. Both patient groups have greater than normal blood pressure responses to angiotensin, a non-adrenergic pressor drug (Polinsky *et al.*, 1981a).   This abnormality is manifested as an increase in the slope of the blood pressure dose–response. Similar evidence of hyperresponsivity in patients with autonomic failure has been described in response to a variety of cardiovascular drugs: NA (Bannister *et al.*, 1979; Polinsky *et al.*, 1981a), isoproterenol (Mathias, Matthews and Spalding, 1977; Bannister *et al.*, 1981a; Baser *et al.*, 1991), dopamine (Wilcox and Aminoff, 1976), tyramine (Bannister *et al.*, 1979; Polinsky *et al.*, 1981a), methoxamine (Parks *et al.*, 1961), vasopressin (Möhring *et al.*, 1980), and somatostatin and its analogues (Hoeldtke, O'Dorisio and Boden, 1986). Such a non-specific increase in pressor and depressor responses reflects defective baroreflex modulation of blood pressure (Figure 6.6). In contrast, the leftward shift of the blood pressure dose–response curve to NA in PAF is attended by low baseline plasma NA levels, a reduced tyramine-induced NA increment and decreased neuronal uptake; this pharmacological pattern is characteristic of denervation that results from the primarily postganglionic involvement in this disorder. Although the normal indices of peripheral sympathetic noradrenergic function in MSA are consistent with preganglionic dysfunction, pharmacological investigations have not further elucidated the exact site of the lesion(s) responsible for decentralization of cardiovascular control. The loss of baroreflex capability to buffer the effects of stimuli that raise or lower blood pressure has important therapeutic consequences: it facilitates the use of drugs with pressor activity that would be supressed in normal subjects. The *in vitro* observation of increased affinity and number of α-receptors on platelets from patients with MSA seems paradoxical, since this change might be expected in a disorder associated with denervation (Davies *et al.*, 1982; Kafka *et al.*, 1984). However, the relationship between platelet α-receptors and those at the vascular neuroeffector junction is not clear.   Despite normal basal levels of plasma NA in MSA, the ambulatory levels throughout the day would be low in comparison with normal subjects. Thus, as indicated by the selective reduction in urinary normetanephrine, a decrease in released NA may induce a slight upregulation in platelet α-adrenoreceptors relative to normals. Although the time course of receptor changes has not been clearly defined, pressor infusion studies are generally carried out in the early morning. Presumably, circulating NA in patients with MSA would be similar to normal subjects during the night.

ADRENAL MEDULLARY DYSFUNCTION

Stressful stimuli elicit a variety of protective compensatory mechanisms including secretion of adrenaline from the adrenal medulla which functions as a specialized ganglion of the sympathetic nervous system. The characteristics of the adrenal medullary response are indicative of its role: rapid release of large amounts of adrenaline effectively produces a bolus injection into the bloodstream after a critical threshold is exceeded by a metabolic or circulatory insult. Although it has very important cardiovascular actions, adrenaline is more closely involved with metabolic control. The neurotransmitter functions as a true hormone; unlike NA which is released at or near its site of action adrenaline is distributed throughout the body via the bloodstream.

Catecholamines affect intermediary metabolism through a variety of mechanisms. The importance of adrenaline as an hypoglycemic counter-regulatory hormone is highlighted by the fact that increases in plasma precede release of glucagon, growth hormone and cortisol during hypoglycemia (Garber *et al.*, 1976).

Detailed studies of the adrenal medullary response to insulin-induced hypoglycemia began during the first quarter of this century (Cannon, McIver and Bliss, 1924). Despite the demonstration by Luft and von Euler (1953) that two patients with postural hypotension failed to increase their urinary catecholamine excretion during hypoglycemia, there have been relatively few studies of this important sympathetic nervous system response in autonomic failure. Investigation of patients with adrenergic insufficiency has yielded valuable insights into normal physiological mechanisms. The glucose recovery curve following the nadir of hypoglycemia is biphasic in normal subjects: an initial rapid rise precedes a slower rate of glucose elevation in restoring euglycemia. Coincident with the fall in plasma glucose after insulin administration is a precipitous rise in catecholamines, consisting primarily of an increase in adrenaline and to a lesser extent NA. Patients who lack the catecholamine responses to hypoglycemia do not have the initial rapid phase of glucose recovery (Polinsky *et al.*, 1980). Thus, these patients with adrenergic insufficiency provide evidence that adrenaline is responsible for this critical aspect of the counter-regulatory mechanism in man. This reflects the teleological function of the adrenal medulla, which rapidly responds with sufficient magnitude during crisis while other more gradually responsive mechanisms are activated. Furthermore, it appears that adrenaline is not required to elicit other counter-regulatory hormones that function normally in patients with adrenergic insufficiency (Polinsky *et al.*, 1981b).   The increment in plasma NA levels observed during insulin-induced hypoglycemia in normal subjects may reflect a compensatory cardiovascular reflex to prevent the hypotensive effect of insulin (Brown *et al.*, 1986; Brown, Polinsky and Baucom, 1989).

Most patients with MSA have deficient catecholamine responses to hypo-glycemia.   Longitudinal studies suggest that adrenal medullary involvement develops with continued progression of the disease.   Although assessment of adrenal medullary function is clinically important, there is no localizing value. The response involves activation of glucose receptors in the hypothalamus or brainstem, transmission of an efferent sympathetic response through the ILC and stimulation of adrenal medullary activity via preganglionic neurons. Thus, a lesion at any point in the central or peripheral pathways can interfere with the response. A central lesion in MSA is supported by the observation that arecoline produces a normal plasma adrenaline increment following pretreatment with a muscarinic blocker.

IMPAIRED HORMONAL MECHANISMS

The autonomic nervous system interacts with many hormonal and peptide systems that control a diversity of functions including cardiovascular parameters, digestion, metabolism, mineral homeostasis and stress responses.   For some of these substances, a specific function has yet to be determined in man. Their importance in clinical investigation is based on a relationship with specific aspects of autonomic nervous system function.   Two examples in this latter class are melatonin and pancreatic polypeptide.

The pineal gland releases melatonin in a diurnal pattern that is under control by an endogenous circadian oscillator located in the suprachiasmatic nucleus. Light appears to play a major role in determining the rhythm and its effects are mediated through retinohypothalamic projections. Pineal innervation is derived from the superior cervical ganglion which is supplied by preganglionic fibres arising from the ILC. Nocturnal stimulation of the β-adrenergic pineal receptors results in enhanced synthesis and release of melatonin. Following rapid metabolism, 6-hydroxymelatonin is conjugated and excreted in the urine. Normal subjects excrete most of the urinary 6-hydroxymelatonin at night and relatively little during the day (Tetsuo *et al.*, 1981). In PAF, the total amount excreted each day is low but the diurnal pattern is preserved. The main abnormality in MSA is a disruption in the pattern: many patients excrete as much 6-hydroxymelatonin in the day as at night. Central nervous system lesions in MSA presumably interfere with the pathways controlling melatonin secretion.

Pancreatic polypeptide is one of several gut peptides released during insulin-induced hypoglycemia. Release of the polypeptide appears to be under a vagally mediated, cholinergic mechanism. Thus, measurement of the plasma level is a biochemical index of vagal function. In combination with catecholamine responses it is possible to simultaneously assess sympathetic, parasympathetic and adrenal medullary activity. Most patients with MSA do not manifest the normal rise in plasma levels of pancreatic polytide during hypoglycemia (Long *et al.*, 1979;

Polinsky *et al.*, 1982). This abnormality provides biochemical evidence of the pandysautonomia in MSA.

Many of the gut peptides are secreted from cells within or near the structures where they exert their predominant effects, unlike typical endocrine hormones whose secretory cells are clustered in specific organs. Through its local effects on antral motility and acid production, gastrin plays an important role in the digestive process. A variety of factors affect gastrin release; the relative contributions of specific influences may vary according to the clinical setting. For example, vagal stimulation increases gastrin release although basal gastrin levels are elevated following vagotomy. Gastric distention and high stomach pH resulting from vagotomy contribute to stimulation of gastrin release in this situation. Adrenergic effects have also been postulated since the gastrin response persists after highly selective vagotomy. Gastrin levels distinguish patients with PAF and MSA (Polinsky *et al.*, 1988b). High basal levels in PAF are consistent with peripheral involvement of vagal function. The gastrin increments during hypoglycemia are respectively greater and less than normal in patients with PAF and MSA. This dichotomy may be explained on the basis of adrenergic receptor supersensitivity which occurs only in PAF. Thus, the diminished catecholamine responses in both groups have different effects. Lower than normal catecholamine levels during hypoglycemia in MSA result in a reduced gastrin increment, since adrenergic receptors have normal sensitivity. The supersensitive gastrin release mechanism in PAF is reflected by exaggerated gastrin responses to intravenous isoproterenol (Polinsky *et al.*, 1989a).

Food ingestion causes post-prandial hypotension in patients with autonomic failure (Robertson, Wade and Robertson, 1981). Inadequate adrenergic compensatory mechanisms may contribute to this phenomenon; however, release of gut peptides after eating may lower blood pressure through their effects on the circulatory system. Normal levels of gastrin, motilin, vasoactive intestinal polypeptide, somatostatin and cholecystokinin have been observed after fluid intake in patients with autonomic failure (Mathias *et al.*, 1986). Although enteroglucagon and pancreatic polypeptide were increased they apparently do not lower blood pressure. In contrast, neurotensin appears to be higher than normal in autonomic failure patients given a test meal and this peptide is a potent vasodilator. Insulin may also play a role since patients with autonomic dysfunction are more sensitive to its hypotensive effects (Brown *et al.*, 1986; Mathias *et al.*, 1987). It must be remembered, however, that measurements of plasma levels may not give an accurate assessment, since local effects of gut peptides may be more relevant.

Several peptide systems related to blood pressure control have also been studied in patients with autonomic failure. The renin–angiotensin system participates in the regulation of haemodynamics and electrolyte balance (for review see Skeggs, 1984). Reductions in blood volume, renal perfusion pressure and plasma sodium activate the cascade of enzymatic reactions that result in the formation and release of angiotensin II, the dominant hormone of this system. Renin, released from juxtaglomerular cells in the kidney, initiates the sequence by its actions on angiotensinogen. Autonomic nervous system activity affects the renal handling of sodium and water through α- and β-adrenergic receptors. The former affect renal blood flow, while the latter modulate secretion of renin. Patients with autonomic failure exhibit excessive nocturnal salt and water excretion (Wilcox, Aminoff and Slater, 1977). Their low supine plasma renin levels fail to increase in response to salt restriction, postural change or isoproterenol (Wilcox et al., 1974; Bannister, Sever and Gross, 1977; Wilcox, Aminoff and Slater, 1977; Baser et al., 1991).

Atrial natriuretic peptide (ANP) is synthesized and stored in atrial muscle cells. Although atrial distention appears to be the primary stimulus for ANP release, the sympathetic nervous system may also affect release of the peptide. Natriuresis, decreased tone in vascular smooth muscle, lowering of blood pressure, and inhibition of renin and aldosterone secretion are among the many cardiovascular effects of ANP. Plasma ANP levels decrease at night in patients with autonomic failure; the peptide seems to respond appropriately to change in posture and intravascular volume (Kaufmann et al., 1990). Urinary ANP excretion in autonomic failure is higher during the day, although natriuresis is maximal at night (Polinsky et al., 1989d). From these studies it appears that autonomic lesions do not interfere with ANP release.

Vasopressin, a potent vasoconstrictor, also plays an integral role in regulating fluid volume. Serum osmolality and activation of thoracic stretch receptors are the primary stimuli that elicit release of vasopressin. Thus, the plasma levels increase in normal subjects during standing (Davis et al., 1976; Zerbe, Henry and Robertson, 1983). This response is blunted in patients with autonomic failure (Zerbe, Henry and Robertson, 1983; Williams et al., 1985). An afferent lesion was suggested by Williams, Lightman and Bannister, (1985) on the basis of a normal vasopressin response to infusion of hypertonic saline. Vasopressin is not suppressed by administration of levodopa or naloxone in MSA, presumably as a result of the widespread central nervous system lesions (Lightman and Williams, 1988).

CENTRAL NERVOUS SYSTEM NEUROTRANSMITTER SYSTEMS

Three general strategies have been used to investigate central nervous system function in patients with MSA: (1) neuroendocrine approach; 2) measurement of neurotransmitter-related substances in CSF; and (3) positron emission tomography (PET). There are limitations to each of these techniques. The validity of the first approach depends on the specificity of the provocative drug; in addition, peripheral effects may complicate interpretation of the results. A variety of factors affect the reliability of CSF measurements; standardized methods for collection, storage and analysis must be employed. Furthermore, the use of metabolites or other substances as valid indices of neuronal function requires justification. The final and most direct approach for studying brain metabolism and function in man has only recently been applied to study MSA. The ^{18}F-fluorodeoxyglucose (FDG) method for studying cerebral glucose metabolism has been used extensively over the last decade. However, recent application of PET scanning coupled with neurotransmitter precursors or receptor ligands depends critically on the demonstration that these compounds behave pharmacologically as expected despite their chemical modification. The above strategies will be discussed in relation to MSA, beginning with the least direct approach.

The hormonal responses to insulin-induced hypoglycemia include release of the pituitary peptides, β-endorphin and adrenocorticotrophic hormone (ACTH). A central cholinergic pathway appears to mediate the release of β-endorphin, since arecoline and atropine respectively stimulate and block its release. Patients with MSA but not PAF have essentially absent β-endorphin/ACTH responses during hypoglycemia (Polinsky et al., 1987). Decreased plasma ACTH responses during arecoline in patients with MSA suggest that the degenerative process in hypothalamic cholinergic neurons affects postsynaptic elements. Another approach for assessing the integrity of central cholinergic pathways is the measurement of CSF levels of neurotransmitter-related substances. Since it is not possible to measure acetylcholine or a metabolite, enzyme markers constitute a less desirable alternative. Unfortunately, attempts to measure choline acetyltransferase, an established presynaptic cholinergic neuronal marker, have been unsuccessful. However, acetylcholinesterase (AChE), the enzyme responsible for breaking down acetylcholine, is released into CSF in response to electrical stimulation and administration of drugs. The AChE level in CSF is lower than normal in patients with MSA, consistent with central cholinergic involvement (Polinsky et al., 1989b).

Monoaminergic systems have been evaluated by measuring their respective metabolite levels in CSF. The major brain metabolite of NA is 3-methoxy-4-hydroxyphenylglycol (MHPG). In order to use CSF MHPG levels as an index of central noradrenergic activity, it is necessary to correct the CSF level for the plasma contribution, because free MHPG readily crosses the blood-brain barrier (Kopin et al., 1983a). Patients with MSA have low total CSF MHPG, but normal plasma levels of the metabolite (Polinsky, Jimerson and Kopin, 1984). Thus, the component of CSF MHPG due to central NA metabolism is reduced in MSA, consistent with central nervous system noradrenergic dysfunction. This contrasts with the findings in PAF where the diminished total CSF MHPG level results from a decreased plasma contribution, a reflection of peripheral sympathetic neuronal involvement.

Two other monoamine systems, dopamine and serotonin, are particularly relevant in MSA since they play important roles in extrapyramidal function and blood pressure regulation. Homovanillic acid (HVA) and 5-hydroxyindoleacetic acid (5-HIAA) are the primary CSF metabolites of dopamine and serotonin respectively. Both metabolites are reduced in patients with MSA (Polinsky et al., 1988a). Those patients with the most severe signs of Parkinsonism have the lowest CSF levels of HVA. However, HVA is lower than normal even in those MSA patients who have a purely cerebellar syndrome (OPCA variant). The lack of clinical evidence of Parkinsonism in this group may be a function of severity, since it has been estimated that 80% of neuronal function must be impaired before symptoms appear in Parkinson's disease. Furthermore, there are central dopaminergic pathways that do not participate in motor control. The increase in CSF HVA following administration of Sinemet® (Polinsky et al., 1988a) and decease in response to bromocryptine (Williams *et al.*, 1979b) in patients with MSA suggest that some dopamine synthesis, release, metabolism and receptor function may remain. This could account for the therapeutic benefit from anti-Parkinsonian medications in the early and middle stages of the illness. Serotonergic involvement is not likely related to the motor dysfunction, since all clinical MSA subgroups manifest similar reductions in 5-HIAA. In addition, the highest concentrations of serotonin in the central nervous system are found in areas concerned with vigilance, blood pressure and other vegetative functions.   The origin of CSF 5-HIAA is less clearly defined than for HVA. A substantial spinal cord contribution has been suggested.

Relatively few studies of peptides in the CSF have been performed in patients with autonomic failure. The lumbar CSF levels of substance P (Nutt *et al.*, 1980), somatostatin (Polinsky, Hooper and Baser, 1989c), and corticotropin releasing factor are lower than normal in patients with MSA. These peptides appear to function as neurotransmitters or neuromodulators in the nervous system, but their widespread regional distributions preclude correlation with specific clinical and pathological aspects of the disease.   There appears to be little correlation between reductions in CSF peptide concentrations and decreases in indices of other transmitter systems in MSA, suggesting that they may be independently affected by the degenerative process.

Cerebral glucose metabolism has been investigated using the PET-FDG method in patients with MSA as well as two related disorders, striatonigral degeneration and OPCA. There appears to be a generalized reduction in glucose utilization; however, the hypometabolism is most severe in the cerebellum, brainstem, striatum and frontal and motor cortices (Figure 6.7) (Fulham *et al.*, 1991). In the study by De Volder *et al.* (1989), two of the patients had autonomic failure; interestingly, cerebellar hypometabolism was only observed in these two individuals who appear likely to have had MSA rather than striatonigral degeneration.   One concern regarding the application of PET technology to the investigation of degenerative disorders relates to partial volume effects, i.e.   apparent reductions in glucose metabolism that result solely from atrophy.   This does not appear to be the case in the MSA studies, since hypometabolism occurred even in patients without structural changes on CT or MRI scanning (Gilman *et al.*, 1988; Fulham *et al.*, 1991).   More recently, PET investigations have focused on the dopaminergic system.   A reduction of ^{18}F-fluorodopa uptake was observed in two of three MSA patients (Bhatt *et al.*, 1990). The patient with normal uptake had a short duration of mild Parkinsonism.   Brooks *et al.*   (1990) assessed striatonigral function using ^{18}F-dopa and S-^{11}C-nomifensine.   The latter compound binds to dopamine reuptake sites in the thalamus.   Influx constants for ^{18}F-dopa were reduced in the caudate and putamen in patients with MSA but not PAF. The putaminal change correlated with locomotor disability.   A similar putaminal abnormality was observed in patients with Parkinson's disease; however, caudate function appeared normal.   The striatal abnormality was also attended by a reduction in nomifensine binding, consistent with a loss of nigrostriatal nerve terminals.

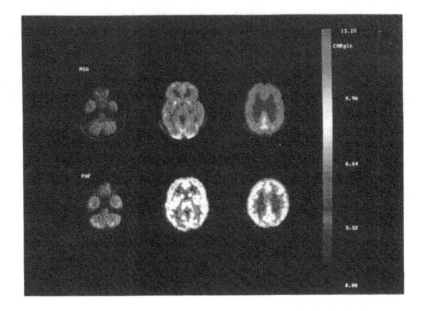

FIGURE 6.7   Normal FDG-PET scan in a patient with PAF contrasts with hypometabolism observed in MSA. See text for detailed observations. MSA = multiple system atrophy, CMRglc = glucose cerebral metabolic rate.

PATHOLOGICAL BASIS OF THE DISEASE

Histological studies have contributed substantially to the development of suitable schemes for classification and clinico-pathological correlation in autonomic failure. Early reports concentrated on descriptive neuropathology. In parallel with the general trends in neurology, these reports were followed by efforts to relate clinical symptoms to morphological abnormalities. The transition from quantitative morphometric studies to neurochemical measurements was facilitated by many technological advances that found application in the neurosciences. Despite significant progress in our understanding of pathophysiology, relatively little achievement is evident regarding the etiology of MSA.

MORPHOLOGICAL CHANGES

Since the first neuropathological description of the findings in a patient with multiple system atrophy attended by autonomic failure (Shy and Drager, 1960), there have been more than 50 cases reported in the literature. Most cases have added little to the original report (for review see Oppenheimer, 1988). The degenerative process is characterized by neuronal loss and gliosis in many central nervous system areas

TABLE 6.4
Neuropathological involvement in multiple system atrophy

Caudate

Putamen

Substantia nigra

Cerebellar cortex

Inferior olives

Anterior horn cells

Pyramidal tract

Locus coeruleus

Dorsal vagal nucleus

Edinger–Westphal nucleus

Intermediolateral column

Onuf's Nucleus

(Table 6.4). The non-specific nature of these morphological changes highlights the importance of documenting their geographic distribution to establish the diagnosis. A relationship between OPCA and SND became evident relatively soon after the neuropathological changes in the latter entity were described. The lesions in OPCA often involve the striatonigral and pyramidal systems. Although the findings in SND resemble Parkinson's disease, Lewy bodies are notably absent. In 1969, Graham and Oppenheimer proposed that the term MSA be used to encompass SND and OPCA. The relationship between MSA and PAF is less clear; those cases labelled as PAF had a short duration of illness, central neurological signs or other atypical neurological features. Neuropathological investigation in an unequivocal case of PAF has not been performed.

The Lewy body has been used to classify patients with autonomic failure. Since Lewy bodies were found in presumed PAF cases, PAF has been lumped together with those patients who have Parkinson's with autonomic insufficiency. On this basis, Oppenheimer (1983) has proposed that PAF is a "form fruste" of Parkinson's disease. The significance of Lewy bodies is unclear because they occur in normals (Forno, 1966) and less frequently in MSA and other neurological disorders (Perry, Irving and Tomlinson, 1990).

Although Shy and Drager (1960) first described the condition subsequently named after them, Johnson *et al.* (1966) were the first authors to attribute the autonomic insufficiency to loss of preganglionic neurons in ILC of the spinal cord. Quantitative studies demonstrate a dramatic reduction in the number of ILC neurons (Low, Thomas and Dyck, 1978; Oppenheimer, 1980). Most importantly, the investigators compared cell counts between patients and normal controls. Of particular interest, however, is the study by Gray, Vincent and Hauw, (1988) who did not find a correlation between cell loss in the lateral horns and severity of autonomic failure. Comparable quantitative investigations of parasympathetic neurons are lacking, though cell loss in the dorsal vagal nuclei has been reported. Sacral parasympathetic outflow, i.e. Onuf's nucleus, also appears to be affected by the degenerative process in MSA (Konno *et al.*, 1986; Chalmers and Swash, 1987).

The question of whether sympathetic ganglia are primarily involved has not been adequately answered. Mild cell loss has been described, but the counts have not been compared with control material. Parasympathetic ganglia have not been examined. Although the title suggests that the patients in the report by Petito and Black (1978) had PAF, at least two (and probably three) had clinical findings consistent with MSA. They observed a number of abnormalities in sympathetic ganglia, including focal neuronal phagocytosis, increased satellite cells and perivascular lymphocytic infiltrates. Unmyelinated axons were abnormal by electron microscopy. Tohgi *et al.* (1982) also found evidence of unmyelinated fibre degeneration in MSA. These observations are consistent with the occasional clinical signs of peripheral nerve involvement. Particularly relevant to the respiratory stridor is the selective atrophy of the posterior cricoarytenoid muscles; the histological changes suggest a myopathic basis for this aspect of MSA (Bannister *et al.*, 1981b).

NEUROCHEMICAL ABNORMALITIES

Despite the large number of pathological studies in MSA, very few have employed neurochemical techniques to further characterize the involvement of various neuronal systems in MSA. One of the earliest studies employed measurements of biochemical markers in sympathetic ganglia. Petito and Black (1978) found decreased dopamine β-hydroylase but normal choline acetyltransferase in sympathetic ganglia. These observations imply postganglionic involvement, since the enzymes are respectively markers of post- and presynaptic neurons in ganglia. Unfortunately, interpretation of this study is complicated by the uncertain clinical aspects discussed above.

Four MSA brains were studied neurochemically by Spokes, Bannister and Oppenheimer (1979). The levels of NA and dopamine were reduced in several brain regions, including the striatum, nucleus accumbens, substantia nigra, septal nuclei, hypothalamus and locus coeruleus. Quik *et al.* (1979) also found that substantia nigra binding sites for dopamine were decreased in MSA. The differentiating feature relative to Parkinson's disease is the more marked depletion of NA in MSA. Regional measurements of choline acetyltransferase are more complex; striatal changes in MSA are similar to Parkinsonian subjects. The inconsistency of these findings raises doubts about their pathophysiological significanance. However, low enzyme activity (choline acetyltransferase) in the hypothalamus, dentate, red nucleus, pons and olive parallel some of the sites exhibiting morphologic signs of degeneration. Although glutamic acid decarboxylase is generally low in MSA brains, interpretation is hampered by lability of the enzyme which is greatly affected by pre- and post-mortem factors. Two other studies are relevant to MSA. Benzodiazepine receptors are normal in cerebellar cortex, but increased in the dentate nucleus from OPCA brains (Whitehouse *et al.*, 1986). Muscimol binding was diminished in the granule layer. Muscarinic cholinergic receptor density was also lower than normal in the molecular and granule cell layers, but increased in the dentate. Malessa *et al.* (1990) found tyrosine-hydroxylase-immunoreactive neurons markedly depleted in the A1 and A2 medullary regions of the brainstem from SND patients. This may be particularly important in relation to blood pressure control. Markedly reduced levels of substance P in the basal ganglia may explain the low CSF levels mentioned previously (Kwak, 1985).

Spinal cord peptides have also been measured (Anand, 1988). Reduced neuropeptide concentrations are most apparent in the dorsal aspect of the cord. This includes substance P, substance K, calcitonin gene-related peptide (CGRP), somatostatin and galanin. Anand *et al.* (1988) also studied the skin flare response in patients with MSA, since substance P and/or CGRP release from unmyelinated sensory fibres may play a role in generating this phenomenon. Histamine-induced skin flares were normal in MSA.

AETIOLOGIC CLUES

The length of this chapter and number of references are testimony to the growing interest and increased knowledge in this disorder. Despite substantial progress we have no convincing evidence to implicate a particular aetiologic mechanism in MSA. Although various inherited patterns of OPCA are well-known, only rarely have familial cases of MSA been reported (Lewis, 1964). Bannister, Mowbray and Sidgwick, (1983) found a significant association between HLA-Aw32 and autonomic failure, suggesting a genetic contribution to the aetiology. However, in a larger study Nee, Brown and Polinsky, (1989) failed to confirm the earlier observations. Another potential mechanism for neuronal degeneration is deficient DNA repair. Cultured lymphoblasts from MSA patients show reduced survival after exposure to mutagenic agents (Robbins *et al.*, 1981). The significance of these findings is not clear, a similar abnormality occurs in several other primary neuronal degenerations. An immunological basis for the degenerative process is suggested by the presence of CSF antibodies in MSA that react specifically with the locus ceruleus, a noradrenergic nucleus consistently affected in the disorder (Polinsky *et al.*, 1990). Further investigation is required to determine whether this is a primary abnormality or an epiphenomenon. A case-control study of 60 patients with MSA revealed that the patients had more exposures to a variety of toxins, generally through occupational exposure (Nee *et al.*, 1991). In addition, clinical symptoms related to MSA were reported more frequently by patients' relatives than control subjects. Nee *et al.* (1991) proposed that MSA might result from a genetically determined selective vulnerability in the nervous system. A larger study will be required to further explore this hypothesis.

CONCLUSION

The Shy–Drager syndrome is a devastating degenerative neurological disorder that severely compromises the quality of life and significantly shortens the lifespan of affected individuals. Improved understanding of pathophysiology has led to many rational therapeutic approaches for treating specific aspects of the disorder. Treatment of autonomic insufficiency is discussed in another chapter of this book. There are, however, several aspects of management that are particularly relevant for care in MSA.

Despite the widespread central nervous system involvement, some of the neurological manifestations can be improved. Rigidity, bradykinesia and tremor do respond to anti-Parkinsonian drugs, although not as completely as in patients with idiopathic Parkinson's disease. Unfortunately, dopaminergic drugs must generally be avoided because they exacerbate the postural hypotension. Anticholinergic drugs seem to work equally well without affecting the blood pressure. This class of medications promotes urinary retention, but many patients view the side-effect as a benefit. There is no effective therapy for the cerebellar dysfunction.

Another consideration relates to anaesthetic management. As mentioned earlier, most patients lack the adrenal medullary response that may be critical during surgery or traumatic injuries. In addition, respiratory control and sensitivity to centrally acting drugs such as sedatives and analgesics may predispose patients to catastrophic reactions. Furthermore, abnormal temperature control may alter their response to infection. In general, patients should avoid surgery. When necessary, however, it is important for the surgeon and anaesthetist to become familiar with the manifestations and management of various aspects of MSA.

REFERENCES

Anand, P. (1988) Neuropeptides in the spinal cord in multiple system atrophy. In *Autonomic Failure*, edited by R. Bannister, pp. 511–520. London: Oxford University Press.

Anand, P., Bannister, R., McGregor, G.P., Ghatei, M.A., Mulderry, P.K. and Bloom, S.R. (1988) Marked depletion of dorsal spinal cord substance P and calcitonin gene-related peptide with intact skin flare responses in multiple system atrophy. *J. Neurol. Neurosurg. Psych.*, **51**, 192–196.

Bannister, R., Ardill, L. and Fentem, P. (1967) Defective autonomic control of blood vessels in idiopathic orthostatic hypotension. *Brain*, **90**, 725–746.

Bannister, R., Ardill, L. and Fentem, P. (1969) An assessment of various methods of treatment of idiopathic orthostatic hypotension. *Quart. J. Med.*, **38**, 377–395.

Bannister, R., Boylston, A.W., Davies, I.B., Mathias, C.J., Sever, P.S. and Sudera, D. (1981a) Beta-receptor numbers and thermodynamics in denervation supersensitiviy. *J. Physiol. Lond.*, **319**, 369–377.

Bannister, R., Crowe, R., Eames, R., Rosenthal, T. and Sever, P. (1979) Defective cardiovascular reflexes and supersensitivity to sympathomimetic drugs in autonomic failure. *Brain*, **102**, 163–176.

Bannister, R., Gibson, W., Michaels, L. and Oppenheimer, D.R. (1981b) Laryngeal abductor paralysis in multiple system atrophy: a report on three necropsied cases, with observations on the laryngeal muscles and the nuclei ambigui. *Brain*, **104**, 351–368.

Bannister, R. and Mathias, C. (1988) Testing autonomic reflexes. In *Autonomic Failure*, edited by R. Bannister, pp. 289–307. London: Oxford University Press.

Bannister, R., Mathias, C. and Polinsky, R. (1988) Clinical features of autonomic failure: a comparison between UK and US experience. In *Autonomic Failure*, edited by R. Bannister, pp. 281–288. London: Oxford University Press.

Bannister, R., Mowbray, J. and Sidgwick, A. (1983) Genetic control of progressive autonomic failure: evidence for an association with an HLA antigen. *Lancet*, **1**, 1017.

Bannister, R., Sever, P. and Gross, M. (1977) Cardiovascular reflexes and biochemical responses in progressive autonomic failure. *Brain*, **100**, 327–344.

Baser, S.M., Brown, R.T., Curras, M.T., Baucom, C.E., Hooper, D.R. and Polinsky, R.J. (1991) Beta-receptor sensitivity in autonomic failure. *Neurology*, **41**, 1107–1112.

Bhatt, M.H., Snow, B.J., Martin, W.R.W., Cooper, S. and Calne, D.B. (1990) Positron emission tomography in Shy-Drager syndrome. *Ann. Neurol.*, **28**, 101–103.

Bradbury, S. and Eggleston, C. (1925) Postural hypotension: a report of three cases. *Am. Heart J.*, **1**, 73–86.

Briebach, T., Laubenberger, J. and Fischer, P.A. (1989) Transcranial Doppler sonographic studies of cerebral autoregulation in Shy-Drager syndrome. *J. Neurol.*, **236**, 349–350.

Brooks, D.J., Salmon, E.P., Mathias, C.J., Quinn, N., Leenders, K.L., Bannister, R., *et al.* (1990) The relationship between locomotor disability, autonomic dysfunction, and the integrity of the striatal dopaminergic system in patients with multiple system atrophy, pure autonomic failure, and Parkinson's disease, studied with PET. *Brain*, **113**, 1539–1552.

Brown, R.T., Polinsky, R.J. and Baucom, C.E. (1989) Euglycemic insulin-induced hypotension in autonomic failure. *Clin. Neuropharmacol.*, **12**, 227–231.

Brown, R.T., Polinsky, R.J., Di Chiro, G., Pastakia, B., Wener, L. and Simmons, J.T. (1987) MRI in autonomic failure. *J. Neurol. Neurosurg. Psych.*, **50**, 913–914.

Brown, R.T., Polinsky, R.J., Lee, G.K. and Deeter, J.A. (1986) Insulin-induced hypotension and neurogenic orthostatic hypotension. *Neurology*, **36**, 1402–1406.

Cannon, W.B., McIver, M.A. and Bliss, S.W. (1924) Studies on the conditions of activity in endocrine glands. XIII. A sympathetic and adrenal mechanism for mobilizing sugar in hypoglycemia. *Am. J. Physiol.*, **69**, 46–66.

Chalmers, D. and Swash, M. (1987) Selective vulnerability of urinary Onuf motoneurons in Shy-Drager syndrome. *J. Neurol.*, **234**, 259–260.

Coccagna, G., Martinelli, P., Zucconi, M., Cirignotta, F. and Ambrosetto, G. (1985) Sleep-related respiratory and haemodynamic changes in Shy-Drager syndrome: a case report. *J. Neurol.*, **232**, 310–313.

Cohen, J., Low, P., Fealey, R., Sheps, S. and Jiang, N.-S. (1987) Somatic and autonomic function in progressive autonomic failure and multiple system atrophy. *Ann. Neurol.*, **22**, 692–699.

Davies, B., Sudera, D., Sagnella, G., Marchesi-Saviotti, E., Mathias, C., Bannister, R., *et al.* (1982) Increased numbers of alpha receptors in sympathetic denervation supersensitvity in man. *J. Clin. Invest.*, **69**, 779–784.

Davis, R., Slater, J.D.H., Forsling, M.L. and Payne, N. (1976) The reponse of arginine vasopressin and plasma renin to postural change in normal man, with observations on syncope. *Clin. Sci. Mol. Med.*, **51**, 267–274.

DeVolder, A.G., Francart, J., Laterre, C., Dooms, G., Bol, A., Michel, C., *et al.* (1989) Decreased glucose utilization in the striatum and frontal lobe in probable striatonigral degeneration. *Ann. Neurol.*, **26**, 239–247.

Dotson, R., Ochoa, J., Marchettini, P. and Cline, M. (1990) Sympathetic neural outflow directly recorded in patients with primary autonomic failure: clinical observations, microneurography, and histopathology. *Neurology*, **40**, 1079–1085.

Esler, M., Jackman, G., Kelleher, D., Skews, H., Jennings, G., Bobik, A., *et al.* (1980) Norepinephrine kinetics in patients with idiopathic autonomic insufficiency. *Circ. Res.*, **46**(Suppl 1), 47–48.

Forno, L.S. (1966) Pathology of Parkinsonism; a preliminary report of 24 cases. *J. Neurosurg.*, **24**, 266–271.

Fulham, M.J., Dubinsky, R.M., Polinsky, R.J., Brooks, R.A., Brown, R.T., Curras, M.T., *et al.* (1991) Computerised tomography, magnetic resonance imaging and positron emission tomography with [^{18}F] fluorodeoxyglucose in mutiple system atrophy and pure autonomic failure. *Clin. Auton. Res.*, **1**, 27–36.

Garber, A.J., Cryer, P.E., Santiago, J.V., Haymond, M.W., Pagliara, A.S. and Kipnis, D.M. (1976) The role of adrenergic mechanisms in the substrate and hormonal response to insulin-induced hypoglycemia in man. *J. Clin. Invest.*, **58**, 7–15.

Gilman, S., Markel, D.S., Koeppe, R.A., Junck, L., Kluin, K.J., Gebarski, S.S., *et al.*, (1988) Cerebellar and brainstem hypometabolism in olivopontocerebellar atrophy detected with Positron Emission Tomography. *Ann. Neurol.*, **23**, 223–230.

Gilmartin, J.J., Wright, A.J., Cartlidge, N.E.F. and Gibson, G.J. (1984) Upper airway obstruction complicating the Shy-Drager syndrome. *Thorax*, **39**, 313–314.

Goldstein, D.S., Polinsky, R.J., Garty, M., Robertson, D., Brown, R.T., Biaggioni, I., *et al.*, (1989) Patterns of plasma levels of catechols in neurogenic orthostatic hypotension. *Ann. Neurol.*, **26**, 558–563.

Goldstein, D.S., Udelsman, R., Eisenhofer, G., Stull, R., Keiser, H.R. and Kopin, I.J. (1987) Neuronal source of plasma dihydroxyphenylalanine. *J. Clin. Endocrinol. Metab.*, **64**, 856–861.

Graham, J.G. and Oppenheimer, D.R. (1969) Orthostatic hypotension and nicotine sensitivity in a case of multiple system atrophy. *J. Neurol. Neurosurg. Psych.*, **32**, 28–34.

Gray, F., Vincent, D. and Hauw, J.J. (1988) Quantitative study of lateral horn cells in 15 cases of multiple system atrophy. *Acta. Neuropathol.*, **75**, 513–518.

Guilleminault, C., Tilkian, A., Lehrman, K., Forno, L. and Dement, W.C. (1977) Sleep apnea syndrome: states of sleep and autonomic dysfunction. *J. Neurol. Neurosurg. Psych.*, **40**, 718–725.

Guindi, G.M., Bannister, R., Gibson, W.P.R. and Payne, J.K. (1981) Laryngeal electromyography in multiple system atrophy with autonomic failure. *J. Neurol. Neurosurg. Psych.*, **44**, 49–53.

Hoeldtke, R.D., O'Dorisio, T.M. and Boden, G. (1986) Treatment of autonomic neuropathy with a somatostatin analogue SMS-201-995. *Lancet*, **2**, 602–605.

Johnson, R.H., Lee, G.deJ., Oppenheimer, D.R. and Spalding, J.M.K. (1966) Autonomic failure with orthostatic hypotension due to intermediolateral column degeneration. *Quart. J. Med.*, **35**, 276–292.

Kafka, M.S., Polinsky, R.J., Williams, A., Kopin, I.J., Lake, C.R., Ebert, M.H., *et al.* (1984) Alpha-adrenergic receptors in orthostatic hypotension syndromes. *Neurology*, **34**, 1121–1125.

Kaufman, H., Oribe, E., Pierotti, A.R., Roberts, J.L. and Yahr, M.D. (1990) Atrial natriuretic factor in human autonomic failure. *Neurology*, **40**, 1115–1119.

Kenyon, G.S., Apps, M.C.P. and Traub, M. (1984) Stridor and obstructive sleep apnea in Shy-Drager syndrome treated by laryngofissure and cord lateralization. *Laryngoscope*, **94**, 1106–1108.

Kirby, R., Fowler, C., Gosling, J. and Bannister, R. (1986) Urethro-vesical dysfunction in progressive autonomic failure with multiple system atrophy. *J. Neurol. Neurosurg. Psych.*, **49**, 554–562.

Konno, H., Yamamoto, T., Iwasaki, Y. and Iizuka, H. (1986) Shy-Drager syndrome and amyotrophic lateral sclerosis: cytoarchitectonic and morphometric studies of sacral autonomic neurons. *J. Neurol. Sci.*, **73**, 193–204.

Kopin, I.J., Gordon E.K., Jimerson, D.C. and Polinsky, R.J. (1983a) Relationship between plasma and cerebrospinal fluid levels of 3-methoxy-4-hydroxyphenylglycol. *Science*, **219**, 73–75.

Kopin, I.J., Polinsky, R.J., Oliver, J.A., Oddershede I.R. and Ebert M.H. (1983b) Urinary catecholamine metabolites distinguish different types of sympathetic neuronal dysfunction in patients with orthostatic hypotension. *J. Clin. Endocrinol. Metab.*, **57**, 632–637.

Kwak, S. (1985) Biochemical analysis of transmitters in the brains of multiple system atrophy. *No Shinkei*, **37**, 691–694.

Lake, C.R., Ziegler, M.G. and Kopin, I.J. (1976) Use of plasma norepinephrine for evaluation of sympathetic neuronal function in man. *Life Sci.*, **18**, 1315–1326.

Lehrman, K.L., Guilleminault, C., Schroder, J.S., Tilkian, A. and Forno, L.N. (1978) Sleep apnea syndrome in a patient with Shy-Drager syndrome. *Arch. Intern. Med.*, **138**, 206–209.

Lewis, P. (1964) Familial orthostatic hypotension. *Brain*, **47**, 719–728.

Lightman, S.L. and Williams, T.D.M. (1988) Hypothalamic function in autonomic failure. In *Autonomic Failure*, edited by R. Bannister, pp. 381–392. London: Oxford University Press.

Long, R.G., Barnes, A.J., Albuquerque, R.H., Prata, A., Bannister, R., Adrian, T.E., *et al.* (1979) Pancreatic hormone release in Chagas' disease and the Shy-Drager syndrome. *Gut*, **20**, A921.

Low, P.A., Thomas, J.E. and Dyck, P.J. (1978) The splanchnic autonomic outflow in Shy-Drager syndrome and idiopathic orthostatic hypotension. *Ann. Neurol.*, **4**, 511–514.

Luft, F. and von Euler, U. (1953) Two cases of postural hypotension showing a deficiency in release of norepinephrine and epinephrine. *J. Clin. Invest.*, **32**, 1065–1069.

Malessa, S., Hirsch, E.C., Cervera, P., Duyckaerts, C. and Agid, Y. (1990) Catecholaminergic systems in the medulla oblongata in parkinsonian syndromes: a quantitative immunohistochemical study in Parkinson's disease, progressive supranuclear palsy, and striatonigral degeneration. *Neurology*, **40**, 1739–1743.

Mann, S., Altman, D.G., Raftery, E.B. and Bannister, R. (1983) Circadian variation of blood pressure in autonomic failure. *Circulation*, **68**, 477–483.

Martinelli, P., Coccagna, G., Rizzuto, N. and Lugaresi, E. (1981) Changes in systemic arterial pressure during sleep in Shy-Drager syndrome. *Sleep*, **4**, 139–146.

Mathias, C.J., da Costa, D.F., Fosbraey, P., Bannister, R. and Christensen, N.J. (1986) Post-cibal hypotension in autonomic failure. In *The Sympatho-adrenal System*, edited by N.J. Christensen, O. Henriksen and N.A. Lassen, Alfred Benson Symposium, No. 23, pp. 402–413. Copenhagen: Munksgaard.

Mathias C.J., da Costa, D.F., Fosbraey, P., Christensen, N.J. and Bannister, R. (1987) Hypotensive and sedative effects of insulin in autonomic failure. *Br. Med. J.*, **295**, 161–163.

Mathias, C.J., Matthews, W.B. and Spalding, J.M.K. (1977) Postural changes in plasma renin activity and responses to vasoactive drugs in a case of Shy-Drager syndrome. *J. Neurol. Neurosurg. Psych.*, **40**, 138–143.

Möhring, J., Glänzer, K., Maciel, Jr., J.A., Düsing, R., Kramer, H.J., Arbogast, R., *et al.* (1980) Greatly enhanced pressor response to antidiuretic hormone in patients with impaired cardiovascular reflexes due to idiopathic orthostatic hypotension. *J. Cardiovasc. Pharmacol.*, **2**, 367–376.

Munschauer, F.E., Loh, L., Bannister, K. and Newsom-Davis, J. (1990) Abnormal respiration and sudden death during sleep in multiple system atrophy with autonomic failure. *Neurology*, **40**, 677–679.

Nee, L.E., Brown, R.T. and Polinsky, R.J. (1989) HLA in autonomic failure. *Arch. Neurol.*, **46**, 758–759.

Nee, L.E., Gomez, M.R., Dambrosia, J., Bale, S., Eldridge, R.J. and Polinsky, R.J. (1991) Environmental-occupational risk factors and familial associations in multiple system atrophy: a preliminary investigation. *Clin. Auton. Res.*, **1**, 9–13.

Nutt, J.G., Mroz, E.A., Leeman, S.E., Williams, A.C., Engel, W.K. and Chase, T.N. (1980) Substance P in human cerebrospinal fluid: reductions in peripheral neuropathy and autonomic dysfunction. *Neurology*, **30**, 1280–1285.

Oppenheimer, D. (1980) Lateral horn cells in progressive autonomic failure. *J. Neurol. Sci.*, **46**, 393–404.

Oppenheimer, D. (1983) Neuropathology of progressive autonomic failure. In *Autonomic Failure*, edited by R. Bannister, pp. 267–283. New York: Oxford University Press.

Oppenheimer D. (1988) Neuropathology and neurochemistry of autonomic failure. A. Neuropathology of autonomic failure. In *Autonomic Failure*, edited by R. Bannister, pp. 451–463. London: Oxford University Press.

Parks, V.J., Sandison, A.G., Skinner, S.L. and Whelan, R.J. (1961) Sympathomimetic drugs in orthostatic hypotension. *Lancet*, **1**, 1133–1136.

Pastakia, B., Polinsky, R., Di Chiro, G., Simmons, J.T., Brown, R. and Wener, L. (1985) Multiple system atrophy (Shy-Drager Syndrome): MR imaging. *Radiology*, **159**, 499–502.

Perry, R.H., Irving, D. and Tomlinson, B.E. (1990) Lewy body prevalence in the aging brain: relationship to neuropsychiatric disorders, Alzheimer-type pathology and catecholaminergic nuclei. *J. Neurol. Sci.*, **100**, 223–233.

Petito, C.K. and Black, I.B. (1978) Ultrastructure and biochemistry of sympathetic ganglia in idiopathic orthostatic hypotension. *Ann. Neurol.*, **4**, 6–17.

Polinsky, R.J., Brown, R.T., Burns, R.S., Harvey-White, J. and Kopin, I.J. (1988a) Low lumbar CSF levels of homovanillic acid and 5-hydroxyindoleacetic acid in multiple system atrophy with autonomic failure. *J. Neurol. Neurosurg. Psych.*, **51**, 914–919.

Polinsky, R.J., Brown, R.T., Lee G.K., Timmers, K., Culman, J., Foldes, O., *et al.* (1987) Beta-endorphin, ACTH, and catecholamine responses in chronic autonomic failure. *Ann. Neurol.*, **21**, 573–577.

Polinsky, R.J., Curras, M.T., Baucom, C.E., Morse, K.L. and Brown, R.T. (1989a) Plasma gastrin responses to isoproterenol in patients with autonomic failure. *J. Auton. Nerv. System*, **27**, 88–89.

Polinsky, R.J., Goldstein, D.S., Brown, R.T., Keiser, H.R. and Kopin, I.J. (1985) Decreased sympathetic neuronal uptake in idiopathic orthostatic hypotension. *Ann. Neurol.*, **18**, 48–53.

Polinsky, R.J., Holmes, K.V., Brown, R.T. and Weise, V. (1989b) CSF acetylcholinesterase levels are reduced in multiple system atrophy with autonomic failure. *Neurology*, **39**, 40–44.

Polinsky, R.J., Hooper, D. and Baser, S.M. (1989c) Reduced CSF somatostatin in multiple system atrophy with autonomic failure. *Neurology*, **39**(Suppl 1), 142.

Polinsky, R.J., Jimerson, D.C. and Kopin, I.J. (1984) Chronic autonomic failure: CSF and plasma 3-methoxy-4-hydroxyphenylglycol. *Neurology*, **34**, 979–983.

Polinsky, R.J., Kopin, I.J., Ebert, M.H. and Weise, V. (1980) The adrenal medullary response to hypoglycemia in patients with orthostatic hypotension. *J. Clin. Endocrinol. Metab.*, **51**, 1401–1406.

Polinsky, R.J., Kopin, I.J., Ebert, M.H. and Weise, V. (1981a) Pharmacologic distinction of different orthostatic hypotension syndromes. *Neurology*, **31**, 1–7.

Polinsky, R.J., Kopin, I.J., Ebert, M.H., Weise, V. and Recant, L. (1981b) Hormonal responses to hypoglycemia in orthostatic hypotension patients with adrenergic insufficiency. *Life Sci.*, **29**, 417–425.

Polinsky, R.J., McRae, A., Baser, S.M. and Dahlstrom, A. (1990) Antibody in the CSF of patients with multiple system atrophy reacts specifically with rat locus ceruleus. *Neurology*, **40**(Suppl. 1), 401.

Polinsky, R.J. and Nee, L.E. (1989) Autonomic failure: nosology, clinical evolution, and prognosis. *Ann. Neurol.*, **26**, 121.

Polinsky, R.J., Taylor, I.L., Chew, P., Weise, V. and Kopin, I.J. (1982) Pancreatic polypeptide responses to hypoglycemia in chronic autonomic failure. *J. Clin. Endocrinol. Metab.*, **54**, 48–52.

Polinsky, R.J., Taylor, I.L., Weise, V. and Kopin, I.J. (1988b) Gastrin responses in patients with adrenergic insufficiency. *J. Neurol. Neurosurg. Psych.*, **51**, 67–71.

Polinsky, R.J., Zamir, N., Baucom, C. and Kopin, I.J. (1989d) Atrial natriuretic peptide in autonomic failure. *Neurol (India)*, **37**(Suppl.), 148.

Prasher, D. and Bannister, R. (1986) Brainstem auditory evoked potentials in patients with multiple system atrophy with progressive autonomic failure (Shy-Drager syndrome). *J. Neurol. Neursurg. Psych.*, **49**, 278–289.

Quik , M., Spokes, E.G., Mackay, A.V.P. and Bannister, R. (1979) Alterations in [^{3}H] spiperone binding in human caudate nucleus, substantia nigra and frontal cortex in the Shy-Drager syndrome and Parkinson's disease. *J. Neurol. Sci.*, **43**, 429–437.

Robbins, J.H., Moshell, A.N., Scarpinata, R.G., Polinsky, R.J., Nee, L.E. and Tarone, R.E. (1981) Hypersensitivity to ionizing radiation in sporadic primary neuronal degenerations. *Clin. Res.*, **29**, 669.

Robertson, D., Wade, D. and Robertson, R.M. (1981) Postprandial alterations in cardiovascular hemodynamics in autonomic dysfunctional states. *Am. J. Cardio.*, **48**, 1048–1052.

Rutledge, J.N., Hilal, S.K., Silver, A.J., Defendini, R. and Fahn, S. (1987) Study of movement disorders and brain iron by MR. *AJR*, **149**, 365–379.

Salinas, J.M., Berger, Y., De La Rocha, R.E. and Blaivas, J.G. (1986) Urological evaluation in the Shy-Drager syndrome. *J. Urol.*, **135**, 741–743.

Savoiardo, M., Strada, L., Girotti, F., D'Incerti, L., Sberna, M., Soliveri., *et al.* (1989) MR imaging in progressive supranuclear palsy and Shy-Drager syndrome. *J. Comput. Assist. Tomogr.*, **13**, 555–560.

Shy, G.M. and Drager, G.A. (1960) A neurological syndrome associated with orthostatic hypotension. *Arch. Neurol.*, **2**, 511–527.

Silverberg, R., Naparstek, Y., Lewis, B.S. and Levy, M. (1979) Angina pectoris with normal coronary arteries in Shy-Drager syndrome. *J. Neurol. Neurosurg. Psych.*, **42**, 910–913.

Singh, N. and Fahn, S. (1980) Electrophysiologic studies in Shy-Drager syndrome. *Neurology*, **30**, 394.

Skeggs, L.T., jr. (1984) Historical overview of the renin-angiotensin system. In *Hypertension and the Angiotensin System: Therapeutic Approaches*, edited by A.E. Doyle and A.G. Bearn, pp. 31–45. New York: Raven Press.

Spokes, E.G.S., Bannister, R. and Oppenheimer, D.R. (1979) Multiple system atrophy with autonomic failure: clinical, histological and neurochemical observations on four cases. *J. Neurol. Sci.*, **43**, 59–82.

Teräväinen, H. and Udd, B. (1982) Vocal cord paralysis in the Shy-Drager syndrome. *Acta. Neurol. Scand.*, **66**, 505–507.

Tetsuo, M., Polinsky, R.J., Markey, S.P. and Kopin, I.J. (1981) Urinary 6-hydroxymelatonin excretion in patients with orthostatic hypotension. *J. Clin. Endocrinol. Metab.*, **53**, 607–610.

Thomas, D.J. and Bannister, R. (1980) Preservation of autoregulation of cerebral blood flow in autonomic failure. *J. Neurol. Sci.*, **44**, 205–212.

Tohgi, H., Tabuchi, M., Tomonaga, M. and Izumiyama, N. (1982) Selective loss of small myelinated and unmyelinated fibers in Shy-Drager syndrome. *Acta. Neuropathol.*, **57**, 282–286.

Trendelenburg, U. (1963) Supersensitivity and subsensitivity to sympathomimetic amines. *Pharmacol. Rev.*, **15**, 225–276.

Uematsu, D., Hamada, J. and Gotoh, F. (1987) Brainstem auditory evoked responses and CT findings in multiple system atrophy. *J. Neurol. Sci.*, **77**, 161–171.

Wallin, B.G., Sundlöf, G., Eriksson, B.-M., Dominiak, P., Grobecker, H. and Lindblad, L.-E. (1981) Plasma noradrenaline correlates to sympathetic muscle nerve activity in normotensive man. *Acta. Physiol. Scand.*, **111**, 69–73.

Whitehouse, P.J., Muramoto, O., Troncoso, J.C. and Kanazawa, I. (1986) Neurotransmitter receptors in olivopontocerebellar atrophy: an autoradiographic study. *Neurology*, **36**, 193–197.

Wilcox, C.S. and Aminoff, M.J. (1976) Blood pressure responses to noradrenaline and dopamine infusions in Parkinson's disease and the Shy-Drager syndrome. *Br. J. Clin. Pharmacol.*, **3**, 207–214.

Wilcox, C.S., Aminoff, M.J. and Slater J.D.H. (1977) Sodium homeostasis in patients with autonomic failure. *Clin. Sci. Mol. Med.*, **53**, 321–328.

Wilcox, C.S., Aminoff, M.J., Kurtz, A.B. and Slater, J.D.H. (1974) Comparison of the renin response to dopamine and noradrenaline in normal subjects and patients with autonomic insufficiency. *Clin. Sci. Mol. Med.*, **46**, 481–488.

Williams, A., Hanson, D. and Calne, D.E (1979a) Vocal cord paralysis in the Shy-Drager syndrome. *J. Neurol. Neurosurg. Psych.*, **42**, 151–153.

Williams, A.C., Nutt, J., Lake, C.R., Pfeiffer, R., Teychenne, P.E., Ebert, M., *et al.* (1979b) Actions of bromocriptine in the Shy-Drager and Steele-Richardson-Olszewski syndromes. In *Dopaminergic Ergots and Motor Control*, edited by K. Fuxe and D.B. Calne, pp. 271–283. New York: Pergamon Press.

Williams, T.D.M., Lightman, S.L. and Bannister, R. (1985) Vasopressin secretion in progressive autonomic failure: evidence for defective afferent cardiovascular pathways. *J. Neurol. Neurog. Psych.*, **48**, 225–228.

Yoshida, K., Tashiro, K. Matsumoto, A., Maruo, Y., Hamada, K., Hamada, T., *et al.* (1987) An autopsy case of Shy-Drager syndrome preceded by a urinary disturbance for over 20 years. *No To Shinkei*, **39**, 59–64.

Zerbe, R.L., Henry, D.P. and Robertson, G.L. (1983) Vasopressin response to orthostatic hypotension: etiologic and clinical implications. *Am. J. Med.*, **74**, 265–271.

Ziegler, M.G., Lake, C.R. and Kopin, I.J. (1977) The sympathetic-nervous-system defect in primary orthostatic hypotension. *N. Engl. J. Med.*, **296**, 293–297.

7 Peripheral Autonomic Neuropathies

O. Appenzeller

New Mexico Health Enhancement and
Marathon Clinics Research Foundation,
1559 Eagle Ridge, N.E. Albuquerque,
New Mexico 87122, USA

The integrative activity of the Autonomic Nervous System is best recognized when it's maintenance of homeostasis fails. The pathology and clinical syndromes of autonomic failure in peripheral neuropathies are described with emphasis on those with predominant autonomic signs.

KEY WORDS: Peripheral nerves, autonomic function, clinical-pathologic aspects

INTRODUCTION

The recognition of the clinical importance of autonomic accompaniments of peripheral nerve disease has been given impetus by the advent of simple objective tests to assess autonomic nervous system involvement. Many of these tests are now considered reliable, reproducible, simple and quick to carry out and, above all, non-invasive. Such criteria for values of tests of autonomic function are mainly fulfilled by those that test cardiovascular reflexes. Whether these tests reflect damage elsewhere in the autonomic nervous system is not definitively established, but suggestive evidence supports this view. It is of clinical and prognostic importance to assess autonomic damage in peripheral nerve disease, particularly in patients with diabetes mellitus, but the tests may be used to diagnose autonomic dysfunction caused by other disorders.

BASIC PRINCIPLES

The integrative activity of the autonomic nervous system is best seen in syndromes of autonomic failure which provide an opportunity to observe clinically the extensive ramifications of autonomic dysfunction. Although most patients with autonomic failure associated with peripheral nerve disease suffer from chronic autonomic insufficiency, there are some rare exceptions in which acute devastation of autonomic functions occurs and the normal adaptability of the nervous system has not yet restored partial function (Appenzeller, 1990).

AUTONOMIC ANATOMY

Autonomic functions are regulated by peripheral and central neurons comprising the autonomic nervous system.

The autonomic nervous system includes all neurons lying outside the central nervous system that innervate viscera, except the sensory neurons in the posterior root ganglia and in some cranial nerve ganglia. Spinal cord, brainstem and brain neurons that connect with peripheral autonomic neurons are also part of the autonomic nervous system, which is then subdivided functionally into sympathetic and parasympathetic divisions.

The endocrine system, which regulates and releases hormones, is also included in the autonomic nervous system by some, and the hypothalamus or "head ganglion" of the autonomic nervous system has a major role in endocrine function. The autonomic dysfunction discussed here relates to autonomic neurons lying outside the central nervous system. Those disorders of central autonomic pathways usually associated with other chronic neurological syndromes are not included.

ACUTE DISORDERS OF AUTONOMIC FUNCTION

Autonomic dysfunction may be acute or chronic. In acute pandysautonomia, autonomic activity throughout the body ceases acutely. Symptoms and signs include postural hypotension, fixed heart rate, absent sweating, dry eyes, nasal and oral mucus membranes, mid-position non-reactive pupils, absent bowel motility and hypotonic bladder. Laboratory studies are normal in this condition. The disease is self-limiting, at least symptomatically, and most patients eventually recover. The glucose tolerance curve may be diabetic and protein may be elevated in the cerebrospinal fluid. Mental function, extra-ocular motility, muscle strength and coordination, sensation, and deep tendon and superficial reflexes remain normal. The condition has been attributed to an immunological insult against peripheral unmyelinated fibres. In the chronic stage of the disorder when clinical recovery is already present, nerve biopsies show many smaller-than-normal unmyelinated axons, suggesting regeneration of such fibres (Appenzeller and Kornfeld, 1973).

CHRONIC DISORDERS OF AUTONOMIC FUNCTION

Cardiovascular deconditioning is an example of chronic, autonomic dysfunction and occurs in persons exposed to prolonged weightlessness or in patients bedridden for time. Symptoms, primarily dizziness, may appear after resumption of normal daily activities and often take several days to disappear.

An affection of autonomic function, perhaps not exclusively related to the peripheral autonomic nervous system, occurs after drug therapy that interferes with autonomic activity. Examples include postural hypotension associated with antihypertensive therapy and the hyperthermia resulting from drugs that interfere with temperature regulation (sweating and vasomotor function).

SECONDARY DISORDERS OF AUTONOMIC FUNCTION

These are associated with other diseases, but symptoms of autonomic failure may overshadow those caused by the underlying condition. Examples include diabetic neuropathy with severe postural hypotension, impotence, sweating abnormalities and changes in pupillary reactivity. Similar changes plus sphincter dysfunction occur in patients with amyloid neuropathy.

POSITIVE AUTONOMIC PHENOMENA

Autonomic dysfunction can also manifest with positive phenomena. In tetanus, for example, hyperhidrosis accompanies hypertension with tachycardia and, in autonomic hyperreflexia due to spinal cord transection, bouts of hypertension, sweating, bradycardia, pupillary changes and outpouring of catecholamines occur in response to stimuli of bladder filling stretch or peristaltic movements. Alternatively, autonomic paralysis may be associated with compensatory hyperactivity in normally innervated areas; for example, facial hyperhidrosis in patients with diabetic neuropathy and the segmental hyperhidrosis in partial peripheral nerve injuries. Autonomic function may be impaired in the paraspinous area when lung tumours invade the pleura, but hyperhidrosis occurs in such areas associated with tumour invasion of intercostal nerves.

NEGATIVE AUTONOMIC PHENOMENA

These are common and include dizziness due to orthostatic hypotension, heat illness or heat stroke, susceptibility to accidental hypothermia, particularly in the elderly, anhidrosis in parts of the body and sphincter dysfunction.

SPECIAL ASPECTS OF AUTONOMIC DYSFUNCTION

Because of the pervasive influence of the autonomic nervous system on all body systems, its failure affects the whole body and may seriously influence disease of other systems. The autonomic nervous system, in turn, can also compensate to some extent for dysfunction of other systems helping to preserve the body's functional integrity. This compensatory capacity declines with advancing age. Thus, thermoregulation may be reasonably intact under ordinary circumstances in patients with anhidrosis due to skin disease, for example. If the system is additionally stressed by excessive ambient temperatures, the compensatory abilities (vasodilatation) may not be sufficient to maintain normal body temperature. Similarly, with advancing age and decreasing cerebral blood flow secondary to atherosclerosis, slight postural falls in blood pressure (orthostatic hypotension) may result in transient (or permanent) neurological deficits.

In the past, medical thinking has completely separated volitional from autonomic activity. More recently, however, it is clear that volitional changes in autonomic activity are possible with training, resulting in the widespread practice of biofeedback conditioning. Body temperature, heart rate, blood pressure and sphincter control, to name a few, can be changed with this method, but its role in modifying activity and in treating disease is uncertain.

PATHOLOGY OF AUTONOMIC NEUROPATHIES

Autonomic nervous system pathology associated with peripheral neuropathies has been studied extensively in patients with diabetic neuropathy, and specific findings are few. Peripheral autonomic structures show change in a variety of patients with peripheral neuropathy and autonomic dysfunction, and the degree of alteration varies from case to case and at different sites. The major autonomic ganglia, the unmyelinated visceral fibres, the vagus nerve, the smooth muscles and the perivascular innervation of vasa nervorum are principally involved. Abnormal neurons, found scattered throughout the sympathetic ganglia, are considerably larger than normal (giant sympathetic neurons), and often have rounded outlines and a peripheral nucleus; empty spaces are seen where ganglion cells have disappeared (Duchen *et al.*, 1980). Abnormal argyrophilic masses are frequently apposed to ganglion cells in the chain ganglia and occasionally continue along the axons and dendritic processes. Their origin is not known. Lymphocytic infiltrations, macrophages and plasma cells are found along autonomic nerve bundles and ganglia. Unmyelinated axon tangles are occasionally found in Schwann cells in perivascular areas or abdominal organs. These nodules are contiguous with nerve bundles, resemble neuromas histologically and may represent aberrant axonal regeneration.

Cell counts of spinal cord intermediolateral cell columns are reduced in patients with diabetic or alcoholic autonomic neuropathy.

The vagus nerve can be severely demyelinated in patients with diabetic and alcoholic neuropathies, attributed to a common demylinating factor previouly demonstrated in diabetic patients with peripheral nerve disease (Low *et al.* 1975a). Because of the heterogeneity of the findings (both degenerative and inflammatory) many etiologic factors must be considered in peripheral autonomic neuropathy, including toxic and immunological insults.

CARDIAC DISCONNECTION IN AUTONOMIC NEUROPATHIES

Cardiac disconnection has important clinical consequences although transplanted (disconnected) hearts function within limits without central autonomic modulation. Heart transplants usually have a high resting heart rate without sinus arrhythmia because of the absence of vagal tone. During exercise, heart rate increases slowly and peak rates are achieved only after about five minutes; slowing of the heart rate on cessation of activity is comparably delayed. The heart rate changes are attributed to alterations in circulating catecholamines with exercise and not to autonomic activity. The disconnected human heart function depends on the Frank Starling mechanism (the force of contraction is related to the stretching of heart muscle), and it therefore functions well enough, but the lack of central sympathetic drive limits exertional increase in stroke volume, and the absence of vagal tone results in resting tachcardia. Most activities undertaken by heart transplant recipients are performed well, and circulatory needs are met by adjustments of stroke volume and by effects mediated by circulating catecholamines (Anonymous, 1980). Cardiac denervation-disconnection due to autonomic neuropathy, particularly in diabetics where it has been most extensively studied, is a similar situation. Measurement of heart rate variation during deep breathing (sinus arrhythmia) has been used diagnostically as a test of autonomic function. In normal subjects the heart rate varies considerably during deep breathing, but these variations are absent in patients with autonomic neuropathy, indicating parasympathetic damage.

The autonomic control of the heart rate in patients with diabetes has been studied with ambulatory 24-hour electrocardiographic monitoring and also during all-night polysomnography. In insulin dependent diabetics (IDDM), the R–R intervals during sleep were longer than during waking, showing a slower heart rate during sleep, which was not, however, different from controls. However, the standard deviation from the mean, which was similar to controls in those with diabetes, was significantly reduced in diabetics with autonomic neuropathy during sleep but not during wakefulness. This is in keeping with a reduction in heart rate variability found by others during deep breathing tests. There was also a significant decrease in sudden changes in heart rate during sleep in patients with diabetic autonomic neuropathy when compared to controls. These sophisticated and long-duration studies give a good understanding of defective parasympathetic control of heart rate in patients with diabetic autonomic neuropathy.

The phasic changes in heart rate were also related to spontaneous body movements during sleep which were not related to sleep apnea. The tachycardia that occurs in response to body movement may be a reaction to physical effort, although it could also result from an internal arousal stimulus because of its relatively short latency of 8 seconds before the onset of movement. In controls, heart rate increases related to body movements during sleep were very marked, always present, and did not show marked intraindividual variability. Heart rate increases during prolonged body movements in IDDM, however, were significantly reduced (Canal, 1989).

Blackouts, faintness, and dizziness or visual obscuration on standing are frequent complaints of patients with autonomic neuropathy and reflect the effects of postural hypotension on brain perfusion. These symptoms are similar to those of hypoglycemia in insulin-treated diabetics but, if blood pressure is measured during the episodes, the postural hypotension leaves little doubt about the cause of the symptoms.

In managing patients with postural hypotension, one needs to consider the possible deleterious effects on blood pressure of a variety of drugs including hypotensive agents, diuretics, tricyclic antidepressants, phenothiazines, vasodilators and glyceryl trinitrate. Insulin can also aggravate postural hypotension, tentatively attributed to decreased venous return, altered capillary endothelial permeability and reduced plasma volume, or a direct insulin effect on the neurovascular junction. The advent of congestive heart failure or the nephrotic syndrome often improves existing postural hypotension.

Postural hypotension is generally attributed to lesions in the efferent part of the baroreflex arcs and the damaged fibers are thought to be in the sympathetic vasoconstrictors of the splanchnic bed, skeletal muscles and skin. Nevertheless, renal denervation may be associated with a diminished plasma renin response to standing and potentiation of the postural fall in blood pressure. Decreased basal and reflex outpouring of plasma noradrenaline on standing is also contributory. Occasional patients with diabetic autonomic neuropathy have elevated noradrenaline levels on standing and postural hypotension. The mechanism of this paradoxic situation is not known; a decrease in receptor sensitivity has been postulated, similar to that found in the elderly with postural hypotension and elevated catecholamine levels (Palmer and Ziegler, 1978).

RESTING TACHYCARDIA

In patients with postural hypotension a resting heart rate of 90–100 beats per minute has often been found, and rates of up to 130 beats per minute occur (Watkins and Mackay, 1980). The explanation of this tachycardia was deduced from heart rate measurements in those with autonomic blockade or in patients with cardiac transplants, suggesting this to be due to parasympathetic denervation.

CARDIOVASCULAR REFLEX IMPAIRMENT IN AUTONOMIC NEUROPATHY

Valsalva's manoeuvre

Valsalva's manoeuvre is associated with tachycardia and peripheral vasoconstriction during the period of increased intrathoracic pressure which is then followed by an overshoot in the blood pressure and bradycardia after release of the high intrathoracic pressure. Measurement of intra-arterial blood pressure has hitherto been the standard method for assessing responses to Valsalva's manoeuvre, but it is invasive. Heart rate changes are also reliable guides to the hemodynamic events during Valsalva's manoeuvre, are not invasive, and can be used to evaluate the response in those with suspected abnormalities of cardiovascular reflex function.

The standard technique for performance of Valsalva's manoeuvre is to ask the subject to blow into a mouthpiece connnected to a manometer and to hold the pressure at 40 mmHg for 10 s. The mouthpiece has a slight leak to make it difficult to maintain inflation with a closed glottis without the pressure being transmitted to the intrathoracic structures. The subject is connected to an electrocardiogram which records continuously before, during, and after the manoeuvre. A 'Valsalva ratio' has been calculated from the ratio of the longest pulse interval (R-R interval) after the manoeuvre, which reflects the overshoot bradycardia, to the shortest R-R interval during the manoeuver, which reflects the tachycardia during the strain. A Valsalva ratio of 1.21 or greater is normal, 1.11 to 1.20 is borderline, and less than 1.10 is abnormal. Accuracy of this test depends on the patient's effort and cheating is possible, but if this is excluded it gives reliable evidence of reflex abnormalities. Other methods of non-invasive assessment of cardiovascular reflex function in autonomic neuropathy include heart rate variation during deep breathing, during standing with measurements of the standard deviation, during lying with measurements of mean square successive differences, and after a single deep breath. Additional measures of cardiovascular reflex function in autonomic disturbances include heart rate and blood pressure responses to standing and to sustained muscular exercise (Table 7.1).

CATECHOLAMINE LEVELS IN POSTURAL HYPOTENSION

Postural hypotension may be a symptom of several disorders that impede blood pressure maintenance against gravity. The specific mechanisms resulting in postural hypotension should be identified for successful therapy, which is usually independent of the pathogenesis, but is effective if properly directed. Among the many possible causes of postural hypotension are abnormalities in catecholamine release to appropriate stimulation, specifically, noradrenaline. The primary neurogenic component of the circulatory response to orthostasis or exercise is an increase in sympathetic tone mediated by noradrenaline release, and this release may be seriously affected in a variety of patients with peripheral autonomic neuropathies. The finding of low circulating noradrenaline levels sometimes allows an educated guess about the pathogenesis of postural hypotension. For example, in patients with pure autonomic failure (PAF) who have isolated autonomic dysfunction, noradrenaline is markedly depleted in peripheral neurovascular bundles. Such patients do not respond to tyramine, known to release noradrenaline from its stores, with the normally expected rise in blood noradrenaline, and during recumbency noradrenaline levels are low, reflecting the depleted tissue stores.

TABLE 7.1

Some useful tests of autonomic function in man.

Test	Methods	Expected normal response	"Lesion"	Clinical signs or symptoms	Abnormal responses
Postural change in BP; Valsalva's manoeuvre; Lower body negative pressure	Sphygomanometer; intra-arterial BP recording	No change in BP; "overshoot" at end of Valsalva's manoeuvre or after release of negative pressure	Afferent, central or efferent baroreceptor reflex pathways	Symptomatic fall in blood pressure in acute stage, only rarely thereafter	Fall in BP on standing or tilt-up; no "over-shoot"
Radiant heating of trunk	Hand or forearm plethysmography or calorimetry	Vasodilatation, increase in blood flow	Skin receptors; afferents; central structures; efferents or blood vessels	?Sensory loss on heated area	No increase in blood flow
Application to cold feet	Plethysmography of hand; BP recording	Reflex fall in blood flow within seconds; it accomodates; increase in BP		?Sensory loss on cooled area	No change in blood flow; no increase in blood pressure
Measurement of oral, rectal or external auditory canal temperature	Heating or appropriate weather conditions	Normal temperature	Central thermoregulatory mechanism, sudomotor or vasomotor fibres; skin disease	Confusion, coma, absence of sweating dry hot skin	Hyperthermia
	Cooling or appropriate weather conditions		Central thermoregulatory structures or efferent vasoconstrictor connections; metabolic	Confusion, coma, cold skin, rigidity, absence of shivering	Hypothermia
	Intravenous pyrogen	Fever	Central thermoregulatory structures or efferent vasomotor connections, ?hypothalamic or cord above sympathetic outflow		No fever

TABLE 7.1
Continued.

Test	Methods	Expected normal response	"Lesion"	Clinical signs or symptoms	Abnormal responses
Prolonged heating of limb	Plethysmography or calorimetry of hand or forearm	Vasodilatation, increase in blood flow	Central thermoregulatory structures, efferent vasomotor connections or abnormalities in vessels		No change in blood flow
Body heating, monitor temperature in external auditory canal	Measurement of electrical skin resistance or observation	Fall in skin resistance, sweating	Central thermoregulatory structures, efferent sympathetic sudomotor connections or sweat glands	?Skin disease; heat intolerance	No change in skin resistance, no sweating
Mental arithmetic, noise, pain	Plethysmography of hand intra-arterial BP recording	Fall in hand blood flow, rise in BP	Efferent constrictor fibres to skin vessels		No change
Passive elevation of legs in recumbent subjects, negative pressure breathing, squatting	Forearm plethysmography	Increase in forearm blood flow (decrease in constrictor tone)	Sympathetic vasoconstrictor fibres to vessels in skeletal muscle or blood vessels		No change in blood flow
Tilting from horizontal to standing, positive pressure breathing, Valsalva's manoeuvre, radial acceleration, hypercapnia, exercise of legs	Same	Decrease in forearm blood flow (increase in constrictor tone)	Same		No change in blood flow

TABLE 7.1
Continued.

Test	Methods	Expected normal response	"Lesion"	Clinical signs or symptoms	Abnormal responses
Emotional stimuli, fright, mental arithmetic, pain	Same	Increase in forearm blood flow	Sympathetic cholinergic vasodilator fibres to skeletal muscle or blood vessels		No change
Observation of hairs on skin	Intradermal injection of acetylcholine	Piloerection	Postganglionic sympathetic nerves		No change
Infusion of ephedrine or methylamphetamine into brachial artery, 50–500 µg/min	Plethysmography of hand	Fall in hand blood flow	Same	Hot, dry hand	No change
Measurement of finger temperature or heat elimination during cooling of fingers	Immersion in cold water	Alternating vasodilatation and constriction, Lewis's shunting reaction	Blood vessels	Evidence of blood vessel disease	No change
Intra-arterial (brachial artery) or intravenous noradrenaline, 30–100 µg/min	Plethysmography of hand or intra-arterial BP	Fall in blood flow; rise in BP			No change
Intra-arterial (brachial artery) acetylcholine, 300–500µg/min	Plethysmography of hand; observation	Rise in blood flow or flush			No change
Heart rate	Intravenous atropine	Increase in heart rate	Vagus nerve or nuclei	Tachycardia	No change
Gastric acidity	0.01 U insulin/kg body weight	Increase in gastric acidity	Same		No change
Resting heart rate	Electrocardiogram	60–80 beats/min	Parasympathetic damage	Tachycardia 90–100 beats/min	No change

TABLE 7.1
Continued.

Test	Methods	"Lesion"	Clinical signs or symptoms	Expected normal response	Abnormal responses
Beat-to-beat (R-R interval) heart rate variation during deep breathing	Deep breaths (6/min) recorded on instantaneous heart rate monitor; difference between max. and min. heart rates	Parasympathetic damage		≥15 beats/min	≤10 beats/min
30/15 Ratio	ECG recording; measure R-R intervals at beat 15 and 30 after standing	Vagus nerve		>1.03	<1.00
Valsalva ratio	Blow into mouthpiece connected to manometer; maintained at 40 mmHg for 10s (tubing has small leak) Longest R–R interval after Valsalva/shortest R–R interval during Valsalva	Afferent, central or efferent baroreceptor reflex pathways	Symtomatic fall in blood pressure on standing acute stage; rare thereafter	≤1.21	1.11–1.20 (borderline) ≤1.10
Diving response (not recommended in endurance-trained subjects)	Facial immersion (cool water)	Trigeminal nerve sensory distribution; carotid bodies; vagal reflexes (pulmonary); brainstem (inspiratory neurons; cardio-inhibitory centre); vagus efferents	Reflex cardiac arrest during intubation or respiratory toilet; asthma; dental procedures; chronic respiratory insufficiency; sleep apnea; Pickwickian syndrome; Ondine's curse; apnea-stridor-bradycardia syndrome	Bradycardia; apnea; increase in BP	Tachycardia or no change in heart rate; no change or fall in BP

From Appenzeller O. (1990)

BP = blood pressure; ECG = electrocardiogram.

On standing, the blood pressure drops; there is no associated increase in heart rate, and plasma noradrenaline is unchanged. Exercise also is not associated with increased circulating noradrenaline in these patients.

In patients with peripheral neuropathy and orthostatic hypotension, noradrenaline secretion is also less than normal in response to orthostasis and low levels are found during recumbency. In some neuropathies, particularly diabetic and uremic, intravascular volume changes are important in producing symptomatic orthostatic hypotension. If plasma volume effects are dominant, then the sympathetic response to standing may be greater than normal, a situation characteristically found in those with orthostatic hypotension due solely to volume depletion. In diabetics and uremics with mild neuropathy, noradrenaline levels may be high in response to orthostasis, and postural hypotension can occur in the presence of considerable tachycardia when the patient stands. In this situation, treatment is aimed at volume expansion only. A similar mechanism of postural hypotension exists in otherwise normal individuals, particularly during or immediately after physical activity in hot ambient conditions. Normal autonomic function results in noradrenaline release on standing and blood pressure falls due to volume depletion. In such individuals, high resting plasma noradrenaline and further increases in noradrenaline to very high values on standing occur without adequate maintenance of blood pressure and brain perfusion. Concurrent thorough neurologic examination, including the autonomic nervous system, gives useful information on the cause of the postural hypotension, whether it is related to drug therapy, diseased peripheral autonomic structures, abnormal autonomic integration, or perhaps volume depletion (Ziegler, 1980). In general, pharmacologic studies have lent support to the clinical and biochemical classification of Pure Autonomic Failure (PAF) as a peripheral (postganglionic) autonomic neuropathy, but pathologic studies have not as yet fully supported this conclusion (Appenzeller, 1990).

THE NEUROPATHIC BLADDER

Pathological alterations in the bladders of diabetics closely resemble those described in sympathetic ganglia of patients with the same disease. Axon thickening and vacuolation of the thickened areas with spindle-shaped beaded appearances visualized by silver impregnation techniques have been described in severe cases of diabetic neuropathy. The changes were unrelated to inflammatory or vascular alterations which sometimes accompany the disease in patients with diabetic bladder dysfunction. Similar abnormalities were seen in the corpora cavernosa in impotent diabetic males and in the myenteric plexuses, particularly of the oesophagus. The bladder wall ganglion cells show changes previously described, including giant sympathetic neurons, shrunken nerve cells, hypochromatic ganglion cells and occasional empty spaces previously occupied by neurons (Mastri, 1980).

Neurogenic bladder involvement in diabetics (diabetic cystopathy) begins insidiously, and symptoms usually appear when the disease is advanced. Diabetic cystopathy and peripheral neuropathy occur together in 75–100% of patients, and some consider the presence of peripheral neuropathy a *sine qua non* for the diagnosis of diabetic cystopathy. On the other hand, in those who have peripheral nerve dysfunction, 83% have diabetic cystopathy on cystometric examination. Symptoms and signs of neurogenic bladder include residual urine from 90 to over 1000 ml (the definition of residual urine depending upon the investigator). The absence of the desire to void when the bladder contains over 500 ml and increased bladder capacity are indications of diabetic cystopathy. An obstructed bladder neck should be ruled out in the presence of the above abnormalities.

Characteristically, the patient with a diabetic neurogenic bladder has residual urine, infection, pyelonephritis, sepsis and azotemia. The patient may present with the complaint of increasing intervals between voiding, and urination may occur once or twice a day only. This is often accompanied by the need for straining to initiate and maintain voiding, weakness of the stream, dribbling and a sensation of incomplete emptying. Many patients are oblivious to the symptoms and a proper history can be obtained only by direct questioning. Some present with abdominal or pelvic tumour, and differential diagnosis includes peritoneal metastasis, prostatic hypertrophy or intestinal obstruction. The diagnosis can be established by abnormal cystometric studies and a large amount of residual urine. The cystometrogram shows a long low pressure curve and no sensation of filling when bladder capacity is tested until bladder distention is greater than normal. Ureteral orifices are often incompetent, showing reflux during radiographic studies.

The treatment of diabetic cystopathy includes use of an indwelling catheter for 10 days together with appropriate antibiotics. Thereafter, the patient should void every 3 h, aided by manual compression of the suprapubic area (Crede manoeuvre) and receive parasympathomimetic drugs. About 40% of patients respond to this therapy, at least temporarily, until urinary tract infection recurs. Transurethral surgery and bladder neck resection in those without obvious mechanical obstruction may also be useful. Initially, the parasympathomimetic drug can be given parenterally, for example, bethanechol twice weekly and may be continued orally in a dose of 40–50 mg every 6 or 8 h. Cholinergic treatment is withdrawn when residual volumes are less than 100 ml for at least a week (Ellenberg, 1980a).

SEXUAL DYSFUNCTION IN PERIPHERAL NEUROPATHIES

The sacral parasympathetic and thoracolumbar sympathetic nerves are efferent vasodilator nerves for penile vessels and are responsible for erection produced by arterial dilatation and increased blood flow to the erectile issues. Erection is reflexly initiated by visual or olfactory stimuli, or by imagery that affect supraspinal centres, but it can also be achieved by gentle stimulation that reflexly activates spinal mechanisms. The parasympathetic nerves also stimulate secretion from the seminal vesicles and prostate, including Cowper's glands, during the plateau phase of the sexual response. Emission of semen, ejaculation and the accompanying sensations occur during orgasm. Emission depends on sympathetic innervation of the urethra, which initiates contraction of smooth muscles in the vas deferens, seminal vesicles and prostate. The rhythmic contractions of striated muscles, including the bulbocavernosus and ischiocavernosus innervated by the pudendal nerves, eject semen from the urethra. It is clear, therefore, that peripheral neuropathies involving the nerves controlling any phase of the sexual response in males or females lead to sexual dysfunction. Diabetic autonomic neuropathy of the pelvic parasympathetic nerves (nervi erigentes) accounts for about 60% of impotence in diabetic males, and diabetic men tend to become impotent at an earlier age than others. Psychological factors are by far the commonest cause of impotence in the general population, but this is usually of sudden onset in contrast to diabetic impotence which progresses slowly. Neurogenic impotence is characterized by absence of erection at any time under any circumstances and is often associated with decreased testicular sensitivity, whereas in psychogenic impotence testicular sensation is unimpaired.

Many medications may cause impotence. These inclulde oestrogens used for the treatment of carcinoma of the prostate, alcohol, phenothiazines, antidepressants, some of the antihypertensive agents and other drugs used in patients with diabetes.

When the diabetes is poorly controlled, impotence can reflect malnutrition, wasting and weakness, and may improve with proper diabetic management. If psychogenic factors are associated with the decreasing potency, these should be appropriately managed and drugs affecting sexual function should be withdrawn. In many cases, however, the neurogenic nature of diabetic sexual dysfunction leaves no useful therapy and the prognosis is poor. Mechanical devices have been proposed and used, but their efficacy is difficult to evaluate (Karakan, 1980).

The diagnosis of erectile impotence can be aided by monitoring nocturnal penile tumescence; this can provide evidence of erectile capacity or lack thereof and establish the organic nature of impotence. Penile blood pressure in organic impotence is low and bulbocavernosus reflex response latencies are delayed, but the plasma concentrations of testosterone and prolactin are normal in such patients. These tests should be performed together with behavioural evaluation before penile prostheses for diabetic autonomic impotence are considered.

In many patients with diabetic neuropathy, impotence is an early symptom; in others, however, despite normal erection and orgasm, no ejaculation occurs. In some such patients it has been shown that the ejaculation was retrograde, that is, the semen was propelled into the bladder (Greene et al., 1963; Ellenberg and Weber, 1966). Normal emission, which implies delivery of seminal fluid into the posterior urethra, is due to the contraction of the smooth muscle of the seminal vesicles.

Ejaculation follows because of forceful contraction of the bulbocavernosus muscle (Whitelaw and Smithwick, 1951). If the vesical neck does not close during ejaculation, as it normally does, the seminal fluid is propelled back into the bladder by so called retrograde ejaculation. The sympathetic nervous system mediates the closure of both the bladder neck and the contraction of the seminal vesicles during emission, but in some diabetic patients the bladder neck does not contract while the autonomic innervation to the rest of the genital system remains apparently unimpaired. Diabetic patients with this disorder usually are aware of frothy material in the urine after intercourse. In such patients, there were signs of severe peripheral neuropathy and evidence of widespread autonomic nervous system involvement manifested by orthostatic hypotension, nocturnal diarrhoea and a large capacity hypotonic bladder with residual urine.

Most patients with this disorder had long-standing insulin-dependent diabetes mellitus with diabetic retinopathy. It has also been shown that the disturbance in ejaculation was of gradual onset, leading to a decrease in the ejaculate (Ellenberg and Weber, 1966) and this might account for the high prevalence of seminal deficiency in diabetic men (Rubin and Babbott, 1958). While there is sufficient evidence for impairment of autonomic function and retrograde ejaculation in patients with diabetes mellitus, it is difficult to understand why emission remains normal whereas the closure of the bladder neck, which is dependent on the function of closely related sympathetic fibres, becomes impaired and allows retrograde ejaculation. Retrograde ejaculation is also seen after transurethral resection, after inflammation, trauma and congenital abnormalities of the posterior urethra. It occurred commonly after bilateral thoracolumbar sympathectomies (Allen and Adson, 1938) and with the use of sympatholytic antihypertensive therapy (Ellenberg and Weber, 1966).

In acute diabetes, impotence may be the presenting symptom and in such patients it usually clears within weeks or months after effective treatment. In patients who had diabetes of insidious onset or who developed impotence after many years of adequate diabetic treatment, impotence was usually permanent (Keen, 1959). Impotence affected one in four diabetic males in the 30–35 year age group, but 75% of diabetic males aged 60–65 were impotent (Rubin and Babott, 1958). Although failure of erection and ejaculation usually occurred together, either may exist independently (Keen, 1959). Because impotence in the diabetic was frequently associated with evidence of peripheral neuropathy and other autonomic dysfunction, it has been attributed to disturbances in parasympathetic innervation causing failure of erection and sympathetic disturbances accounting for the lack of ejaculation. Moreover, experimentally an analogous functional disturbance was produced in animals by section of the 2nd to the 4th sacral nerves and of the nervi erigentes (Learmonth, 1931).

The sympathetic nervous system in patients with diabetic neuropathy was studied clinically and pathologically (Low et al., 1975a). Clear evidence of demyelination in the splanchnic nerves was found and degeneration of unmyelinated fibres in these structures was also present. Clinical correlation of autonomic dysfunction with pathological evidence for demyelination and axonal degeneration in autonomic nerves was obtained. These studies show a morphological basis for the widespread impairment of autonomic function so commonly found in diabetics. A clinical study of outpatient male diabetic patients which included 175 subjects showed that 49% were impotent, 2% had premature ejaculation and 1% had retrograde ejaculation. The duration of diabetes was approximately 6 years and the mean age of subjects who were impotent was 53. Diabetics who were not impotent had a mean age of 45 and a mean duration of diabetes of 5 years. The most common progression of impotence was one of a decrease in firmness of the erection and a gradual onset of impotence over a period of about 6 months, though the level of sexual interest and erotic stimulation was sustained in almost all impotent diabetic subjects. The impotence could not be correlated with assurance with the duration of diabetes or with insulin administration nor with the use of oral anti-diabetic agents. Those with sexual dysfunction had increased clinical evidence of peripheral neuropathy. Examination of plasma testosterone levels in the impotent group excluded androgen deficiency as a cause (Kolodny *et al.*, 1974).

While secondary impotence is well recognized in diabetic patients the incidence of diabetes in sexually impotent men has also been looked for. Secondarily impotent men were examined by 3 h glucose tolerance tests and compared with men of similar age with normal sexual function or with premature ejaculation. All individuals were apparently non-diabetic. Subsequently, the diagnosis of diabetes was based on impaired glucose tolerance tests and on serum glucose levels during the tests as recommended by the National Diabetes Data Group. The results were corrected for weight and age differences, but the mean glucose levels in patients with secondary impotence implies that localized neuropathy or perhaps penile ischemia may produce impotence in otherwise asymptomatic subjects (Deutsch and Sherman, 1980).

The majority of men who have secondary impotence have always been thought to have this on a psychogenic basis. However, examination of serum testosterone levels in 105 consecutive patients uncovered 37 with previously unsuspected disorders of hypothalamic–pituitary gonadal function. Some had hypogonadotropic-hypogonadism; in others, hypergonadotropic-hypogonadism was found and yet others had hyperprolactinemia and a few were found to be hyperthyroid. The correction of the endocrine deficits results in restoration of potency. Thus, screening of serum testosterone in large numbers of impotent males might disclose a significant group with treatable endocrine disorders and the concept that 95% of patients with impotence have this on the basis of psychogenic factors should not be accepted (Spark *et al.*, 1980).

Illness of any sort, whether or not it affects sexual organs or their innervation, does modify the entire life pattern and it removes the patient from the usual personal and social interaction. The effect of illness on sexual competence is, however, often forgotten and remains unrecognized. For example, patients with chronic alcoholism are often impotent and clinically deficits of autonomic function have been demonstrated in alcoholics, and demyelination in paravertebral sympathetic myelinated fibres has been found (Appenzeller and Ogin, 1973). However, the splanchnic nerves in patients with alcoholic neuropathy failed to show marked changes in this particular nerve (Low *et al.*, 1975b). One must, therefore, conclude that though some changes in autonomic structures have been found in alcoholics, impotence which is so widespread amongst chronic sufferers from alcoholism must, in part at least, be based on the illness itself. This may lead to a loss of the sense of intactness because of abnormalities in body image which can cause great anxiety about the capacity to function sexually in those who are afflicted by chronic alcoholism. Their depression, often associated with feelings of worthlessness and loss of attractiveness to others and including the loneliness of addiction to alcohol, plays an important role in their impotence (Labby, 1975). While the autonomic nervous system is similarly involved in females with diabetes mellitus, there is little to suggest that sexual dysfunction in females is common or troublesome. Careful studies indicate that diabetes has no effect on sexual performance in women. So far, no anatomical, neurological or physiological differences have been identified to explain the observed differences in sexual function between the two sexes. Perhaps it is a psychological aspect of sexual behaviour that accounts for the remarkable differences in the effect of autonomic neuropathy in the two sexes (Ellenberg, 1980b).

SWEATING

Two types of sweating are recognized: (1) Thermoregulatory sweating, which occurs over the whole body, in response to environmental temperature changes or exercise; (2) Emotional sweating, which is confined to palms, axillae, soles of feet and some parts of the face.

It is not entirely clear whether thermoregulatory sweating always depends on a rising blood temperature that activates central structures responsible for thermoregulation or whether a peripheral heat-sensitive receptor reflexly activates these central structures. Under certain circumstances, either of these mechanisms alone or combined may produce sweating and may be differentially affected in disease. An example of reflex peripheral regulation of sweating is the temporary inhibition of sweat secretion in a hot environment, achieved by applying ice to the skin, but sweating induced by the action of pilocarpine on sweat glands is not inhibited by cooling.

Microelectrode recordings from human cutaneous nerves show that sudomotor impulses occur in periods with intervals of about 0.6 s. This rhythmic sudomotor nerve activity corresponds to changes in electrodermal responses that measure skin conductance which, in turn, depends upon sweat secretion. Therefore, sweat periodicity, well-documented in humans, is governed by sympathetic nervous activity. Evaporation rates correlated closely with discharge periodicity from cutaneous nerves.

Normal postural influences related to pressure on the skin affect the sweating rate reflexly. When a subject lies on his back in a hot environment, sweating occurs equally over the body. Lying on one side, however, increases sweating remarkably on the uppermost side and inhibits sweat gland activity on the other. Standing after recumbency induces sweating on the upper part of the body and inhibits it in the legs (Takagi and Sakurai, 1950). These hemihidrotic reactions can be activated readily by pressing the axillae or adjacent pectoral region, and the responses are limited either to the upper or lower half of the body separated by an ill-defined border adjacent to the iliac crests.

The relationship of sweat secretion to cutaneous blood flow is complicated and varies from one body part to another. In some parts of the body, vasodilatation of skin vessels may depend on vasodilator substances in the sweat, and both of these thermoregulatory modalities occur concurrently. However, cutaneous vasodilatation in the hand in response to increased central temperature may be abolished in patients with the Landry-Guillain-Barré syndrome and yet sweating occurs normally in response to heat loads (Appenzeller and Marshall, 1963). In freshly amputated limbs without circulation, direct electrical stimulation of peripheral nerves produces sweating for some time after amputation.

In essential hyperhidrosis, the constituents of sweat and sweat gland morphology are normal. The aetiology of essential hyperhidrosis is unknown and pharmacological blockage of sweat gland activity has been an unsatisfactory treatment for this condition. Sympathectomy cures the socially embarrassing sweating that usually starts in childhood, but does not become burdensome until adult life. If not treated surgically, hyperhidrosis lasts throughout life, and it is most troublesome when it affects the hands, feet and axillae. Recently, abnormal vasomotor function has also been noted in some of these patients, and baroreflex activity becomes normal after surgical extirpation of the relevant sympathetic ganglia. Results of biofeedback therapy for essential hyperhidrosis are conflicting and, in cases in which it has been successful, the permanence of reduction in sweating has not been established.

Essential hyperhidrosis must be distinguished from secondary hyperhidrosis which occurs in hypothalamic disorders, due to ingestion of cholinergic agents, thyrotoxicosis or fever. Sweating associated with cold skin occurs with hypoglycemia, the dumping syndrome, withdrawal from alcohol or other drugs, during shock or syncope, and with intense pain. Endocrine abnormalities such as diabetes mellitus or gout have been associated with hyperhidrosis. Patients with familial dysautonomia have hyperhidrosis that affects mainly the limbs, and patients with causalgia have hyperhidrosis confined to the affected extremity.

Sweating abnormalities produced by human nervous system lesions give information about functional pathways in this thermoregulatory activity. Most such studies have involved peripheral nerve disorders associated with complete sympathetic paralysis leading to symptom grouping into syndromes. Horner's syndrome usually occurs after interruption of cervical sympathetic pathways, and pupillary constriction, dilatation of conjunctival vessels, eyelid drooping, anhidrosis and dilatation of facial vessels ipsilateral to the lesions occur. A partial Horner's syndrome occurs with centrally placed lesions.

Sweating in sympathetically denervated areas of the face during eating has been attributed to facial sweat fibres joining the trigeminal nerve distal to its sensory root for peripheral distribution rather than traversing the paravertebral sympathetic chain and external carotid plexuses to reach the face. These fibres arise in the brainstem and are called accessory sweat secretory fibres. Like other fibres innervating sweat glands, they are cholinergic.

Hyperhidrosis of the face is common. It may be confined to one-half of the face and be associated with tearing and nasal discharge, or sweat may occur only on the cheeks. Emotional or gustatory stimuli provoke this type of sweating. Most exaggerated sweat responses are mediated by accessory sweat fibres except the sweating confined to the cheeks which occurs after injury or parotid gland surgery; this may be related to diffusion of acetylcholine or it is released by stimulation of parotid secretory nerves.

Anhidrosis is the usual effect of destruction of sympathetic supply to the face. However, in about 35% of patients with sympathetic denervation of the face, accessory fibres (reaching the face through the trigeminal system) become hyperactive and hyperhidrosis occurs, occasionally causing the interesting phenomenon of alternating hyperhidrosis and Horner's syndrome (Ottomo and Heimburger, 1980).

The auriculotemporal syndrome is paradoxical reflex gustatory sweating; it occurs after nerve injury in the face. Sweating and flushing of the skin supplied by the auriculotemporal nerve occur during eating, particularly of spicy or sour foods. Because the auriculotemporal nerve carries sympathetic postganglionic fibres to blood vessels and sweat glands, and parasympathetic preganglionic secretomotor fibres to the parotid gland, it may be that reflex sweating during eating is due to cross-excitation between parasympathetic and sympathetic fibres. Successful treatment of this syndrome by division of the ninth cranial nerve has been reported, suggesting that the impulses initiating the abnormal sweating were carried in parasympathetic fibres. The auriculotemporal syndrome has also been attributed to injury of the auriculotemporal nerve by forceps during delivery.

The chorda tympani syndrome is submental gustatory sweating. It occurs after surgical trauma and is attributed to cross-excitation of sympathetic fibres that are in close proximity to parasympathetic secretory fibres to the submaxillary gland.

Diabetic anhidrosis is a condition in which sweating is absent in the lower limbs or trunk of patients with diabetic neuropathy (Goodman, 1966). Patients with this disorder are intolerant to heat, but may experience excessive perspiration of the head, face and neck, or compensatory hyperhidrosis that is particularly profuse on the face at mealtimes (enough to be socially unacceptable). Gustatory sweating in patients with diabetic neuropathy is usually confined to the territory of the superior cervical ganglion. All patients with diabetic anhidrosis or hyperhidrosis of the face have abnormal autonomic function in other systems and organs. Gustatory sweating is attributed to sprouting or cross-innervation of fibres, and suggests that in diabetic autonomic neuropathy, despite the persistence of the metabolic abnormality presumably responsible for autonomic disturbances, regeneration of axons is possible (Watkins, 1973).

In diabetic patients, the degree of involvement of the autonomic nervous system is often uneven and tests of autonomic function must, therefore, include many systems to show autonomic deficits that might not be symptomatic. Of all tests, the assessment of thermoregulatory sweating has been neglected because of the cumbersomeness of the procedures. In a comprehensive study of abnormal patterns of sweating in patients with diabetes, Fealey et al. (1989) showed that damage to the autonomic innervation of sweat glands is similar to damage to somatic innervation in diabetes. A pattern of length-dependent damage producing distal hypo- or anhidrosis or focal, multifocal or regional lesions of nerves that show restricted areas of sudomotor dysfunction.

Sweating depends on small fibre activity and, in diabetic neuropathy, the assessment of peripheral nerve function by standard techniques, such as nerve conduction and electromyography assesses only large-diameter myelinated fibre function. Thus, one would expect that sweat testing could be important in delineating the extent of small-diameter fibre damage in diabetic neuropathy. Thermoregulatory sweating, moreover, is a test that stresses all aspects of sweating, whereas another test, the quantitative sudomotor axon reflex test, tests only postganglionic sudomotor function. Thus, using the two tests together might be helpful in distinguishing pre- from postganglionic lesions. However, heat chambers, space and time required for the performance of thermoregulatory sweat tests make it difficult to recommend this as a standard autonomic diagnostic procedure. Nevertheless, in diabetic truncal neuropathies, as opposed to focal neuropathies affecting major nerves in the limbs, thermoregulatory sweat testing may be superior and offer the only way to demonstrate autonomic dysfunction in the former, whereas nerve conduction and electromyography may give reliable results in the latter more quickly and with less effort (Stewart, 1989).

Hyperhidrosis may be a feature of traumatic peripheral nerve lesions and occasionally occurs in other peripheral neuropathies. It indicates incomplete interruption of peripheral nerve bundles. When it is associated with causalgia, it occurs most commonly in median or sciatic nerve injuries and in those associated with a swollen, cyanotic, extremely painful limb. Sympathectomy in patients with causalgia relieves the pain, excessive sweating and other symptoms. The hyperhidrosis of peripheral nerve lesions usually disappears in those with progressive disease and, with advancing disease, the distal extremities eventually become anhidrotic. In some patients with a cervical rib, segmental hyperhidrosis of the ipsilateral limb occurs; it responds to surgical resection of the rib. Nonthermal hyperhidrosis also occurs in involved skin segments of occasional patients with mononeuropathies or mononeuritis multiplex and often manifests as strips of sweating when patients are emotionally stressed, suggesting heightened susceptibility of partially denervated sweat glands to circulating stress hormones or perhaps a sweat gland denervation supersensitivity. This condition occasionally helps diagnose mononeuritis multiplex by pointing to the focal nature of the peripheral nerve involvement.

Excessive sweating of the extremities is the rule in familial dysautonomia in which peripheral nerve dysfunction and morphological abnormalities also occur. In such patients, the threshold for sweating in response to local heat is much lower than in controls, causing the patient to sweat most of the time. This, together with other evidence, suggests that the hyperhidrosis in familial dysautonomia is due to hyperexcitability of control sudomotor centers.

Sympathectomy causes anhidrosis in appropriate cutaneous segments that is usually permanent, whereas the vasomotor paralysis induced by this surgery may recover after a time. Nevertheless, "escape areas" of preserved sweating, the result of intermediate ganglia in the rami communicantes may occur, particularly if the limb has been sympathectomized.

Abnormal sweating in patients with collagen disease most often relates to concomitant involvement of peripheral nerves. This occurs in patients with rheumatoid arthritis and polyarteritis nodosa, a well-known cause of peripheral neuropathy.

TABLE 7.2
Some tests for the clinical assessment of sweating.

Test	Methods	Expected normal response	"Lesion"	Clinical signs or symptoms	Abnormal responses
Body heating	Heat cradle or two limbs in hot water (43°C) until central temperature 1°C above normal, dust body with quinizarin powder purple when wet	Sweating	Central; postganglionic; skin or sweat glands	General or regional hypo- or anhidrosis, heat intolerance	No sweating
Number of functioning sweat glands per unit area	Starch containing paper impregnated with iodine (paper + solid iodine crystals in closed vessel) apply paper to skin–count number of dark spots	Number of normally active glands varies with area of skin examined	Central; postganglionic; skin or sweat glands	General or regional hypo- or anhidrosis, heat intolerance	Reduced number of dots per unit area
Psychogalvanic skin reflex	Voltage or current flow change between indifferent area (earlobe) and test skin	Change in voltage or current flow (can be continuously recorded)	Central; postganglionic; skin or sweat glands	General or regional hypo- or anhidrosis, heat intolerance	No change in voltage or current flow
Axon reflex [Quantitative Sudomotor and Axon Reflex Test (QSART)]	Faradism; intradermal injection of 5–10 mg acetylcholine solution	Local piloerection; sweating	Sympathetic ganglionic; postganglionic; skin or sweat glands	General or regional hypo- or anhidrosis, heat intolerance	No piloerection; reduced or absent sweating
Sweat gland function	Pilocarpine iontophoresis	Sweating (can be quantitatively assessed)	Sweat glands (normal function if denervated)	General or regional hypo- or anhidrosis, heat intolerance	No sweating; reduced quantity of sweat
Skin biopsy	Skin punch biopsy	Normal histological and histochemical appearance of sweat glands and the skin	Sweat glands or skin	General or regional hypo- or anhidrosis, heat intolerance, skin lesions	Abnormal sweat composition, morphological changes

From Appenzeller, O. (1990)

Anhidrosis corresponding to the territory of small cutaneous sensory branches of peripheral nerves is a feature of some types of leprosy. It is of great diagnostic value when the small anhidrotic patches on the skin do not follow the cutaneous sensory distribution of major peripheral nerves.

Tests for the clinical assessment of sweating are listed in Table 7.2.

NEUROGENIC IMPAIRMENT OF THE CIRCULATION

Afferent or efferent impairment of baroreflex causes circulatory disturbance. In tabes dorsalis, the Valsalva manoeuvre is not associated with a normal "overshoot", but vasodilatation occurs normally in the hands in response to heating the feet, and vasoconstriction following a gasp is preserved. These tests suggest that postural hypotension in tabes dorsalis results from interruption of the afferent side of the baroreflex arc. Similar reasoning also suggests baroreflex afferent involvement in diabetics with autonomic neuropathy. The point should be made that autonomic neuropathy may not be symptomatic even when cardiovascular reflexes are clearly abnormal when assessed by special tests. Autonomic responses may be abnormal in diabetics during ketoacidosis without clinical autonomic neuropathy. Additionally, parasympathetic function tends to be more widely involved and more commonly abnormal than sympathetic function in patients with diabetes mellitus; progression occurs with time in both parasympathetic and sympathetic deficits. Baroreflexes are markedly abnormal in haemodialyzed uremic patients and are different from those in patients with primary hypertension. Some uremic patients, however, may have a neurogenic component to their hypertension similar to that found in experimental baroreceptor afferent denervation (Levy and Lilley, 1978). Moreover, in such patients Valsalva's manoeuvre and the haemodynamic adaptation after amyl nitrite inhalation show severe vasomotor failure, particularly in those with clinical and electrical evidence of peripheral neuropathy.

Reflex vasoconstriction is impaired in patients with chronic hypoxemia because of failure of sympathetic function. However, vasoconstrictor responses in such patients improve promptly when the hypoxemia is temporarily corrected. The abnormal sympathetic function is probably based on both central and peripheral disturbances; patients with chronic hypoxemia may develop peripheral neuropathy (Appenzeller, Parks and Macgee, 1968). Impaired autonomic neurotransmission rather than structural autonomic pathology probably accounts for the vasomotor abnormalities. In an experiment when healthy males inhaled carbon monoxide, vasoconstrictor responses were also impaired even when arterial oxygen tension was normal (Heistad and Wheeler, 1972).

The orthostatic hypotension of primary amyloidosis may suggest the correct diagnosis of this type of peripheral neuropathy prior to nerve biopsy. Abnormal sweating also occurs. These disturbances are attributed to invasion of somatic and autonomic peripheral nerves by the disease process. Clinical and pathological studies of patients with both primary and secondary amyloidosis show that structural involvement of the autonomic nervous system does not necessarily produce autonomic dysfunction. However, when autonomic structures are severely involved, as they often are in secondary amyloidosis, autonomic disturbances are usually disabling (Nordborg *et al.*, 1973). In dominantly inherited amyloidosis, (Andrade, 1952), postural hypotension and widespread sweating abnormalities are common. Similar vasomotor disturbances also occur in alcoholic, porphyric and several other neuropathies. Abnormal vascular reflexes in patients with Charcot–Marie–Tooth disease have been confirmed repeatedly, and the finding of giant nerve bundles in bowel myenteric plexuses of these patients supports the notion that the autonomic nervous system is involved (Brooks, 1980). Charcot and Marie (1886) noted the bluish or reddish discolouration and mottling of the feet and legs in their patients with this disease, recognizing vasomotor abnormalities commonly found in such individuals.

Features of diabetic autonomic neuropathy, apart from the vasomotor abnormalities, include gastrointestinal symptoms which resemble those found after surgical vagotomy, and the term "autovagotomy" has been used to describe this condition. In these patients, vagal stimulation or hypoglycaemia may not produce gastric acid secretion, but acid secretion still occurs in response to administration of humoral agents such as pentagastrin. Pancreatic polypeptide secretion in response to insulin hypoglycaemia is decreased in juvenile diabetics with autonomic neuropathy when compared with those without clinical neuropathy. A significant correlation exists between the duration of diabetes and the response to insulin-induced hypoglycaemia. Abnormal responses are attributed to a vagal neuropathy, since pancreatic polypeptide release depends normally upon intact vagal innervation of the pancreas.

HORMONAL RESPONSES IN DIABETES

Immersion of humans to the neck in water produces hydrostatic pressure gradients in the sitting subject, which redistributes blood from the limbs into the central thoracic compartment. It is well established that volume expansion produced in this fashion causes hormonal changes, including suppression of renin–aldosterone, catecholamines and antidiuretic hormone secretion. There is also an enhanced release of atrial natriuretic peptide. The study of diabetic subjects with and without neuropathy and normal age-matched controls by immersion allows a comparison of these volume expansion-induced changes in hormonal and peptide secretions and an assessment of the effect of autonomic denervation on these responses. Fourteen insulin-dependent diabetic subjects, seven with and seven without autonomic neuropathy, were studied. All diabetic subjects showed impaired natriuretic responses to volume expansion, but the presence of autonomic neuropathy had no effect on this impaired response. The suppression of plasma renin or aldosterone during immersion was no different in the groups studied. A suppression of plasma catecholamine secretion was found in all diabetics on immersion, but basal values were lowest in those who had autonomic neuropathy. On the other hand, atrial natriuretic peptide levels were significantly increased by immersion in all groups, and diabetic autonomic involvement had no effect on this increase. These results show cardiac denervation by autonomic dysfunction in diabetes does not affect the volume expansion-induced increase in circulating atrial natriuretic peptide. Similar findings have been reported in patients with disconnected transplanted hearts (Singer *et al.*, 1986).   Though the kidney has an important autonomic innervation in diabetic subjects irrespective of the presence of neuropathy, retention of sodium during volume expansion occurs and this increased retention of sodium is not attributable to differences in circulating hormones that are known to be important in natriuresis and is not influenced by autonomic neuropathy (O'Hare *et al.*, 1989).

IMMUNOLOGICAL ABNORMALITIES

In insulin dependent diabetics (IDDM), anti-adrenal medullary antibodies have previously been demonstrated and anti-sympathetic ganglia antibodies have also been found.   The catecholamine responses to standing have now been evaluated in IDDM patients (Cryer, 1989) with and without complement-fixing anti-adrenal medullary and anti-sympathetic ganglia antibodies.   Four IDDM patients were antibody-negative and three were antibody-positive.   The baseline mean noradrenaline and adrenaline levels were not significantly different in the two groups. However, significantly higher levels were found in noradrenaline after 5 min of standing in antibody-negative subjects and similar significant differences were also present in the levels of adrenaline.   Thus, a decreased response of catecholamines to changes in posture is associated with complement-fixing anti-adrenal medullary and anti-sympathetic ganglia antibodies.

The suggestion that the immune system contributes to the development of autonomic neuropathy seems to be supported by this study and also by the well-known presence of lymphocytic perivascular infiltrates in sympathetic ganglia of patients dying with IDDM (Appenzeller, 1986).   The contribution of an immunological response to autonomic neuropathy in IDDM is perhaps due to the abnormalities in preganglionic sympathetic nerve fibre innervation of the adrenal medulla, which might account for a decrease in adrenaline secretion in response to the stress of standing. Though none of the subjects in this study had symptoms of autonomic neuropathy, it is possible that the immune reaction had not progressed far enough or for a long enough time, and further follow-up might reveal the appearance of symptomatic dysfunction in antibody-positive subjects (Brown *et al.*, 1989).

A number of other disorders which affect the peripheral nerves and roots and which are clinically associated with autonomic dysfunction, have been recognized. These include, amongst others, amyloidosis, chronic alcoholism, chronic renal failure and congenital insensitivity to pain (Thrush, 1973). All of these conditions have been excluded on clinical and pathogenetic grounds in patients with acute pandysautonomia (Young *et al.*, 1969). The presence of underlying malignancy seems unlikely in view of the long follow-up of some patients and their clinical recovery without specific treatment. The results of autonomic function tests in acute pandysautonomia suggest, in most patients, a postganglionic autonomic dysfunction and the morphology of the sural nerve is indicative of regeneration. The pathogenesis of such complete sympathetic and parasympathetic postganglionic denervation in acute pandysautonomia is not clear, but the clinical features together with the documented recovery of function and regeneration of unmyelinated fibres suggest an immunological basis of the disorder. The high cerebrospinal fluid protein level without an accompanying cellular reaction in several cases implies a relationship to the Landry–Guillain–Barré syndrome and further supports this interpretation. The association of pandysautonomia with malignant disease and the production of experimental autonomic neuropathy are all in keeping with an immunologically produced autonomic neuropathy (Appenzeller, 1983).

A number of diseases of peripheral nerves might present with severe autonomic failure causing the most disabling symptoms. In one such patient, who recovered incompletely after a period of observation extending over 13 months, the cerebrospinal fluid protein level was elevated and the clinical course was complicated by an almost global loss of sensation including corneal ulceration. A sural nerve biopsy showed degeneration of both myelinated and unmyelinated fibres. The condition was attributed to an acute ganglionopathy, but the cause was unknown, toxins and viral aetiologies having been excluded (Colan et al., 1980). Three similar cases have subsequently been reported (Fagius, Westerberg and Olsson, 1983; Singer et al., 1987; Kanda et al., 1990).

Fabry disease is characterized by cell storage of glycolipid ceramide trihexoside within the nervous system, including neurons of the autonomic nervous system, and a dying-back neuropathy, most prominent in small myelinated and unmyelinated fibres. Autonomic dysfunction has been demonstrated in this disease (Cable, Kolodny and Adams, 1980). Nine males with the disorder were examined and showed impaired sweating, absent corrugation of the skin induced by warm water immersion (a sign of autonomic denervation), and an abnormal flare component of the triple response of Lewis. Abnormal pupillary responses to pilocarpine, reduced saliva production, tear formation and abnormal cardiovascular responses, including decreased reflex rises in plasma noradrenaline, were also present. Sympathetic and parasympathetic function was impaired and was thought to represent small nerve fibre dysfunction in peripheral nerves. Autonomic nervous system involvement is rarely as incapacitating in this disease as in acute pandysautonomia.

MULTIPLE MUCOSAL NEUROMAS

Rarely, patients with multiple mucosal neuromas have dysautonomia and an abnormal triple response to intradermal injection of histamine (Horstink et al., 1974). Clinically, the syndrome is characterized by painless greyish yellow tumours on the conjunctiva, tongue, in the oral cavity, pharynx and larynx. They are usually distributed symmetrically about the midline and a typical site is along the anterior dorsal area of the tongue. Such patients also have a marfanoid habitus, thick everted lips and multiple endocrine tumours such as medullary thyroid carcinoma and pheochromocytomas. Pes cavus, funnel chest, myelinated fibres in the cornea, skin pigmentation, hypertrophic nerves and abnormalities of Auerbach and Meissner plexuses and a myopathy have been described. The clinical features start in early childhood, but endocrine tumours make their appearance only in adolescence. Autosomal dominant heredity is probable. The syndrome has been attributed to a disturbance of neural crest formation, because several crest-derived structures are involved, including dental abnormalities and the autonomic nervous system. Clinically impaired lacrimation, parasympathetic denervation supersensitivity of the pupils with intact sympathetic responses, orthostatic hypotension with an intact overshoot in response to the Valsalva manoeuvre, normal responses to vagal stimulation (eyeball pressure and carotid sinus massage), impaired pilomotor activity, absent reflex vasodilatation with preservation of reflex vasodilatation in blood vessels of the hand, abnormal glucose levels in response to intravenous insulin administration and dermographia have been documented. Sweating and salivary gland function were normal. A sural nerve biopsy showed degeneration and regeneration of unmyelinated fibres similar to the findings in some patients with acute pandysautonomia.

BOTULISM

Autonomic dysfunction occurs in acute botulism. A minor variant of this condition due to intoxication with botulism B with preponderant effects upon cholinergic autonomic function has also been delineated (Jenzer et al., 1975). This condition is characterized by a benign but protracted course, with paresis of accommodation, impaired salivary and lacrimal secretion, constipation, pupillary abnormalities and disturbances of micturition, all due to peripheral cholinergic dysfunction. Orthostatic hypotension and apathetic behaviour are also present, but could not be definitely attributed to botulinus toxin. A striking feature was the duration of symptoms, varying from 38 to 80 days. Neurological examination was otherwise normal and electromyography showed no defect in neuromuscular transmission. No specific treatment was given and administration of antitoxin was not recommended. It is important to recognize this condition and distinguish it from acute pandysautonomia, which on clinical and morphological grounds is an entirely different entity.

ACRODYNIA

Acrodynia (pink disease) is a syndrome with characteristic clinical features, the aetiology of which is not fully understood. The disease has become rare but was in the past particularly common in England and Australia. In Europe, it was described by Feer who called the disorder "infantile vegetative neurosis" (Feer, 1923)

Acrodynia is practically confined to infancy and appears to be associated in some cases, with repeated ingestion or contact with mercury. The number of infants exposed to sufficient amounts of mercury and for long periods is, however, far greater than the number of patients with the disease. It remains, therefore, uncertain whether the disorder is due to poisoning with mercury.

The clinical manifestations include listlessness, progressing to restlessness and irritability. Associated upper respiratory tract infections are often seen. Recurring widespread rashes are found. The tips of the fingers and toes become pink and, later, a dusky colour spreads to the palms and soles. The extremities become cold and clammy and the cheek and nose may appear bright red. These changes are the most distinctive features of the disease and are responsible for the name "pink disease". When the syndrome is established, there is profuse and persistent sweating. Desquamation of the skin of the palms and soles may re-occur at intervals. The fingers swell, due to hyperplasia and hyperkeratosis of the skin. Fever appears only with intercurrent infections. The children have extremely painful hands and feet and older children will complain of burning sensations in the extremities. Photophobia is very common and characteristically children bury their heads in the pillow to shield their eyes. There is hypotonia of limbs and this allows the patients to assume unusual postures for hours. Loss of normal teeth is a characteristic feature and may be accompanied by necrosis of the mandible. The gums become swollen and inflamed and there is excessive salivation. Fingernails are discoloured and may drop off and occasionally gangrene of fingers and toes has been observed. The hair falls out and is often pulled out by the child. Large quantities of water are drunk, anorexia is frequent and diarrhoea is the rule. Prolapse of the rectum is a common complication. There is accompanying tachycardia and hypertension. Tendon reflexes usually disappear in the course of the illness and the child refuses to walk although there is no paralysis. Characteristic pathological features have not been found at necropsy, but judging from the symptomatology the disorder seems to affect terminal sprouts of peripheral nerves. Laboratory investigations are not helpful.

The disease can be avoided by shunning contact with compounds containing mercury. Treatment is difficult and must be prolonged because the disease lasts from a few months up to a year. Management is mostly directed at relieving distress. Dimercaprol has been used with success, particularly when started early in the disorder. Ganglion-blocking agents have also been employed and are effective in abolishing the hyperhidrosis, the peripheral vasospasm and hypertension, but usually aggravate the photophobia. The mortality is higher in hospitalized children than in those kept at home because of the risk of cross-infection and the hospitalized children should be managed during their long illness by their parents.

The link between pink disease and mercury is complicated by the absence of this disorder in recent epidemics of mercury poisoning in Japan and other parts of the world. An exception is an epidemic that resulted from exposure of 7000–10000 babies to phenylmercury fungicide, which was used to treat the cloth of diapers. 509 exposed infants were matched to 166 control infants. Among those exposed to diapers treated with phenylmercury fungicide, three infants developed pink disease, but the urinary excretion of mercury in exposed infants showed approximately 20 times greater concentrations of mercury than non-exposed controls. This recent epidemic, due to mercury exposure not related to teething powders, supports other mounting evidence that a threshold mechanism for toxicity of several metals including mercury exists (Gotelli et al., 1985).

INFLAMMATORY NEUROPATHIES

Autonomic dysfunction may be prominent in acute and rarely in chronic inflammatory neuropathy. Heart rate variability changes, orthostatic hypotension and changes in sweating and sphincter disorders result from demyelination of fibres in the vagus and glossopharyngeal nerves and in the innervation of the sphincters. Preganglionic sympathetic efferent fibres are also involved. Clinically syncope, sometimes with irreversible damage to the brain and sudden death due to cardiac arrhythmia, has been reported associated with autonomic disorders (Appenzeller and Marshall, 1963; Lichtenfeld, 1971).

Chronic inflammatory demyelinating neuropathies are rarely associated with clinically important autonomic dysfunction, but this can be demonstrated on special tests. This difference is related to the rapidity of the process, particularly the sudden conduction block, in the acute disease.

TOXIC AND MISCELLANEOUS CAUSES

Autonomic dysfunction occurs in alcoholic peripheral neuropathy, although postural hypotension is seldom disabling. Chronic alcoholics have evidence of vagal denervation, shown by abnormal heart rate responses, and also sympathetic dysfunction. Pathologically, giant sympathetic neurons and a relative preservation of the density and fibre diameters of myelinated nerves in the splanchnic, vagus and sympathetic efferents have been found but, in severe cases, degeneration in the vagus and sympathetic nerves occurs (Novak and Victor, 1974). The vagus is involved earlier and there is a relative preservation of sympathetic vasomotor control, in keeping with the predominant dying back features of alcoholic neuropathy where short and smaller diameter myelinated fibres in the sympathetic nervous system are relatively spared until later in the illness. There is a disturbance of oesophageal motility in chronic alcoholics with peripheral neuropathy as a result of damage to the vagus nerve (Winship et al., 1968).

Autonomic dysfunction has been reported in peripheral nerve disorders associated with vincristine (McLeod and Penny, 1969) perhexiline (Fraser, Campbell and Miller, 1977) heavy metals such as thallium (Bank, et al., 1972), arsenic (LeQuesne and McLeod, 1977) organic solvents (Matikainen and Juntunen, 1985), mercury (Kark, 1979) acrylamide (Auld and Bedwell, 1967), intermittent acute and variegate porphyrias (Stewart and Hensley, 1981), vitamin B6 intoxication, where peripheral vasomotor failure leads to the "burning hand, burning feet" syndrome, Charcot–Marie–Tooth disease (Appenzeller, 1963), rheumatoid arthritis (Edmonds et al., 1979), malignancy (Park et al., 1972), leprosy (Appenzeller, 1963), systemic lupus erythematosus and mixed connective tissue disease (Gudesblatt et al., 1985) and in some sensory neuropathies (Okajima et al., 1983).

A review of 86 patients with peripheral neuropathy showed autonomic disturbances in 45%. This excluded peripheral neuropathy due to diabetes mellitus, toxic causes and the Guillain–Barre' syndrome, where the frequency of autonomic disturbances was 41%. On the other hand, in familial amyloid polyneuropathy and in thallium neuropathy, autonomic dysfunction was present in all patients and in 80% of those with causalgia (Kuroiwa *et al.*, 1986).

Autonomic neuropathy has also been reported in chronic renal insufficiency (Appenzeller and Leuker, 1972). Evaluation of normal subjects and comparison with those with renal insufficiency before and after dialysis and of patients who have been dialyzed for some time, including those with diabetic endstage renal disease, showed autonomic abnormalities in uremia independent of other complications. Dialysis transiently improves autonomic deficits, particularly in non-diabetic endstage renal disease. In contrast, the condition of diabetic patients with chronic renal failure tends to deteriorate in spite of adequate dialysis and shows deteriorating sympathetic disorders in addition to parasympathetic dysfunction, which predominates in the non-diabetic uremic patients (Heidbreder, Schafferhans and Heidland, 1985).

In central autonomic failure of the Shy-Drager type (multiple system atrophy) several reports have implicated the peripheral nerves as well. Pathologically, lesions are present in the peripheral nervous system but not in all cases. However, only half the reported cases of autonomic failure with multiple system atrophy have had their peripheral nerves and ganglia examined at autopsy. About 8 showed changes in sympathetic ganglia, usually mild non-specific cell loss with some proliferation of supporting cells, but these reports remain unconvincing in the absence of comparisons with control material. There is, however, one case with definite evidence of damage to sympathetic ganglia (Rajput and Rozdilsky, 1976). There is no persuasive evidence of lesion in parasympathetic ganglia nor in visceral nerve plexuses, but occasional observations of normal innervation of gut, bladder, cardiac and peripheral ganglia, including the sweat glands in the skin, suggest that more sophisticated methods of histopathologic study are necessary to recognize abnormalities. In one patient with an additional severe sensory neuropathy, the 9th and 10th cranial nerves were also involved, including a marked loss of fibres of the tractus solitarius (Oppenheimer, 1983).

REFLEX SYMPATHETIC DYSTROPHIES

Numerous painful syndromes affecting limbs have been described, but the discussion here is limited to those which are accompanied by vasomotor and sudomotor dysfunction, and atrophy. Sometimes these syndromes are relieved by autonomic nervous system surgery and they may be related to injury to peripheral nerves.

Pain syndromes associated with overt or suspected peripheral nerve lesions are particularly common in areas of incomplete sensory loss due to traumatic transection or after surgery for repair of a peripheral nerve with only partial recovery or in conjunction with neuromas. There is usually a heightened threshold to ordinary stimuli in association with pain of a particularly unpleasant quality confined to the cutaneous distribution of the injured nerve. It is thought that the sensory abnormalities are due to incomplete regeneration of pain fibres in the cutaneous sensory network (Weddell, Sinclair and Feindez, 1948). Numerous clinical examples of this condition have been found many years after injury to the limbs (White and Sweet, 1955). The majority of patients eventually recover and do not require surgical intervention, but excision of neuromas or neurolysis is not usually effective. When the dysesthetic area becomes sensitive to cold, relief may be obtained by sympathectomy. This condition has to be distinguished from causalgia in which the disorder is not confined to a single nerve distribution.

CAUSALGIA

Causalgia is characterized by hyperpathia, trophic changes and autonomic phenomena. It occurs after peripheral nerve injury and usually involves either the arm or the leg. There is a partial nerve injury as a rule and the hyperpathia is most marked in the fingers and toes. Trophic changes in the skin are prominent, but are not different from those seen after peripheral nerve lesions without causalgia. The pain and dysesthesias are not confined to denervated areas, but involve a large part of the affected limb. Excessive sweating and vasodilatation occurs and the limb may be either hot or cold depending mainly on the degree of sweating. Cool limbs are particularly common in the later stages of the disorder (White, Heroy and Goodman, 1948). Because movement causes pain, the limb is kept immobile and the joints become stiff. Contact with clothing may be exceedingly unpleasant and patients may wrap their limbs in cotton wool or moistened towels. Many ordinarily well-tolerated stimuli cause excruciating pain in the affected limb.

It has been suggested that causalgia is a form of psychoneurosis. This assumption has, however, never been well documented. On the contrary, psychological studies of patients who have recovered showed no abnormal personality traits (Speigel and Milowsky, 1945). Pain is the most characteristic feature of this syndrome and the vasomotor and sudomotor abnormalities are thought to be secondary to the pain. The true pathogenesis of causalgia is not understood and continues to be a matter of contention (Janig, 1989).

Examination of paravertebral sympathetic ganglia showed different abnormalities from those associated with chromatolysis consistent with increased neurotransmitter synthesis (McLellan and Duckett, 1987).

SUDECK ATROPHY

Sudeck (1900) described a post-traumatic dystrophy in which after minor traumas, usually in the region of the wrist or ankle joint, atrophy and patchy decalcification of bone occurs. Pain may be severe and abnormal vasomotor and sudomotor activity is seen. Many changes in Sudeck atrophy are due to disuse because of pain. This condition can also be successfully treated by sympathectomy.

Sudeck atrophy has been reported associated with pelvic or lumbar spine lesions (Cayla and Rondier, 1974). It is more common in females and is often bilateral, nearly always involving the regions of both the foot and the knee. Though radiologically the hip joint may show osteoporosis, this is usually clinically silent. Pain of a causalgic nature is often present and brings the patient to the attention of the physician. Almost all patients had had surgical procedures on the back, but some were found after hip joint surgery and a number of cases occurred after infectious spondylitis and metastatic carcinoma in the pelvis. Two patients were pregnant and in another the condition developed after pelvic surgery. Stiffness of the joints, especially the ankle, is prominent, though reflex vasomotor abnormalities are less marked than in Sudeck atrophy. Radiology shows rarefaction of the bone, but this may not appear until several weeks after the onset of symptoms. The aggravation of symptoms by stress is not very prominent in this type of Sudeck atrophy, but precipitating traumatic factors are almost universally present. The best treatment for this type of reflex sympathetic dystrophy is lumbar sympathetic block, with additional local anaesthetic infiltration of the femoral artery perivascular nerve plexuses.

OTHER REFLEX SYMPATHETIC DYSTROPHIES

Other reflex sympathetic dystrophies occur in limbs without osteoporosis and are characterized by sweating, pain and vasomotor changes after minor trauma (Holden, 1948). In the early stages there is hyperaemia accompanied by oedema and rapid muscle atrophy. Pain on slight movement is very severe and the limb is kept immobile. In the later stages, oedema persists but the limb becomes cyanotic, cold and clammy. Vasomotor abnormalities have been found in unaffected limbs of these patients. The diagnosis should be made early and treatment by sympathectomy carried out so that movement of joints becomes possible again and secondary atrophy and demineralization of bones is prevented.

Some patients develop swelling of the legs, oedema and severe pain on movement after a considerable interval following back surgery. The oedema is often responsive to diuretics and there may be no trophic changes in the affected limb. On occasion, Raynaud phenomenon develops in the hands. The exact mechanism of this has not been determined, but it could result from an abnormal firing of the whole of the postganglionic sympathetic nervous system induced by the initial trauma to the back. Lumbar sympathectomy in such patients is also effective in removing the symptoms if carried out soon after the onset.

Another form of reflex sympathetic dystrophy is the shoulder–hand syndrome. This occurs commonly after myocardial infarction or hemiplegia. The shoulder becomes painful on movement, the hand may be sweaty with vasomotor changes and finally atrophy, particularly of hand muscles, occurs. The palmar aponeurosis is often thickened in longstanding cases (Bayles, Judson and Potter, 1950). Thickening of the skin of the palm is also seen in patients with longstanding hemiplegia, but the painful shoulder often develops within days of the onset of the weakness and cannot as a rule be attributed to adhesions of the shoulder joint. The cause of this syndrome is not known and sympathectomy is rarely necessary.

A case of causalgia complicating meningococcal meningitis has also been reported. This patient developed severe burning pain over the left foot and lower leg with paresthesia and this increased in severity associated with wasting and weakness in spite of rapid improvement of the meningitis with appropriate treatment. No local cause for the extensive denervation and pain was found on full investigation and the patient made a remarkable recovery after guanethidine blocks, the technique being similar to that of a Bier block. The good response to guanethidine blockade may result from blocking the larger fibre afferent input to the spinal cord (McLelland and Ellis, 1986).

On rare occasions, lumbar sympathectomy itself may lead to neuralgic pain. This usually occurs about 2 weeks after surgery. The pain is localized to the thighs and is deep and boring in character and disappears spontaneously, as a rule, within 10 days or 2 weeks. A few patients, however, have severe pain which is not relieved by narcotics. Several patients were markedly improved by the administration of phenytoin or carbamazepine (Raskin et al., 1974). The most common indication for sympathectomy in these patients was vascular intermittent claudication and leg ulcers. The mechanism has not been elucidated, but the effectiveness of the drugs, which alter excitability of nerve membranes and depress polysynaptic responses, suggests abnormal excitability in postsympathectomy neuralgia.

A motor paresis accompanying a clinical syndrome similar to reflex sympathetic dystrophy that included severe pain, swelling, coldness of limbs and muscle atrophy, was dramatically relieved by sympathetic blockade and exacerbated by sympathetic stimulation using adrenaline, noradrenaline, or isoproterenol hydrochloride (Yokota, Furukawa and Tsukagoshi, 1989). A dramatic improvement in four patients in strength after sympathetic blockade occurred even in the absence of pain, the paresis usually was of gradual onset and preceded pain in some of the patients. Day-to-day fluctuations in pain intensity were observed and precipitating factors for the reflex sympathetic dystrophy in these patients were not obvious in two and may have been an upper respiratory infection in one. Only one patient had spondylitic changes in cervical vertebrae and lumbar disc herniation. The weakness was primarily monoparetic and involved limbs only, though in one patient the facial musculature was also affected. Examination of the peripheral circulation using radionuclide angiograms showed no abnormality of peripheral blood perfusion. When reserpine, a sympathetic blocking agent was infused intravenously, the paresis was also improved. In these patients the paresis was thought to be due to abnormalities at the neuromuscular junction or in skeletal muscle receptors, but the paresis was different from that observed in myasthenia gravis, which was excluded on repetitive stimulation and single fibre electromyography and by lack of improvement of weakness with edrophonium. Paresis related to sympathetic dysfunction and improved by sympathetic blockade, termed by the author's sympathetic motor paresis, is important to recognize because of marked therapeutic benefit obtained by permanently blocking the paravertebral sympathetic ganglia related to autonomic function in the affected paralyzed limb.

MECHANISMS OF AUTONOMIC DISORDERS ASSOCIATED WITH PAIN

Numerous pain mechanisms associated with the sympathetic nervous system have been recognized. It is clear that under normal physiological conditions the efferent sympathetic nervous system is not involved in pain perception. Nevertheless, the sympathetic nervous system is involved in response to threatening, damaging signals from the periphery or viscera. These reactions have biological significance.   After injury, particularly that involving limbs with or without direct lesions to nerves, additional reactions to injury which involve a number of mechanisms bring the efferent sympathetic nervous system into play as a causative or permissive component in pain generation and its associated processes. Evidence that postganglionic noradrenergic fibres influence nociceptors that supply the extremities has emerged. This mechanism depends on several different conditions in the periphery that may operate singly or together. Chemical coupling is thought to be a well-established pathophysiological mechanism for some form of neuropathic pain. Ephaptic coupling between postganglionic and somatic afferent axons is still only hypothetical, but indirect coupling through alterations in the microenvironment of nociceptors, for example, causing a sensitization, is thought to be a likely mechanism by which the sympathetic nervous system acts as a promoter of afferent pain. Components that together change the neurovascular transmission are thought to be important in this mechanism (Janig, 1989).

Pain is a physiological or a pathological signal that induces a number of inter-related and complex events that affect the whole organism, including the somatic, the visceral sensory systems, the motor system, both somatic and sympathetic, and neuroendocrine regulation. Coordinated reactions in unanaesthetized organisms follow at various intervals the perception of pain.

The observations in animals and humans that the sympathetic nervous system is important in the generation and particularly in maintenance of neuropathic pain, whether in response to injury, chronic inflammation or even after central lesions, and after prolonged excitation from viscera and perhaps deep somatic tissue rests on the empiric success of sympathetic blockade in relieving pain and its associated clinical phenomena.   This method may be effective also in abolishing "pain behaviour" of appropriately instrumented animals.   There are numerous ways in which the sympathetic nervous system may influence the perception of acute pain and promote the persistence of chronic pain after peripheral nerve trauma or inflammation. These can be classed anatomically into primary afferent neuron, postganglionic sympathetic neuron, and into changes in spinal cord activity and direct effects of the sympathetic nervous system on target organs in the periphery. This includes muscle spindles, cutaneous mechanoreceptors, cutaneous nociceptors and some thermal receptors.

Peripheral nerve lesions or central lesions in the nervous system (rarely) and even chronic visceral stimuli may cause a clinical syndrome characterized by pain, disregulation of blood flow and sudomotor activity, abnormalities in motor activity, and trophic changes in the skin and its appendages, the subcutaneous tissues, bones and joints. The whole syndrome, although not always present, is now lumped into the category of reflex sympathetic dystrophy. Even temporary blockade may abolish these symptoms and this has been a major argument for the involvement of the sympathetic supply to the limb in the generation of pain. Since sympathetic activity is removed from the periphery by adequate blockade, the implication is that some coupling occurs between sympathetic outflow and afferent fibres in the generation of pain. What remains a puzzling aspect of this syndrome is that not all patients with appropriate lesions develop reflex sympathetic dystrophy and that, at present, there are no distinguishing features that separate those that develop this disabling syndrome from those whose lesions remain without this reflex sympathetic dystrophy. It is also difficult to understand that similar clinical phenomena are produced with and sometimes without involvement of the sympathetic nervous system, and in other situations the sympathetic nerves may only be permissive in the generation of pain.

The coupling of afferent myelinated and unmyelinated fibres after release of noradrenaline and following peripheral nerve lesions has been convincingly demonstrated in experimental animals and extensively reviewed by Blumberg and Janig, 1984.

Ephaptic transmission between postganglionic sympathetic and afferent axons in somatic nerves has not been experimentally observed. However, it is possible that such mechanisms may be instrumental even in the absence of either macroscopically or microscopically recognizable lesions in the generation of causalgic reflex sympathetic dystrophy. The envisioned mechanism is an interruption of nerve fibres by minor trauma which then causes sprouting into distal nerves and to the periphery between nerve fibres that have remained unaffected by the minor injury. In this way, coupling between postganglionic and afferent axons may occur.

In reflex sympathetic dystrophy, the blood vessels are clearly involved and modulate the environment of afferent nociceptor fibres. This environment includes the modulation of myogenic activity of blood vessel walls by the sympathetic nervous system, the vasoconstrictor and vasodilator supply, and also the release of vasoactive substances, including peptides, from many sensory endings and from cells, for example, mast cells and endothelial cells. In addition, the responses of these mechanisms to environmental influences so important in the generation of pain in reflex sympathetic dystrophy, for example, low temperatures, may also contribute to the reactivity and influence of autonomic fibres and their effect on nociceptor afferents.

Noradrenergic postganglionic activity causes vasoconstriction and the degree of constriction depends on tissue (endothelial cells) and ambient temperatures and on the metabolic state, for example, muscle contraction. When nociceptive afferents are activated, there is, in addition to the orthodromic nerve impulse traffic, also a release of a variety of substances, for example, substance P and other peptides that have a direct effect on blood vessel diameter, mainly precapillary vasodilation and an increase in vascular permeability, which is predominantly postcapillary, with the release of substances that may include kinins. These effects can be directly mediated but they also can be the result of indirect activation of mast cells with the release of their vasoactive substances, for example, histamine or serotonin. Experimentally, excitation of unmyelinated afferents causes vasodilation and extravasation, whereas thin myelinated afferent activation causes only vasodilation (Janig and Lisney, 1989).

The ambient conditions of nociceptive receptors and afferents are seriously disrupted during and after trauma of any kind, particularly when there is degeneration and regeneration of nerve fibres that result then in indirect effects on sympathetic nervous activity, on afferent fibre terminals and abnormal blood vessel reactions to environmental and local influences (Janig, 1985).

TROPHIC DISORDER

Trophic phenomena are related to the formation of connections between a nerve and its target organ. The maintenance of the integrity of target cells by their nerves is exemplified by the close dependence of muscles on their nerve supply and the regulation of some properties of target cells by innervating neurons which adapt the target to an intended task which can be changed by changing its innervation. Until the true mechanisms of trophic interactions has been clarified, this definition remains tentative.

The role of the nervous system in the maintenance of muscle is obvious and well documented. Acetylcholine, which is an established neurotransmitter at the neuromuscular junction, is necessary for trophic functions in the maintenance of muscle. But it is not clear whether an acetylcholine-mediated neuromuscular transmission alone is sufficient for trophic function *in vitro* or *in vivo*. In addition, there are other non-cholinergic factors that play a role in trophic activities, including axoplasmic flow. Some nerve trophic activity is closely related to adequate muscle contractions, as stimulation experiments show that usage alone can almost be sufficient to maintain muscle even if it is denervated. The spontaneous release of acetylcholine alone, however, is not sufficient for normal trophic function, but the possibility that it may contribute cannot be excluded.

The wide scope of neurotrophic relations that includes maintenance of connections and the development and regeneration of parts of the nervous system and other tissues of the body, has led to the suggestion that the term be restricted to functional manifestations necessary for the long-term maintenance and regulation of structures by their innervation but which are independent of nerve impulses. Since a number of different components are involved in intercellular relationships, including cell to cell contact, it cannot be expected that all trophic relations are based on the same mechanisms.

After peripheral nerve damage by a variety of injurious agents resulting in peripheral neuropathies, trophic changes occur and are part of the autonomic manifestations of peripheral nerve diseases. The characteristic atrophy of muscles is well known and is attributed to lack of trophic influence of motor fibres, since only minor changes are observed in muscles that remain innervated from experimentally isolated cord segments. The trophic changes in the skin are obvious and usually most marked in hands and feet, particularly when pain is a feature of the nerve lesion or when the affected parts are subjected to continuous trauma or applications of excessive heat or cold, which may not be appreciated as such because of sensory loss. The skin becomes tight and smooth, later shiny and transparent and the subcutaneous tissues may also be atrophic. Pigmentation, common in chronically denervated areas, may occur and eczema-like changes may be seen sharply localized to denervated areas. Minor mechanical or thermal trauma leads to indolent ulcers, particularly in fingers and toes, which may be painful even in the presence of extensive sensory loss of the affected parts. Fibrosis of subcutaneous tissues may occur particularly in partial nerve injury, the overlying skin being elevated into thick folds by the contraction of the fibrous tissue. Clubbing of the fingers is occasionally observed. The fingernails in areas of severe sensory loss become transversely striped, thickened, ridged, brittle and often claw-like. Nail growth, however, is not always retarded, except when the arterial supply to the limb has also become impaired. The hair is thinned or disappears from denervated areas of limbs but occasionally hypertrichosis is seen, as in neuropathies with immunoglobulin abnormalities and liver failure. Most marked trophic changes usually occur in traumatic nerve lesions and in chronic sensory neuropathies where trophic ulceration of the feet is very common. This is also a feature of diabetic neuropathy with marked sensory loss (Appenzeller, 1990).

VASCULAR AXON REFLEXES AND NEUROGENIC COMPONENTS OF THE INFLAMMATORY RESPONSE TO INJURY

Lewis (1942) published a monograph on clinical observations and the inflammatory response to injury of the skin. The vascular axon reflex (flare) is one component of Lewis's triple response and is important in the neurogenic inflammatory activity, though anatomically this function is not related to the autonomic nervous system but rather to unmyelinated peripheral nerves, the nociceptive C-fibres, it nevertheless, by affecting vascular smooth muscle through neurogenic mechanisms, is often discussed in the setting of the autonomic nervous system.

The mechanism of the flare component of the triple response is thought to be closely related to substance P and histamine. Intradermal substance P induces a flare and wheal similar to that produced by histamine, but it is 100 times more active on a molar basis. Histamine H1 and H2 antagonists are capable of inhibiting the flare, and it is thought that histamine is the direct effector of flare initiation and that it is released from mast cells by substance P which, in turn, is released from sensory nerve endings. Indirect support for this proposition is the close apposition of mast cells with substance P-containing fibres to form neuroeffector junctions around blood vessels. Mast cell degranulation with release of histamine has been demonstrated in response to antidromic stimulation and vasodilatation induced by such stimulation is prevented experimentally by pretreatment with a mast cell degranulating agent. In addition to substance P, other peptides are probably also important in the flare component of the triple response. The peptides somatostatin and neurokinin-A are derived from the same precursor (neurokinin-K), as is the peptide substance P. Calcitonin gene-related peptide is also a potent vasodilator and is colocalized with substance P in the same nerve fibres. Its role in neurogenic vasodilatation is as yet not fully established. This unique neurovascular system and its neurotransmitters and neuropeptides, though not clearly defined at the time of Lewis, nevertheless was recognized by him as protective, and he called this unique system the nocifensor system. While the clinical impact of this system has not been fully assessed, the overall response to injury of the skin is strongly influenced by its innervation. It is well known that denervated, but otherwise normal, skin is slow to heal. In such cases, the flare response provides evidence of neurogenic impairment of the inflammatory response and of the integrity of the nociceptive C-afferent fibres. Indirectly, therefore, because both afferent and efferent function is subserved by the same nerve, the flare response provides an indirect measure of afferent nerve function. From the aforementioned discussion, it is clear that the flare component may also be affected by local mechanisms such as mast cells and dysfunction of large and small vessels in the skin. Thus, an interpretation of abnormal responses implies knowledge of normal vascular function, and this presupposes evidence of normal vasodilation independent of neurogenic blood vessel innervation.

A new standard method of eliciting the flare has been used. This produces reproducible results, is non-invasive and uses electrophoresis of 10% acetylcholine for 5 min to the skin from a small capsule to which a laser Doppler flowmeter is attached in the centre, which measures blood flow. This method of measuring the response to acetylcholine iontophoresis is independent of the extent of the flare, which is often irregular and difficult to measure. The vasodilation produced by acetylcholine electrophoresis has been shown to be due to an axon reflex (Parkhouse and LeQuesne, 1988). The flare produced by this method is likely to be due to the known stimulation of unmyelinated C-fibres by acetylcholine and clinically has the spread and irregular outline recognized after stimulation of C-fibres by histamine, and the flux recorded by the laser Doppler flowmeter is very similar to that found after intradermal histamine injection and induction of the flare (LeQuesne and Parkhouse, 1989).

AUTONOMIC DYSFUNCTION IN POSTPOLIO SUBJECTS

The temperature is often abnormally low in limbs affected by poliomyelitis and this may occur in mildly cold ambient conditions. Severe burning pain is also often present and hyperesthesia and colour changes are frequent, ranging from cyanosis, mottling to reddish violet or deep blue. The colour changes and discomfort persist for hours and lengthy rewarming to relieve pain is often necessary. Many patients also report that during these periods, the dexterity of the limb and the muscle strength decreases considerably. This cold-related impairment is exaggerated with advancing age.

The hypothesis that these manifestations represent a reflex sympathetic dystrophy is supported by histological damage to the intermediolateral cell columns after polio virus infection. This damage has been said to cause a decrease in inhibition of sympathetic outflow to the affected limbs, one consequence of which would be an increase in tonic sympathetic vasoconstrictor tone. Mild ambient temperatures and particularly cold might produce exaggerated vasospastic phenomena, resulting in a marked decrease in limb temperature. However, a recent re-examination of spinal cords in some postpolio patients failed to find damage in the intermediolateral cell columns (Pezeshkpour and Dalakas, 1988).

This interpretation of the postpolio syndrome does not account for the trouble-some pain and dysesthesia. Moreover, no consistently lower skin temperature or decreases in muscle blood flow in affected limbs in thermoneutral ambient temperatures have been found (Abramson *et al.*, 1943). Inconsistent findings of temperature changes in affected limbs during the day and the violet hue of the extremity, which suggest venous dilatation and not arterial constriction, also argue against an increase in sympathetic tone due to a fall in inhibitory output from the intermediolateral cell columns.

Further study of postpolio subjects showed that equal decreases in blood flow with cooling occurred in affected limbs and in controls, suggesting that the sympathetic vasoconstrictor outflow was normal and not increased. But mental arithmetic, which causes an increase in sympathetic tone under normal conditions, initiated by cortical activity, was associated with a decrease and often paradoxical vasomotor response, suggesting that stress-induced vasoconstriction is decreased. This may perhaps result from both pre- and postganglionic sympathetic neuron damage. Mean blood flow in affected limbs in 30°C ambient temperature was lower, perhaps due to damaged sympathetic vasoconstrictor neurons, resulting in passive dilatation of the cutaneous venous (capacitance) vascular beds. This would amount for the venous congestion that can cause a decrease in arterial inflow to the skin. These two conditions will promote an increase in heat loss from the skin and result in the very cold limbs often found in such patients.

Nerve conduction and latencies are delayed in the affected limbs, but this could be the result of the colder limb temperature. The difficulty in rewarming cold postpolio limbs is probably related to an increase in alpha-adrenergic receptor sensitivity with decreasing temperatures. This would accentuate vasoconstriction and further reduce blood flow and promote stasis in the limb.

Cold-induced pain of postpolio limbs remains to be elucidated, but may be related to local hypersensitivity or change in peptide or catecholamine content of nerves or of perivascular nerve plexuses. The cold-associated decrease in conduction velocity has been implicated in the reduced dexterity of affected limbs with cold exposure (Bruno, Johnson and Berman, 1985).

FAMILIAL DYSAUTONOMIA (RILEY–DAY SYNDROME)

This congenital condition is characterized by diminished lacrimation, hyperhidrosis, blotching of the skin, which may be transient, abnormal swallowing reflexes, impaired vestibular reflexes, labile blood pressure, disturbed temperature control, emotional instability associated with episodes of severe vomiting, poor coordination, relative insensitivity to pain and diminished deep tendon reflexes; all suggest peripheral nervous system involvement which has indeed been documented morphologically. Most cases occur in Jewish children, and extensive reviews of the subject have been published (Riley, 1957; McKusick *et al.*, 1967; Aguayo, Nair and Bray, 1971; Pearson, Axelrod and Dancis, 1974).

A combination of pain and temperature sensation loss with extensive autonomic failure is the hallmark of his autosomal recessive disorder. Early manifestations, such as poor sucking, swallowing incoordination, gastroesophageal reflux and irritability, and autonomic "crises" that may last several days may occur. Tachycardia, hypertension and hyperhidrosis, including skin blotching, have been tentatively attributed to an increased sensitivity to circulating catecholamines. In between crises, postural hypotension without an increase in heart rate is usually found. There is lack of lachrymation, even during emotional crying, but the pupillary responses are normal, yet methacholine does cause an exaggerated pupillary constriction consistent with denervation supersensitivity. An early recognizable hallmark of the condition is an absence of fungiform papillae of the tongue, which fail to develop. High-altitude exposure of these patients fails to elicit the normal response to hypoxia and hypercarbia is also not associated with an increase in ventilatory rate. Some patients hold their breath for prolonged periods without discomfort. Though sensory loss is initially mainly in cutaneous pain and temperature discrimination, touch is well preserved, as is visceral pain. Later, a loss of joint position sense is accompanied by absent tendon jerks and hypotonia. The sural nerve shows an extensive loss of unmyelinated axons and less severe depletion of small myelinated and of the largest myelinated fibres, and the number of neurons in the dorsal root ganglia, and in the autonomic and trigeminal ganglia are reduced.

Clinical–pathological correlation is now possible for some of the symptoms. Impaired taste results from reduced numbers of, or defective, lingual taste buds. If sural nerve changes are representative of those in the rest of the peripheral nervous system, postural hypotension, altered catecholamine excretion, adrenergic denervation supersensitivity, skin blotching and impaired thermoregulation can be accounted for. Unmyelinated C-fibre loss in peripheral nerves is consistent with impaired pain and temperature perception, and absent corneal and axonal reflexes. Because large myelinated fibres are rare, the tendon stretch reflexes dependent on muscle spindle afferents subserved by these fibres are also impaired or abolished. The pathogenesis of familial dysautonomia is not known. It has been suggested that nerve growth factor is involved and that developmental arrest of neuronal migration from the neural crest explains the absence of unmyelinated and large myelinated fibres, including muscle spindle afferents.

An allied condition, described by Nordborg et al., 1981 in three patients with a congenital non-progressive sensory neuropathy with dysautonomia, is clinically reminiscent of familial dysautonomia. However sural nerve biopsies showed a severe loss of all myelinated fibres and a reduction of unmyelinated axons. This morphological pattern is akin to that found in the recessively inherited sensory neuropathy type II of Dyck et al., 1983.

A mutilating acropathy in a family with autosomal recessive inheritance of the condition and accompanied by loss of pain and temperature sensitivity and impaired sudomotor function and keratitis has also been reported (Donaghy et al., 1987). In these patients there is a selective loss of small myelinated fibres. Some unmyelinated axons may also have been lost but this could not be well documented. These patients have some resemblance to those reported as having congenital sensory neuropathy (Low, Burke and McLeod, 1978).

Other inherited diseases can produce the clinical manifestations of familial dysautonomia. If the diagnosis is to be made in acutely ill infants, this possibility should be kept in mind.   Hyperammonemia, due to propionyl-coenzyme-A carboxylase deficiency, is an example of metabolic abnormality that can produce phenocopies of familial dysautonomia. (Harris *et al.*, 1980) The diagnosis of familial dysautonomia in infants depends on abnormal responses to intracutaneous histamine and methacholine (absence of flare component of triple response of Lewis that depends on intact small myelinated pain afferents) and abnormal urinary catecholamine excretion. In the infants with hyperammonemia, however, appropriate dietary manipulation results in reversal of the clinical symptoms.

LESCH–NYHAN SYNDROME

The Lesch–Nyhan syndrome, a disorder of purine metabolism, is characterized by hyperuricemia and excessive uric acid production. It is X-linked and manifests profound neurological dysfunction, including spasticity, choreoathetosis, self-mutilation and mental retardation. There is an almost total absence of the enzyme hypoxanthine-guanine-phosphoribosyl transferase (HGPRT). The pathogenesis of behavioural and neurological disturbances in patients with HGPRT deficiency is not known. Because of self-mutilating behaviour induced in animals by caffeine or amphetamine administration, it has been suggested that this part of the Lesch–Nyhan syndrome results from altered function in central nervous system pathways affected by these agents. A study of patients with the HGPRT deficiency and self-mutilating behaviour shows a unique pattern of adrenergic dysfunction. Plasma dopamine beta-hydroxylase, an enzyme which catalyzes noradrenaline formation from dopamine, is elevated. This enzyme is released simultaneously into the synaptic cleft with noradrenaline. Though this enzyme was elevated in patients with Lesch–Nyhan syndrome, none of them had autonomic manifestations of excessive adrenergic activity such as hypertension, tachycardia or mydriasis. But clinical evaluation of adrenergic responsiveness such as the cold pressor test showed that the normally expected rise in blood pressure (due to vasoconstriction) was absent in those with the HGPRT deficiency syndrome who also exhibited self-mutilation. The mechanism of this failure of vasoconstriction is not known. Any implied relationship to high plasma dopamine beta-hydroxylase activity remains conjectural. The experimentally induced self-mutilation of animals by caffeine injection may link this behavioural abnormality to the endorphin system, since caffeine is a potent stimulator of plasma, but not cerebrospinal fluid beta-endorphin release.

HEREDITARY AUTONOMIC NEUROPATHIES

Two distinct categories are recognizable, the first with known metabolic disorders, examples being porphyria and inherited amyloid neuropathies, and in the latter category without metabolic abnormalities, which usually also include symptomatic sensory impairment occurring together with autonomic dysfunction. This group is now referred to as hereditary sensory and autonomic neuropathies.

Peripheral neuropathy with autonomic involvement occurs in acute intermittent, variegate porphyria and in hereditary coproporphyria. These are autosomal dominantly inherited disorders. The autonomic dysfunction in these disorders may precede, accompany or follow the predominantly motor disability and include abdominal pain, constipation, and painless and persistent vomiting. The bladder involvement is characterized by difficulty in initiating micturition and retention of urine. Sympathetic and parasympathetic effects on the cardiovascular system are noteworthy, and include tachycardia and hypertension during the acute disturbance and an increase in circulating catecholamines. A baroreceptor denervation has also been recognized and chronic hypertension may result, but the devastating acute autonomic dysfunction during attacks often improves when attacks subside (Laiwah *et al.*, 1985). Sudden death has been reported with acute attacks and is thought to be due to cardiac arrhythmias. The eye shows pupillary disturbances with impaired accommodation in acute attacks and hyperhidrosis on the face is often found. A dying-back peripheral neuropathy with chromatolysis in cell bodies is also found in the coeliac ganglion cells, but the metabolic disturbances leading to axonal degeneration remain to be determined.

HEREDITARY AMYLOID NEUROPATHIES

At least seven varieties of inherited amyloidosis affecting the peripheral nerves have been reported and are related to an abnormal transthyretin, which has been shown to have an isovaline substitution at position 30 in the molecule. Autonomic symptoms are often a very early feature of these patients and include postural hypotension, anhidrosis, micturition problems, constipation, diarrhoea, impotence and pupillary abnormalities. Many patients develop perforating ulcers in the feet, related to but not entirely the result of the loss of pain sensation. Neuropathic joints also are frequently found, and these might be associated not only with a loss of deep pain sensitivity, but also an abnormality of the perivascular autonomic innervation of the blood supply to the joints. Amyloid deposits are found not only in the peripheral nerves but also in sensory and autonomic ganglia. The neuropathy is of "small fibre" type and affects initially unmyelinated and small myelinated axons.

In the Iowa form of familial amyloidotic polyneuropathy and other hereditary amyloid neuropathies, autonomic involvement is less prominent than in the Portuguese variety.

HEREDITARY SENSORY AND AUTONOMIC NEUROPATHIES (HSAN)

These disorders are not fully characterized because responsible genes have not been as yet identified. Their classification is based on inheritance and clinical features. The dominantly inherited sensory neuropathy, first recognized by Denny-Brown (1951), has not been fully investigated to delineate the extent of autonomic involvement, though the perforating ulcers and distal mutilation and neuropathic joints suggest autonomic involvement, while clinically, in addition, bladder dysfunction is common. This type has been classified as HSAN Type 1.

HSAN Type 2, in the classification of Dyck *et al.*, 1983, is an autosomal recessively inherited sensory neuropathy. In this type, bladder dysfunction and impotence have been reported and autonomic tests carried out by Johnson and Spalding (1964) were negative.

A rare autosomal recessive disorder accompanied by motor retardation and unexplained episodes of fever, together with failure to normally react to painful stimuli, has been reported. In this condition, loss of pain and temperature sensation and anhidrosis, together with cutaneous ulceration, painful fractures and self-mutilation, have been found. A selective loss of small myelinated fibres occurs and unmyelinated axons are almost totally absent. Morphologically, this hereditary anhidrotic sensory neuropathy closely resembles the reported neuropathy exclusively found in Navajo children (Johnson and Johnsen, 1987).

ETHNIC AUTONOMIC SYNDROMES

Navajo Neuropathy

Navajo neuropathy was first reported by Appenzeller, Kornfeld and Snyder, 1976. It is a unique sensory motor neuropathy with corneal ulceration and acral mutilation hitherto confined to one ethnic grip, Navajo children. Some 20 cases have been identified, and more patients are being recognized. The onset is between 5 days to 4 1/2 years of age (mean 1.2 years), and the presenting symptom is usually an infection (in 28%), corneal ulceration (in 20%), liver disease (in 24%) and failure to thrive (in 24%). All cases have had weakness with sensory deficits (80%), and an abnormal sural nerve biopsy (12 out of 13 showed decreases in large and small myelinated fibres and relative preservation of unmyelinated axons). Pneumonia, meningitis and other sepsis are common. Eventually, 75% of patients develop corneal ulceration, and liver disease causes morbidity and mortality in a large proportion. The level of protein in the cerebrospinal fluid may be elevated without a cellular response. Magnetic resonance imaging (five out of seven) showed early cerebellar white matter lesions (the "matutine cerebellar butterfly"), progressive cerebral white matter lesions and cervical cord thinning. The patients eventually become wheelchair-bound, and death has occurred between the ages of 6 months and 25 years after onset (mean 9.9 years). The oldest living case is now 27 years old. Affected siblings have been identified, and the condition is unrelated to Cree encephalitis or leukoencephalopathy, except that all these diseases seem restricted to Athapaskan North American Indians (Singleton *et al.*, 1990). Autonomic dysfunction, though present, is overshadowed by the severe and mutilating neuropathy and the innervation of blood vessels in the sural nerve (vasa nervorum), suggests a partial depletion of noradrenergic perivascular nerve fibres.

One infant with the disease became symptomatic at 2 months of age with respiratory difficulties. By 7 months he was ventilator dependent, paretic and hypotonic with barely obtainable tendon jerks. No peroneal nerve conduction could be obtained. Sural nerve biopsy showed axonal degeneration with a decrease in myelinated and preservation of unmyelinated axons similar to the findings in other cases of Navajo neuropathy. This case extends the age range of onset of the disease, implying intrauterine insult to the nervous system or its development or degeneration of the myelinated nerve fibres. This case also supports the proposition that corneal ulceration present in the older children, but not found in this infant, results from corneal anaesthesia and repeated injury rather than from corneal dystrophy, as reported in the patients of Donaghy *et al.*, 1987.

Navajo Neuropathic Arthropathy

Eight patients with this condition have been reported (Johnson and Johnsen, 1987). This disease is confined to Navajo children, is also familial, and is characterized by hyperkeratosis of the palms, Charcot joints, unsuspected fractures, with clinically intact neurological examination, except for heat intolerance. Electromyography and nerve conduction studies are normal. Deep pain, however, has not been systematically examined in these patients. Sural nerve biopsies showed a mild loss of small myelinated and more prominent reduction of unmyelinated fibres and fibrosis. This contrast to the biopsy findings in Navajo neuropathy, where the unmyelinated fibre population is almost entirely preserved, with marked loss of myelinated nerve fibres. In contrast to Navajo neuropathy, the development of these children is usually normal until the second decade of life, when painless fractures were present in all, primarily in weight-bearing bones. A progressive, severe arthropathy developed mostly in the knees and ankles associated with recurrent joint swelling. Fusion and fixation of the affected joints were required, and the history of heat intolerance and lack of sweating without noted hyperpyrexia was present in all cases. In contrast to patients with Navajo neuropathy, there was no mutilation and there were no clinical somatic neurological deficits, in particular, no corneal ulceration; but pupillary reflexes were difficult to elicit in some patients and in others were entirely absent.

Extensive autonomic testing of one patient with this condition [Appenzeller, unpublished observations] showed sympathetic denervation of the postganglionic type and, in addition, an abnormal Lewis's shunting reaction, suggestive of a primary abnormality of blood vessels as well. The pupil cycle time was 1720 ms (upper limit of normal, 1000 ms). In this disease, from which no recovery has yet been reported, extensive autonomic deficits occur in a setting of relative preservation of somatic neurological function, including normal electrophysiological tests, although the latter do not assess small-fibre function. The central nervous system in these patients has not as yet been imaged. Neither the vasa nervorum and their innervation in the sural nerve, nor the innervation of major blood vessels have yet been evaluated (Appenzeller, 1989).

AUTONOMIC NERVOUS SYSTEM DYSFUNCTION IN HUMAN IMMUNODEFICIENCY VIRUS INFECTION

Both the central and peripheral nervous systems have been shown to be directly involved by the human immunodeficiency virus.   Whether direct autonomic nervous system invasion occurs by the virus remains controversial.   Testing of autonomic function in patients with AIDS showed abnormality in four out of five (Craddock *et al.*, 1987).   In a series of 10 patients who were sero-positive (HIV-positive), five had autonomic nervous system involvement, determined by quantitative sudomotor axon reflex testing for assessment of postganglionic sympathetic sudomotor function, orthostatic blood pressure recording and heart rate variability to deep breathing and Valsalva's manoeuvre. All five patients with involvement of both parasympathetic and sympathetic nervous systems were in Group 4 of the Center for Disease Control classification; that is, they had overt AIDS rather than acute asymptomatic or persistent generalized lymphadenopathy. The dysfunction of the autonomic nervous system in these patients may have been the result of central or peripheral involvement. However, autonomic abnormalities were not associated with clinical evidence of peripheral somatic neuropathy in this series, and one patient with peripheral neuropathy and AIDS did not have autonomic nervous system dysfunction. It is also important to realize that none of the patients had dysfunction of the autonomic nervous system as the first manifestation of HIV infection (Cohen and Laudenslager, 1989).

VASA NERVORUM: POSSIBLE ROLE IN AUTONOMIC NEUROPATHY

The term "vasa nervorum" is used here collectively to denote all blood vessels found in the endoneurium. A distinction can also be made between blood vessels found in various fibrous compartments of peripheral nerves and the terms "endoneurial", "perineurial", and "epineurial" vasa nervorum are used. It is assumed that the vasa nervorum play a crucial role in nutritional maintenance and in the normal function of peripheral nerves. This assumption is based on the fact that the vasa nervorum change caliber, depending on physiologic demands, and in pathologic conditions. The mechanism by which these vessels control blood flow or are themselves controlled remains obscure.   The proposition that innervation of blood vessels plays an important role in the regulation of blood vessels and in the regulation of blood flow to nerves under normal and pathologic conditions is supported by analogy of the role of these nerves in other tissues, particularly the brain (Dhital and Appenzeller, 1988). Experimentally, stimulation of the lumbar sympathetic chain causes general vasoconstriction of intraneural vessels and the presence of adrenergic nerve terminals in the walls of vasa nervorum has also been demonstrated histochemically (Amenta, Mione and Napoleone, 1983).

   The permeability of vasa nervorum has been studied experimentally using a number of different techniques in a variety of diseases.   It has been examined also after experimental nerve injury produced by trauma, toxins, crush lesions, transections, or ligature. All these manipulations disrupt the blood-nerve barrier initially, and a second wave of increased permeability of vasa nervorum and disruption of the blood-nerve barrier occurs 14 days after injury. The delayed response is a result of distal and distant blood vessel abnormalities. The role of the innervation of vasa nervorum and permeability changes in the responses to toxins and mechanical manipulations away from the site of injury may be important but has not as yet been defined.

INNERVATION OF VASA NERVORUM AND THE NERVI NERVORUM IN HUMAN SURAL NERVES

The vasa nervorum and nervi nervorum were studied in sural nerves at the level of the ankle in normal controls and patients with chronic alcoholism, diabetes, and some with non-alcoholic, non-diabetic neuropathy (NAND). Normal controls and the majority of neuropathic nerves were obtained at autopsy after sudden death, whereas 6 were obtained at biopsy. Plexuses of dense catecholamine-containing perivascular nerves of extraneurial nutrient arteries were found. Veins were not innervated. Epineurial arteries had a similar innervation. Perineurial arteries showed scant parallel noradrenergic fibres and only occasional veins were sparsely innervated. The endoneurial blood vessels showed a scanty innervation, including the veins, with adrenergic, perivascular nerve plexuses. The nervi nervorum containing noradrenaline were present in the epineurium, less commonly in the perineurium and not in the endoneurium. Neuropeptide Y (NPY) was perivascular with occasional nervi nervorum containing this putative neurotransmitter. Vasoactive intestinal polypeptide (VIP) was found in perivascular plexuses of nutrient arteries and in occasional nervi nervorum. Calcitonin gene-related peptide (CGRP), cholescystokinin (CCK) and 5-hydroxytryptamine (5-HT) were found in myelinated and unmyelinated nervi nervorum. Similar studies in sural nerves of aborted fetuses showed no noradrenergic perivascular nerve plexuses, but in advanced neuropathies (alcoholic, diabetic, and NAND) a denervation of the vasa nervorum was found; conversely, adrenergic and VIP-ergic nervi nervorum were relatively preserved in such nerves. In acute inflammatory neuropathy, the perivascular innervation and nervi nervorum appeared normal in distribution and transmitter content. In minimally symptomatic diabetic neuropathy and in Navajo neuropathy, a reduction in the perivascular innervation of vasa nervorum was found (Appenzeller, Lincoln and Zumwalt, 1989). The autonomic innervation of the vasa nervorum and the nervi nervorum needs further study in larger numbers of patients with neuropathies and more controls to delineate a possible role of neurotransmitters of the autonomic perivascular nerves and of nervi nervorum in the genesis of such neuropathies.

Quantitative analysis of catecholamines in human sural nerves showed a preservation of catecholamine content of the nerves and the epineurium in alcoholic neuropathies, when compared to controls. However, diabetic and NAND showed a significant decrease in catecholamine content, and this was independent of age or postmortem time.

APPROACH TO PATIENTS WITH SYMPTOMS SUGGESTIVE OF AUTONOMIC DYSFUNCTION

It is important to establish the presence or absence of autonomic dysfunction in patients with peripheral neuropathy for diagnostic and prognostic purposes. Certain peripheral neuropathies have prominent, if not exclusive autonomic dysfunction (amyloid neuropathy, diabetic neuropathy, and acute pandy-autonomia). In other neuropathies, autonomic dysfunction may be symptom-less, but may become important and life-threatening (Landry–Guillain–Barré syndrome). Delineation of autonomic deficits (Table 7.1) may, therefore, help with diagnosis, prognosis, or to anticipate and avoid serious complications (Burgos *et al.*, 1989).

PATHOGENESIS OF AUTONOMIC FAILURE IN PERIPHERAL NERVE DISEASE

Cholinergic postganglionic sympathetic unmyelinated fibres form part of peripheral nerves and innervate sweat glands. In peripheral neuropathy these fibres may degenerate or there may be demyelination of preganglionic sympathetic efferent fibres. This accounts for the loss of sweating, but hyperhidrosis may be seen after partial nerve injuries, as part of causalgia or in malignant and toxic neuropathies. Compensatory sweating may appear when the degree of sweating loss on the extremities and trunk (particularly in diabetic autonomic neuropathy) is widespread.

Postural hypotension is caused by damage or loss of small diameter myelinated and unmyelinated fibres in afferent and efferent nerves of the baroreflex arcs. This is most often seen in diabetes mellitus, in amyloidosis, and also in tabes dorsalis, a primary disease of roots. Small fibres are also affected in acute inflammatory neuropathies with segmental demyelination that involve myelinated autonomic fibres in the vagus and sympathetic pathways. When the splanchnic vascular bed is extensively denervated, orthostatic hypotension is marked since these blood vessels lose tone and are of crucial importance in maintaining blood pressure by neurogenic vasoconstriction. The impairment of heart rate control (heart rate variability) is predominantly due to parasympathetic denervation and is seen in those with autonomic neuropathy including diabetes mellitus and respiratory neuropathy (Appenzeller, Parks and Macgee, 1968).

Disturbances of bladder function, impotence, and pupillary abnormalities occur in peripheral nerve disease if autonomic fibres to these structures are involved. However, autonomic dysfunction, particularly orthostatic hypotension, is not common in dying back neuropathies, which affect mostly long peripheral nerve fibres with a large diameter. There is, therefore, a relative sparing of shorter autonomic fibres with preservation of autonomic function.

TREATMENT

General Principles

The most disabling symptoms of autonomic failure are due to cardiovascular dysfunction. Orthostatic hypotension with decreased perfusion of the brain is the most troublesome. It is of particular importance, however, not to be too concerned about low standing blood pressures if the patients remain asymptomatic. With chronic autonomic failure, patients may tolerate very low standing blood pressures without dizziness or syncope. This is attributed to maintenance of cerebral blood flow by remarkable cerbrovascular autoregulation leading to considerable vasodilatation and preservation of perfusion in the face of low blood pressures. Whatever the correct explanation may be clinically, patients with autonomic failure have great tolerance to low blood pressures without developing symptoms of cerebral ischemia. Because of defective baroreflexes, orthostatic hypotension may be associated with paradoxic recumbent hypertension which may complicate the treatment of the postural fall in blood pressure.

While the loss of baroreflex function is of primary importance in the immediate response of the blood pressure to standing, the control of blood volume regulated by low pressure receptors in the kidney through the release of antidiuretic hormone and the renin-angiotensin-aldosterone system is important in the long-term adjustment of patients with autonomic failure to postural hypotension.

There are two principles of treatment of postural hypotension. One involves the reduction of the volume which is available for blood pooling on standing, and the second is to increase the volume of blood available for pooling. Drugs which decrease the capacity for blood to pool below the heart, however, have a tendency to increase the occurrence of recumbent hypertension, and those that increase available blood volume may overload the circulation, causing cardiac failure and edema.

Any measure which temporarily restores the patient to his feet irrespective of the actual standing blood pressure will enhance homeostatic mechanisms which are triggered by standing. For example, an increase in extracellular fluid volume will occur and improved myogenic tone of blood vessels can be expected. Thus, a continued symptomatic improvement in postural hypotension may be erroneously attributed to specific treatment where, under controlled conditions, it may be possible to withdraw a dangerous drug and replace it by mechanical supports of circulation.

In principle, it is advisable to use combination therapy since most patients with autonomic failure have defects at various levels of the baroreflex pathways. Drugs with central or ganglionic and postganglionic effects may have synergistic effects and require lower doses. Other drugs, which increase noradrenaline release, can be combined with those which reduce its re-uptake or which increase receptor sensitivity. In general, however, it is best to attempt therapy by mechanical methods before complicating matters with the administration of powerful pharmacologic agents.

It is possible to increase blood volume by a head-up position at night. This method of treatment, tried by Bannister, Ardill and Fentem, (1969), evidenced by an increase in body weight overnight, presumably due to increases in extracellular fluid volume. Many patients require nothing more than persistent head-up tilting while asleep. Studies have shown that this position prevents the increased sodium and water loss during the night which may explain the retention of fluid and peripheral edema observed in normal subjects who are confined to a head-up position in crowded aircraft on trans-Atlantic flights.

The head-up position also promotes renin release because of reduced renal arterial pressure. Consequently, angiotension II formation and aldosterone stimulation occur and this, in turn, increases blood volume.

Drugs

Fludrocortisone is commonly used; it has many pharmacologic effects. The initial dose is 0.1 mg per day. The drug increases effective vasoconstriction because it augments the action of noradrenaline release by normal sympathetic efferent activity; it does not usually aggravate recumbent hypertension. Fludrocortisone increases the sensitivity of vascular receptors to pressors and it may increase fluid content of blood vessel walls, therefore decreasing their distensibility. Because patients with postural hypotension are sensitive to sodium intake it is necessary to support all methods of treatment with a high sodium intake provided this is not contraindicated for other reasons.

Simple external supports to reduce the volume for blood pooling in the legs and abdomen are useful, but because they are cumbersome they are not usually accepted for long. Pressor drugs, which cause improvement in postural hypotension, include phenylephrine, which has a direct sympathomimetic action and, less effectively, ephedrine, an indirect sympathomimetic drug. These drugs have a tendency to aggravate recumbent hypertension. Midodrine (Schirger *et al.*, 1981) has been used. The constrictor effects of this drug are attributed to its alpha-agonist activity on both arterioles and veins.

Monamine oxidase inhibitors and tyramine have also been used (Diamond, Murray and Schmid, 1970). The effects of tyramine are attributed to release and re-uptake of noradrenaline at sympathetic endings. This drug requires an intact sympathetic supply to blood vessels for effectiveness. When tyramine is given with a monoamine oxidase inhibitor in the presence of adrenergic denervation supersensitivity and baroreflex block, severe hypertension can occur. Moreover, careful trials with pure tyramine and monoamine oxidase inhibitors have shown erratic blood pressure responses and aggravation of recumbent hypertension (Bannister, 1983).

Dihydroergotamine has been advocated because of its direct vasoconstrictor effect on smooth muscles of capacitance vessels (veins). It acts as a direct alpha-agonist and increases central blood volume but, in autonomic failure, it causes constriction of resistance vessels as well. After intravenous injection, dihydroergotamine, though highly effective in abolishing postural hypotension, causes severe recumbent hypertension. The oral dose, which may be needed for control of postural hypotension, may be 30-35 mg daily. It is more effective given intramuscularly in small doses during the day to avoid recumbent hypertension at night.

The considerations which led to the recommendation of indomethacin for postural hypotension are based on its antiprostaglandin activity. Prostaglandins are potent vasodilators, and the decreased levels of prostaglandins after indomethacin therapy are helpful in autonomic failure. Indomethacin has a number of other effects, however. It increases sensitivity to infused noradrenaline and angiotensin II and its effectiveness is now attributed to these actions rather than to prostglandin synthesis inhibition.

Propranolol, a beta-blocker, is used in treatment of postural hypotension because beta-agonist-induced vasodilatation might contribute to symptomatic orthostatic hypotension. Beneficial effects have been reported in patients with progressive autonomic failure using doses of 40-240 mg per day, but these patients were also taking fludrocortisone and added salt to their diet. It may, however, be useful in patients with postural hypotension due to tachycardia leading to a falling cardiac output and, eventually, syncope.

Other drugs used with varying success are pindolol, given to patients with diabetic autonomic neuropathy in whom there was a demonstrated denervation supersensitivity to noradrenaline (Frewin et al., 1980). Some patients have an improved cardiac output but unchanged vascular resistance while taking this drug. However, subsequent reports have not been enthusiastic and pindolol, like metoclopramide, a dopamine agonist, have now been abandoned. Clondine, a partial alpha-receptor agonist, acts centrally and peripherally, causes hypotension normally, and is used in the treatment of hypertension. It causes an inhibition of parasympathetic tone by its primary action on the nucleus of the tractus solitarius. Recently, a report in four patients with autonomic failure showed that an oral dose of 0.4 mg twice daily caused long-term improvement in those with low plasma noradrenaline levels and denervation supersensitivity but no further systematic studies in patients with peripheral autonomic failure have been reported.

The use of L-threo-3,4-dihydroxyphenylserine (L-threo-DOPS), a precursor of noradrenaline has been found beneficial in one patient with multiple system atrophy (Kachi et al., 1988), but has not as yet been reported to influence orthostatic hypotension in peripheral autonomic failure.

Thermoregulatory Activity

Because many patients with peripheral autonomic failure have abnormal sweating and vasomotor responses, heat tolerance may be severely impaired. Proper advice for regulation of microclimate (clothing) and exposure to heat should be given to avoid serious complications resulting from excessive heatloads or hypothermia (Appenzeller, 1990).

CONCLUSION

The treatment of postural hypotension and of autonomic peripheral failure, in general, have been disappointing. In those conditions in which the deficits are transient, all efforts should be made to control the patient's internal and external environment either mechanically or with the aid of drugs which mimic normal modulatory activity of the autonomic nervous system. Conditions in which recovery is not expected also need treatment, but the long-term effects and, particularly, side effects of therapy should be carefully balanced with the expected benefits. In general, it is better to tolerate minor disability than cause serious disabling side effects.

REFERENCES

Abramson, D.I., Flachs, K., Feiberg, J. and Mirsky, A. (1943) Blood flow in extremities affected by anterior poliomyelitis. *Archives of Internal Medicine*, **71**, 391–396.

Aguayo, A.J., Nair, C.P.V. and Bray, G.M. (1971) Peripheral nerve abnormalities in the Riley-Day syndrome. *Archives of Neurology*, **24**, 106–116.

Allen, E.V. and Adson, W.A. (1938) Physiologic effects of extensive sympathectomy for essential hypertension: further observations. *Annals of Internal Medicine*, 2151–2171.

Amenta, F., Mione, M.C. and Napoleone, P. (1983) The autonomic innervation of vasa nervorum. *Journal of Neurological Transmission*, **58**, 291–297.

Andrade, C. (1952) A peculiar form of peripheral neuropathy: familial atypical generalized amyloidosis with special involvement of peripheral nerves. *Brain*, **75**, 408–427.

Anonymous (1980) Function of the transplanted heart. *British Medical Journal*, **281**, 529.

Appenzeller, O. (1963) The neurogenic control of vasomotor function in the hand: a study of patients with lesions at various levels of the nervous system. Thesis, London.

Appenzeller, O. (1983) Immune autonomic neuropathies: experimental studies in autonomic function. In *Autonomic Failure: A Textbook of Clinical Disorders of the Autonomic Nervous System*, edited by R. Bannister, pp. 640–649 Oxford: Oxford University Press.

Appenzeller, O. (1986) *Clinical Autonomic Failure: Practical Concepts*. Amsterdam: Elsevier.

Appenzeller, O. (1989) Autonomic neuropathies: clinical syndromes. *New Issues of Neuroscience*, **1**, 291–310.

Appenzeller, O. (1990) *The Autonomic Nervous System: An Introduction to Basic and Clinical Concepts*, 4th revised and enlarged edn. Amsterdam: Elsevier Biomedical Press.

Appenzeller, O. and Kornfeld, M. (1973) Acute pandysautonomia: clinical and morphologic study. *Archives of Neurology*, **29**, 334–339.

Appenzeller, O. and Leuker, R.D. (1972) Autonomic dysfunction in patients on chronic hemodialysis. *Neurology*, **97**, 245–246.

Appenzeller, O. and Marshall, J. (1963) Vasomotor disturbances in the Landry–Guillain–Barré syndrome. *Archives of Neurology*, **9**, 368–372.

Appenzeller, O. and Ogin, G. (1973) Myelinated fibers in the human paravertebral sympathetic chain: quantitative studies on white rami communicantes. *Journal of Neurology, Neurosurgery and Psychiatry*, **36**, 777–785.

Appenzeller, O., Parks, R. and Macgee, J. (1968) Peripheral neuropathy associated with chronic respiratory tract disease. *American Journal of Medicine*, **44**, 873–880.

Appenzeller, O., Kornfeld, M. and Snyder, R. (1976) Acromutilating paralyzing neuropathy with corneal ulceration in Navajo children. *Archives of Neurology*, **33**, 733–738.

Appenzeller, O., Lincoln, J. and Zumwalt, R. (1989) Vasa nervorum and nervi nervorum: adrenergic and peptidergic innervation of normal human and neuropathic sural nerves. *Annals of Neurology*, **26**, 185.

Appenzeller, O., Lincoln, J., Milner, P. and Qualls, C. (1991) Innervation des vasa nervorum dans le nerf sural humain normal et pathologique. E'tude immunocytochimique et quantitative. *L'Expansion Scientifique Francaise*, **1**, 27–31.

Auld, R.B. and Bedwell, S.F. (1967) Peripheral neuropathy with sympathetic activity from industrial contact with arcylamide. *Canadian Medical Association Journal*, **96**, 652–654.

Bank, W.J., Pleasure, D.E., Suzuki, K., Nigro, N. and Katz, R. (1972) Thallium poisoning. *Archives of Neurology*, **26**, 456–464.

Bannister, R. (1983) *Autonomic Failure: A Textbook of Clinical Disorders of the Autonomic Nervous System.* Oxford: Oxford University Press.

Bannister, R., Ardill, L. and Fentem, P. (1969) An assessment of various methods of treatment of idiopathic orthostatic hypotension. *Quarterly Journal of Medicine*, **38**, 377–395.

Bayles, T.B., Judson, W.E. and Potter, T.A. (1950) Reflex sympathetic dystrophy of the upper extremity (hand-shoulder syndrome). *Journal of American Medical Association*, **144**, 537–542.

Blumberg, H. and Janig, W. (1984) Discharge pattern of afferent fibers from a neuroma. *Pain*, **20**, 335–353.

Brooks, A.P. (1980) Abnormal vascular reflexes in Charcot-Marie-Tooth disease. *Journal of Neurology, Neurosurgery and Psychiatry*, **43**, 348–350.

Brown, F.M., Brink, S.J., Freeman, R. and Rabinowe, S.L. (1989) Anti-sympathetic nervous system autoantibodies. *Diabetes*, **38**, 938–941.

Bruno, R.L., Johnson, J.C. and Berman, W.S. (1985) Motor and sensory function with changing ambient temperature in post-polio subjects: autonomic and electrophysiological correlates. In *Late Effects of Poliomyelitis*, edited by L.S. Halstead and D.O. Wiechser, pp. 95–101. Miami: Symposia Foundation.

Burgos, L.G., Ebert, T.J., Assiddao, C., Turner, L.A., Pattison, C.Z., Wang-Cheng, R., *et al.* (1989) Increased intraoperative cardiovascular morbidity in diabetics with autonomic neuropathy. *Anesthesiology*, **70**, 591–597.

Cable, W.J.L., Kolodny E.H. and Adams, R.D. (1980) Fabry disease: a clinical demonstration of impaired autonomic function. *Neurology*, **3**, 1352.

Canal, N. (1989) Autonomic dysfunction in diabetes. *New Issues in Neuroscience*, **1**, 415–423.

Cayla, J. and Rondier, J. (1974) Algodystrophies reflexes des membres inferieurs d'origine vertebro-pelvienne: a propos 23 cas. *Semaine Des Hopitaux* (Paris), **50**, 275–286.

Charcot, J.M. and Marie, P. (1886) Sur une form particulière d'iatrophie musculaire, progressive souvent familiale debutant par les pieds et les jambes et atteignant plus tard les mains. *Rev. Med.*, **6**, 97–138.

Cohen, J.A. and Laudenslager, M. (1989) Autonomic nervous system involvement in patients with human immunodeficiency virus infection. *Neurology*, **39**, 1111–1112.

Colan, R.V., Carter Snead, O., Oh, S.J. and Kashlan, M.B. (1980) Acute autonomic sensory neuropathy. *Annals of Neurology*, **8**, 441–444.

Craddock, C., Pasvol, G., Bull, R., Protheroe, A. and Hopkin, J. (1987) Cardiorespiratory arrest and autonomic neuropathy in AIDS. *Lancet*, **2**, 16–18.

Cryer, P.E. (1989) Decreased sympathochromaffin activity in IDDM. *Diabetes*, **38**, 405–409.

Denny-Brown, D. (1951) Hereditary sensory radicular neuropathy. *Journal of Neurology, Neurosurgery and Psychiatry*, **14**, 237–240.

Deutsch, S. and Sherman, L. (1980) Previously unrecognized diabetes mellitus in sexually impotent men. *Journal of American Medical Association*, **244**, 2430–2432.

Dhital, K.K. and Appenzeller, O. (1988) Innervation of vasa nervorum. In *Non-adrenergic Innervation of Blood Vessels*, edited by G. Burnstock and S.G. Griffith, pp. 191–211. Boca Raton: CRC Press.

Diamond, M.A., Murray, R.H. and Schmid, P.G. (1970) Idiopathic postural hypotension: physiologic observations and report of a new mode of therapy. *Journal of Clinical Investigations*, **49**, 1341–1348.

Donaghy, M., Hakin, R.N., Bamford, J.M., Garner, A., Kirkby, A. and Thomas, P.K. (1987) Hereditary sensory neuropathy with neurotrophic keratitis: description of an autosomal recessive disorder with a selective reduction of small myelinated nerve fibres and a discussion of the classification of the hereditary sensory neuropathies. *Brain*, **110**, 181–193.

Duchen, L.W., Anjorin, A., Watkins, P.J. and MacKay, J.D. (1980) Pathology of autonomic neuropathy in diabetes mellitus. *Annals of Internal Medicine*, **92**, 301–303.

Dyck, P.J., Mellinger, J.F., Reagan, T.J., Horowitz, S.J., McDonald, J.W., Litchy, W.J., *et al.* (1983) Not indifference to pain but varieties of hereditary and autonomic neuropathy. *Brain*, **106**, 373–390.

Edmonds, M.E., Jones, T.C., Saunders, W.A. and Sturrock, R.D. (1979) Autonomic neuropathy in rheumatoid arthritis. *British Medical Journal*, **2**, 173–175.

Ellenberg, M. (1980a) Development of urinary bladder dysfunction in diabetes mellitus. *Annals of Internal Medicine*, **92**, 321–323.

Ellenberg, M. (1980b) Sexual function in diabetic patients. *Annals of Internal Medicine*, **92**, 331–333.

Ellenberg, M. and Weber, H. (1966) Retrograde ejaculation in diabetic neuropathy. *Annals of Internal Medicine*, **65**, 1237–1246.

Fagius, J., Westerberg, C.E. and Olsson, Y. (1983) Acute pandysautonomia and severe sensory deficity with poor recovery. A clinical, neurophysiological and pathological study. *Journal of Neurology, Neurosurgery and Psychiatry*, **46**, 725–733.

Fealey, R.D., Low, P.A. and Thomas, J.E. (1989) Thermoregulatory sweating abnormalities in diabetes mellitus. *Mayo Clinic Proceedings*, **64**, 617–628.

Feer, E. (1923) Eine eigenartige Neurose des vegetativen Systems beim Kleinkinde. *Ergebnisse der Inneren Medizen Und Kinderheilkunde*, **24**, 100–122.

Fraser, D.M., Campbell, I.W. and Miller, H.C. (1977) Peripheral and autonomic neuropathy after treatment with perhexiline maleate. *British Medical Journal*, **2**, 675–676.

Frewin, D.B., Leonello, P.P., Pentall, R.K., Hughes, L. and Harding, P.E. (1980) Pindolol in orthostatic hypotension: possible therapy? *Medical Journal of Australia*, **1**, 128.

Goodman, J.F. (1966) Diabetic anhidrosis. *American Journal of Medicine*, **41**, 831–855.

Gotelli, C.A., Astolfi, E., Cox, C., Cernichiari, E. and Clarkson, T.W. (1985) Early biochemical effects of an organic mercury fungicide on infants: "dose makes the poison". *Science*, **227**, 638–640.

Greene, L.F., Kelalis, P.O. and Weeks, R.E. (1963) Retrograde ejaculation of semen due to diabetic neuropathy: report of 4 cases. *Fertility and Sterility*, **14**, 1237–1246.

Gudesblatt, M., Goodman, A.D., Rubenstein, A.E., Bender, A.N. and Choi, H.S. (1985) Autonomic neuropathy associated with autoimmune disease. *Neurology*, **35**, 261–264.

Harris, D.J., Yang, B.I.Y., Wolf, B. and Snodgrass, P.J. (1980) Dysautonomia in an infant with secondary hyperammonemia due to propionylcoenzyme A carboxylase deficiency. *Pediatrics*, **65**, 107–110.

Heidbreder, E., Schafferhans, K. and Heidland, A. (1985) Autonomic neuropathy in chronic renal insufficiency. *Nephron*, **41**, 50–56.

Heistad, D.D. and Wheeler, R.C. (1972) Effect of carbon monoxide on reflex vasoconstriction in man. *Journal of Applied Physiology*, **32**, 7–11.

Holden, W.D. (1948) Sympathetic dystrophy. *Archives of Surgery*, **57**, 373–384.

Horstink, M.W.I.M., Gabreels, F.J.M., Joosten, E.M.G., Gabreels-Feston, A.W.M., Jaspar, H.H.J., van Haelst, V.J.G., *et al.* (1974) Multiple mucosal neuromas, dysautonomia and abnormal intradermal histamine reaction. *Clinical Neurology and Neurosurgery*, **3**, 212–224.

Janig, W. (1985) Causalgia and reflex sympathetic dystrophy: in which way is the sympathetic nervous system involved? *Trends in Neuroscience*, **8**, 471–477.

Janig, W. (1989) Mechanisms of pain associated with abnormalities in the sympathetic nervous system. *New Issues in Neuroscience*, **1**, 369–387.

Janig, W. and Lisney, S.J. (1989) Small diameter myelinated afferents produce vasodilatation but not plasma extravasation in rat skin. *Journal of Physiology*, **415**, 477–486.

Jenzer, G., Mumenthaler, M., Ludin, H.P. and Robert, F. (1975) Autonomic dysfunction in botulism B: a clinical report. *Neurology*, **25**, 150–153.

Johnson, P.C. and Johnsen, S.D. (1987) Hereditary neuropathies in Navajo children. *Neurology*, **37**, 255.

Johnson, R.H. and Spalding, J.M.K. (1964) Progressive sensory neuropathy in children. *Journal of Neurology, Neurosurgery and Psychiatry*, **27**, 125–127.

Kachi, T., Iwase, S., Mano, T., Saito, M., Kunimoto, M. and Sobue, I. (1988) Effect of L-threo-3, 4-dihydroxy-phenylserine on muscle sympathetic nerve activities in Shy-Drager syndrome. *Neurology*, **38**, 1091–1094.

Kanda, F., Uchida, T., Jinnai, K., Tada, K., Shiozawa, S., Fujita, T., *et al.* (1990) Acute autonomic and sensory neuropathy: a case report. *Journal of Neurology*, **237**, 42–44.

Karacan, I. (1980) Diagnosis of erectile impotence in diabetes mellitus. An objective and specific method. *Annals of Internal Medicine*, **92**, 334–337.

Kark, R.A.P. (1979) Clinical and neurochemical aspects of inorganic mercury intoxication. In *Handbook of Clinical Neurology: Intoxication of the Nervous System, Part I*, edited by P.J. Vinken and G.W. Bruyn, pp. 147–197. Amsterdam: North Holland Publishing Company.

Keen, H. (1959) Autonomic neuropathy in diabetes mellitus. *Postgraduate Medical Journal*, **35**, 272–280.

Kolodny, R.C., Kahn, C.B., Goldstein, H.H. and Barnett, D.M. (1974) Sexual dysfunction in diabetic men. *Diabetes*, **23**, 306–309.

Kuroiwa, Y., Goto, I., Itoyama, Y., Tobinatsu, S., Ochiai, J., Furuya, H., et al. (1986) Autonomic disorders in patients with peripheral neuropathies. *Functional Neurology*, **1**, 291–296.

Labby, D.H (1975) Sexual concomitants of disease and illness. *Postgraduate Medicine*, **58**, 103–111.

Laiwah, A.C., MacPhee, G.J., Boyle, P., Moore, M.R. and Goldberg, A. (1985) Autonomic neuropathy in acute intermittent porphyria. *Journal of Neurology, Neurosurgery and Psychiatry*, **48**, 1025–1030.

Learmonth, J.R. (1931) A contribution to the physiology of the urinary bladder in man. *Brain*, **54**, 147–176.

LeQuesne, P.M. and Parkhouse, N. (1989) Cutaneous vascular axon reflexes. *New Issues of Neuroscience*, **1**, 359–367.

LeQuesne, P.M. and McLeod, J.G. (1977) Peripheral neuropathy following a single exposure to arsenic. *J. Neurol. Sci.*, **32**, 437–451.

Levy, S.B. and Lilley, J.J. (1978) Baroreflex function in uremic and hypertensive man. *American Journal of Medical Science*, **276**, 57–66.

Lewis, T. (1942) *Pain*, New York: MacMillan.

Lichtenfeld, P. (1971) Autonomic dysfunction in the Guillain-Barré syndrome. *American Journal of Medicine*, **59**, 772–780.

Low, P.A., Walsh, J.C., Huang, C.Y. and McLeod, J.G. (1975a) The sympathetic nervous system in diabetic neuropathy: a clinical and pathological study. *Brain*, **98**, 341–356.

Low, P.A., Walsh, J.C., Huang, C.Y. and McLeod, J.G. (1975b) The sympathetic nervous system in alcoholic neuropathy: a clinical and pathological study. *Brain*, **98**, 357–364.

Low, P.A., Burke, W.J. and McLeod, J.G. (1978) Congenital sensory neuropathy with selective loss of small myelinated nerve fibers. *Annals of Neurology*, **3**, 179–182.

Mastri, A.R. (1980) Neuropathology of diabetic neurogenic bladder. *Annals of Internal Medicine*, **92**, 239–245.

Matikainen, E. and Juntunen, J. (1985) Autonomic nervous system dysfunction in workers exposed to organic solvents. *Journal of Neurology, Neurosurgery and Psychiatry*, **48**, 1021–1024.

McKusick, V.A., Norum, R.A., Farkas, H.J., Brunt, P.W. and Mahloudji, M. (1967) The Riley-Day syndrome, observations on genetics and survivorship. *Israel Journal of Medical Sciences*, **3**, 372–379.

McLellan, T.L. and Duckett, S. (1987) Pathology of sympathetic ganglia in reflex sympathetic dystrophy. *Neurology*, **37**(Suppl. 1), 254.

McLelland, J. and Ellis, S.J. (1986) Causalgia as a complication of meningococcal meningitis. *British Medical Journal*, **292**, 1710.

McLeod, J.C. and Penny, R. (1969) Vincristine neuropathies — an electrophysiological and histological study. *Journal of Neurology, Neurosurgery and Psychiatry*, **32**, 297–304.

Nordborg, C., Kristensson, K., Olsson, Y. and Sourander, P. (1973) Involvement of the autonomic nervous system in primary and secondary amyloidosis. *Acta Neurologica Scandinavica*, **49**, 31–38.

Nordborg, C., Conradi, N. Sourander, P. and Westerberg, B. (1981) A new type of non-progressive sensory neuropathy in children with atypical dysautonomia. *Acta Neuropathologica*, **55**, 135–141.

Novak, D.J. and Victor, M. (1974) The vagus and sympathetic nerves in alcoholic neuropathy. *Neurology*, **30**, 273–284.

O'Hare, J.P., Anderson, J.V., Millar, N.D., Dalton, N., Tymms, D.J., Bloom, S.R., et al. (1989) Hormonal response to blood volume expansion in diabetic subjects with and without autonomic neuropathy. *Clinical Endocrinology*, **30**, 571–579.

Okajima, T., Yamamura, S., Hamada, K., Kawasaki, S., Ueno, H. and Tokuomi, H. (1983) Chronic sensory and autonomic neuropathy. *Neurology*, **33**, 1061–1064.

Oppenheimer, D. (1983) Neuropathology of progressive autonomic failure. In *Autonomic Failure: A Textbook of Clinical Disorders of the Autonomic Nervous System*, edited by R. Bannister, pp. 267–283. Oxford: Oxford University Press.

Ottomo, M. and Heimburger, R.F. (1980) Alternating Horner's syndrome and hyperhidrosis due to dural adhesions following cervical spinal cord injury. *Journal of Neurosurgery*, **53**, 97–100.

Palmer, G.J., Ziegler, M.G. and Lake, C.R. (1978) Response of noradrenaline and blood pressure to stress increases with age. *Journal of Gerontology*, **33**, 482–487.

Park, D.M., Johnson, R.H., Crean, G.P. and Robinson, J.F. (1972) Orthostatic hypotension with recovery after radiotherapy in a patient with bronchial carcinoma. *British Medical Journal*, **3**, 510–511.

Parkhouse, N. and LeQuesne, P.M. (1988) Quantitative objective assessment of peripheral nociceptive C fibre function. *Journal of Neurology, Neurosurgery and Psychiatry*, **51**, 28–34.

Pearson, J., Axelrod, F. and Dancis, J. (1974) Trophic functions of the neuron in familial dysautonomia. Current concepts of dysautonomia. Neuropathological defects. *Annals of the New York Academy of Sciences*, **228**, 288–300.

Pezeshkpour, G.H. and Dalakas, M.C. (1988) Long-term changes in the spinal cord of patients with old poliomyelitis: signs of continuous disease activity. *Archives of Neurology*, **45**, 505–508.

Rajput, A.H. and Rodzilsky B. (1976) Dysautonomia in parkinsonism: a clinico-pathological study. *Journal of Neurology, Neurosurgery and Psychiatry*, **39**, 1092–1100.

Raskin, N.H., Levinson, S.A., Hoffman, P.M., Pickett, J.B.E. and Fields, H.L. (1974) Postsympathectomy neuralgia: amelioration with diphenylhydantoin and carbamazepine. *American Journal of Surgery*, **128**, 75–78.

Riley, C.M. (1957) Familial dysautonomia. *Advanced Pediatrics*, **9**, 157–190.

Rubin, A. and Babbott, D. (1958) Impotence in diabetes mellitus. *Journal of the American Medical Association*, **168**, 498–500.

Schirger, A., Sheps, S.G., Thomas, J.E. and Fealey, R.O. (1981) Midodrine—a new agent in the management of idiopathic orthostatic hypotension and Shy-Drager syndrome. *Mayo Clinic Proceedings*, **56**, 429–433.

Singer, D.R.J., Buckley M. and MacGregor, G.A. (1986) Raised concentrations of plasma atrial natriuretic peptide in cardiac transplant recipients. *British Medical Journal*, **293**, 1391–1392.

Singer, B., Parc, R., Levy, E., Pinta, P., Bianco, P., Dancea, S., *et al.* (1987) Un cas de neuropathie aigue autonome progressive. *Annales de Medecine Interne*, **138**, 41–44.

Singleton, R., Helgerson, S.D., Snyder, R.D., O'Conner, P.J., Nelson S., Johnsen, S.D., *et al.* (1990) Neuropathy in Navajo children: clinical and epidemiologic features. *Neurology*, **40**, 363–367.

Spark, R.F., White, R.A. and Connolly, P.B. (1980) Impotence is not always psychogenic. Newer insights into hypothalamic-pituitary-gonadal dysfunction. *Journal of American Medical Association*, **43**, 750–755.

Speigel, I.L. and Milowsky, J.L. (1945) Causalgia: a preliminary report of nine cases successfully treated by surgical and chemical interruption of sympathetic pathways. *Journal of American Medical Association*, **127**, 9–15.

Stewart, J.D. (1989) Sweating abnormalities and other autonomic disorders in diabetes mellitus. *Mayo Clinic Proceedings*, **64**, 712–715.

Stewart, P.M. and Hensley, W.J. (1981) An acute attack of variegate porphyria complicated by severe autonomic neuropathy. *Australian and New Zealand Journal of Medicine*, **11**, 82–83.

Sudeck, P. (1900) Über die akute enzundliche Knochenatrophie. *Archiv Fur Klinische Chir*, **62**, 147–156.

Takagi, K. and Sakurai, T. (1950) A sweat reflex due to pressure on the body surface. *Japanese Journal of Physiology*, **1**, 22–28.

Thrush, D.C. (1973) Autonomic dysfunction in four patients with congenital insensitivity to pain. *Brain*, **96**, 591–600.

Watkins, P.J. (1973) Facial sweating after food: a new sign of diabetic autonomic neuropathy. *British Medical Journal*, **1**, 583–587.

Watkins, P.J. and MacKay, J.D. (1980) Cardiac denervation in diabetic neuropathy. *Annals of Internal Medicine*, **92**, 304–307.

Weddell, G., Sinclair, D.E. and Feindez, W.H. (1948) An anatomical basis for alterations in quality of pain sensibility. *Journal of Neurophysiology*, **11**, 99–109.

White, J.C. and Sweet, W.H. (1955) *Pain: Its Mechanisms and Neurosurgical Control*, Springfield: C.C. Thomas.

White, J.E., Heroy, W.W. and Goodman, E.N. (1948) Causalgia following gunshot injuries of nerves: role of emotional stimuli and surgical cure through interruption of diencephalic efferent discharge by sympathectomy. *American Journal of Surgery*, **128**, 161–183.

Whitelaw, G.P. and Smithwick, R.H. (1951) Some secondary effects of sympathectomy with particular reference to disturbance of sexual function. *New England Journal of Medicine*, **245**, 121–130.

Winship, D.M., Caflisch, C.R., Zboralski, E.F. and Hogan, W.J. (1968) Deterioration of esophageal peristalsis in patients with alcoholic neuropathy. *Gastroenterology*, **55**, 173–178.

Yokota, T., Furukawa, T. and Tsukagoshi, H. (1989) Motor paresis improved after sympathetic block. A motor form of reflex sympathetic dystrophy? *Archives of Neurology*, **46**, 683–687.

Young, R.R., Asbury, A.K., Adams, R.D. and Corbett, J.L. (1969) Pure pan-dysautonomia with recovery. *Transactions of the American Neurological Association*, **94**, 355–357.

Ziegler, M. (1980) Postural hypotension. *Annual Review of Medicine*, **31**, 239–245.

8 Genetic Disorders of the Autonomic Nervous System

David Robertson

Departments of Medicine, Pharmacology, and Neurology,
Clinical Research Center, Vanderbilt University,
Nashville, TN 37232-2195, USA

Major strides in the molecular biology of autonomic disorders are currently underway. Early studies are defining the clinical presentation of severe genetic deficiencies but future work may focus on the recognition of milder phenotypes. This review highlights the four major diseases due to abnormalities in catecholamine enzymes. It is likely that other abnormalities with hypotensive phenotypes will also be found to be due to altered gene products that are normally critcal for the function of the autonomic nervous system.

KEY WORDS: Dopamine-β-hydroxylase, dopa decarboxylase, autonomic failure, hypotension, dopamine

INTRODUCTION

With the advent of molecular biology techniques, there has been an explosion in our understanding of autonomic disorders. Careful clinical investigations supplemented by this new information have led to identification of two previously unrecognized disorders and promise to lead to the recognition of many others in the future.

DOPAMINE β-HYDROXYLASE DEFICIENCY

BACKGROUND AND PATHOPHYSIOLOGY

Dopamine β-hydroxylase (DBH) (EC 1.17.14.1) is required for the conversion of dopamine to noradrenaline. However, tyrosine hydroxylase rather than DBH is the

197

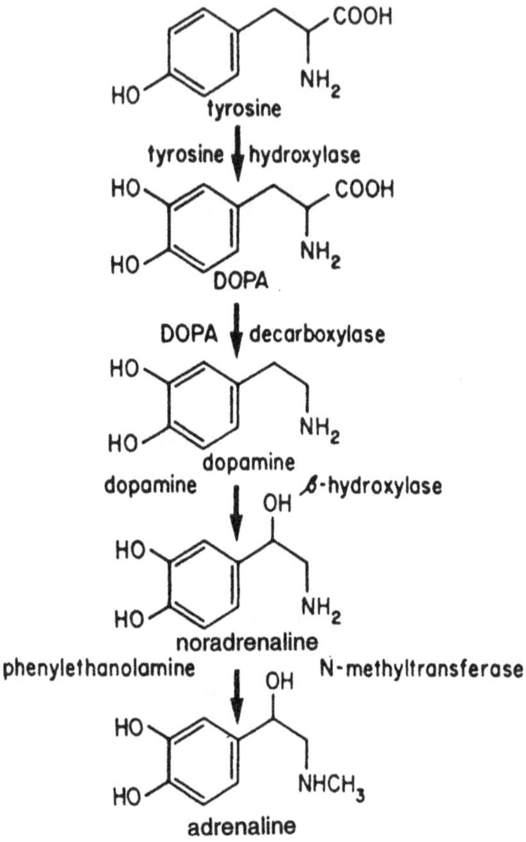

FIGURE 8.1   Metabolic pathway of catecholamine synthesis

rate-limiting step in noradrenaline synthesis under almost all circumstances in man (Figure 8.1). Even in situations of high sympathetic activation, such as prolonged treadmill exercise, noradrenaline and adrenaline remain the predominant circulating catecholamines, with minimal step-up in plasma dopamine levels (Robertson *et al.*, 1979). Thus, in healthy subjects under ordinary circumstances, DBH activity is sufficient for the needs of autonomic cardiovascular regulation.

There are occasional patients, however, in whom neuronal DBH activity is inadequate (Robertson *et al.*, 1986a; Man in't Veld *et al.*, 1987a). These individuals have the syndrome of severe DBH deficiency (Robertson *et al.*, 1986a). The survival of individuals with essentially a complete absence of noradrenaline into adulthood strongly suggests that individuals with partial enzyme deficiency will also be found. Unlike earlier recognized forms of autonomic failure, this disorder was localized to a discrete enzymatic defect, which enabled investigators to approach its treatment more rationally than had heretofore been possible.

DBH is unique among the catecholamine synthesizing enzymes in being located almost exclusively in the chromaffin granules of the adrenal medulla and the large dense-core synaptic vesicles of noradrenergic neurons (Kaufman and Friedman, 1965; Goldstein, 1966; Axelrod, 1972). It is found in both peripheral and central noradrenergic and adrenergic neurons. DBH exists both in the dimeric and tetrameric forms, with two copper atoms per monomeric subunit (Sabban and Goldstein, 1984; Sabban, Kuhn and Levin, 1987). The four subunits are linked by disulphide bridges into two dimers, which are joined to each other by non-covalent bonds. The copper is essential for enzyme activity. DBH also requires molecular oxygen and ascorbic acid for enzyme activity. DBH is not substrate-specific, since it oxidizes almost any phenylethylamine to its corresponding phenylethanolamine (including the hydroxylation of tyramine into octopamine) and converts the α-methyldopa metabolite, α-methyldopamine, to α-methylnoradrenaline. The K_m of this enzyme for dopamine is approximately $5 \times 10^{-3}M$ (Nagatsu, 1986).

DBH occurs in both a soluble and a membrane-bound form (Sokoloff, Frigon and O'Connor, 1985). These are present in approximately equal amounts in the vesicle. The soluble enzyme is released into the synaptic cleft at the time of vesicular exocytosis and is presumably the source of the enzyme present in blood. Much recent study has gone into the identification of the differences between these two forms (O'Connor, Frigon and Stone, 1979). Current evidence suggests that both forms of DBH originate from a single gene and that the soluble form is derived from the membrane bound form (Dhawan et al., 1987). There is evidence that neither glypiation (Stewart and Klinman, 1988) nor retained signal peptide (Taylor, Kent and Fleming, 1989) can account fully for the membrane-binding characteristic of the enzyme.

The sequence of DBH cDNA was reported by Lamouroux (1987). The cDNA was cloned from a human pheochromocytoma expressing high levels of DBH activity. The corresponding polypeptide chain contained 603 amino acids corresponding to an unmodified protein of 64,862 daltons, preceded by a cleaved signal peptide of 25 residues. Kobayashi and coworkers subsequently show that there is a single DBH gene of approximately 23 kb and that it is composed of 12 exons, with exon 12 providing two alternative polyadenylation sites (Kobayashi et al., 1989; Nagatsu, 1991). Restriction analysis of the positive DBH clones revealed two types of DBH cDNA, type A (2.7 kb) and type B (2.4 kb). The ratio of type A and B mRNAs in the pheochromocytoma was 5:1. Four clones were selected for further analysis. Sequencing of the cDNA inserts demonstrated that type A cDNA (DBH-1) differed from type B cDNA (DBH-2) by an additional 300 nucleotides in the 3' untranslated region, the former set of clones being 2.7 kb. The clones in each set differed from each other at six nucleotides found in various portions of the cDNA. This was the first published data for polymorphism at the DBH locus at the molecular level, although other restriction endonuclease polymorphisms have subsequently been reported. Transcription regulatory sites such as TATA, CCAAT, CACCC and GC boxes were identified in the 5' flanking region as were sequences homologous to glucocorticoid and cyclic AMP response elements (Kobayashi *et al.*, 1989). It is not known if the primary structure of DBH reported by Lamouroux is Type A or B (Lamouroux *et al.*, 1987), but it cannot be grouped with either set of cDNAs published by Kobayashi *et al.* Furthermore, the nucleotide difference reported at position 910 would cause an amino acid change (Ala versus Ser), and there is also a change from Arg to Cys at position 1642, which corresponds to Kobayashi's position 1603.

Transgenic studies of the human DBH promotor region employing lacZ n (nlacZ) reporter genes have shown expression in the sympathetic ganglia, the adrenal gland, the locus coeruleus, and in early development, also in a number of other CNS sites, and the developing palate (Mercer *et al.*, 1991).

Assays of DBH activity in blood have been widely employed during the past 25 years (Weinshilboum and Axelrod, 1971).   They are based on the enzymatic conversion of a substrate (e.g. tyramine into a corresponding product octopamine) by DBH, taking advantage of the non-substrate specificity of the enzyme. Inhibitors of DBH are normally present in serum but can be inactivated by N-ethylmaleimide or by copper sulphate (Nagatsu, 1986).   In addition to serum DBH activity measurements, the actual amount of enzyme present can be determined by radioimmunoassay (O'Connor, Levine and Frigon, 1983). When antibodies to homologous protein are used, there is usually an excellent correlation between immunoreactive DBH and DBH enzymatic activity (Weinshilboum, 1989).

Genetic studies of the 1970's and early 1980's using these assays documented polymorphism in DBH in healthy individuals. Serum DBH enzymatic activity and immunoreactive protein increase from low levels in young children to high levels in most adults, but levels of both remain low in a minority (Weinshilboum *et al.*, 1975). Fifty to 70% of the variance in basal human serum DBH activity results from this genetic variation in DBH levels. In linkage studies this trait has been mapped to chromosome 9q34, the region containing the gene for DBH (Goldin *et al.*, 1981; Craig *et al.*, 1988; Wilson *et al.*, 1988). In addition, approximately 8% of a randomly selected population was found to have a thermolabile form of serum DBH which exhibited familial aggregation (Dunnette and Weinshilboum, 1982). This thermolability is a characteristic of the DBH molecule itself and depends on an interaction with oxygen. Although one would expect such a characteristic to derive from the structural DBH gene, thermolability has not been mapped to 9q34.

While in most persons there is a good correlation between immunoreactive serum and enzymatic DBH levels, a few individuals have a disparity (Weinshilboum, 1989). These individuals have much higher amounts of immunoreactive material than enzymatic DBH activity. This disparity may reflect changes in the active site and could indicate subclinical impaired function. There is clearly a familial aggregation of this trait.

CLINICAL AND BIOCHEMICAL FEATURES

A syndrome characterized by severe orthostatic hypotension, noradrenergic failure and ptosis of the eyelids was reported in 1986 (Robertson *et al.*, 1986b).

TABLE 8.1

Clinical Features of Dopamine-β-Hydroxylase Deficiency

Feature	Frequency (%)
Severe orthostatic hypotension	100
Retrograde ejaculation (n=2)	100
Ptosis	67
Complicated perinatal course	67
Nocturia	67
Nasal stuffiness	50
Seizures (with hypotension)	33
Atrial fibrillation	33

This disorder was simultaneously recognized in the Netherlands (Man in't Veld *et al.*, 1987a). Based on a battery of biochemical and physiological tests, it was determined that this disorder was due to a deficiency of DBH. The characteristics of patients with DBH deficiency are distinct from previously recognized forms of autonomic dysfunction (see Table 8.1) and in some cases the anamnesis is so characteristic as to provide the diagnosis. DBH deficiency differs from familial dysautonomia (Axelrod and Pearson, 1984) and various other autonomic disorders seen in adults (Robertson and Biaggioni, 1993) in that the peripheral defect can be localised to the noradrenergic and adrenergic tissues. There is virtual absence of noradrenaline and adrenaline, coupled with greatly increased dopamine in plasma, cerebrospinal fluid and urine (Robertson *et al.*, 1986b; Man in't Veld *et al.*, 1987a). Furthermore, there is no evidence of other neurological defects, either central or peripheral (Laragh and Brenner, 1990). The full clinical spectrum of DBH deficiency is still not known, because of the limited number of patients who have been reported. The description here is based primarily on the data in the first six published cases (Robertson *et al.*, 1986a; Biaggioni, Hollister and Robertson, 1987; Man in't Veld *et al.*, 1987a, 1988a; Biaggioni *et al.*, 1990; Mathias *et al.*, 1990). It is likely that many features not currently recognized will ultimately be found to be associated with the disorder as the number of reported cases increases. Conversely, some abnormalities found in individual patients may ultimately prove to be fortuitous associations.

Parents of DBH-deficient patients have appeared normal (Robertson *et al.*, 1986a; Mathias *et al.*, 1990), but a history of spontaneous abortions and stillbirths has been noted among siblings of affected patients (Man in't Veld *et al.*, 1987). The perinatal period in DBH-deficient subjects has sometimes been particularly difficult (Robertson *et al.*, 1986a; Man in't Veld *et al.*, 1987; Biaggioni *et al.*, 1990). Delay in opening of the eyes (2 weeks in one case) may occur (Biaggioni *et al.*, 1990) and ptosis of eyelids has occurred in most infants (Robertson *et al.*, 1986a; Man in't Veld *et al.*, 1987a, 1988a; Biaggioni *et al.*, 1990). The infants have occasionally been so sickly at birth that parents were advised their survival was unlikely (Robertson *et al.*, 1986a). Although records are incomplete in some cases, it appears that hypotension, hypoglycemia, and hypothermia have occurred primarily in the first year of life (Man in't Veld *et al.*, 1987a). The causes of hypoglycemia and hypothermia are not fully understood at present, but adrenaline has a well-characterized calorigenic effect in animals, and excessive dopamine may reduce temperature in animals. Vomiting occurred four times in the first year of life in one patient (Man in't Veld *et al.*, 1987a). Sometimes seizures have occurred, probably because of hypoglycemia or hypotension (Man in't Veld *et al.*, 1988a; Biaggioni *et al.*, 1990). DBH-deficient patients have had markedly reduced ability to exercise because of postural hypotension occurring with exertion (Robertson *et al.*, 1986a; Mathias *et al.*, 1990). The syncope associated with this postural hypotension has led to trials of anticonvulsive medications (Robertson *et al.*, 1986a), even though the electroencephalogram did not suggest a seizure disorder and efficacy was not observed.

Symptoms have generally worsened considerably in late adolesence and early adulthood (Biaggioni *et al.*, 1990; Mathias *et al.*, 1990; Robertson *et al.*, 1991). Patients complain of profound orthostatic hypotension, especially early in the day and during hot weather or after alcohol ingestion. There is greatly reduced exercise tolerance, ptosis of the eyelids, nasal stuffiness (Robertson *et al.*, 1986a; Biaggioni *et al.*, 1990), and prolonged or retrograde ejaculation (Biaggioni *et al.*, 1990; Mathias *et al.*, 1990); the retrograde ejaculation is recognized by the presence of semen in the post-ejaculation urine void. Presyncopal symptoms include dizziness, blurred vision, dyspnea, nuchal discomfort, and, occasionally, chest pain. Some patients have adopted novel strategies for maintaining upright posture. One patient crossed his legs at a 30° angle and leaned his torso 30° forward, placing his right hand on his right anterior thigh for support (Biaggioni *et al.*, 1990). Sexual maturation has been normal, with menarche occurring at ages 12–14 (Robertson *et al.*, 1986a; Man in't Veld *et al.*, 1988b; Mathias *et al.*, 1990).

TABLE 8.2

Catecholamines in DBH Deficiency

Noradrenaline	↓↓↓↓
Adrenaline	↓↓↓↓
Dopamine	↑↑↑↑
Dopa	↑↑
Normetanephrine	↓↓↓↓
Metanephrine	↓↓↓↓
Vanillylmandelic Acid	↓↓↓↓

On physical examination, patients have a low normal supine blood pressure and a normal heart rate but an upright blood pressure less than 80 mmHg systolic. Heart rate rises on standing, but certainly inadequately when one considers the magnitude of the hypotension in the upright posture. Patients are usually unable to stand motionless for more than 30 secs. Pupils are somewhat small but respond to light and accommodation. Parasympatholytics usually dilate the eye appropriately, but in two patients homatropine has failed to do so (Man in't Veld et al., 1987a, 1988b). There is usually ptosis of the eyelids. Joints may be hyperflexible (Man in't Veld et al., 1988b) or hyperextensible (Biaggioni et al., 1990). In particular, sweating, a sympathetic non-noradrenergic function, is normal.

These patients are easily differentiated from those with familial dysautonomia (Riley–Day syndrome). Cholinergic sensitivity as assessed by the ophthalmic response to conjunctival administration of 2.5% methacholine was normal in that there was no response. Intradermal histamine evoked a typical flare reaction, whereas this does not occur in familial dysautonomia. These patients are further distinguished from familial dysautonomia in that the DBH-deficient patients have: (a) normal tearing; (b) intact corneal and deep tendon reflexes; (c) normal sensory function; and (d) normal senses of taste and smell. Also, subjects thus far recognized have not been of Ashkenazi Jewish extraction.

There have been other clinical abnormalities in these patients that bear a still uncertain relationship to the pathology as we understand it. Two of six subjects have evidence of mild renal failure (Biaggioni et al., 1990; Mathias et al., 1990) and at least two patients have experienced recurrent hypomagnesemia (Biaggioni et al., 1990; Laragh and Brenner, 1990). One patient developed atrial fibrillation at age 40 (Biaggioni et al., 1990), which proved remarkably resistant to therapy.

Patients with DBH deficiency have had such striking abnormalities in catecholamine metabolism that they are readily distinguishable from patients with all other known disorders. The combination of minimal or undetectable plasma noradrenaline with a 5- to 10-fold elevation of plasma dopamine is probably pathognomonic of the disorder (Table 8.2). Indeed, perhaps the only other disorder in which plasma dopamine exceeds plasma noradrenaline is Menkes kinky hair disease, a dramatic illness associated with profound mental retardation (Hoeldtke et al., 1988).

Menkes syndrome is an X-linked recessive disorder characterized by early growth retardation, stubby and white hair, hypopigmentation; arterial rupture and thrombosis, urinary tract diverticulae and focal cerebral and cerebellar degeneration. Survival beyond 10 years is rare, and brain damage is usually severe.

DBH deficiency would probably have been recognized earlier were it not for the fact that most medical centres tend to measure noradrenaline and adrenaline but not dopamine in the evaluation of patients with autonomic dysfunction. Without comparative details about the levels of noradrenaline and dopamine and the patterns of their respective metabolites, the special nature of the enzymatic defect in this disorder can be entirely missed. Such patients were probably considered to have an atypical form of the Bradbury–Eggleston syndrome or idiopathic orthostatic hypotension (Bradbury and Eggleston, 1925). In addition, commonly used radioenzymatic methods for catecholamine determinations have the disadvantage of a small, but significant, crossover of dopamine into adrenaline. Because of normally low levels of dopamine, this is usually of minor practical importance. However, in a setting of elevated dopamine levels, as present in DBH deficiency, a proportion of dopamine may be erroneously measured as adrenaline (Robertson et al., 1986a).

Plasma dopamine levels in DBH-deficient subjects approximate plasma noradrenaline levels in normal subjects, but with greater variability. This is believed to occur because dopamine, rather than noradrenaline, is being stored and released by noradrenergic neurons in DBH-deficient subjects. For this reason, plasma dopamine levels respond to various stimuli which would elicit an increase in plasma noradrenaline levels in normal subjects. For example, a change from supine to upright posture will double or triple the plasma dopamine level. Likewise, the administration of a central suppressant of sympathetic activity such as clonidine (Robertson et al., 1986b) will greatly reduce the plasma dopamine level. Plasma dopamine levels have thus been shown to be greatly elevated by insulin hypoglycemia (Man in't Veld et al., 1987a), edrophonium (Man in't Veld et al., 1987), tyramine (Robertson et al., 1986a), tilt (Man in't Veld et al., 1987a; Mathias et al., 1990), and upright posture (Robertson et al., 1986a). Perhaps because of high levels of dopamine, plasma prolactin is low in this disorder (Man in't Veld et al., 1987a; Laragh and Brenner, 1990). It is noteworthy that plasma dopa levels are also raised two to three-fold, while the enzyme dopa decarboxylase is also normal in plasma (Man in't Veld et al., 1987a).

Noradrenaline metabolites have been low or absent in plasma, urine and cerebrospinal fluid (CSF). Conversely, dopamine metabolites such as homovanillic acid and 3-methoxytyramine are raised. Determination of whether or not noradrenaline exists at all in patients with DBH deficiency must await further investigations and improvements in assay methodology. A low, but apparently detectable, level of vanillylmandelic acid was found in the urine of three patients (Robertson et al., 1986a; Mathias et al., 1990), and a low, but detectable, level of 3-methoxy-4-hydroxyphenylglycol was found in the CSF of another patient (Man in't Veld et al., 1987a). In other patients, these metabolites have been beneath the limits of detection of the assay. Whether these reflect genuine differences in pathology or the limitations of the respective assays remains to be seen.

Skin biopsies in three subjects have not shown staining for DBH, but tyrosine hydroxylase was present in all (Robertson et al., 1986a; Mathias et al., 1990). In the two subjects in whom data have been reported, neuropeptide Y, calcitonin gene-related peptide, substance P and vasoactive intestinal peptide have all been present (Mathias et al., 1990).

Tests of autonomic function also provide diagnostic information of great specificity. In DBH deficiency, autonomic tests that measure sympathetic noradrenergic and adrenergic function are uniformly abnormal. Cold pressor testing (immersion of a hand in ice water for 1 min) causes either a fall or no change in blood pressure. Isometric handgrip exercise (sustained handgrip for 3 min) fails to increase blood pressure significantly. The Valsalva manoeuvre results in a profound fall in blood pressure together with an increase in heart rate, reflecting parasympathetic withdrawal. The phase IV overshoot of the Valsalva manoeuvre does not occur. Hyperventilation causes a fall in blood pressure, as is also the case in patients with the Bradbury–Eggleston syndrome. In contrast to the absence of sympathetic activation, the presence of sweating underscores the integrity of sympathetic cholinergic fibres. Moreover, parasympathetic function is preserved, since these patients have normal sinus arrhythmia. This selective sympathetic noradrenergic impairment is quite characteristic of DBH deficiency. Other forms of autonomic failure have both sympathetic and parasympathetic involvement (Bradbury and Eggleston, 1925; Shy and Drager, 1960; Schatz, 1986).

DBH deficiency shares many pharmacological features of other forms of autonomic failure. There is a several-fold hypersensitivity to α_1-adrenoreceptor agonists and β-adrenoreceptor agonists. This is also found in other forms of autonomic failure and represents a compensatory receptor up-regulation as a result of the chronic relative depletion of catecholamines (Robertson et al., 1984). This phenomenon is analogous to other forms of "denervation hypersensitivity". Tyramine is an indirectly acting pressor amine that will induce noradrenaline release from adrenergic nerve terminals. Tyramine, in intravenous doses of 2–3 mg, will raise plasma noradrenaline and blood pressure in normal subjects and in patients with other types of autonomic failure (Robertson et al., 1984), but no blood pressure elevation occurred even with 6–8 mg of tyramine in DBH-deficient subjects. Plasma dopamine, instead of noradrenaline, is increased following the administration of tyramine in these patients.

Propranolol, a β-adrenoreceptor antagonist, does not lower the basal heart rate in these patients, but pindolol, a β-antagonist with sympathomimetic properties, raises heart rate significantly. Intravenous atropine raises heart rate by 40–60 beats per min. The respiratory arrhythmia which occurs in the baseline state in DBH deficiency disappears with the administration of atropine. Taken together, these observations imply normal parasympathetic, but defective sympathetic, control of heart rate. It is also of interest that atropine elicits a much more pronounced pressor effect in DBH-deficient subjects than in normal subjects (Robertson et al., 1986a; Man in't Veld et al., 1987a).

Clonidine acts on α_2-adrenoreceptors or imidazoline receptors in the brainstem to reduce sympathetic outflow and lower blood pressure (Robertson et al., 1986b). It can also exert peripheral pressor effects by stimulation of vascular α_2-adrenoreceptors (Robertson et al., 1983). DBH-deficient patients have no fall in seated mean arterial pressure following the administration of clonidine, probably reflecting the fact that in these patients, blood pressure is not maintained by sympathetic tone. On the contrary, dramatic increases in blood pressure are seen with higher doses of this agent. It is noteworthy that heart rate decreases in DBH-deficient patients following the administration of clonidine, even though blood pressure does not fall, consistent with the postulated central parasympathetic component in clonidine-induced bradycardia.

Finally, direct measurements of sympathetic nerve traffic to the vasculature of the skeletal muscle have been carried out using microneurography in a patient with

DBH deficiency (Rea *et al.*, 1990). They confirm that sympathetic neural traffic is present and regulated in a qualitatively normal fashion.

THERAPY

DBH-deficient patients have been difficult to treat using standard therapeutic approaches for autonomic failure.   Most have failed empirical therapy with anticonvulsant agents prior to diagnosis. Fludrocortisone, at dosages of 0.1–0.8 mg daily, has been used to raise blood pressure with some benefit (Robertson *et al.*, 1986a), but marked orthostatic hypotension still occurs. Likewise, indomethacin (50 mg four times daily) has been of limited benefit in raising blood pressure in these subjects; furthermore, one patient had aggressive ideation on this drug (Biaggioni *et al.*, 1990). Monoamine oxidase inhibition (tranylcypromine) has produced paranoid ideation (Biaggioni *et al.*, 1990). There is some pressor response to phenylpropanolamine (25 and 50 mg), presumably owing to the denervation hypersensitivity of the patients' vascular α-adrenoreceptors (Biaggioni *et al.*, 1990).

Because both plasma dopa and dopamine levels were elevated in DBH-deficient patients, the vasodepressor effects of dopamine, either through direct vasodilatation or by means of a diuretic effect at the level of the kidney, were proposed as possible explanations for the striking severity of low blood pressure in these patients (Kuchel, Debinsky and Larochelle, 1986; DiBona, 1986). It was hypothesized that if dopa and dopamine were reducing blood pressure in DBH-deficient subjects, the administration of metyrosine (α-methyl-para-tyrosine), might prove therapeutic (Biaggioni, Hollister and Robertson, 1987).

Metyrosine blocks tyrosine hydroxylase, the enzyme leading to the synthesis of dopa. This results in reduced levels of dopamine and noradrenaline, and therefore blood pressure falls in normal subjects, particularly when the subject is in the upright posture. We hypothesized that our patients might have such high dopamine levels that paradoxical pressor effects might occur. Since metyrosine is a depressor in healthy individuals, a failure of metyrosine to affect blood pressure, or a reduction in blood pressure with metyrosine, would not support a contribution of dopamine to the low blood pressure in our patients. On the other hand, a rise in blood pressure with metyrosine would suggest that dopa and dopamine were indeed exerting depressor effects and that these effects could be attenuated by an agent which reduced manufacture and release of dopamine. In the event, metyrosine given in doses used to treat pheochromocytoma exerted a dramatic pressor effect, which appeared to correlate with the metyrosine-associated reduction in urinary dopamine excretion (Biaggioni, Hollister and Robertson, 1987).

In spite of this initially favourable response to metyrosine, much more experience with it will be required before it can be recommended for treatment. Patients receiving metyrosine experienced significant sedation; one patient experienced a dystonic reaction (Biaggioni, Hollister and Robertson, 1987), but fortunately responded promptly to a 10 mg intravenous dose of diphenhydramine.

The most effective therapy in these patients so far has been dihydroxyphenylserine (DOPS) (Biaggioni and Robertson, 1987). We administered DOPS in the hope that it would result in an endogenous conversion (by dopa decarboxylase) of the drug to noradrenaline. This might occur because DBH is not needed for the conversion of DOPS to noradrenaline and, thus, this enzyme could be bypassed in the patients in whom it is defective. There would be an increase in plasma noradrenaline following the administration of DOPS.

The administration of DOPS to patients with DBH deficiency has resulted in dramatic increases in blood pressure and concomitant restoration of plasma and urinary levels of noradrenaline toward normal (Biaggioni and Robertson, 1987; Man in't Veld *et al.*, 1987b). There has been an associated modest decline in dopamine levels, as though the novel provision of noradrenaline to intraneuronal sites might be reducing the activity of tyrosine hydroxylase through feedback inhibition. The increase in plasma noradrenaline was highly correlated with the increase in mean arterial blood pressure. Standing time was greatly increased following DOPS.

We could not be certain whether *de novo* synthesis of noradrenaline from DOPS occurred in neuronal tissues or in extraneuronal tissues, since dopa decarboxylase activity is present in many extraneuronal tissues. However, long-term treatment with DOPS in this disorder is associated with intraneuronal restoration of nora-drenaline, which is released upon assuming the upright posture. Thus, DOPS in DBH deficiency appears to be far more effective than any other therapy for any form of autonomic dysfunction (Biaggioni and Robertson, 1987; Man in't Veld *et al.*, 1987b, 1988b; Mathias *et al.*, 1990).

DBH deficiency is clearly a rare disease in adults, but it could be more common in the perinatal period. Medical histories of DBH-deficient patients include near fatal illness during the neonatal period due to hypotension, hypoglycemia and hypothermia. We suspect that many DBH-deficient infants succumb undiagnosed at this point, never reaching childhood and adulthood.

Prior to recognition of DBH deficiency, it was assumed that humans could not live without noradrenaline. Yet, stretching current assay methodology to the limit, it is not certain that any noradrenaline at all is present in the severely affected individuals we have studied; if it is present in plasma, it is less than 1% of normal (Goldstein *et al.*, 1989). Since noradrenaline and its receptor sites have long been postulated to play a role in a number of psychiatric disorders, the generally normal (Man in't Veld *et al.*, 1988b; Biaggioni *et al.*, 1990) or near-normal (Mathias *et al.*, 1990) mood and mental status of DBH-deficiency subjects so far encountered has elicited great interest among investigators in the area of depression and schizophrenia.

Shortly after DBH was recognized as an important step in catecholamine syn-thesis, attempts were made to treat hypertension with DBH inhibitors. Disulfiram (Antabuse), a copper chelator, was early recognized to inhibit DBH (Hoeldtke *et al.*, 1988). Early clinical studies also demonstrated that fusaric acid and its precur-sor bupicomide could lower blood pressure in hypertensive subjects and decrease serum DBH activity (Nagatsu, 1986). However, tachycardia and increased excre-tion of urinary catecholamines were observed. This apparent contradiction can be explained by the fact that fusaric acid apparently stimulates the release of cate-cholamines from the adrenal gland. More specific and potent DBH inhibitors are currently being tested as antihypertensive agents. As in our patients, inhibition of DBH following the administration of SKF 102698 to rats results in a decrease in plasma and tissue noradrenaline associated with an increase in dopamine levels (Ohlstein *et al.*, 1987). Also, as our results with metyrosine suggest, the hypotensive effects of specific DBH inhibitors may be related to both a decrease in noradrenaline and an increase in dopamine with its attendant vasodilatory and natriuretic effects.

The presence of such a severe deficit in neurotransmitter synthesis encourages us to continue to search for other disorders of neurotransmitter synthesis. It also provides us with the unique opportunity to study the physiological role of dopamine. It provides a model that may help us to determine, in general, dopamine's role in cardiovascular control in humans (DiBona, 1986). In normal subjects, the presence of noradrenaline would obscure the interpretation of any intervention aimed at modulating dopamine synthesis or action. Our preliminary results suggest that increased endogenous dopamine is not only a simple marker of the enzymatic defect, but that it exerts a tonic depressor effect, perhaps in relation to its vasodilatory effect, or more importantly, due to its natriuretic properties.

Most importantly, perhaps, DBH deficiency and its successful treatment by DOPS encourages us to hope that other autonomic disorders may one day also yield to genuinely effective therapeutic interventions.

MONOAMINE OXIDASE DEFICIENCY

Monoamine oxidase (MAO) (EC1.4.3.4) is a flavin-containing enzyme involved in the breakdown of a number of biogenic amines, including noradrenaline, dopamine and serotonin (Murphy, 1978). It is located in the outer mitochondrial membrane of most cell types throughout the body. Of the two forms of the enzyme, monoamine oxidase A (MAOA) was originally defined by its sensitivity to inhibition by low concentrations of clorgyline (Johnston, 1968) and monoamine oxidase B (MAOB) by its sensitivity to inhibition by low concentrations of selegiline (Knoll and Magyar, 1972). MAOA selectively deaminates noradrenaline and serotonin, while MAOB selectively deaminates phenylethylamine and benzylamine (Garrick and Murphy, 1982). MAOA is often localized in association with catecholamine neurons, while MAOB is often associated with serotonin-containing neurons (Thorpe *et al.*, 1987).

MAO deficiency, unlike other autonomic disorders, was discovered by "reverse genetics". First, individuals with a specific chromosome deletion (Norrie's syndrome) lacking chromosomal DNA that included the genes encoding for MAOA and MAOB were identified, and then efforts to extrapolate the clinical consequences of this defect were undertaken. These patients have severe chromosomal deletions in addition to the gene encoding MAO. There is still no information about the clinical presentation of the enzyme defect in isolation.

MAOA and MAOB are encoded in separate genes which have been localized to the Xp21–p11 region (Ozelius *et al.*, 1988). The Norrie syndrome locus was mapped to Xp11.3–11.2 by linkage to locus DXS7 recognized by the anonymous

DNA probe L1.28 (Gal *et al.*, 1985). The apparent closeness of the Norrie gene and the genes for MAOA an MAOB encouraged Sims *et al.* (1989) to study the DNA of Norrie syndrome families to see if the Norrie gene deletion might extend into the DNA loci of MAOA and MAOB; two cousins in one family found to lack MAOA and MAOB were studied further.

The Norrie syndrome is an X-linked recessive disorder of unknown cause which includes retinal dysplasia (with consequent congenital blindness) and bilateral retrolental ocular masses (Norrie, 1927). About half of the affected individuals also have mental retardation and/or progressive sensorineural hearing loss (Warburg, 1975). The family possessing the extended deletion had a number of additional features, including profound mental retardation, disruptive behaviour, somatic growth failure, abnormal sexual maturation, atonic seizures, hypotonia, repetitive head rolling, myoclonic contractions, sleep disturbance and hyperreflexia without spasticity.

These patients also have clinical findings that could be related to disturbances of the autonomic nervous system, including supine hypotension, respiratory irregularities and flushing. Current understanding does not permit one to say that these last abnormalities are definitely related to MAO deficiency, and clearly, it is not certain that the other "atypical Norrie" features are unrelated to the MAO deficiency.

Preliminary analysis of urinary catecholamine metabolites reveals a several-fold increase in metanephrine and normetanephrine, a halving of vanillylmandelic acid and homovanillic acid and levels of 3-methoxy-4-hydroxyphenylglycol that are only about 2% of normal (Sims *et al.*, 1989).

AROMATIC L-AMINO ACID DECARBOXYLASE DEFICIENCY

Recently, Hyland and coworkers (Hyland and Clayton, 1990, 1992; Hyland *et al.*, 1992) have reported an inborn error of dopa decarboxylase. Male identical twins with birth weights of 2.85 and 2.80 kg were born following a normal pregnancy and delivery. There were no immediate neonatal problems. At the age of 2 months, they presented with developmental delay, generalized hypotonia, and paroxysmal movements consisting of a cry followed by extension of arms and legs, eyes rolling up, and cyanosis. The movements occurred once every 2 to 3 days. Anticonvulsant therapy was ineffective. Social smiling had developed at 3 months and babbling speech had developed at 9 months. They were perceived to be floppy, with poor head control and a paucity of spontaneous movements; they cried incessantly but were able to visually fixate and follow, and to chew and swallow. The parents were first cousins, and an elder boy had died of a similar disease at 9 months.

Examination at 9 months showed the twins to be within the normal range of length, weight and head circumference. They were irritable and had skin pallor. Muscle bulk was normal, but they had central and peripheral hypotonia with very few spontaneous movements and flattening of the occiput. Direct tendon reflexes were normal and superficial reflexes were present.

When awake, they had a fine chorea of the distal limbs. On occasion, two distinct paroxysmal movements would occur. The first started with a cry and was followed by extension of the legs and ankles, extension of the arms with internal rotation at the shoulder, flexion, and ulnar deviation at the wrists, finger extension, and abdominal rigidity. The second consisted of a cry followed by deviation of the eyes either up or down for up to 30 sec. The eye movements were conjugate when upward but were convergent when downward; doll's head manoeuvres did not alter the position of the eyes; the children were awake throughout the movements (but could not respond), and the electroencephalogram was normal.

The abnormal eye movements disappeared on treatment with bromocriptine. For these reasons, the eye movements were thought to be extrapyramidal in nature and are termed "oculogyric crises". Both eye and limb movements could occur together. The children were noted to sweat excessively and to have normal tear production. There was miosis and ptosis and a reverse Argyll Robertson pupil. Cocaine and adrenaline instilled into the eyes had no effect; phenylephrine caused raising of the eyelids with no effect on the pupil, but mydriasis followed cyclopentolate. At 9 months old, the twins showed no postural drop in blood pressure, but this was noted at 1 year. An electrocardiogram showed normal beat-to-beat variation. Temperature instability was present in both children. There were full extraocular movements and the ocular fundi were normal.

Concentrations of biogenic amines and their metabolites were reduced considerably both centrally and peripherally. Pterin and phenylalanine metabolism were normal. Activity of aromatic L-amino acid decarboxylase was virtually absent in a liver biopsy sample and greatly reduced in plasma. Concentrations of L-dopa, 3-methoxytyrosine and 5-hydroxytryptophan were elevated in CSF, plasma, and urine. CSF S-adenosylmethionine concentrations were reduced. Pyridoxine treatment had no clinical effect but led to a fall in CSF L-dopa and 3-methoxytyrosine and a rise in S-adenosylmethionine. Treatment with either bromocriptine or tranylcypromine stopped the abnormal eye movements; tranylcypromine treatment also improved muscle tone and lead to a rise in plasma noradrenaline and whole blood serotonin. Combined treatment with pyridoxine, bromocriptine, and tranylcypromine produced sustained improvement in tone and voluntary movements. The twins' parents were asymptomatic but had reduced plasma aromatic L-amino acid decarboxylase activity, consistent with heterozygosity.

TETRAHYDROBIOPTERIN DEFICIENCY

Phenylketonuria (1–3% residual activity of phenylalanine hydroxylase) with benign variants (6–30% residual activity) has been recognized for several years. Plasma phenylalanine is greater than 20 mg % and tyrosine less than 5 mg % in severe cases (Okano et al., 1991). Patients are normal at birth, but developmental delay is seen after 2 months. By 2 years, mental retardation (IQ < 50), pigment dilution (blond hair, blue eyes), aggressive behaviour, tremor, hypertonia and spasticity have begun. Skin is eczemoid and tonic–clonic seizures may be seen. The ferric chloride urine test is green.

About 2% of patients with persistent hyperphenylalaninemia have deficient tetrahydrobiopterin, the cofactor of tyrosine hydroxylase, phenylalanine hydroxylase and tryptophan hydroxylase. This may result from inadequate biopterin synthesis or decreased regeneration by dihydropteridine reductase (Kaufman et al., 1975, 1978). The resulting impaired hydroxylation of tyrosine and tryptophan may account for the marked developmental delay and seizures in these infants. Plasma dopamine, noradrenaline and adrenaline are low.

PHENYLETHANOLAMINE N-METHYLTRANSFERASE DEFICIENCY

No cases of deficiency in phenylethanolamine N-methyltransferase (PNMT) have yet been reported. The major reason to suspect that some individuals have this disorder is that individuals with DBH deficiency, with absence of noradrenaline and adrenaline, have now been recognized. Presumably, if individuals can survive in the absence of both these neurotransmitters, they should also be able to survive in the absence of adrenaline alone. This presupposes that PNMT's sole action is to convert noradrenaline into adrenaline, a traditional view that has been challenged by the finding that some PNMT-containing neurons in the medulla oblongata do not contain DBH; this could mean that in some sites it might act on a different substrate to produce a hitherto unrecognized product.

PNMT catalyzes the N-methylation of noradrenaline using S-adenosylmethionine as methyl donor (Kirshner and Goodall, 1957; Axelrod, 1962). In addition to its localization in the brainstem sites alluded to above (Hökfelt et al., 1973), it is also found in the adrenal medulla, the amygdala, hypothalamus and retina. It is the smallest of the catecholamine-synthesizing enzymes with a molecular weight of 30,853 daltons and is the only one that functions primarily as a monomer. Its synthesis is dramatically stimulated by glucocorticoid (Bohn, 1986).

In 1986, the nucleotide sequence of bovine cDNA encoding adrenal PNMT (Baetge, Suh and Joh, 1986) and, 3 years later, rat adrenal cDNA (Weisberg et al., 1989) were reported. Kaneda and coworkers (1988) elucidated the complete nucleotide sequence of human PNMT cDNA and deduced the amino acid sequence of the enzyme, which was shown to consist of 283 amino acids. They subsequently assigned PNMT to chromosome 17. Homology was 88% between human and bovine enzymes and 85–90% between human and rat enzymes.

The genomic structures of human (Sasaoka et al., 1989) and bovine (Batter et al., 1988) PNMT were subsequently elucidated. There appear to be two types of PNMT mRNA deriving from this single gene, a major form reported by Kaneda et al. (1988), and a minor form which carries an approximately 700 nucleotide-long untranslated region in the 5' terminus. This suggests that the two types of mRNA are produced from a single gene through the control of alternate promoters. Upstream of the cap site of type A mRNA is a TATA box (−30 bp), three Sp1 binding sites (−643, −275 and −263 bp) and three glucocorticoid response elements (−856, −672 and −723). Similar promotor elements were detected in the rat gene for PNMT (Ross *et al.*, 1990). Transgenic rats with the human PNMT coding region and either 2 or 8 Kb of 5' flanking sequence resulted in expression of human PNMT mRNA in the adrenal gland and the eye (Baetge *et al.*, 1988).

It is possible that deficiency of PNMT would be a subtler abnormality than DBH deficiency, but predictions are difficult. Possible presentations would include enhanced perinatal morbidity and mortality, labile hypertension, asthma and β-adrenoreceptor hypersensitivity to exogenous stimuli.

REFERENCES

Axelrod, F.B. and Pearson, J. (1984) Congenital sensory neuropathies: diagnostic distinction from familial dysautonomia. *Am. J. Dis. Child*, **138**, 947–954.

Axelrod, J. (1962) Purification and properties of phenylethanolamine N-methyltransferase. *J. Biol. Chem.*, **237**, 1657–1660.

Axelrod, J. (1972) Dopamine-β-hydroxylase: regulation of its synthesis and release from nerve terminals. *Pharmacol. Rev.*, **24**, 233–243.

Baetge, E.E., Behringer, R.R., Messing, A., Brinster, R.L. and Palmiter, R.D. (1988) Transgenic mice express the human phenylethanolamine N-methyltransferase gene in adrenal medulla and retina. *Proc. Natl. Acad. Sci. USA*, **85**, 3648–3652.

Baetge, E.E., Suh, Y.H. and Joh, T.H. (1986) Complete nucleotide and deduced amino acid sequence of bovine phenylethanolamine N-methyltransferase: partial amino acid homology with rat tyrosine hydroxylase. *Proc. Natl. Acad. Sci. USA*, **83**, 5454–5458.

Batter, D.K., D'Mello, S.R., Turzai, L.M., Hughes, H.B., III, Gioio, A.E. and Kaplan, B.B. (1988) The complete nucleotide sequence and structure of the gene encoding bovine phenylethanolamine N-methyltransferase. *J. Neurosci. Res.*, **19**, 367–376.

Biaggioni, I., Goldstein, D.S., Atkinson, T. and Robertson, D. (1990) Dopamine-beta-hydroxylase deficiency in humans. *Neurology*, **40**, 370–373.

Biaggioni, I., Hollister, A.S. and Robertson, D. (1987) Dopamine in dopamine-beta-hydroxylase deficiency. *New Engl. J. Med.*, **317**, 1415.

Biaggioni, I. and Robertson, D. (1987) Endogenous restoration of noradrenaline by precursor therapy in dopamine-beta-hydroxylase deficiency. *Lancet*, **2**, 1170–1172.

Bohn, M.C. (1986) Expression and development of phenylethanolamine N-methyltransferase (PNMT): role of glucocorticoids. *Neurol. Neurobiol. (NY)*, **16**, 245–271.

Bradbury, S. and Eggleston, C. (1925) Postural hypotension: a report of three cases. *Am. Heart J.*, **1**, 73–75.

Craig, S.P., Buckle, V.J., Lamouroux, A., Mallet, J. and Craig, I.W. (1988) Localization of the human dopamine beta hydroxylase (DBH) gene to chromosome 9q34. *Cytogenet. Cell Genet.*, **48**, 48–50.

Dhawan, S., Duong, L.E., Ornberg, R.L. and Fleming, P.J. (1987) Subunit exchange between membranous and soluble forms of bovine dopamine-beta-hydroxylase. *J. Bio. Chem.*, **262**, 1869–1875.

DiBona, G.F. (1986) Neural mechanisms in body fluid homeostasis. *Fed. Proc.*, **45**, 2871–2877.

Dunnette, J. and Weinshilboum, R. (1982) Family studies of plasma dopamine-beta-hydroxylase thermal stability. *Am. J. Hum. Genetics*, **34**, 84–99.

Gal, A., Bleeker-Wagemakers, L.M., Wienker, T.F., Warburg, M. and Ropers, H.H. (1985) Localization of the gene for Norrie disease by linkage to the DXS7 locus. *Cytogenet. Cell Gen.*, **40**, 633.

Garrick, N.A. and Murphy, D.L. (1982) Monoamine oxidase type A: differences in selectivity towards nora-drenaline compared to serotonin. *Biochem. Pharmacol.*, **31**, 4061–4066.

Goldin L.R., Gershon, E.S., Lake, C.R., Murphy, D.L., McGinniss, M. and Sparkes, R.S. (1981) Segregation and linkage studies of plasma DBH: possible linkage between the ABO locus and a gene controlling DBH activity. *Am. J. Hum. Gen.*, **34**, 250–262.

Goldstein, D.S., Polinsky, R.J., Garty, M., Robertson, D., Brown, R.T., Biaggioni, I. *et al.* (1989) Patterns of plasma levels of catechols in idiopathic orthostatic hypotension. *Ann. Neurol.*, **26**, 558–563.

Goldstein M. (1966) Inhibition of noradrenaline biosynthesis at the dopamine-beta-hydroxylation stage. *Pharmacol. Rev.*, **18**, 77–91.

Hoeldtke, R.D., Cavanaugh, S.T., Hughes, J.D., Mattis-Graves, K., Hobnell, E. and Grover, W.D. (1988) Catecholamine metabolism in kinky hair disease. *Pediatr. Neurol.*, **4**, 23–26.

Hökfelt, T., Fuxe, K., Goldstein, M. and Johansson, O. (1973) Evidence for adrenaline neurons in the rat brain. *Acta. Physiol. Scand.*, **89**, 286–288.

Hyland, K. and Clayton, P.T. (1990) Aromatic L-amino acid decarboxylase deficiency in twins. *J. Inherit. Metab. Dis.*, **13**, 301–304.

Hyland, K. and Clayton, P.T. (1992) Aromatic L-amino acid decarboxylase deficiency: diagnostic methodology. *Clin. Chem.*, **31**, 2405–2410.

Hyland, K., Surtees, R.A.H., Rodeck, C. and Clayton, P.T. (1992) Aromatic L-amino acid decarboxylase deficiency: clinical features, diagnosis, and treatment of a new inborn error or neurotransmitter amine synthesis. *Neurology*, **42**, 1980–1988.

Johnston, J.P. (1968) Some observations upon a new inhibitor of monoamine oxidase in human brain. *Biochem. Pharmacol.*, **17**, 1285–1297.

Kaneda, N., Ichinose, H., Kobayashi, K., Oka, K., Kishi, F., Nakazawa, A., *et al.* (1988) Molecular cloning of cDNA and chromosomal assignment of the gene for human phenylethanolamine N-methyltransferase, the enzyme for epinephrine biosynthesis. *J. Bio. Chem.*, **263**, 7672–7677.

Kaufman, S. and Friedman, S. (1965) Dopamine-β-hydroxylase. *Pharmacol. Rev.*, **17**, 71–100.

Kaufman, S., Berlow, S., Summer, G.K., Milstein, S., Schulman, J.D., Orloff, S. *et al.* (1975) Hyperphenylalanine due to a deficiency of biopterin: A variant form of phenylketonuria. *N. Engl. J. Med.*, **293**, 785–790.

Kaufman, S., Holtzman, N.A., Milstien, S., Butler, I.J. and Krumholz, A. (1978) Hyperphenylalaninemia due to a deficiency of biopterin. A varient form of phenylketonuria. *N. Engl. J. Med.*, **299**, 673–679.

Kobayashi, K., Kurosawa, Y., Fujita, K. and Nagatsu, T. (1989) Human dopamine beta-hydroxylase gene: two mRNA types having different 3'-terminal regions are produced through alternative polyadenylation. *Nucleid Acids Research*, **17**, 1089–1102.

Kirshner, N. and Goodall, M. (1957) The formation of adrenaline from noradrenaline. *Biochim. Biophys. Acta.*, **24**, 658–659.

Knoll, J. and Magyar, K. (1972) Some puzzling pharmacological effects of monoamine oxidase inhibitors. *Adv. Biochem. Psvchopharmacol.*, **5**, 393–408.

Kuchel, O., Debinski, W. and Larochelle, P. (1986) Isolated failure of autonomic noradrenergic neurotransmission (Letter). *N. Engl. J. Med.*, **315**, 1357–1358.

Lamouroux, A., Vigny, A., Faucon Biguet, N., Darmon, M.C., Franck, R., Henry, J.-P., *et al.* (1987) The primary structure of human dopamine-beta-hydroxylase: insights into the relationship between the soluble and the membrane-bound forms of the enzyme. *EMBO J.*, **6**, 3931–3937.

Laragh, J. and Brenner, B.M. (eds). (1990) *Hypertension: Pathophysiology, Management, and Diagnosis.* New York: Raven Press.

Man in't Veld, A.J., Boomsma, F., Julien, C., Lenders, J., van den Meiracker, A.H., Tulen, J., *et al.* (1988a) Patients with dopamine-B-hydroxylase deficiency: a lesson in catecholamine physiology. *Am. J. Hypertension*, **1**, 231–238.

Man in't Veld, A.J., Boomsma, F., Moleman P. and Schalekamp, M.A.D.H. (1987a) Congenital dopamine-β-hydroxylase deficiency: a novel orthostatic syndrome. *Lancet*, **1**, 183–187.

Man in't Veld, A.J., Boomsma, F., van den Meiracker, A.H., and Schalekamp, M.A.D.H. (1987b) Effect of an unnatural noradrenaline precursor on sympathetic control and orthostatic hypotension in dopamine-β-hydroxylase deficiency. *Lancet*, **2**, 1172–1175.

Man in't Veld, A.J., Boomsma, F., van den Meiracker, A.H., Julien, C., Lenders, J. and Schalekamp, M.A. (1988b) D,L-threo-3,4-dihydroxyphenylserine restores sympathetic control and cures orthostatic hypotension in dopamine-beta-hydroxylase deficiency. *J. Hypertension*, **6** (Suppl 4), 547–549.

Mathias, C.J., Bannister, R.B., Cortelli, P., Heslop, K., Polak, J.M., Raimbach, S. *et al.* (1990) Clinical autonomic and therapeutic observations in two siblings with postural hypotension and sympathetic failure due to an inability to synthesize noradrenaline from dopamine because of a deficiency in dopamine beta hydroxylase. *Quart. J. Med.*, **75**, 617–633.

Mercer, E.H., Hoyle, G.W., Kapur, R.P., Quaife, C.J., Brinster, R.L. and Palmiter, R.D. (1991) Use of the dopamine-β-hydroxylase and phenylethanolamine-N-methyltransferase gene promotors for expression of transgenes in mice. In *Neurotransmitter Regulation of Gene Transcription*, edited by E. Costa and T.H. John, pp. 57–68. New York: Thieme Verlag.

Murphy, D.L. (1978) Substrate-selective monoamine oxidases: inhibitor, tissue, species and functional differences. *Biochem. Pharmacol.*, **27**, 1889–1893.

Nagatsu T. (1986) Dopamine beta-hydroxylase. In *Neuromethods I* edited by R.R. Boulton and G.B. Baker, pp. 79–116. Clifton, New Jersey: Humana.

Nagatsu, T. (1991) Genes for human catecholamine-synthesizing enzymes. *Neurosci. Res.*, **12**, 315–345.

Norrie, G. (1927) Causes of blindness in children. *Acta. Ophthalmol.*, **5**, 357–386.

O'Connor, D.T., Frigon, R.P. and Stone, R.A. (1979) Human pheochromocytoma dopamine-beta-hydroxylase: purification and molecular parameters of the tetramer. *Molecular Pharmacol.*, **16**, 529–538.

O'Connor, D.T., Levine, G.L. and Frigon, R.P. (1983) Homologous radioimmunoassay of human plasma dopamine-β-hydroxylase: analysis of homospecific activity, circulating plasma pool and intergroup differences based on race, blood pressure and cardiac function. *J. Hypertens.*, **1**, 227–233.

Ohlstein, E.H., Kruse, L.I., Ezekiel, M., Sherman, S.S., Erickson, R. and DeWolf, W.E. (1987) Cardiovascular effects of a new potent dopamine-β-hydroxylase inhibitor in spontaneously hypertensive rats. *J. Pharmacol. Exp. Ther.*, **241**(2), 554–559.

Okano, Y., Eisensmith, R.C., Güttler, F., Lichter-Konecki, U., Koneck, D.S., Trefz, F.K., *et al.*, (1991) Molecular basis of phenotypic heterogeneity in phenylketonuria. *N. Engl. Med.*, **324**, 1232–1983.

Ozelius, L., Hus, Y.-P.P., Bruns, G., Powell, J.F., Chen, S., Weyler, W., *et al.* (1988) Human monoamine oxidase gene (MAOA): chromosome position (Xp21–p11) and DNA polymorphism. *Genomics*, **3**, 53–58.

Rea, R.F., Biaggioni, I., Robertson, R.M., Haile, V. and Robertson, D. (1990) Reflex control of sympathetic nerve activity in dopamine β-hydroxylase deficiency. *Hypertension*, **15**, 107–112.

Robertson, D. and Biaggioni, I. (eds). (1994) *Disorders of the Autonomic Nervous System*. London: Harwood, in press.

Robertson, D., Goldberg, M.R., Hollister, A.S., Onrot, J., Wiley, R., Thompson, J.G., *et al.* (1986a) Isolated failure of autonomic noradrenergic neurotransmission: evidence for impaired beta-hydroxylation of dopamine. *N. Engl. J. Med.*, **314**, 1494–1497.

Robertson, D., Goldberg, R.M., Hollister, A.S., Wade, D. and Robertson, R.M. (1983) Clonidine raises blood pressure in severe idiopathic orthostatic hypotension. *Am. J. Med.*, **74**, 193–200.

Robertson, D., Goldberg, M.R., Tung, C.S., Hollister, A.S. and Robertson D. (1986b) Use of alpha₂-adrenoreceptor agonists and antagonists in the functional assessment of the sympthetic nervous system. *J. Clin. Invest.*, **78**, 576–581.

Robertson, D., Hollister, A.S., Carey, E.L., Tung, C.S., Goldberg, M.R. and Robertson, R.M. (1984) Increased vascular beta₂-adrenoreceptor responsiveness in autonomic dysfunction. *J. Am. Coll. Cardiol.*, **3**, 850–856.

Robertson, D. Johnson, G.A., Robertson, R.M., Nies, A.S., Shand, D.G. and Oates, J.A. (1979) Comparative assessment of stimuli that release neuronal and adrenomedullary catecholamines in man. *Circulation*, **59**, 637–643.

Robertson, D., Perry, S.E., Hollister, A.S., Robertson, R.M. and Biaggioni, I. (1991) Dopamine β-hydroxylase deficiency: a genetic disorder of cardiovascular regulation. *Hypertension*, **18**, 1–8.

Ross, M.E., Evinger, M.J., Hyman, S.E., Carroll, J.M., Mucke, L., Comb, M., *et al.* (1990) Identification of a functional glucocorticoid response element in the phenylethanolamine N-methyltransferase promoter using fusion genes introduced into chromaffin cells in primary culture. *J. Neurosci.*, **10**, 520–530.

Sabban, E.L. and Goldstein, M. (1984) Subcellular site of biosynthesis of the catecholamine biosynthetic enzymes in bovine adrenal medulla. *J. Neurochem.*, **43**, 1663–1668.

Sabban, E.L., Kuhn, L.J. and Levin, B.E. (1987) *In vivo* biosynthesis of two subunit forms of dopamine beta-hydroxylation in rat brain. *J. Neurosci.*, **7**, 192–200.

Sasaoka, T., Kaneda, N., Kurosawa, Y., Fujita, K. and Nagatsu, T. (1989) Structure of human phenylethanolamine N-methyltransferase gene: existence of two types of mRNA with different transcription initiation sites. *Neurochem. Int.*, **15**, 555–565.

Schatz, L.J. (1986) *Orthostatic Hypotension*, pp. 1–128. Philadelphia: FA Davis.

Shy G.M. and Drager, G.A. (1960) A neurological syndrome associated with orthostatic hypotension. *Arch. Neurol.*, **2**, 511–527.

Sims, K.B., de la Chapelle, A., Norio, R., Sankila, E.-M., Hus, Y.-P.P., Rinehart, W.B., *et al.* (1989) Monoamine oxidase deficiency in males with an X chromosome deletion. *Neuron*, **2**, 1069–1076.

Sokoloff, R.L., Frigon, R.P. and O'Connor, D.T. (1985) Dopamine-β-hydroxylase: structural comparisons of membrane-bound versus soluble forms from adrenal medulla and pheochromocytoma, *J. Neurochem.*, **44**, 411–420.

Stewart, L.C. and Klinman, J.P. (1988) Bovine membranous dopamine beta-hydroxylase is not anchored via covalently attached phosphatidylinositol. *J. Bio. Chem.*, **263**, 12183–12186.

Taylor, C.S., Kent, U.M. and Fleming, P.J. (1989) The membrane-binding segment of dopamine beta-hydroxylase is not an uncleaved signal sequence. *J. Bio. Chem.*, **264**, 14–16.

Thorpe, L.W., Westlund, K.N., Kochersperger, L.M., Abell, C.W. and Denney, R.M. (1987) Immunocytochemical localization of monoamine oxidase A and B in human peripheral tissues and brain. *J. Histochem. Cytochem.*, **35**, 23–32.

Warburg, M. (1975) Norrie's disease-differential diagnosis and treatment. (1975) *Acta. Ophthalmol.*, **53**, 217–236.

Weinshilboum, R. (1989) Catecholamine biochemical genetics. In *Catecholamines II*, edited by U. Trendelenburg and N. Weiner, pp. 391–426. Berlin: Springer-Verlag.

Weinshilboum, R. and Axelrod, J. (1971) Serum dopamine-beta-hydroxylase activity. *Circulation Res.*, **28**, 307–315.

Weinshilboum, R.M., Schrott, H.G., Raymond, F.A., Weidman, W.H. and Elveback, L.R. (1975) Inheritance of very low serum dopamine-beta-hydroxylase activity. *Am. J. Human Genetics*, **27**, 573–585.

Weisberg, E.P., Baruchin, A., Stachowiak, M.K., Stricker, E.M., Zigmond, M.J. and Kaplan, B.B. (1989) Isolation of a rat adrenal cDNA clone encoding phenylethanolamine N-methyltransferase (PNMT), mRNA and protein. *Mol. Brain Res.*, **6**, 159–166.

Wilson, A.F., Elston, R.C., Siervogel, R. and Tran, L.D. (1988) Linkage of a gene regulating dopamine-beta-hydroxylase activity and the ABO blood group locus. *Am. J. Hum. Gen.*, **42**(1), 160–166.

9   Familial Dysautonomia

Felicia B. Axelrod

*New York University School of Medicine,
530 First Avenue, New York, NY 10016, USA*

Familial dysautonomia (FD) is an autosomal recessive disease affecting the development and survival of sensory, sympathetic and some parasympathetic neurons. Diagnostic tests and consistent pathological findings are described. Treatment is primarily supportive but has resulted in improved survival. Some treatments are applicable to other congenital sensory neuropathies.

KEY WORDS: Congenital sensory neuropathy, hypertension, tachycardia, sweating, hypotonia, vomiting, constipation, cornea, ischaemia

INTRODUCTION

Familial dysautonomia (FD), also termed the Riley-Day syndrome, is a rare inherited neurologic disease affecting the development and survival of sensory, sympathetic and some parasympathetic neurons (Pearson, Axelrod and Dancis, 1974). Since the original report in 1949 (Riley *et al.*, 1949), knowledge of the disorder has expanded so that genetic transmission has been studied, diagnostic tests devised and consistent pathological findings described. In addition, treatment programs have been developed resulting in improved survival (Axelrod and Abularrage, 1982).

Despite its low incidence, FD is the most extensively described of a group of disorders known as congenital sensory neuropathies (Axelrod and Pearson, 1984). Although some of the other congenital sensory neuropathies are phenotypically very similar, we assume they involve different genetic abnormalities. Many of the treatment modalities for FD individuals, however, can be applied to the other disorders.

217

GENETICS AND DIAGNOSIS

FD is an autosomal recessive disorder which currently appears confined to individuals of Ashkenazi Jewish extraction (Moses *et al.*, 1967; Brunt and McKusick, 1970). In this population, the latest estimation of carrier rate is 1 in 30, with a disease frequency of 1 in 3600 live births (Maayan *et al.*, 1987). The gene for FD has been recently been localized to chromosome 9q (Blumenfeld *et al.*, 1993). Using linkage analysis techniques, carrier and prenatal tests can be offered to families in which there is already one FD individual. However, screening for the general population is not yet available.

Although penetrance is complete, there is marked variability in expression of the disease. Diagnosis is based upon clinical recognition of both sensory and autonomic dysfunction. It is expected that, in addition to a history of Ashkenazi Jewish extraction, the following criteria be present in every affected individual.

(i)   Absence of fungiform papillae on the tongue and decreased taste (Smith, Farbman and Dancis, 1965b)

(ii)   Absence of flare after intradermal histamine (Smith and Dancis, 1963)

(iii)   Decreased or absent deep tendon reflexes (Axelrod, Nachtigall and Dancis, 1974)

(iv)   Absence of overflow tears (Smith, Dancis and Breinin, 1965a)

(v)   Intraocular hypersensitivity to parasympathomimetic agents (Smith, Dancis and Breinin, 1965a)

ABSENT FUNGIFORM PAPILLAE

Absent fungiform papillae are an example of deficiency in the sensory pathway involving neurons of the geniculate ganglion. Fungiform papillae normally are concentrated towards the tip of the tongue. To the naked eye, they should appear as pinhead sized red projections interspersed among the grey-white filiform papillae (Figure 9.1). The red colour results from their being highly vascularised. In the patient with familial dysautonomia the fungiform papillae are sparse and rudimentary so that the tongue often appears smooth and pale (Figure 9.2).

ABSENCE OF FLARE AFTER INTRADERMAL HISTAMINE

In the normal individual the intradermal injection of histamine phosphate 1:10,000 produces pain and local erythema followed within minutes by the development of a central wheal surrounded by the axon flare, a diffuse poorly demarcated zone of erythema which is best appreciated 15 minutes post injection (Figure 9.3). The flare is usually sustained for over 30 minutes and slowly resolves with infiltration of areas of pallor. In the dysautonomic, although the initial local area of erythema is noted, the pain is not appreciated and the flare does not appear. The central wheal is only surrounded by a narrow violaceous areola which is sharply demarcated due to vascular congestion (Figure 9.4).

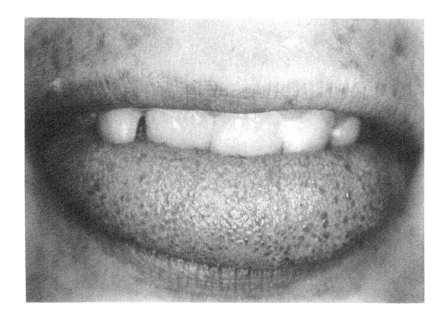

FIGURE 9.1   Normal tongue with fungiform papillae present on the tip.

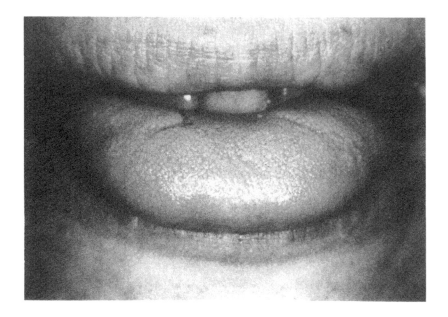

FIGURE 9.2   Dysautonomic tongue.

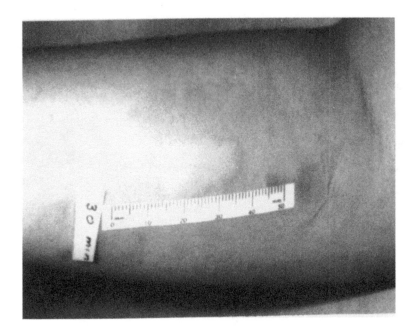

FIGURE 9.3   Normal histamine test. Reaction displays diffuse axon flare around central wheal.

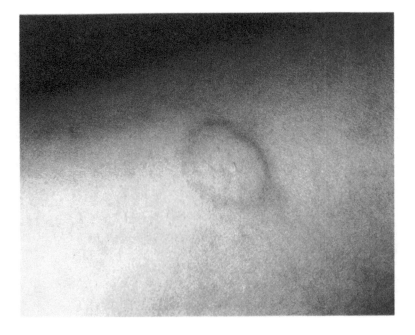

FIGURE 9.4   Dysautonomic histamine test. Only a narrow areola surrounds the wheal.

DECREASED DEEP TENDON REFLEXES

As a result of impairment of the sensory part of the reflex arc, patellar reflexes are depressed, as are corneal reflexes.

ABSENCE OF OVERFLOW TEARS

Alacrima or lack of overflow tears, which is very unusual in the normal infant after the first few months of life, is almost always present in the child with FD. Thus lack of overflow tears with emotional crying is one of the diagnostic criteria. History is usually sufficient to document this finding but semiquantitation of the deficiency can be obtained by the Schirmer filter paper test or by trying to provoke tearing emotionally or with an irritant.

The Schirmer test is performed with the patient in a moderately lighted room. A piece of filter paper, 5mm wide and 35mm long, is partially folded 5 mm from one end. The folded short end is placed over the lateral one-third of a lower eyelid. After five minutes the degree of filter paper wetting is assessed. If there is less than 10mm of wetting, then both baseline and reflex tear secretions are diminished. If there is greater than 10mm of wetting, then there is either normal secretion or pseudoepiphora (Jones, 1966).

INTRAOCULAR HYPERSENSITIVITY TO PARASYMPATHOMIMETIC AGENTS

The instillation of methacholine 2.5% or pilocarpine 0.0625% causes miosis in almost all patients with familial dysautonomia and has no observable effect on the normal pupil. Both pupils are carefully examined for symmetry and then the dilute parasympathomimetic agent is instilled in one eye, with the other serving as a control. The pupils are compared at 5 minute intervals for 30 minutes. There is usually at least a 20 minute delay before miosis appears.

The combination of the five "cardinal" criteria, i.e. alacrima, absent fungiform papillae, depressed patellar reflexes, abnormal histamine test and an abnormal methacholine test, in an individual of Ashkenazi Jewish extraction is usually sufficient to make the diagnosis. Further supportive evidence is provided by findings of decreased response to pain and temperature, orthostatic hypotension, periodic erythematous blotching of the skin, and increased sweating. In addition, cine-esophagrams may reveal delay in cricopharyngeal closure, tertiary contractions of the esophagus, gastroesophageal reflux and delayed gastric emptying. Sural nerve biopsy is rarely required unless one of the five "cardinal" criteria are not present or the patient is not of Jewish extraction.

NEUROPATHOLOGY

FD appears to be a disorder in which there is development arrest of the sensory and autonomic systems, with sympathetic development more widely affected than parasympathetic.

Consistent neuropathologic findings have now been described.

SENSORY NERVOUS SYSTEM

Sural nerve

The sural nerve, a predominantly sensory peripheral nerve, is reduced in its transverse fascicular area and contains markedly diminished numbers of non-myelinated axons, as well as diminished numbers of small diameter myelinated axons (Aguayo, Nair and Bray, 1971; Pearson *et al.*, 1975). Even in the youngest subject extensive pathology has been evident, as might be expected from the fact that clinical symptoms are present at birth. Catecholamine-containing fibers are missing. The findings indicate a peripheral neurological deficit sufficiently characteristic for familial dysautonomia to differentiate it from other sensory neuropathies (Axelrod and Pearson, 1984).

Spinal cord

Consistent with decreased peripheral sensory neurons, intrauterine development and postnatal maintenance of dorsal root ganglion neurons are abnormal in familial dysautonomia (Pearson *et al.*, 1978). The dorsal root ganglia are grossly reduced in size. The number of neurons is markedly diminished. Within the spinal cord, lateral root entry zones and Lissauer's tracts are severely depleted of axons. As evidence of slow progressive degeneration, there is a definite trend with increasing age for further depletion of the number of neurons in dorsal root ganglia and an increase in the abnormal numbers of residual nodules of Nageotte in the dorsal root ganglia. In addition, loss of dorsal column myelinated axons becomes evident in older patients. Neuronal depletion in dorsal root ganglia and the progressive pattern of cord changes correlate well with the clinical observations of worsening pain and vibration sense with age (Axelrod *et al.*, 1981).

Diminution of primary substance P axons in the substantia gelatinosa of spinal cord and medulla has been demonstrated using immunohistochemistry (Pearson, Brandeis and Cuello, 1982). Because substance P may be involved in sensory neuron synaptic transmission, the immunoreactive findings support the electron microscopic findings.

AUTONOMIC NERVOUS SYSTEM

Sympathetic nervous system

The sympathetic nervous system has similar findings as the peripheral sensory nervous system. In adult patients with FD, the mean volume of superior cervical sympathetic ganglia is reduced to 34% of the normal size reflecting an actual severe decrease in numbers of neurons (Pearson and Pytel, 1978a). The anatomical defect in the ganglion cells extends to preganglionic neurons as the intermediolateral gray columns of the spinal cord also contain low numbers of neurons. This may be a retrograde effect of the depletion of ganglionic neurons.

Tyrosine hydroxylase, as measured by immunocytochemical techniques, can identify catecholaminergic neurons which produce dopamine (Pearson, Goldstein and Brandeis, 1979a). Although clinical, anatomic, biochemical and pharmacologic data indicate diminution in the numbers of sympathetic neurons in FD, there was more intense staining for tyrosine hydroxylase in FD neurons from sympathetic ganglia than in the controls (Pearson, Brandeis and Goldstein, 1979b).

Ultrastructural study of peripheral blood vessels demonstrated the absence of autonomic nerve terminals (Grover-Johnson and Pearson, 1976). Lack of innervation would explain orthostatic hypotension, as well as exaggerated responses to sympathomimetic and parasympathomimetic agents (Smith and Dancis, 1964a; Smith et al., 1965c) in terms of denervation hypersensitivity.

Parasympathetic nervous system

Although patients with FD do not produce overflow tears and there is pharmacological evidence suggesting denervation supersensitivity in effector tissues normally supplied by postganglionic parasympathetic nerve terminals (Smith, Dancis and Breinin, 1965a; Smith et al., 1965c), parasympathetic ganglia are not as abnormal as one might expect. The sphenopalatine ganglia are consistently reduced in size with low total neuronal counts, but the neuronal population is only questionably reduced in the ciliary ganglia (Pearson and Pytel, 1978b). The paucity of neurons in the sphenopalatine ganglion would explain the supersensitivity of the lacrimal gland to infused methacholine (Smith et al., 1965c), but the defect in the ciliary ganglion is too slight to account for the pupillary hypersensitivity (Smith, Dancis and Breinin 1965a).

BIOCHEMICAL DATA

As one would expect from the decreased numbers of sympathetic neurons present, norepinephrine (NE) metabolite excretion is diminished (Goodall, Gi How and Alton, 1971). There is a 60 percent diminution in NE synthesis, yet dopamine products continue to be excreted in normal amounts resulting in abnormal 3-methoxy-4-hydroxymandelic acid (VMA) to 3-methoxy-4-hydroxyphenylacetic acid (HVA) ratios. In addition, there is no appropriate increase in plasma levels of NE and dopamine-beta-hydroxylase (DBH) when the FD patient goes from supine to standing position (Ziegler, Lake and Kopin, 1976). Yet, during emotional crises plasma NE and dopamine are markedly elevated and there are lesser rises in epinephrine. During such crises, vomiting usually coincides with high dopamine levels. The high NE may appear through peripheral conversion of dopamine by DBH. Diazepam sedates patients in crises and relieves vomiting, possibly by blocking the release of dopamine.

Because the neuropathology suggests that FD may be due to a deficit in a trophic substance essential to the development of sensory and sympathetic neurons, nerve growth factor (NGF) became a natural candidate for investigation. Cultured fibroblasts from FD patients produce slightly less radioimmunoassayable NGF than those in healthy controls (Schwartz and Breakefield, 1980). Using cloned DNA probes, the structural gene regions which encode for beta-NGF and the beta-NGF receptor were excluded as possible causes for FD (Breakefield *et al.*, 1984). However, this analysis did not eliminate other genes involved in beta-NGF action, such as those coding for processing enzymes, or other subunits of the NGF complex.

CLINICAL SYMPTOMS AND TREATMENT

Although the primary abnormality in familial dysautonomia is due to anatomical depletion of sensory and autonomic neurons, the clinical manifestations are the concern of the treating physician. The pervasive nature of the autonomic nervous system results in protein functional abnormalities (Table 9.1). Signs of the disorder are present from birth and neurological function slowly deteriorates with age so that symptoms and problems will vary with time.

The disease process cannot be arrested. Treatment is preventative, symptomatic and supportive. It must be directed towards specific problems within each system which vary considerably among patients and with different ages.

NEUROLOGIC

Sensory system

Pain loss in FD typically is not universal and spares hands, soles of feet, neck and genital areas. Lower extremities are more affected than upper and older individuals have greater losses than younger. Temperature appreciation is also more abnormal on the trunk and lower extremities. Patellar reflexes are depressed as are corneal reflexes.

In the older individual, evidence of progressive sensory deficiencies appear as vibration sense and joint position become abnormal and the Romberg test may be positive.

The ability to taste is deficient, especially in recognition of sweet (Smith and Dancis, 1964b). This corresponds to the absence of fungiform papillae on the tip of the tongue, and can be assumed to reflect a sensory deficiency involving the neurons of the geniculate ganglion. Although the speech is frequently characterized as dysarthric and nasal, hearing is normal.

TABLE 9.1
CLINICAL MANIFESTATIONS

Neurologic	Sensation	Decreased pain preception Decreased temperature appreciation Decreased deep tendon reflexes Decreased corneal reflex Deficient taste discrimination
	Motor	Hypotonia Ataxia gait
	Autonomic	Excessive sweating Blotching of skin Cold mottled extremities
Gastrointestinal	Oropharyngeal	Feeding difficulties Drooling
	Esophageal and gastric dysmotility	Misdirected swallows Gastroesophageal reflux Episodic vomiting
	Bowel dysmotility	Constipation
Respiratory	Peripheral receptors (Insensitivity to hypoxia and hypercapnia)	Problems with high altitudes, air travel and underwater swimming
	Aspiration	Repeated pneumonias Bronchiectasis and atelectasis Hyperreactive airways
Blood Pressure	Orthostatic hypotension	Dizzy spells Syncope Difficulty with anesthetics
	Hypertension	Headaches
Orthopedic	Spine Joints	High frequency of curvature Aseptic necrosis Neuropathic joints
	Long bones	Unrecognized fractures
Ophthalmologic	Cornea	Excessive dryness Corneal ulcerations
	Optic nerve	Pallor
Renal	Glomeruli	Progressive ischemic sclerosis Rising blood urea nitrogen and creatinine with age

Motor system

Varying degrees of hypotonia will be noted in the infant and young child contributing to delay in motor milestones. In the ambulatory individual, the gait is often broad based and mildly ataxic with special difficulties in performing rapid movements or turning. Many patients walk listing forward with a compensatory increased stiffness in shoulders and neck leading to protracted shoulders. Physical therapy is useful in avoiding pathological movements, decreasing contracture, and promoting more normal alignment. It will not prevent spinal curvature but it is a useful adjunct to brace therapy. Deterioration of gait with progressive ataxia is a common complaint of adult patients. Many have had to resort to use of walkers or wheelchairs when outside the home.

Autonomic dysfunction

Autonomic dysfunction becomes apparent during the examination especially if the child is agitated, excessive sweating and blotching can then be noted. Distal vasoconstriction in the peripheral skin causes cutis marmorata accompanied by hypertension and tachycardia. In addition, the autonomic dysfunction is so generalized that it is easier to think of its effects within the various other systems.

Intelligence

Patients with familial dysautonomia usually have normal intelligence. In a study by Welton *et al.* (1979), 62% of children had IQs over 90. Verbal skills were better than motor skills. Most children are able to attend regular school but perceptual problems may require provision for special education.

GASTROINTESTINAL SYSTEM

Oropharyngeal incoordination is one of the earliest signs. Poor suck or discoordinated swallow is observed in 60% of infants in the neonatal period (Axelrod, Porges and Sein, 1987). Feeding problems are treated with various maneuvers to improve feeding and nutrition and to avoid aspiration. If thickening of formula and different nipples are ineffective, then gastrostomy may be necessary (Axelrod, 1990).

Esophageal dysmotility can be diagnosed by cine-esophagrams. Gastroesophageal reflux is also a common problem and furthers risk of aspiration. If medical management with prokinetic agents, H2 antagonists, thickening of feeds and position are not successful, then surgical intervention, fundoplication, is performed (Axelrod *et al.*, 1982). Failure of medical management would include persistence of pneumonias, hematemesis or apnea.

Vomiting crises occur in approximately 40% of FD patients. The crises are a systemic reaction to stress, either physical or emotional. They have been managed most effectively with a combination of intravenous or rectally administered diazepam (0.2 mg/kg q3h) and chloral hydrate rectal suppositories (30 mg/kg q6h) and avoidance of dehydration and aspiration (Axelrod, 1990). Chlorpromazine is used only when hypertension is refractory to above treatment.

RESPIRATORY SYSTEM

Because the major cause of lung infections is aspiration, many of the respiratory problems are avoided when gastrointestinal dysfunction is well-managed. For those individuals who have had pneumonias and have developed chronic lung disease, daily chest physiotherapy is recommended. This consists of nebulization, bronchodilators and postural drainage. Suctioning is also required in the individual who has an ineffective cough.

Respiratory control is abnormal. Patients with FD do not have an appropriate ventilatory response to hypercapnia and hypoxia. Filler *et al.* (1965) demonstrated that rebreathing of 12% oxygen for a period of minutes caused dramatic falls in oxygen saturation resulting in extreme cyanosis, syncope, and even convulsions. When Edelman *et al.* (1970) reinvestigated respiratory control, they noted that the response to hypercapnia was normalized if the patient were maintained hyperoxic. However, if FD subjects were hypoxic, there were profound cardiovascular effects. Hypoxia in FD patients caused the heart rate and blood pressure to fall which is opposite to the normal response. This reaction is presumed due to sympathetic insufficiency and secondary decreased circulation to the respiratory center.

Clinical symptoms referable to these abnormal respiratory control responses include drowning when swimming underwater, as well as syncope and convulsions during air travel. In addition, FD patients can become cyanotic and assume decerebrate posturing as a result of breathholding. The breathholding episodes are usually self-limited and decrease in frequency as the child matures.

BLOOD PRESSURE LABILITY

Abnormal orthostatic changes in blood pressure is a cardinal sign of autonomic dysfunction and is present in all patients with FD. Clinical manifestations of orthostatic hypotension include episodes of lightheadedness or dizzy spells. Some patients complain of "weak legs". On occasion, there may be syncope. Orthostatic hypotension is treated by maintaining adequate hydration as monitored by blood urea nitrogen levels. Lower extremity exercises are encouraged to increase muscle tone and promote venous return. Elastic stockings and fludrocortisone have also been of some benefit.

When patients are agitated or in the supine position, blood pressures are often in the hypertensive range, but patients will still demonstrate hypotension without compensatory tachycardia when the erect position is assumed. Hypertension can be quite striking during periods of stress. Because blood pressure is so labile, treatment should be directed to factors precipitating the hypertension rather than use of blocking agents.

General anesthesia has caused profound hypotension and cardiac arrest (Kritchman, Schwartz and Papper, 1959). With greater attention to stabilization of the vascular bed by hydrating the patient prior to surgery and titrating the anesthetic, the risk of these problems was greatly reduced (Axelrod *et al.*, 1988).

ORTHOPEDIC

Spinal curvature

Over 50% of FD patients will develop significant spinal curvature by 16 years of age (Yaslow *et al.*, 1971). Onset can occur very early. Annual examination of the spine will allow early diagnosis of scoliosis and permit appropriate institution of brace and exercise therapy. The latter is helpful in correcting or preventing secondary contractures in shoulders and hips. Extreme care is required in fitting of braces as decubiti may develop on the insensitive skin at pressure points. Braces may also inhibit respiratory excursion and induce gastroesophageal reflux if there is a high epigastric projection. Spinal fusion is recommended if there is rapid progression or severe curvature at time of presentation.

Extremities

Fractures are frequent and are often unrecognized. The child may not complain or indicate only mild discomfort. However, any swelling of an extremity should raise the possibility of a fracture. Pain insensitivity also results in inadvertent trauma to joints. Charcot joints and aseptic necrosis have been reported (Brunt, 1967; Mitnick *et al.*, 1982).

OPHTHALMOLOGIC

The corneal hypesthesia and alacrima predispose the cornea to neurotropic corneal ulcerations due to undetected trauma and excessive dryness. Corneal complications have been decreasing with regular use of artificial tear solutions containing methylcellulose and maintenance of normal body hydration. Artificial tears are instilled three to six times daily, depending on the child's own baseline eye moisture, environmental conditions and whether or not the child is febrile or dehydrated. Moisture chamber spectacle attachments help to maintain eye moisture and protect the eye from wind and foreign bodies. Tarsorraphy has been reserved for unresponsive and chronic situations. Soft contact lenses are also beneficial in promoting corneal healing. Corneal transplants have had limited success.

RENAL PROBLEMS

Renal function appears to deteriorate with advancing age, as indicated by slowly rising serum urea and creatinine (Pearson *et al.*, 1980). Pathological studies reveal excess glomerulosclerosis. On ultrastructural examination of renal biopsies, vascular innervation is deficient in FD patients as compared to controls. Although the cause of the progressive renal disease is not certain, hypoperfusion of the kidney seems a likely explanation. Hypoperfusion could occur because of dehydration, orthostatic hypotension, or vasoconstriction of renal vessels as a result of sympathetic supersensitivity during vomiting crises.

REPRODUCTION

Sexual maturation is frequently delayed, but primary and secondary sex characteristics eventually develop in both sexes (Axelrod, Nachtigall and Dancis, 1974). Women with dysautonomia have conceived and delivered normal infants (Porges, Axelrod and Richards, 1978). Pregnancies were tolerated well. At time of delivery, blood pressures were labile. One male has fathered six children.

PROGNOSIS

Familial dysautonomia can no longer be considered only a disease of childhood. With greater understanding of the disorder and development of treatment programs, survival statistics have markedly improved so that increasing numbers of patients are reaching adulthood. Survival statistics prior to 1960 reveal that 50% of patients died before 5 years of age (Brunt and McKusick, 1970). Current survival statistics indicate that a newborn with FD has a 50% probability of reaching 30 years of age (Axelrod and Abularrage, 1982).

Despite physical and emotional developmental lags, intelligence is normal and many of the adults have been able to achieve independent function. Patients have even married and reproduced with all offspring being phenotypically normal despite their obligatory heterozygote state.

Causes of death are still predominantly pulmonary indicating more aggressive treatment is still needed in this area. Another large group has succumbed to unexplained deaths which may have been the result of unopposed vagal stimulation. A few adult patients have died of renal failure.

REFERENCES

Aguayo, A.J., Nair, C.P.V. and Bray, G.M. (1971) Peripheral nerve abnormalities in the Riley-Day syndrome, findings in sural nerve biopsy. *Arch. Neurol.*, **24**, 106–116.

Axelrod, F.B. (1990) Familial dysautonomia. In *Current Pediatric Therapy*, 12th edn., edited by S.C. Gellis and B.M. Kagen, pp. 92–94. Philadelphia: Saunders.

Axelrod, F.B. and Abularrage, J.J. (1982) Familial dysautonomia. A prospective study of survival. *J. Pediatr.*, **101**, 234–236.

Axelrod, F.B. and Pearson, J. (1984) Congenital sensory neuropathies. Diagnostic distinction from familial dysautonomia. *Am. J. Dis. Child*, **138**, 947–954.

Axelrod, F.B., Nachtigall, R. and Dancis, J. (1974) Familial dysautonomia; Diagnosis Pathogenesis and Management. In *Advances in Pediatrics*, vol. 21, edited by I. Schulman, pp. 75–96. Chicago: Yearbook.

Axelrod, F.B., Iyer, K., Fish, I., Pearson, J., Sein, M.E. and Spielholz, N. (1981) Progressive sensory loss in familial dysautonomia. *Pediatrics*, **65**, 517–522.

Axelrod, F.B., Schneider, K.M., Ament, M.E., Kutin, N.D. and Fonkalsrud, E.W. (1982) Gastroesophageal fundoplication and gastrostomy in familial dysautonomia. *Ann. Surg.*, **195**, 253–258.

Axelrod, F.B., Porges, R.F. and Sein, M.E. (1987) Neonatal recognition of familial dysautonomia. *J. Pediatr.*, **110**, 946–948.

Axelrod, F.B., Donenfeld, R.F., Danziger, F. and Turndorf, H. (1988) Anesthesia in familial dysautonomia. *Anaesthesiology*, **68**, 631–635.

Blumenfeld, A., Slaugenhaupt, S.A., Axelrod, F.B., Lucente, D.E., Maayan, Ch., Lieberg, C.B., Ozelius, L.J., Trofatter, J.A., Haines, J.L., Breakefield, X.O. and Gusella, J.F. (1993) Localization of the gene for familial dysautonomia on Chromosome 9 and definition of DNA markers for genetic dianosis. *Nature Genetics*, **4**, 160–164.

Breakefield, X.O., Orloff, G., Castiglione C., Coussens, L., Axelrod, F.B. and Ullrich, A. (1984) Structural gene for beta nerve-growth-factor not defective in familial dysautonomia. *Proc. Nat. Acad. Sci.*, **81**, 4213–4216.

Breakefield, X.O., Ozelius, L., Bothwell, M.A., Chao, M.V., Axelrod, F.B., Kramer, P., *et al.* (1986) DNA polymorphisms for the nerve growth factor receptor gene exclude its role in familial dysautonomia. *Mol. Biol. Med.*, **3**, 483–494.

Brunt, P.W. (1967) Unusual case of Charcot joints in early adolescence (Riley-Day syndrome). *Br. Med. J.*, **4**, 277–278.

Brunt, P.W. and McKusick, V.A. (1970) Familial dysautonomia. A report of genetic and clinical studies with a review of the literature. *Medicine*, **48**, 343–374.

Edelman, N.H., Cherniak, N.S., Lahiri, S., Richards, E. and Fishman, A.P. (1970) The effects of abnormal sympathetic nervous function upon the ventilatory response to hypoxia. *J. Clin. Invest.*, **49**, 1153–1165.

Filler, J., Smith, A.A., Stone, S. and Dancis, J. (1965) Respiratory control in familial dysautonomia. *J. Pediatr.*, **66**, 509–516.

Goodall, G., Gitlow, S.E. and Alton, H. (1971) Decreased noradrenalin synthesis in F.D. *J. Clin. Invest.*, **50**, 2734–2740.

Grover-Johnson, N. and Pearson, J. (1976): Deficient vascular innovation in familial dysautonomia, an explanation for vasomotor instability. *Neuropathol. Appl. Neurobiol.*, **2**, 217–224.

Jones, L.T. (1966) The lacrimal system and its treatment. *Am. J. Ophthalmol.*, **62**, 47–60.

Kritchman, M.M., Schwartz, H. and Papper, E.M. (1959) Experiences with general anesthesia in patients with familial dysautonomia. *J. Am. Med. Assoc.*, **170**, 259–263.

Maayan, Ch., Kaplan, E., Shachar, Sh., Peleg O. and Godfrey, S. (1987) Incidence of familial dysautonomia in Israel 1977–1981. *Clin. Genet.*, **32**, 106–108.

Mitnick, J., Axelrod, F.B., Genieser, N. and Becker, M. (1982) Aseptic necrosis in familial dysautonomia. *Radiology*, **142**, 89–91.

Moses, S.W., Rotem, Y., Jogoda, N., Talmor, N., Eichorn, F. and Levin, S. (1967) A clinical genetic and biochemical study of familial dysautonomia. *Isr. J. Med. Sci.*, **3**, 358–371.

Pearson, J. and Pytel, B. (1978a) Quantitative studies of sympathetic ganglia and spinal cord intermediolateral gray columns in familial dysautonomia. *J. Neurol. Sci.*, **39**, 47–59.

Pearson, J. and Pytel, B. (1978b) Quantitative studies of ciliary and sphenopalatine ganglia in familial dysautonomia. *J. Neurol. Sci.*, **39**, 123–130.

Pearson, J., Axelrod, F.B. and Dancis, J. (1974) Current concepts of dysautonomia: neurological defects. *Ann. NY. Acad. Sci.*, **228**, 288–300.

Pearson, J., Dancis, J., Axelrod, F.B. and Grove-Johnson, N. (1975) The sural nerve familial dysautonomia. *J. Neuropathol. Exp. Neurol.*, **34**, 413–424.

Pearson, J., Pytel, Grover-Johnson, N., Alexrod, F.B. and Dancis, J. (1978) Quantitative studies of dorsal root ganglia and neuropathologic observations on spinal cords in familial dysautonomia. *J. Neurol. Sci.*, **35**, 77–97.

Pearson, J., Goldstein, M. and Brandeis, L. (1979a) Tyrosine hydroxylase immunohistochemistry in human brain. *Brain Res.*, **165**, 333–337.

Pearson, J., Brandeis, L. and Goldstein, M. (1979b) Tyrosine hydroxylase immunohistoreactivity in familial dysautonomia. *Science*, **206**, 71–72.

Pearson, J., Gallo, G., Gluck, M. and Axelrod, F. (1980) Renal disease in familial dysautonomia. *Kidney Int.*, **17**, 102–112.

Pearson, J., Brandeis, L. and Cuello, A.C. (1982) Depletion of substance P-containing axons in substantia gelatinosa of patients with diminished pain sensitivity. *Nature*, **295**, 61–63.

Porges, R.F., Axelrod, F.B. and Richards, M. (1978) Pregnancy in familial dysautonomia. *Am. J. Obstet. Gynecol.*, **132**, 485–488.

Riley, C.M., Day, R.L., Greeley, D.McL. and Langford, W.S. (1949) Central autonomic dysfunction with defective lacrimation. Report of 5 cases. *Pediatrics*, **3**, 468–477.

Schwartz, J. and Breakefield, X. (1980) Altered nerve-growth-factor in fibroblasts from patients with familial dysautonomia. *Proc. Natl. Acad. Sci.*, **77**, 1154–1158.

Smith, A.A. and Dancis, J. (1963) Response to intradermal histamine in familial dysautonomia–a diagnostic test. *J. Pediatr.*, **63**, 889–894.

Smith, A.A. and Dancis, J. (1964a) Exaggerated response to infused norepinephrine in familial dysautonomia. *N. Engl. J. Med.*, **270**, 704–707.

Smith, A.A. and Dancis, J. (1964b) Taste discrimination in familial dysautonomia. *Pediatrics*, **33**, 441–443.

Smith, A.A., Dancis, J. and Breinin, G. (1965a) Ocular responses to autonomic drugs in familial dysautonomia. *Invest. Ophthalmol.*, **4**, 358–361.

Smith, A.A., Farbman, A. and Dancis, J. (1965b) Absence of taste bud papillae in familial dysautonomia. *Science*, **147**, 1040–1041.

Smith, A.A. Hirsch, J.I. and Dancis, J. (1965c) Responses to infused methacholine in familial dysautonomia. *Pediatrics*, **36**, 225–230.

Welton, W., Clayton, D., Axelrod, F. and Levine, D. (1979) Intellectual development in familial dysautonomia. *Pediatrics*, **63**, 708–712

Yaslow, W., Becker, M., Bartels, J. and Thompson, W. (1971) Orthopedic defects in familial dysautonomia. *J. Bone Joint Surg.*, **53**, 1541–1550.

Ziegler, M.G., Lake, R.C. and Kopin, I.J. (1976) Deficient sympathetic nervous system response in familial dysautonomia. *New Engl. J. Med.*, **294**, 630–633.

10 Cardiovascular Disorders in High Spinal Cord Lesions

Christopher J. Mathias, and Hans L. Frankel

Cardiovascular Medicine Unit, Department of Medicine
St. Mary's Hospital and Medical School/
Imperial College of Science, Technology and Medicine,
University of London, UK

Autonomic Unit, University Department of Clinical Neurology,
Institute of Neurology and National Hospital for Neurology
and Neurosurgery, Queen Square, London, UK

National Spinal Injuries Centre,
Stoke Mandeville Hospital, UK

In patients with cervical and high thoracic spinal cord lesions, a major portion of the sympathetic outflow to the heart and blood vessels is impaired. This results in a number of cardiovascular abnormalities. In the acute stages when recently injured tetraplegics are in spinal shock, blood pressure is low and there are no cardiovascular responses to stimuli working through the isolated spinal cord. There is pressor supersensitivity to noradrenaline. The cardiac vagi may be overactive, and in certain situations this may result in severe bradycardia and cardiac arrest.

In chronic stages, the basal blood pressure in high spinal cord lesions remains low with dependence on other compensatory factors. Postural hypotension, an abnormal Valsalva manoeuvre and enhanced vasodepressor responses to vasodilator drugs result from the separation of the brain from peripheral sympathetic pathways. Severe hypertension, as part of the syndrome of autonomic dysreflexia, also occurs during cutaneous, visceral and skeletal muscle stimulation, as a result of an increase in isolated spinal cord reflex activity. There is pressor supersensitivity to a range of drugs with widely different properties.

The pathophysiological processes responsible for the resultant cardiovascular dysfunction in patients with high spinal cord lesions is discussed, along with how this has aided preventive and management strategies.

KEY WORDS: Catecholamines, tetraplegia, hypertension, hypotension, autonomic nervous system

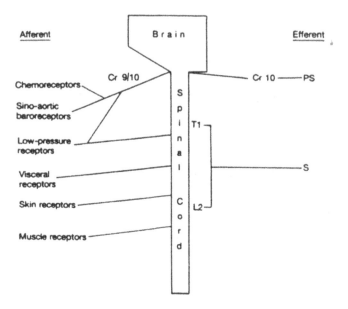

FIGURE 10.1   Schematic outline of the major autonomic pathways controlling the circulation. The major afferent input into the central nervous system is through the glossopharyngeal (Cr 9) and vagus (Cr 10) nerves by activation of baroreceptors in the carotid sinus and aortic arch. Chemoreceptors and low pressure receptors also influence the efferent outflow. The latter consists of the cranial parasympathetic (PS) outflow to the heart via the vagus nerves, and the sympathetic (S) outflow from the thoracic and upper lumbar segments of the spinal cord. Activation of visceral, skin and muscle receptors, in addition to cerebral stimulation, influences the efferent outflow. (From Mathias and Frankel, 1988.)

INTRODUCTION

In patients with cervical or high thoracic spinal cord lesions, a major proportion of the sympathetic outflow to the cardiovascular system is disrupted (Figure 10.1). This can result in abnormalities ranging from bradycardia and cardiac arrest, to profound hypotension and even severe paroxysmal hypertension. The cardiovascular abnormalities may vary depending upon the level of the lesion and its completeness. We will mainly concentrate on disorders of autonomic function which involve the cardiovascular system in tetraplegics with cervical cord transection, and in paraplegics with high thoracic lesions above the level of T5. These patients usually have the most severe disorders affecting cardiovascular function and the pathophysiological basis for these abnormalities, together with possible management approaches, will be described.

There are differences in disorders of autonomic control in the stages after a spinal cord lesion. Patients in the acute stage are often initially in spinal shock, a term used by Hall (1841) to denote a transient state of hypoexcitability of the isolated cord soon after transection. In spinal shock there is flaccid paralysis of the skeletal musculature with a lack of tendon reflexes. Other aspects of spinal cord function, which include those served by the sympathetic and sacral parasympathetic nervous system, appear inactive. Following this stage of spinal cord depression, which may last from a few days to a few weeks, activity of the isolated spinal cord often returns. In the later chronic phase, this contributes to further abnormalities which result not only from separation of efferent sympathetic pathways from cerebral control, but additionally from spinal cord sympathetic reflexes acting independently of higher control.

ACUTE HIGH SPINAL CORD LESIONS

BLOOD PRESSURE

A low basal supine blood pressure is often seen in the early stages while in spinal shock, despite the absence of other traumatic complications, which may lower blood pressure further. Skeletal muscle flaccidity does not contribute, as patients who are tetraplegic due to poliomyelitis do not usually have a lower supine blood pressure. The diastolic blood pressure is usually 25% lower than in normal subjects (Mathias et al., 1979a) and appears to be related to the reduction in sympathetic neural activity, which normally accounts for about 20% of vascular tone. This is consistent with the low levels of plasma noradrenaline and adrenaline observed soon after spinal cord injury.

From the earliest stages, patients with high lesions are susceptible to either hypotension or hypertension because of impairment of the efferent sympathetic pathways of the baroreceptor reflex. Head-up posture lowers blood pressure markedly. Drugs with vasodilatatory properties may cause profound hypotension. Similar effects may occur with drugs that deplete intravascular volume. The reverse, an elevation in blood pressure, occurs if either fluid or blood is transfused rapidly.

In spinal shock, sympathetic reflexes occurring below the level of the lesion, as in chronic lesions, cannot be activated. This has been observed after cutaneous cold, urinary bladder stimulation (Figure 10.2), or attempts to activate skeletal muscles in segments below the level of the lesion. The lack of change in blood pressure is not the result of subsensitivity or refractoriness of the target organs, as infusion of adrenoreceptor agonists such as noradrenaline raises blood pressure substantially to levels seen in chronic tetraplegics who have pressor supersensitivity (Mathias et al., 1979a). Activity within the isolated spinal cord often returns in a few weeks, with increased tendon reflexes and skeletal muscle spasticity. This is usually associated with the ability to activate spinal autonomic reflexes. The reasons for the initial depression in spinal shock, and the later reversal, are not known.

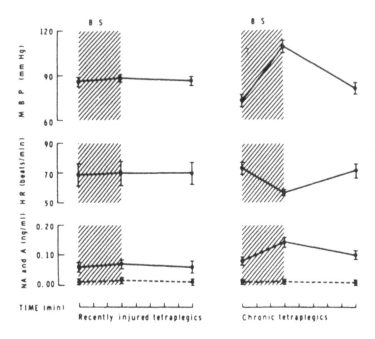

FIGURE 10.2   Average levels of mean blood pressure (MBP), heart rate (HR) and plasma noradrenaline (NA continuous line) and adrenaline (A, interrupted line) in recently injured and chronic tetraplegics, before, during and after bladder stimulation (BS). The bars indicate ± SEM. No changes occur in recently injured tetraplegics, unlike the chronic tetraplegics in whom MBP and plasma NA levels rise and HR falls. There are no changes in plasma A levels. (From Mathias *et al.*, 1979a.)

HEART RATE

The supine basal heart rate in high lesions in the early stages is usually lower than in patients with low thoracic and lumbar spinal cord lesions. The reasons for this are unclear, but probably include the lack of sympathetically mediated neural influences on the heart rate. The afferent and the efferent vagal limbs of the baroreflex are operative in high lesions, as an elevation of blood pressure, following infusion of a pressor agent such as noradrenaline, lowers heart rate (Mathias *et al.*, 1979a). A relative increase in vagal tone in high lesions may further contribute to the low resting heart rate. In patients in spinal shock, and in particular in those with high cervical lesions involving the 4th and 5th segments which supply the phrenic nerves, there is an increased tendency to bradycardia and cardiac arrest (Frankel, Mathias and Spalding, 1975; Mathias, 1976a). These patients need artificial respiration because of diaphragmatic paralysis. Tracheal suction and toilet needs to be performed at regular intervals. If patients are hypoxic, severe

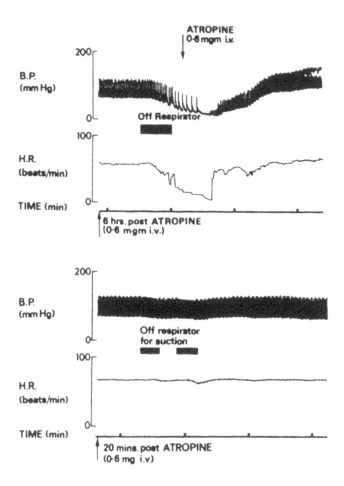

FIGURE 10.3   (a) The effect of disconnecting the respirator (as required for aspirating the airways) on the blood pressure (BP) and heart rate (HR) of a recently injured tetraplegic patient (C 4/5 lesion) in spinal shock, 6 h after the last dose of intravenous atropine. Sinus bradycardia and cardiac arrest (also observed on the electrocardiograph) were reversed by reconnection, intravenous atropine and external cardiac massage. (From Frankel, Mathias and Spalding, 1975.) (b) The effect of tracheal suction, 20 min after atropine. Disconnection from the respirator and tracheal suction did not lower either heart rate or blood pressure. (From Mathias, 1976b.)

bradycardia and cardiac arrest may result (Figure 10.3). This can be prevented by atropine, indicating the major role of vagal efferent pathways.

In tetraplegics, hypoxia may occur because of either respiratory infection or pulmonary emboli. Despite added oxygen, it may be difficult to maintain normoxia. In normal subjects, the primary heart rate response to hypoxia is bradycardia, which is opposed by factors that include the pulmonary inflation vagal reflex; this raises the heart rate, as in spontaneously breathing tetraplegics. In patients on respirators this cannot occur, and the primary response of bradycardia therefore cannot be reversed. During tracheal stimulation vagal afferents are additionally activated, which increases vagal efferent activity. This is normally opposed by either the sympathetic nervous system or by increased respiratory activity. The combination of vagal afferent stimulation and the effects of hypoxia in patients incapable of spontaneous breathing who are likely to have impaired spinal sympathetic activity is, therefore, likely to result in cardiac arrest.

To prevent bradycardia and cardiac arrest it is important to ensure adequate oxygenation. Despite this, however, some patients may have a severe bradycardia during tracheal suction. In these patients intravenous atropine should be administered in a dose of 0.3 or 0.6 mg either subcutaneously or intramuscularly at 4 hourly intervals. Parasympathomimetic drugs such as neostigmine and carbachol, which are used to reverse bladder and bowel atony in spinal shock, should be avoided or used with caution. The beta-1 adrenoreceptor agonist action of isoprenaline has been used to raise heart rate, but it has the potential to lower blood pressure because of its actions on vasodilatatory beta-2 adrenoreceptors. Temporary demand cardiac pacemakers may also be used. With the return of spontaneous breathing and adequate oxygenation, the likelihood of bradycardia and cardiac arrest recedes. Cardiac arrest may, however, also occur in chronic tetraplegics during intubation when vagal afferents are stimulated while respiration is prevented (Welply, Mathias and Frankel, 1975). It is important that such patients are given an adequate amount of atropine as premedication prior to intubation.

CHRONIC HIGH SPINAL CORD LESIONS

BASAL BLOOD PRESSURE

In chronic spinal cord lesions, the basal supine blood pressure is inversely related to the level of the lesion, being lowest in high lesions and progressively rising towards normal levels in patients with lesions in the lower thoracic and the lumbar regions (Frankel et al., 1972). Cervical and high thoracic lesions have low basal plasma levels of noradrenaline and adrenaline, consistent with a diminution of tonic supraspinal sympathetic impulses. Reduced skin and muscle sympathetic discharge in the basal state have been demonstrated using microneurography (Wallin and Stjernberg, 1984; Stjernberg, Blumberg and Wallin, 1986).

The subnormal systolic and diastolic blood pressure in high spinal cord lesions may be lowered further by factors which are unlikely to influence the level of blood pressure in normal subjects. In tetraplegics, recumbency causes a brisk diuresis, as is often observed at night. This is similar to the pronounced nocturnal polyuria seen in patients with chronic primary autonomic failure (Mathias et al., 1986), in whom recumbency causes both diuresis and natriuresis. This is different from the tetraplegics, in whom natriuresis does not occur (Kooner et al., 1987). Recumbency induced diuresis is likely to reduce intravascular volume and, as in the primary autonomic failure patients, may lower blood pressure particularly in the early hours of the morning.

In tetraplegics, the lack of natriuresis during recumbency may be related to their ability to secrete renin and activate the renin–angiotensin–aldosterone (R-A-A) system more effectively than patients with chronic primary autonomic failure (Di Bona and Wilcox, 1992). Basal levels of plasma renin activity are often higher in tetraplegics than in normal subjects (Mathias *et al.*, 1975a, 1980), and through the direct pressor effects of angiotensin-II and the salt-retaining effects of aldosterone are likely to raise intravascular volume and thus blood pressure. The role of the R-A-A system is further emphasised by antagonists of the system. In tetraplegics, the angiotensin converting enzyme inhibitor captopril, which prevents the formation of angiotensin-II, lowers blood pressure substantially even in the supine position (Alam *et al.*, 1992). Reduction in the intravascular fluid volume by even small doses of diuretics often causes a marked fall in supine blood pressure. In tetraplegics, basal blood pressure is also lowered by a low salt diet, despite their ability to reduce salt excretion to the same extent as normal subjects (Sutters *et al.*, 1991).

HYPOTENSION

Patients with high spinal cord lesions are prone to hypotension because of impairment of the sympathetic component of the baroreflex. A variety of stimuli, some physiological and seen during daily life, and others which are pharmacological, may substantially lower blood pressure. The commonest cause is head-up postural change.

HEAD-UP POSTURAL CHANGE

Postural hypotension is often seen in high lesions, especially in the early phases of rehabilitation. At this stage the patients have usually been in bed for a few weeks and the postural fall in blood pressure is mainly due to the inability to activate sympathetic reflexes because of the lesion. Prolonged recumbency, with the lack of activation of various compensatory mechanisms, may also contribute.

Postural hypotension can cause a variety of symptoms that are largely related to cerebral ischaemia. These include dizziness, buzzing and ringing in the ears, a feeling of lightheadedness, and impaired vision. The latter may take on a number of variations, with blacking out or greying out occurring frequently. In some patients the fall in blood pressure may culminate in syncope, although in many it may reach extremely low levels without loss of consciousness. This, however, may not be so, especially in the early stages after an injury or after a period of prolonged recumbency. The reasons for this difference are not entirely clear. It may be that with time there is a greater tolerance to a low cerebral perfusion pressure because of more effective cerebrovascular autoregulation, as observed in patients with chronic primary autonomic failure. With repeated head-up tilt, which appears to reduce both the fall in blood pressure and symptoms, a number of vasoactive substances are locally produced and released into the circulation; whether these beneficially affect the cerebral vasculature is not known.

In the early stages after an injury, the blood pressure on head-up postural changes often continues to fall until the patients are put back to the horizontal (Figure 10.4a). Plasma noradrenaline levels do not rise, consistent with the inability to activate peripheral sympathetic pathways (Figure 10.5) (Mathias *et al.*, 1975a). If head-up tilt is prolonged, especially in the chronic stage, the blood pressure tends to partially recover, often with oscillations (Figure 10.4b). This slow recovery may be related to activation of the R-A-A system. In tetraplegics, renin release in response to head-up tilt is often exaggerated. This occurs independently of sympathetic stimulation (Figure 10.6), and is probably secondary to afferent glomerular dilatation and renal baroreceptor stimulation resulting from the fall in renal perfusion pressure (Mathias *et al.*, 1980). Renin is the enzyme that results in the formation of the octapeptide angiotensin-II, which can raise blood pressure by both central and peripheral actions. (Mathias, May and Taylor, 1984a), the latter being less likely to contribute in high spinal cord lesions. In the periphery, angiotensin-II may raise blood pressure in a number of ways; by acting directly upon the vasculature to cause constriction, by facilitating the release and actions of noradrenaline and as a powerful stimulator of aldosterone release from the adrenal cortex. The secondary rise in aldosterone, with its salt and water retaining effects, increases intravascular volume and buffers the fall in blood pressure. This may explain the reasons why head-up tilt at night and frequent postural change is often of considerable benefit in raising supine blood pressure and diminishing postural hypotension in high lesions.

A further factor in chronic high lesions which helps raise blood pressure during postural change is the activation of spinal reflexes. Movement may thus stimulate the skin, initiate skeletal muscle spasms or cause bladder activity, each of which can elevate blood pressure rapidly. This is used by some patients in wheelchairs who, by tapping suprapubically, can activate the urinary bladder, raise blood pressure and thus overcome the symptoms of postural hypotension (Figure 10.7). In addition, local sympathetic reflexes (veno-arteriolar reflexes) have been described in patients with high lesions during postural change (Skagen *et al.*, 1982) and may play a role in preventing blood pressure from falling even further.

During head-up postural change, the lack of sympathetically induced vasoconstriction causes considerable venous pooling which accounts for the fall in central venous pressure, stroke volume and cardiac output (Corbett *et al.*, 1975). Patients who spend long periods in a wheelchair may have both ankle oedema and cyanotic discolouration of the legs. Urine volume in such patients is often reduced, at times to extremely low levels. This is due to a reduction in renal plasma flow and additionally due to the elevation in levels of the anti-diuretic hormone vasopressin, which can rise to higher levels in tetraplegics than in normal subjects (Sved, McDowell and Blessing, 1985; Poole *et al.*, 1987).

During head-up postural change there is an elevation in heart rate which is due predominantly to withdrawal of vagal tone. As a consequence, the heart rate does not rise above 100 beats per minute, despite severe hypotension.

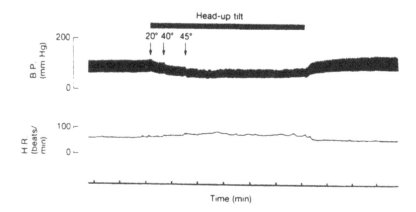

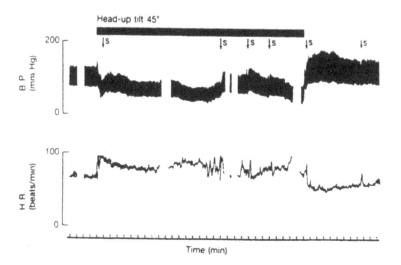

FIGURE 10.4   (a) Blood pressure (BP) and heart rate (HR) in a tetraplegic patient before and after head-up tilt, in the early stages of rehabilitation. This patient had minimal skeletal muscle spasticity and autonomic dysreflexia. (From Frankel and Mathias, 1976.) (b) Blood pressure (BP) and heart rate (HR) in a tetraplegic patient before, during and after head-up tilt to 45°. Blood pressure promptly falls with partial recovery which, in this case, is linked to skeletal muscle spasms (S), including spinal sympathetic activity. Some of the later oscillations are probably due to the rise of renin, measured where there are interruptions in the intra-arterial record. In the later phases, muscle spasms occur more frequently and further elevate blood pressure. On return to the horizontal, blood pressure rises rapidly above the previous basal level and slowly returns to the horizontal. Heart rate tends to move in the opposite direction. There is a transient increase in heart rate during muscle spasms. (From Mathias and Frankel, 1988.)

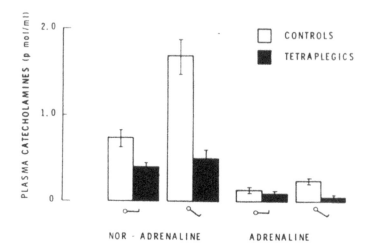

FIGURE 10.5   Plasma noradrenaline and adrenaline level in controls (normal subjeccts) and chronic tetraplegic patients at rest and during head-up tilt to 45° for 10 min. There is a rise in plasma noradrenaline in the control subjects, but no change in the tetraplegics. The bars indicate ± SEM. (From Mathias and Frankel, 1992.)

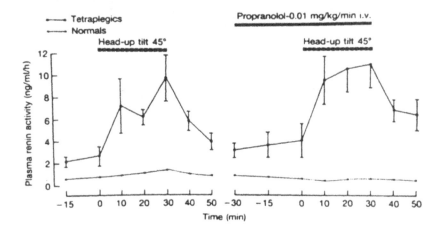

FIGURE 10.6   Levels of plasma renin activity in tetraplegic patients (continuous lines) and in normal subjects (interrupted lines) before, during and after head-up tilt to 45°. Levels in the left panel are before, and levels in the right panel after administration of propranolol in a dose which would have effectively blocked beta-adrenergic receptors. The renin response in the tetraplegics is considerably greater than in the normal subjects. Blood pressure fell in the tetraplegics during head-up tilt. Beta-adrenoreceptor blockade with propranolol had no effect on renin response to head-up tilt in the tetraplegics. This differs from the response in normal subjects. Taken in conjunction with other evidence, this suggests that renin release in tetraplegics during head-up tilt occurs independently of sympathetic activation and stimulation of beta-adrenergic receptors. (From Mathias and Frankel, 1991.)

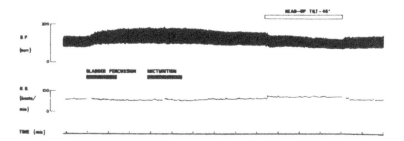

FIGURE 10.7   The effect of bladder percussion and subsequent micturition on the blood pressure (BP, in mmHg) and heart rate, (HR, beats/min) of a chronic tetraplegic. Head-up tilt to 45° caused the blood pressure to fall but it did not reach the low levels seen in the same patient (Figure 10.4a) during a similar degree of head-up tilt. (From Frankel and Mathias, 1976.)

Postural hypotension in high spinal lesions is often a problem in the early stages of rehabilitation, and in the chronic stages after prolonged recumbency. It does not, however, usually cause as many problems as in patients with chronic primary autonomic failure. The management of postural hypotension in high spinal lesions consists of tilting of the head-end of the bed at night, together with frequent postural changes. This is likely to maximally activate the R-A-A system and reduce recumbency induced diuresis. In addition, physical methods such as abdominal binders and thigh cuffs have been used. A range of drugs, as used in autonomic failure (Mathias, 1991), may be needed in extreme cases. Of these, ephedrine appears to be the most beneficial and a dose of 15 mg half an hour before postural change usually helps. Other vasopressor agents such as dihydroergotamine may be used, but these may raise blood pressure further when pressor inducing stimuli are activated. They should, therefore, be used with caution.

ELEVATION IN INTRATHORACIC PRESSURE

The Valsalva manoeuvre is abnormal in high spinal cord lesions. In Phase II (with raised intrathoracic pressure) when the blood pressure falls, there is a rise in heart rate because of intact baroreceptor afferent and vagal efferent pathways (Figure 10.8). Blood pressure, however, can continue to fall while intrathoracic pressure is maintained, and in Phase IV (when intrathoracic pressure is reduced) there is no blood pressure overshoot and, therefore, no heart rate undershoot, indicative of the inability to activate sympathetic pathways. Many patients can lower their blood pressure with small elevations of intrathoracic pressure of even 10 mmHg. A further elevation, as has been noted in a tetraplegic during singing, can result in severe hypotension and syncope (van Lieshout et al., 1991).

Raising intrathoracic pressure has been used beneficially to lower blood pressure in patients who have severe hypertension while undergoing surgery, especially to the urinary tract. This is readily introduced by increasing positive pressure ventilation (Welply, Mathias and Frankel, 1975).

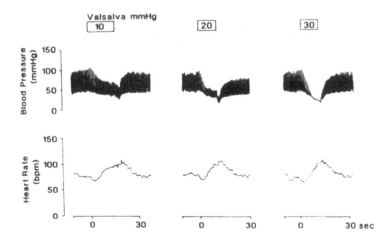

FIGURE 10.8 Blood pressure and heart rate responds to the Valsalva manoeuvre in a tetraplegic patient. The intrathoracic pressure has been raised from 10 to 30 mmHg. The blood pressure during the Valsalva Phase II drops progressively without recovery as the pressure is increased, to the extent that the pulse pressure is virtually obliterated when intrathoracic pressure is raised to 30 mmHg. (From van Lieshout *et al.*, 1991.)

FOOD

Food ingestion in tetraplegics causes only a small fall in supine blood pressure, with a modest elevation in heart rate. This is unlike patients with chronic primary autonomic failure, in whom there are marked falls in supine blood pressure (Mathias, 1990) and a considerable accentuation of post-prandial postural hypotension (Mathias *et al.*, 1991). The latter has not been formally tested in tetraplegics. The reasons for the differences are not clear and may include, in tetraplegics, a brisker pressor hormonal response and the activation of intestinal reflexes as demonstrated in spinal animals (Stein and Weaver, 1988).

HYPOGLYCAEMIA

In tetraplegics, hypoglycaemia lowers blood pressure (Mathias *et al.*, 1979b), although not to the same extent as in patients with chronic primary autonomic failure (Mathias *et al.*, 1987). In neither group is there a rise in plasma noradrenaline or adrenaline levels, unlike normal subjects in whom plasma adrenaline levels rise markedly. The clinical manifestations of anxiety, tremulousness, hunger, sweating and tachycardia do not occur in tetraplegics. The majority are sedated, and this can be quickly reversed with an intravenous injection of hypertonic glucose. If rapidly injected, as is the normal practice, this can cause a marked, albeit transient, fall in blood pressure, similar to observations in patients with autonomic failure (Mathias *et al.*, 1987).

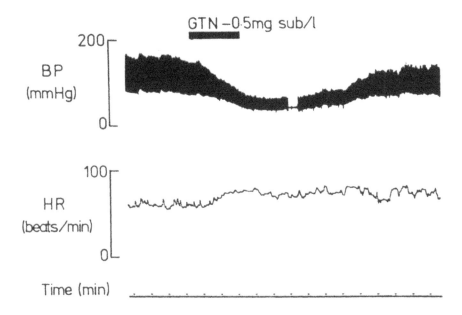

FIGURE 10.9 Changes in blood pressure (BP) and heart rate (HR) in a chronic tetraplegic patient before, during and after the standard dose of sublingual glyceryl trinitrate (GTN), when there is a prompt and substantial fall in blood pressure with the patient lying flat. (From Mathias, 1987.)

DRUGS

Vasodilator drugs can profoundly lower blood pressure in high lesions. This is presumably because their effects are not opposed by activation of the baroreflex pathways and a subsequent increase in sympathetic activity, as a consequence of the spinal lesion. Directly acting vascular smooth muscle dilators, such as glyceryl trinitrate can, therefore, cause a severe reduction in blood pressure, even in the supine position (Figure 10.9). There is a rise in heart rate but this does not prevent the fall in blood pressure. Other drugs, such as isoprenaline, may also lower blood pressure because of their actions on vascular beta-2 adrenoreceptors. This can be demonstrated when the drug is given intravenously either as a bolus or by infusion (Mathias, 1976b).

Other drugs that can lower blood pressure markedly are agents that interfere with components of the many systems that help maintain blood pressure. Blockade of the R-A-A system therefore, will lower blood pressure, as is seen after captopril which prevents angiotensin-II formation, but may have additional effects, such as increasing levels of bradykinin. Small doses of diuretics, such as the thiazides, may also cause marked hypotension, probably by reducing intravascular volume.

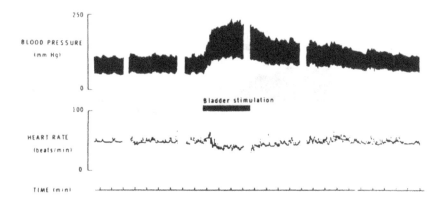

FIGURE 10.10 Intra-arterial blood pressure and heart rate in a chronic tetraplegic patient before, during and after stimulation of the urinary bladder by suprapubic tapping of the anterior abdominal wall. With stimulation there is a marked rise in blood pressure, with an initial transient rise followed by a fall in heart rate. The blood pressure gradually falls to basal levels. The breaks in the record indicate where blood has been withdrawn. (From Mathias and Frankel, 1983.)

HYPERTENSION

AUTONOMIC DYSREFLEXIA

Chronic cervical and high thoracic spinal cord lesions are unique, as they may have severe paroxysms of hypertension following activation of spinal sympathetic reflexes. These responses do not occur in the stage of spinal shock, when stimulation of the skin, urinary bladder or skeletal muscle does not change blood pressure or heart rate (Figure 10.2) (Mathias *et al.*, 1979a). In the chronic stage, however, such stimulation results in a marked rise in blood pressure (Figure 10.10). There is initially a transient rise in heart rate which soon falls, presumably because of activation of the vagal efferent component of the baroreflex. There are a number of additional responses involving target organs supplied by sympathetic and parasympathetic nerves. These include sweating around and above the level of the lesion, evacuation of the urinary bladder and rectum, penile erection and seminal fluid emission (in the male), and skeletal muscle spasms; these were first described by Head and Riddoch in 1917 and referred to as the "mass reflex". The cardiovascular changes during urinary bladder stimulation were described by Guttmann and Whitteridge in 1947.

The cardiovascular responses to cutaneous, visceral and skeletal muscle stimulation include constriction of both resistance and capacitance vessels (Corbett, Frankel and Harris, 1971a,b). This may markedly reduce the peripheral blood flow and cause cold limbs, hence the term poikilopiesis spinalis (Koster and Bethlem, 1960). Stroke volume and cardiac output rise during

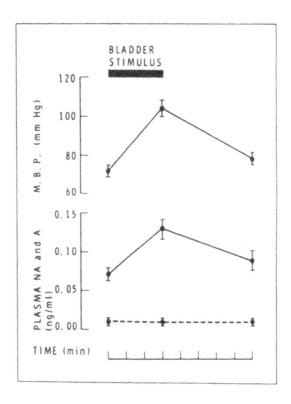

FIGURE 10.11 Mean blood pressure (MBP) and plasma noradrenaline (NA) (●——●) and adrenaline (A) (●–––●) before, at the peak of 3 min of bladder stimulation and 5 min later in a group of chronic tetraplegic patients. (From Mathias *et al*, 1976a.)

stimulation, suggesting the activation of spinal cardiac reflexes (Corbett *et al.*, 1975). These changes occur rapidly, indicating that they are of neurogenic origin and due to reflex sympathetic activity through the isolated spinal cord. Further evidence has been obtained by measurement of plasma noradrenaline levels (Figure 10.11), which rise with stimulation and are closely correlated with blood pressure changes (Mathias *et al.*, 1976a). Plasma adrenaline levels are unchanged, excluding adrenomedullary secretion as being contributory. Levels of plasma dopamine beta-hydroxylase (which is coreleased with noradrenaline) also rise (Figure 10.12), but the rise is slower and levels remain elevated for longer, reflecting the slower entry of dopamine beta-hydroxylase into the circulation (presumably because of its large molecular weight), and also its long half life (Mathias *et al.*, 1976b). Levels of other vasoconstrictor substances, such as renin and angiotensin-II, remain unchanged (Mathias *et al.*, 1981). Potential vasodilator substances such as prostaglandin E-2 rise in plasma (Mathias *et al.*,

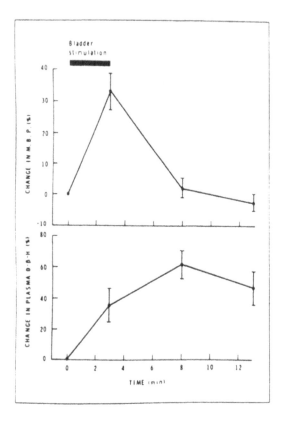

FIGURE 10.12   Change in mean blood pressure (MBP) and in plasma dopamine beta-hydroxylase (DβH) during and after bladder stimulation for 3 min in a group of chronic tetraplegic patients. (From Mathias *et al.*, 1976b.)

1975b); whether these or allied substances help buffer the blood pressure rise is unclear. Drugs preventing prostaglandin formation, such as indomethacin, may raise supine blood pressure, and have the potential to increase the degree of hypertension during autonomic dysreflexia.

Plasma noradrenaline levels in tetraplegics are normally a third lower than basal levels in normal subjects. During autonomic dysreflexia, they are significantly elevated, with a two- or three-fold increase (Mathias *et al.*, 1976a). Despite this, however, the levels are only moderately above the resting basal levels of normal subjects. This differs markedly from the grossly elevated levels often seen during hypertensive crises in patients with phaeochromocytoma. These observations suggest that in high spinal lesions, "sympathetic hyperreflexia" does not occur. This has been confirmed using sympathetic microneurography (Stjernberg, Blumberg and Wallin, 1986). In such patients the rise in blood pressure during autonomic dysreflexia is preceded by only a moderate and transient rise in muscle sympathetic

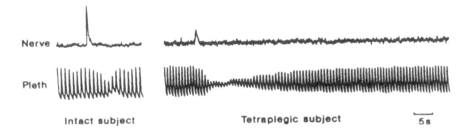

FIGURE 10.13   Relationship between reflex sympathetic discharge in skin nerves (nerve) and cutaneous vasoconstriction (pleth) in a normal subject and in a tetraplegic patient. There is pronounced vasoconstriction in the tetraplegic. (From Stjernberg, 1986.)

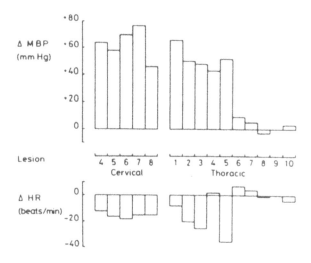

FIGURE 10.14   Changes in mean blood pressure (MBP) and heart rate (HR) in patients with spinal cord lesions at different levels (cervical and thoracic) after bladder stimulation induced by suprapubic percussion. In the cervical and high thoracic lesions, there is a marked elevation in blood pressure and a fall in heart rate. In patients with lesions below T5, there are minimal changes. (From Mathias and Frankel, 1986.)

nerve activity (Figure 10.13). Furthermore, there is no association between the cardiac cycle and muscle sympathetic nerve discharge, as would occur normally. These neurohormonal and neurophysiological studies, therefore, exclude excessive sympathetic efferent activity and indicate hyperactivity of target organs innervated by the autonomic nervous system. This raises the question of supersensitivity of adrenoreceptors, among other mechanisms (Mathias *et al.*, 1976c; Davies *et al.*, 1982).

In patients with lesions below the 5th thoracic segment, autonomic dysreflexia and increased pressor responses to cutaneous, visceral and skeletal muscle stimuli do not occur (Figure 10.14). The neural outflow above the T5 level is clearly

TABLE 10.1

Causes of autonomic dysreflexia. (From Mathias and Frankel, 1992.)

Abdominal or pelvic visceral stimulation

Ureter
    Calculus

Urinary bladder
    Distension by blocked catheter or discoordinated bladder
    Infection
    Irritation by calculus, catheter, or bladder washout

Rectum and anus
    Enemata
    Faecal retention
    Anal fissure

Gastrointestinal organs
    Gastric dilatation
    Gastric ulceration
    Cholecystitis or cholelithiasis

Uterus
    Contraction during pregnancy
    Mensturation, occasionally

Cutaneous stimulation

Pressure sores
Infected ingrowing toenails
Burns

Skeletal Muscle Spasms

Especially in limbs with contractures

Miscellaneous
    Intrathecal neostigmine
    Electro-ejaculatory procedures
    Ejaculation
    Vaginal dilatation
    Urethra-insertion of catheter or abscess
    Fractures of bones

of major importance in the maintenance of blood pressure, probably because it controls the large splanchnic circulation, which accounts for 20 - 30% of the total blood volume. In lesions above T5, stimulation unmasks primary cutaneovascular, viscerovascular and somatovascular reflexes, which raise blood pressure without the ability to compensate and maintain blood pressure homeostasis. These compensatory mechanisms are probably normally modulated by the brain through the descending sympathetic outflow, and in particular through the neural outflow to the splanchnic circulation.

TABLE 10.2

Some of the drugs used in the management of autonomic dysreflexia classified according to their major site of action on the reflex arc and target organs. (From Mathias and Frankel, 1992.)

Afferent		Topical lignocaine
Spinal cord		Clonidine[*]
		Reserpine[*]
		Spinal anaesthetics
Efferent	Sympathetic ganglia	Hexamethonium
	Sympathetic nerve terminals	Guanethidine
	α-adrenoreceptors	Phenoxybenzamine
Target organs	Blood vessels	Glyceryl trinitrate
		Nifedipine
	Sweat glands	Probanthine

[*] Clonidine and reserpine have multiple effects, some of which are peripheral.

The prevention and management of autonomic dysreflexia is of major clinical importance (Mathias and Frankel, 1992). Although minor episodes of autonomic dysreflexia undoubtedly occur at frequent intervals, severe episodes may occur in certain situations as outlined in Table 10.1. Severe hypertension can result in complications ranging from visual deficits and epileptic seizures, to intracerebral haemorrhage and strokes. Sweating over the face and neck may be severe, cause much discomfort and contribute to skin breakdown. A key factor is the prevention of autonomic dysreflexia. It is important to consider common factors, especially the urinary bladder and large bowel. At times it may be difficult to pin down the precipitating cause. An anal fissure, for instance, can be missed because of the absence of pain.

The therapeutic approaches in the management of autonomic dysreflexia are broadly directed to the lowering of blood pressure. The actions of drugs used are related to the pathophysiological abnormalities and these are outlined in Table 10.2, which indicates how a variety of approaches can be used to reduce autonomic activity. The afferents to the spinal cord can be blocked, and an example is the use of a local anaesthetic, such as lignocaine into the urinary bladder. The spinal cord itself may be the site of action; spinal anaesthetics, are particularly effective in preventing autonomic dysreflexia especially when associated with pregnancy. The ganglionic blocker, hexamethonium, was successfully used in the past. Like most peripherally acting drugs, it has the capacity to cause profound postural hypotension. Drugs such as clonidine do not appear to do this (Reid et al., 1977; Mathias et al., 1979c), which may reflect its predominantly central actions in lowering blood pressure. Drugs acting directly upon blood vessels, such as glyceryl trinitrate or the calcium channel blockers, which include nifedipine, can also be used sublingually, but have the potential to cause severe hypotension. Probanthine reduces sweating but may have other side effects.

Alpha-adrenoreceptor blockers, which include phenoxybenzamine and more specific blockers such as prazosin, have been successfully used in autonomic dysreflexia, particularly in situations associated with bladder neck obstruction, where they relax the smooth muscle of the urinary sphincter. In other situations they may not be as effective and possible reasons include their inability to prevent secretion of non-adrenergic substances from sympathetic terminals. These include neuropeptide Y or adenosine triphosphate, which have the potential to act on target organs and induce vasoconstriction independently of alpha-adrenoreceptors. Their effects may also be enhanced in the presence of pressor supersensitivity. Drugs such as reserpine and guanethidine probably deplete sympathetic nerve endings of these substances, which may explain why they are of value in severe autonomic dysreflexia, especially where other agents have failed.

During surgery of the urinary bladder and large bowel, autonomic dysreflexia can be a problem. The use of a spinal anaesthetic, or a general anaesthetic such as halothane in combination with the use of intermittent positive pressure ventilation, is often successful in controlling the hypertension (Welply, Mathias and Frankel, 1975). Short-acting ganglionic blockers which include trimethaphan, may be used successfully in such patients.

DRUGS

Patients with high spinal cord lesions have an enhanced pressor sensitivity to a range of drugs, which include the neurotransmitter noradrenaline (Figure 10.15), more specific alpha-adrenoreceptor agonists such as phenylephrine, and drugs with diverse chemical structures, from the octapeptide angiotensin-II and its antagonist saralasin, (Figure 10.16) to the prostaglandin F-2-alpha (Mathias, 1976; Mathias et al., 1984b). The initial supposition was that pressor supersensitivity was a clinical manifestation of Canon's law, that denervated organs were supersensitive to their neurotransmitter. In high lesions the denervation is preganglionic and there is histochemical evidence of intact adrenergic nerve terminals (Norberg and Normell, 1974). Furthermore, during autonomic dysreflexia there is definite activity of postganglionic sympathetic nerves. Basal supine plasma levels of noradrenaline in tetraplegics are considerably lower than normal, and up-regulation of adrenoreceptors may be contributory. Indirect evidence from platelet binding studies of alpha-2 adrenoreceptors, however, do not favour up-regulation (Davies et al., 1982). Furthermore, this should not apply to angiotensin-II, as basal supine levels of plasma renin activity and angiotensin-II are often either normal or elevated in such patients.

In true denervation supersensitivity there should be impairment of neuronal uptake of noradrenaline resulting in a greater concentration at the postsynaptic site. This does not seem to occur in tetraplegics (Mathias et al., 1976c). It may be that the impairment of the neural pathways that normally descend in the spinal cord and buffer the rise in blood pressure are of particular importance. This may

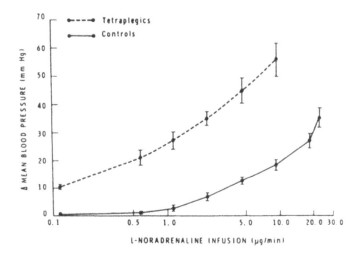

FIGURE 10.15   Changes (Δ) in average mean blood pressure during different dose infusion rates of noradrenaline in five chronic tetraplegics (●– – –●), and 10 control subjects (●———●). The bars indicate ± SEM. There is an enhanced pressor response to noradrenaline in the tetraplegics over the entire dose range studied. (From Mathias *et al.*, 1976c.)

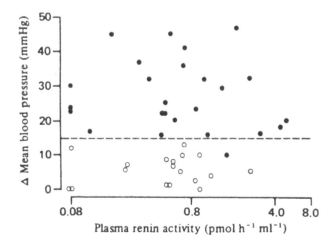

FIGURE 10.16   Maximum change (Δ) in mean blood pressure (MBP) in the early period after infusion of the angiotensin-II receptor antagonist, saralasin on the ordinate, plotted against plasma renin activity on a log scale of the abscissa, for tetraplegic patients (●) and normal subjects (controls, ○). An arbitrary line has been drawn at the 15 mmHg mark. There is a greater pressor response in the tetraplegics, except on one occasion where the responses lie below the line, along with the normal subjects. This immediate pressor response to all doses of saralasin was enhanced in the tetraplegic patients regardless of the prevailing levels of plasma renin activity. (From Mathias *et al.*, 1984b.)

explain why there is a sharp cutoff between the level of the lesion and hypertensive responses to stimulation of the bladder and autonomic dysreflexia, and also to infusion of pressor agents. This level is in the region of T5, and it is likely that in spinal lesions below this level there is preservation of cerebral control of the large splanchnic and renal vascular beds, which maintains blood pressure. This would not be possible in the cervical and higher thoracic lesions.

Patients with high spinal cord lesions, therefore are prone to pressor or depressor supersensitivity to a whole range of vasoactive drugs. This should be carefully borne in mind even with drugs which normally have only minimal vasoactive effects.

EPILOGUE

Patients with high spinal cord lesions exhibit a range of clinical and autonomic problems that provide valuable information on those neural pathways that contribute to cardiovascular regulation. Studies in such patients also emphasise the role of secondary hormonal and locally produced substances that contribute to cardiovascular homeostasis in man.

ACKNOWLEDGEMENTS

We thank our many colleagues at each of the centres who have contributed to many of the studies described. The work we have been involved in was supported by the Wellcome Trust, the International Spinal Research Trust, the Medical Research Council and the Lawson Tait Medical and Scientific Research Trust.

REFERENCES

Alam, M., Unwin, R., Frankel, H.L., Peart, W.S. and Mathias, C.J. (1992) Cardiovascular and hormonal effects of captopril in tetrapiegia. *Clinical Autonomic Research*, **2**, 59.

Corbett, J.L., Frankel, H.L. and Harris, P.J. (1971a) Cardiovascular reflex responses to cutaneous and visceral stimuli in spinal man. *Journal of Physiology (London)*, **215**, 395–405.

Corbett, J.L., Frankel, H.L. and Harris, P.J. (1971b) Cardiovascular changes associated with skeletal muscle spasm in tetraplegic man. *Journal of Physiology, (London)*, **215**, 395–409.

Corbett, J.L., Debarge, O., Frankel, H.L. and Mathias, C.J. (1975) Cardiovascular responses in tetraplegic man to muscle spasm, bladder percussion and head-up tilt. *Clinical Experiments in Pharmacology and Physiology*, Suppl. 2, 189–193.

Di Bona, G.F. and Wilcox, L.S. (1992) The kidney and the sympathetic nervous system. In: *Autonomic Failure. A Textbook of Clinical Disorders of the Autonomic Nervous System*, edited by R. Bannister and C.J. Mathias. 3rd Edition. Oxford: Oxford University Press.

Davies, I.B., Mathias, C.J., Sudera, D. and Sever, P.S. (1982) Agonist regulation of alpha-adrenergic receptor responses in man. *Journal of Cardiovascular Pharmacology*, **4**, S139–S144.

Frankel, H.L., Michaelis, L.S., Golding, D.R. and Beral, V. (1972) The blood pressure in paraplegia-1. *Paraplegia*, **10**, 193–198.

Frankel, H.L., Mathias, C.J. and Spalding, J.M.K. (1975) Mechanisms of reflex cardiac arrest in tetraplegic patients. *Lancet*, **2**, 1183–1185.

Frankel, H.L. and Mathias, C.J. (1976) The cardiovascular system in tetraplegia and paraplegia. In *Handbook of Clinical Neurology*, vol. 26, edited by P.J. Vinken and G.W. Bruyn, pp. 313–333. Amsterdam: Elsevier.

Guttmann, L. and Whitteridge, D. (1947) Effects of bladder distension on autonomic mechanisms after spinal cord injury. *Brain*, **70**, 361–404.

Hall, M. (1841) *New Memoirs on the Nervous System*, London: Balliere.

Head, H. and Riddoch, G. (1917) The autonomic bladder, excessive sweating and some other reflex conditions in gross injuries of the spinal cord. *Brain*, **40**, 188–263.

Kooner, J.S., da Costa, D.F., Frankel, H.L., Bannister, R., Peart, W.S. and Mathias, C.J. (1987) Recumbency induces hypertension, diuresis and natriuresis in autonomic failure but diuresis alone in tetraplegia. *Journal of Hypertension*, **5**(Suppl.), 327–329.

Koster, M. and Bethlem, J. (1960) Paroxysmal hypertension in patients with spinal cord lesions (poikilopiesis spinalis). *Archives Psychiatry & Neurological Scandinavica*, **36**, 347–368.

Mathias, C.J., Christensen, N.J., Corbett, J.L., Frankel, H.L., Goodwin, T.J. and Peart, W.S. (1975a) Plasma catecholamines, plasma renin activity and plasma aldosterone in tetraplegic man, horizontal and tilted. *Clinical Science and Molecular Medicine*, **49**, 291–299.

Mathias, C.J., Hillier, K., Frankel, N.L. and Spalding, J.M.K. (1975b) Plasma prostaglandin E during neurogenic hypertension in tetraplegic man. *Clinical Science and Molecular Medicine*, **49**, 625–628.

Mathias, C.J. (1976a) Bradycardia and cardiac arrest during tracheal suction – mechanisms in tetraplegic patients. *European Journal of Intensive Care Medicine*, **2**, 147–156.

Mathias, C.J., Christensen, N.J., Corbett, J.L., Frankel, H.L. and Spalding, J.M.K. (1976a) Plasma catecholamines during paroxysmal neurogenic hypertension in quadriplegic man. *Circulation Research*, **39**, 204–208.

Mathias, C.J. (1976b) Neurological disturbances of the cardiovascular system. *D. Phil. Thesis*, University of Oxford, Oxford, UK.

Mathias, C.J., Smith, A.D., Frankel, H.L. and Spalding, J.M.K. (1976b) Release of dopamine β-hydroxylase during hypertension from sympathetic overactivity in man. *Cardiovascular Research*, **10**, 176–181.

Mathias, C.J., Frankel, H.L., Christensen, N.J. and Spalding, J.M.K. (1976c) Enhanced pressor response to noradrenaline in patients with cervical spinal cord transection. *Brain*, **99**, 757–770.

Mathias, C.J., Christensen, N.J., Frankel, H.L. and Spalding, J.M.K. (1979a) Cardiovascular control in recently injured tetraplegics in spinal shock. *Quarterly Journal of Medicine, New Series*, **48**, 273–287.

Mathias, C.J., Frankel, H.L. Turner, R.C. and Christensen, J.N. (1979b) Physiological responses to insulin hypoglycaemia in spinal man. *Paraplegia*, **17**, 319–326.

Mathias, C.J., Reid, J.L., Wing, L.M.H., Frankel, H.L. and Christensen, N.J. (1979c) Antihypertensive effects of clonidine in tetraplegic subjects devoid of central sympathetic control. *Clinical Science*, **57**, 425s–428s.

Mathias, C.J., Christensen, N.J., Frankel, H.L. and Peart, W.S. (1980) Renin release during head-up tilt occurs independently of sympathetic nervous activity in tetraplegic man. *Clinical Science*, **59**, 251–256.

Mathias, C.J., Frankel, H.L., Davies, I.B., James, V.H.T. and Peart, W.S. (1981) Renin and aldosterone release during sympathetic stimulation in tetraplegia. *Clinical Science*, **60**, 399–604.

Mathias, C.J. and Frankel H.L. (1983) Autonomic failure in tetraplegia. In *Autonomic Failure. A Textbook of Clinical Disorders of the Autonomic Nervous System*, 2nd edn, edited by R. Bannister, pp. 453–488. Oxford: Oxford University Press.

Mathias, C.J., May, C.N. and Taylor, G.M. (1984a) The renin-angiotensin system and hypertension – basic and clinical aspects. In *Molecular Medicine*, Vol. 1, edited by A.D.B. Malcolm, pp. 177–208. Oxford: IRL Press.

Mathias, C.J., Unwin, R.J., Pike, F.A., Frankel, H.L., Sever, P.S. and Peart, W.S. (1984b) The immediate pressor response to saralasin in man: evidence against sympathetic activation for intrinsic angiotensin-II-like myotropism. *Clinical Science*, **66**, 517–524.

Mathias, C.J. and Frankel, H.L. (1986) The neurological and hormonal control of blood vessels and heart in spinal man. *Journal of Autonomic Nervous System*, Suppl., 457–464.

Mathias, C.J., Fosbraey, P., da Costa, D.F., Thornley, A. and Bannister, R. (1986) Desmopressin reduces nocturnal polyuria, reverses overnight weight-loss and improves morning postural hypotension in autonomic failure. *British Medical Journal*, **293**, 353–354.

Mathias, C.J., da Costa, D.F., Fosbraey, P., Christensen, N.J. and Bannister, R. (1987) Hypotensive and sedative effects of insulin in autonomic failure. *British Medical Journal*, **295**, 161–163.

Mathias, C.J. (1987) Autonomic dysfunction. *British Journal of Hospital Medicine*, **38**, 238–243.

Mathias, C.J. and Frankel, H.L. (1988) Cardiovascular control in spinal man. *Annual Review of Physiology*, **50**, 577–592.

Mathias, C.J. (1990) Cardiovascular effects of food intake in patients with impaired autonomic function. *Journal of Neuroscience Methods*, **34**, 193–200.

Mathias, C.J. (1991) Disorders of the autonomic nervous system. In *Neurology in Clinical Practice*, edited by W.G. Bradley, R.B. Daroff, G.M. Fenichel and C.D. Mardsen pp. 1661–1685, Massachusetts: Butterworth Publishers.

Mathias, C.J. and Frankel, H.L. (1991) Management of cardiovascular abnormalities caused by autonomic dysfunction in spinal injury. In *Spinal Cord Dysfunction: Intervention and Treatment*, edited by L.S. Illis, pp. 101–120. Oxford: Oxford University Press.

Mathias, C.J., Holly, E., Armstrong, E., Shareef, M. and Bannister, R. (1991) The influence of food on postural hypotension in three groups with chronic autonomic failure: clinical and therapeutic implications. *Journal of Neurology, Neurosurgery and Psychiatry*, **54**, 726–730.

Mathias, C.J. and Frankel, H.L. (1992) Autonomic disorders in spinal cord lesions. In *Autonomic Failure. A Textbook of Disorders of the Autonomic Nervous System*, edited by R. Bannister and C.J. Mathias. 3rd edition. pp. 839–883. Oxford: Oxford University Press.

Norberg, K.A. and Normell, L.A. (1974) Histochemical demonstration of sympathetic adrenergic denervation in human skin. *Acta Neurologica Scandinavica*, **50**, 261–271.

Poole, C.J.M., Williams, T.D.M., Lightman, S.L. and Frankel, H.L. (1987) Neuroendocrine control of vasopressin secretion and its effect on blood pressure in subjects with spinal cord transection. *Brain*, **110**, 727–735.

Reid, J.L., Wing, L.M.H., Mathias, C.J., Frankel, H.L. and Neill, E. (1977) The central hypotensive effect of clonidine: studies in tetraplegic subjects. *Clinical Pharmacology and Therapeutics*, **21**, 375–381.

Skagen, K., Jensen, K., Henriksen, O. and Knudsen, L. (1982) Sympathetic reflex control of subcutaneous blood flow in tetraplegic man during postural changes. *Clinical Science*, **62**, 605–609.

Stein, R.B. and Weaver, L.C. (1988) Multi- and single-fibre mesenteric and renal sympathetic responses to chemical stimulation of intestinal receptors in cats. *Journal of Physiology (London)*, **396**, 155–172.

Stjernberg, L., Blumberg, H. and Wallin, B.G. (1986) Sympathetic activity in man after spinal cord injury: outflow to muscle below the lesion. *Brain*, **109**, 695–715.

Stjernberg, L. (1986) Neural and hormonal vasomotor control and temperature regulation in spinal man. Studies based on microneurographic recordings. Comprehensive summaries of Uppsala dissertations from the Faculty of Medicine, Uppsala.

Sutters, M., Wakefield, C., O'Neil, R., Appleyard, M., Frankel, H.L., Mathias, C.J., *et al.* (1992) The cardiovascular, endocrine and renal responses in tetraplegic and paraplegic subjects to dietary sodium restriction. *Journal of Physiology (London)*, **457**, 515–523.

Sved, A.F., McDowell, F.H. and Blessing, W.W. (1985) Release of antidiuretic hormone in quadriplegic subjects in response to head-up tilt. *Neurology*, **35**, 78–82.

van Lieshout, J.J., Imholz, B.P.M., Wesseling, K.H., Speelman, J.D. and Wieling, W. (1991) Singing-induced hypotension: a complication of high spinal cord lesion. *Netherlands Journal of Medicine*, **38**, 75–79.

Wallin, B.G. and Stjernberg, L. (1984) Sympathetic activity in man after spinal cord injury. Outflow to skin below the lesion. *Brain*, **107**, 183–198.

Welply, N.C., Mathias, C.J. and Frankel, H.L. (1975) Circulatory reflexes in tetraplegics during artificial ventilation and general anaesthesia. *Paraplegia*, **13**, 172–182.

11 Baroreflex Failure

David Robertson

Departments of Medicine, Pharmacology, and Neurology,
Clinical Research Center, Vanderbilt University,
Nashville, TN 37232-2195, USA

Baroreflexes maintain blood pressure within the narrow normal range. Afferent input for the baroreflex travels primarily in the glossopharyngeal and vagal nerves. Bilateral damage to these nerves or their central connections in the brainstem results in baroreflex failure. In its fully expressed form, baroreflex failure causes severe episodic hypertension, tachycardia, and sweating, punctuated by periods of normal or even low blood pressure.

KEY WORDS: Baroreflex failure, hypertension, tachycardia, sweating, neuralgia

INTRODUCTION

The syndrome of baroreflex failure in human subjects is still poorly defined in spite of many decades of animal and clinical studies. The principal reason for this is that there have been few studies able to correlate anatomic lesions with clinical consequences in significant numbers of patients. This situation may be changing with improved techniques for assessing autonomic regulation in man (Sprenkle *et al.*, 1986; Sanders, Ferguson and Mark, 1988), greater understanding of the interrelationships of central structures regulating the cardiovascular system (Reis and Doba, 1972; Doba and Reis, 1974) and enhancement of our capacity to visualize brainstem structures through magnetic resonance and other imaging techniques. The development of specialized centers for the referral and evaluation of patients with autonomic disorders has also been important in that it has allowed individual investigative groups to develop experience in the recognition of the clinical spectrum of neurogenic hypertension (Robertson *et al.*, 1984). In this chapter, an effort is made to review the role of baroreceptors in neurogenic hypertension by giving particular emphasis to the results of clinical investigations in patients with impairment in baroreflex function.

ANATOMY AND PHYSIOLOGY OF BAROREFLEX FUNCTION

THE GLOSSOPHARYNGEAL (IX) NERVE

The glossopharyngeal (IX) and vagus (X) nerves are closely related and appreciation of their anatomy is essential for the recognition and understanding of baroreflex and chemoreceptor function. Moreover, their nonautonomic functions must also be recognized in order to evaluate the integrity of the nerve on the physical exam.

The glossopharyngeal nerve is of mixed function; it has both visceral and somatic sensory function. The sensory nerves supply the mucous membranes of the pharynx, the fauces, the palatine tonsils, and the sensation and taste buds of the posterior part of the tongue. The somatic afferents convey pain and temperature sensation from the external auditory meatus and skin of the ear (these central processes ultimately terminate in the nucleus of the spinal tract of the trigeminal (V) nerve). Parasympathetic fibers innervate the parotid gland and small glands in the posterior tongue and neighbouring pharynx. There are also efferent fibers that innervate the stylopharyngeus muscles and perhaps some other striated muscle of the pharynx (Brodal, 1957).

The glossopharyngeal nerve is formed at the level of the brainstem by three to six fiber bundles arising immediately rostral to the vagus nerve fibers. The assembled nerve then goes ventrally and leaves the skull in the anteromedial portion of the jugular foramen. After exiting the jugular foramen, it courses in an arch inferiorly and posteriorly and enters the wall of the throat, lateral to the internal carotid artery (Brodal, 1957). At the base of the tongue, it then splits into its terminal branches.

The glossopharyngeal nerve has two sensory ganglia. The *superior ganglion* is a small swelling of the nerve at the level of the jugular foramen, while the somewhat larger *petrosal ganglion* lies in the *fossula petrosa*, just beneath the jugular foramen. The *otic ganglion* is situated just below the foramen ovale and receives parasympathetic preganglionic fibers from the tympanic branch of the glossopharyngeal nerve where the former courses anteriorly as the lesser superficial petrosal nerve (Brodal, 1957). Postganglionic parasympathetic fibers then enter the auriculotemporal nerve (a branch of the trigeminal nerve) and innervate the parotid gland.

The somatic efferent fibers of the glossopharyngeal nerve arise in the rostral portion of the *nucleus ambiguus*; the visceral efferent ones in the *inferior salivatory nucleus* (immediately rostral to the dorsal motor nucleus of the vagus); and the visceral afferent ones join the *nucleus of the solitary tract* in its rostral portion.

THE VAGUS (X) NERVE

The vagus nerve has the most extensive distribution of any cranial nerve, and embryologically represents the nerves of the fourth, fifth and sixth branchial arches. It has fibers of all types. Its rootlets emerge in a series of bundles from the medulla just dorsal to the inferior olive. The bundles coalesce and leave the skull through the jugular foramen where an enlargement of the nerve is the *jugular ganglion*.

Below this the nerve again thickens to form the much larger *nodose ganglion*. The vagus then lies in close proximity to the accessory (XI) and the glossopharyngeal nerves, and the internal jugular vein. The vagus continues to descend between the internal (then common) carotid and the internal jugular vein (Brodal, 1957).

In the neck and thorax the vagus nerve gives off many branches. The meningeal branch arises in the jugular ganglion then enters the skull through the jugular foramen to innervate the dura mater in the posterior cranial fossa. The auricular branch (Arnold's nerve) also arises at the jugular ganglion, communicates with the glossopharyngeal and facial nerves, and ultimately provides sensory innervation to the skin of the back of the ear and the posterior part of the ear canal. The pharyngeal branches leave the nodose ganglion and form the pharyngeal plexus (together with glossopharyngeal branches and cervical sympathetics) to supply efferent nerves to constrictor muscles of the pharynx, the levator palati muscles, and the muscles of the palatal arch. Sensory fibers innervate the mucous membranes from the pharynx to the upper surface of the epiglottis.

The superior laryngeal nerve exits just below the nodose ganglion and, through its internal and external branches, provides motor innervation to the cricothyroid muscle and sensory innervation of the mucous membrane of the larynx down to the vocal cords. The recurrent laryngeal nerve is the principal motor nerve to the larynx. It leaves the right vagus at a higher level than the left vagus, because it passes around and behind the right subclavian artery, whereas it loops around the aortic arch on the left. Some branches of the recurrent laryngeal nerve pass through the *inferior cardiac rami* to the heart.

The *superior cardiac rami* are the most important parasympathetic innervation of the heart. They exit the vagus between the origins of the superior laryngeal and recurrent laryngeal nerves, and, following the common carotid artery (innominate on the right), travel to the heart. Together with the *inferior cardiac rami*, they form the *cardiac plexus*, the branches of which terminate on the postganglionic parasympathetic neurons in the cardiac ganglia. Many vagal fibers are visceral afferent, and course in a special upper branch of the superior cardiac rami called the *aortic depressor nerve* (Brodal, 1957).

The thoracic portion of the vagus nerve gives off bronchial branches which innervate the bronchi via plexuses to transmit visceral efferent information to the glands and smooth muscles of the airways.

The right vagus then continues anterior to the subclavian artery and posterior to the subclavian vein. It descends posterolateral to the trachea and then behind the right main bronchus, breaking into several branches which form the *plexus oesophagicus* with branches of the left vagus on the dorsal side of the esophagus, where terminal branches pierce the diaphragm and are then disseminated over the posterior wall of the stomach, forming the *posterior gastric plexus*. The plexus gives rise to the *rami coeliaci* which go to the *coeliac ganglion*, passing through this ganglion without interruption to follow the superior mesenteric artery and its branches to the small intestine and the ascending and transverse colon. Other branches go to the kidney and spleen.

The left vagus descends anterior to the subclavian artery and the aortic arch, then posterior to the left main bronchus. It divides on the anterior surface of the esophagus and forms anastomosing branches with the right vagus to constitute the *plexus oesophagicus*, the terminal branches of which pierce the diaphragm anterior to the esophagus and form the *ventral gastric plexus*, giving off branches which accompany the hepatic artery to the liver.

The medullary origin of the vagal fibers reflects the diversity of function that characterizes this nerve. The somatic efferent fibers originate in the *nucleus ambiguus* of the ventrolateral medulla. The visceral efferent fibers originate in the *dorsal motor nucleus of the vagus*. The visceral afferent fibers have cell bodies in the nodose ganglion, from which their central processes pass to the *nucleus of the solitary tract*. The cell bodies of the somatic afferent fibers lie in the jugular ganglion and their central processes terminate principally in the dorsal part of the *nucleus of the spinal trigeminal tract*.

PHYSIOLOGY OF THE ARTERIAL BAROREFLEX NERVES

Afferent impulses arising from mechanoreceptors in many cervical and thoracic structures impinge on the nucleus tractus solitarii and nearby nuclei which govern sympathetic and parasympathetic regulation of the circulation (de Castro, 1931; Asteroth and Kreuziger, 1951; Page and McCubbin, 1967; Page, 1987). This negative reflex feedback system is most importantly localized in the carotid sinus at the bifurcation of the internal and external carotid arteries and in specialized structures in the wall of the aortic arch (Hering, 1927; Heymans, Bouckaert and Regniers, 1933; Möller, 1942).

The carotid sinus (Hering's) nerve endings are located diffusely in the adventitia and areas adjacent to it. Altogether there are usually about 700 neuron fibers in Hering's nerve, with activation occurring primarily during systole but with a smaller burst accompanying the dicrotic notch.

CLINICAL INVESTIGATIONS

In 1954, Kezdi devised a direct approach to assess human carotid sinus function. It was more invasive than had hitherto been applied (Lampen, Kezdi and Kaufmann, 1949a; Lampen *et al.*, 1949b; Lampen, Kezdi and Kopperman, 1949c), consisting of infiltration of the region of the carotid sinus with the anaesthetic procaine (Kezdi, 1954a). The procaine was applied bilaterally to 14 normotensive subjects, 9 patients with essential (Volhard's "red") hypertension (Volhard, 1948), 6 patients with malignant hypertension, and 4 patients with the hypertension of chronic nephritis were assessed. In each case, the right carotid was blocked first, after which there was usually a modest (10–40 mmHg) increment in blood pressure. However, after the block of the second carotid sinus, marked pressor responses were uniformly recorded. Mean values of these variables are shown in Table 11.1. Resistance and cardiac output also increased in these subjects (Wezler and Böger, 1939).

TABLE 11.1

Blood Pressure and Heart Rate Following Carotid Sinus Anaesthesia

Subjects	BP before	BP after	HR before	HR after
Normotensive	125/75	205/130	77	133
Essential hypertensive	192/102	274/147	73	124
Malignant hypertensive	220/134	321/187	86	136
Chronic nephritis	200/122	282/165	76	136

These studies clearly showed the powerful control exerted by baroreflex nerves in both normotensive and hypertensive subjects, but the selectivity of glossopharyngeal block was not demonstrated.

In a series of investigations in the 1960s, Guz and coworkers (1964, 1966) studied the effect of lidocaine block on the vagal and glossopharyngeal nerves. Although these investigations were carried out primarily to address control of respiration, cardiovascular variables were also monitored. Two healthy volunteers were pretreated with atropine 2 mg. Following this, a solution of 1% lidocaine with 1:200,000 adrenaline was injected around the glossopharyngeal and vagal nerves bilaterally at the base of the skull. Evidence of block included (1) total inability to swallow (IX and X); (2) an almost complete abolition of phonation (X); and (3) the development of a bovine cough (X). The presence of bilateral Horner's syndrome indicated that block of cervical sympathetic fibers around the carotid was also effected. One subject experienced transient bilateral facial (VII) nerve block but neither had paralysis of the tongue (XII).

In addition to the studies in the two normal subjects, a presumably selective bilateral vagal block was carried out in a 24 year-old patient with extensive lung involvement due to sarcoidosis. The vagus nerves were exposed under local anaesthesia after pretreatment with atropine 2 mg. The nerves were bathed in a 2% lidocaine solution following which there was disappearance of phonation.

In the normal subjects with bilateral block of IX and X, blood pressure rose from normal to 170/115 mmHg in one subject and to 210/130 mmHg in the second. The heart rate, already high following atropine, did not further increase during the nerve block. The cardiovascular effects of the selective vagal block in the patient, on the other hand, did not alter heart rate or blood pressure (Guz et al., 1964). Similar results were described in the earlier study (Guz et al., 1964). Electrical stimulation of the vagus increased blood pressure slightly in one of two subjects.

During the era when carotid body removal and consequent carotid sinus denervation were believed to be helpful in asthma, Holton and Wood (1965) studied respiratory and cardiovascular effects associated with this intervention. In both subjects, the response to 10% oxygen was hyperpnea before and hypoventilation two weeks later. The procedure caused immediate and marked hypertension that was still raised by 55/35 and 50/20 mmHg several weeks after surgery. Heart rate rose 15–30 bpm. There was considerable lability but even at 61 weeks in one subject and 43 weeks in the other (the last occasions reported), blood pressure remained elevated.

Holton and Wood (1965) noted only subtle changes in their patients in response to tilt. The pre-operative rise in diastolic pressure and fall in pulse pressure was absent for several weeks post-op but normalized ultimately. Orthostatic fall in blood pressure was not noted before or after surgery.

The role of the carotid body in cardiovascular control has always been difficult to assess in man. Nevertheless, there are several lines of evidence, including histological structural features of the artery supplying blood to the carotid body, that support a role beyond chemoreception alone (Heath et al., 1989; Heath, 1991).

The dramatic effects on absolute blood pressure level noted by Holton and Wood (1965) were not observed when special care was used during surgery to preserve the integrity of the carotid sinus and its nerve during carotid body removal (Lugliani et al., 1971; Winter, 1972; Swanson et al., 1978). Adenosine administered to healthy human subjects increased blood pressure due at least in part to carotid body stimulation (Biaggioni et al., 1987), a response which was absent in five subjects who underwent carotid endarterectomy (Griffiths et al., 1990). The large Japanese experience in carotid body removal for asthma provided patients who could be studied 20 years after surgery; these individuals had regained much hypoxic drive, but the reason for this, suboptimal initial surgical resection or assumption of the chemoreceptor role by other chromaffin tissue in the great vessels, could not be ascertained (Honda et al., 1979). Blood pressure effects were said to be minor in those individuals with selective carotid body lesions. Because the carotid body afferent input may be particularly important in control of hypoglossal motor neurons that innervate the genioglossus, obstructive sleep apnea may occur following carotid body resection (Parisi et al., 1987), and it is possible that this may induce secondary effects on blood pressure. This phenomenon deserves further study.

CLINICAL SYNDROMES OF ABNORMAL BAROREFLEX FUNCTION

CAROTID SINUS SYNCOPE (WEISS-BAKER SYNDROME)

In 1933, Soma Weiss and James Baker described three presentations attributable to a hypersensitive carotid sinus (Weiss and Baker, 1933). The first was slowing of heart rate or sinus arrest. The second was a fall in blood pressure without slowing of heart rate. The third is a group of symptoms in the absence of either heart rate or blood pressure changes; pallor, dizziness, syncope, homonymous hemianopsia, hemiplegia, and occasional convulsions. The first two presentations have been well established as manifestations of carotid sinus hypersensitivity, while the third is now known to be due to impaired cerebral blood flow and ought not to be considered part of carotid sinus syncope (Tuckman, Slater and Mendlowitz, 1965).

Treatment of patients with carotid sinus syncope is often difficult. It often presents in older individuals with significant carotid atherosclerosis, and its clinical manifestations are brought out by α_2-agonist therapy with clonidine or α-methyldopa. Sometimes discontinuation of the precipitating or exacerbating drug is sufficient to relieve symptoms, but often the problem persists. Atropine-like agents have been disappointing, especially because of side effects, but in the severely affected patient may be worth a trial. Greater success has been achieved with α_2-antagonists like yohimbine (Onrot et al., 1987). When bradycardia dominates the clinical presentation, a pacemaker may be helpful. About 20% of patients will require surgical decompression or cutting of the involved glossopharyngeal nerve, but hypertension can be a complication (McSwain and Spencer, 1947; Ripley, Hollifield and Nies, 1977).

GLOSSOPHARYNGEAL NEURALGIA (WEISENBURG SYNDROME)

Glossopharyngeal neuralgia differs from carotid sinus syncope in its painful episodic presentation, seemingly unrelated to stretch or twisting of the neck. Severe paroxysmal pain beginning in the tonsillar region, lateral pharynx, or base of tongue radiates deeply to the ear. It may be elicited by eating, talking, or touching the trigger area (Weisenburg, 1910). The syndrome results in sinus bradycardia, hypotension, and sometimes syncope. It is most commonly due to tumor, primary or secondary, (Dykman et al., 1981) and is often treated by carbamazepine. When this is not completely successful, some of the same measures listed as treatment for carotid sinus syncope may be undertaken.

PULSATILE MEDULLARY COMPRESSION

It has long been recognized that, as part of the aging process, arteriosclerosis and ectasia can cause abnormal vascular contacts with neural tissue at the base of the brain. With this background and the recognition that hypertension had been induced in two patients in whom medullary decompression was carried out in an effort to treat glossopharyngeal neuralgia, Jannetta and Gendell (1979) began looking for such abnormalities in patients being operated on for other cranial nerve lesions.

Over the years, retromastoid craniectomy and decompression surgery was performed by this group in more than 700 patients with trigeminal neuralgia, 450 with hemifacial spasm, 70 with tinnitus and/or vertigo, 17 with glossopharyngeal neuralgia, and more than 300 patients with other miscellaneous cranial nerve problems. Special observations were made in 53 consecutive hypertensive and 50 normotensive patients undergoing left retromastoid craniectomy and microvascular decompression and 25 normotensive and 7 hypertensive patients undergoing the same procedures on the right side (Jannetta, Segal and Wolfson, 1985). The primary problems of the 53 hypertensive patients were hemi-facial spasm in 26, trigeminal neuralgia in 13, and various other lesions in the remaining 14. Work-up of hypertension pre-operatively is not described in detail, but all subjects were said to have essential hypertension or neurogenic hypertension.

At surgery, following microvascular decompression of the symptomatic nerve, the cerebellum was gently elevated from the glossopharyngeal and vagal nerves, so that these nerves were exposed to the level of the brainstem in order to visualize whether or not vascular compression of the lateral medulla was present. Microvascular decompression of the lateral medulla was attempted if this could be performed easily, rapidly, and safely. The relevant vessel was mobilized away from the brainstem, and held away with an implant of plastic sponge, muscle, or multiple soft rolled pieces of shredded teflon felt. This was attempted in 42 of the 53 patients (Jannetta, Segal and Wolfson, 1985).

Fifty-one of 53 hypertensive patients had apparent compression of the left an-terolateral medulla by arterial loops. The typical configuration of the compressing artery was almost always a loop, with the outside of the loop pressing onto the medulla, usually between the inferior olive anteriorly and the glossopharyngeal and vagal nerves posteriorly (Jannetta et al., 1985). Only two of the 50 normoten-sive patients had comparable lesions. Thirty-two of 36 patients in whom surgery was judged to be technically successful at operation experienced significant lower-ing of blood pressure. The overwhelming preponderance of left-sided as opposed to right-sided lesions remains unexplained.

Efforts to duplicate these results in cats were not completely successful (Segal et al., 1982), but use of a pulsatile balloon catheter in baboons very effectively duplicated the observations in human subjects (Jannetta et al., 1985b).

Although the application of Jannetta's technique to the management of unusual cases of hypertension has been carried out occasionally, there is virtually no other literature at the present that amplifies the significance and prevalence of the original reports (Jannetta, Segal and Wolfson, 1985a). There is a great need for further study of this phenomenon.

THE SYNDROME OF BAROREFLEX FAILURE

Baroreflex failure, in spite of the dramatic presentation of its fully expressed form, probably often presents with enough subtlety that the vast majority of patients with it are unknowingly followed as patients with essential hypertension, albeit a form that is extremely refractory to therapy. It is easiest to establish the clinical features of baroreflex failure by examining the presentation of patients who prospectively undergo baroreflex denervation in the course of treatment for some other disorder. Observations at the bedside and with physiological and pharmacological testing of such patients have led to the understanding we have developed concerning this clinical syndrome (Table 11.2).

TABLE 11.2
Baroreflex Failure: Clinical Features

- Chronic labile hypertension

- Pressor crises:

    Duration 3–20 minutes
    Severe headache
    Systolic BP to 200–300 mmHg
    Tachycardia to 130–160 bpm
    Hot pale flushing
    Diaphoresis
    [NA] of 1000–3000 pg/ml

- Emotional volatility

- Normal [NA] between crises

- Therapy:

    Biofeedback ($\downarrow$ number $\downarrow$ severity)
    Phenoxybenzamine ($\downarrow\downarrow\downarrow$ severity)
    Clonidine ($\downarrow\downarrow\downarrow$ number $\downarrow$ severity)

The hallmark of baroreflex failure is chronic labile hypertension and tachycardia alternating with periods of hypotension and bradycardia. Following surgical denervation, the labile hypertension tends to be remarkably severe, with systolic blood pressures rising to the 200–300 mmHg range with associated tachycardia of 130–160 bpm. These episodes are usually of short duration, often 3–20 minutes (Figure 11.1). They are frequently attended by a headache of great severity, often the most severe headache ever experienced. These episodes usually respond to antihypertensive therapy, but they are often of sufficiently short duration that many might have resolved without this intervention.

During the pressor crises, the patient may experience a sensation of hot flushing, but pallor is more characteristic than redness. Diaphoresis is marked. In almost all respects, these episodes may resemble those of pheochromocytoma and the differential of baroreflex failure from pheochromocytoma is an exceedingly difficult one.

Another feature which is quite pronounced in the first two to four weeks of baroreflex failure is emotional volatility. The cause of this volatility is not known with certainty. It is possible that it relates to loss of afferent nervous influences from thoracic and abdominal sensory inputs which have an overall calming effect under normal circumstances. Another potential explanation is that the severe pressor crisis may damage the blood-brain barrier or cause other subtle abnormalities in brain function that lead to this emotional volatility. Finally, it is possible that the emotional volatility is simply the patient's behavioral response to the increasing

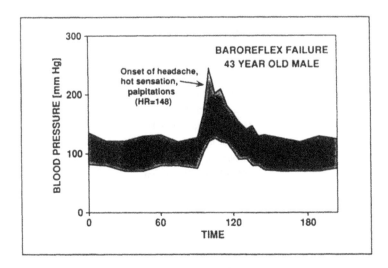

FIGURE 11.1   Blood pressure monitoring over a 200 minute period in a 43 year old man approximately two weeks following surgical removal of a second carotid body tumor, five years after removal of the initial (contralateral) carotid body tumor. While blood pressure was being monitored at normal baseline levels, a cold pressor test with immersion of the hand in ice water for 60 seconds was performed. The blood pressure immediately rose, and continued to rise for several minutes following discontinuation of the cold stimulus. The symptoms appeared during this time and resolved as blood pressure and heart rate returned to normal over the succeeding half-hour. On some occasions, paroxysms of similar magnitude occurred without obvious exogenous causative stimuli.

recognition that his blood pressure is critically dependent upon his emotional state. When the patient is calm or sedated, his blood pressure is usually normal or even low with a directly proportional heart rate. However, in response to even minor excitatory events, blood pressure rises precipitously, and in some cases pressor crises of great severity may ensue. It is possible that the patient's recognition of this normally hidden relationship between inner feelings and external expression becomes a matter of obvious clinical significance and leads to the more overt expression of emotional volatility. Whatever the reason for the emotional volatility, it usually can be attenuated by agents such as benzodiazepines or α_2-adrenoreceptor agonists such as clonidine or α-methyldopa.

Between crises, plasma noradrenaline levels are often normal, although they may be low during the relatively low blood pressures accompanying sedation. In contrast, plasma noradrenaline concentration during attacks is dramatically elevated, often to the 1,000–3,000 pg/ml range that is consistent with pheochromocytoma. Once again, the differentiation from pheochromocytoma requires careful attention to the history and the clinical presentation, and may not be an easy differentiation to make even in the best of circumstances.

Most typically, patients' pressor crises tend to moderate over time, in part due to the patients' learning to control the onset of the pressor crisis (manifested by severe headache) by exerting calming self-control. Thus, the patient may develop a spontaneous biofeedback treatment of his disorder that may result in a reduction in both the number and the severity of attacks.

Physiological testing of patients with baroreflex failure would usually lead to supranormal pressor responses to the handgrip test, the cold pressor test, and especially the mental arithmetic test.

Pharmacological testing will reveal an excess pressor response to phenylephrine without a compensatory fall in heart rate. Comparably, the depressor response to intravenous nitroprusside is pronounced, but again heart rate may rise little or not at all. The response to atropine is most typically no change in blood pressure, and during a relatively pressor phase of the disorder, no increase in heart rate. Likewise, propranolol's effect on heart rate will reflect the prevailing activity of the sympathetic nervous system. Perhaps the most helpful agent in defining baroreflex failure is clonidine. A small dose of clonidine (for example, 0.1–0.2 mg po) will often lower blood pressure precipitously in a baroreflex failure patient during the pressor phase of his illness.

There is a need for greater recognition of baroreflex failure in order for the frequency of various etiologies to be determined with certainty. Table 11.3 illustrates the etiology of unambiguous baroreflex failure in 11 patients studied at Vanderbilt's Autonomic Dysfunction Center. Four of 11 patients had the familial paraganglioma syndrome bilaterally leading to loss of both glossopharyngeal nerves and, in some cases, minor damage to vagal fibers as well. The relatively high incidence of this syndrome is related to the presence in the Cumberland Valley of Tennessee of a large family with this disorder. The second largest etiology in our experience was neck irradiation, usually for throat tumor; three such patients were noted. One of our patients had surgical section of a glossopharyngeal nerve unilaterally that led to baroreflex failure even though the glossopharyngeal nerve appeared to be normal. We are unable to fully explain the reason for this syndrome developing in the face of unilateral baroreflex damage. That patient had a more benign course than others described in our series and it is possible that some subclinical damage to the other side was present. Finally, one patient developed the syndrome in response to a central nervous system disorder, presumably of toxic etiology, that was associated with bilateral infarctions of the nuclei of the solitary tract. Finally, in two individuals, with well-characterized baroreflex failure, no cause could be ascertained.

The treatment of baroreflex failure is exceedingly difficult. The biofeedback procedures that patients themselves develop are of considerable help over the long haul, but clonidine, especially in the form of the clonidine patch, may be a necessary supplement for most of them. The pressor crises can be completely controlled with phenoxybenzamine, if one is willing to give this drug in very high dosages (up to 10 mg po qd). This can then be reduced as the patient is switched over to clonidine. In a single patient, we have also had good results with the use of α-methyldopa, but only when dosages of 3,750 mg po qd were employed

TABLE 11.3

Etiology of Baroreflex Failure: Vanderbilt Experience (n= 11)

Familial Paraganglioma Syndrome	4
Neck Irradiation for Tumor	3
Surgical Section	1
Bilateral NTS Infarctions	1
Unknown Cause	2

(in 5 divided doses). Unfortunately, none of these approaches help the patient deal with the significant hypotension which occasionally occurs. With clonidine and α-methyldopa, orthostatic hypotension may be severe. However, in patients receiving clonidine or α-methyldopa alone, sometimes blood pressure is actually higher in the upright posture than supine. A practical point in the management of patients in the first few weeks after the development of baroreflex failure is caution of medical personnel and family in regard to causing the patient emotional upset. Quite commonly, even the appearance of the physician in the patient's room has significant effect to elevate blood pressure and pressor crises under these circumstances are by no means rare. For this reason, benzodiazepines at relatively low doses have sometimes been employed, and are particularly favoured by the patients. More experience is needed before clear recommendations concerning treatment of baroreflex failure can be made.

REFERENCES

Asteroth, H. and Kreuziger, H. (1951) Dehnungsstudien am isolierten Sinus caroticus von verstorbenen Normotonikern und Hypertonikern. *Ztsch Kreislaufforsch*, **40**, 11–15.

Biaggioni, I., Olafsson, B., Robertson, R.M., Hollister, A.S. and Robertson, D. (1987) Cardiovascular and respiratory effects of adenosine in conscious man. Evidence for carotid body chemoreceptor activation. *Circ. Res.*, **61**, 779–786.

Brodal, A. (1957) *Hjernenervene*, pp. 1–31. Copenhagen: Munksgaard.

deCastro, F. (1931) Die reflektorische Selbststeuerung des Kreislaufes. In *Ergebnisse der Kreislaufforschung* edited by E. Koch, (Dresden: Steinkopff).

Doba, H. and Reis, D.J. (1974) Role of central and peripheral adrenergic mechanisms in neurogenic hypertension produced by brainstem lesions. *Circ. Res.*, **34**, 293–298.

Dykman, T.R., Montgomery, E.B., Jr., Gerstenberger, P.D., Zeiger, H.E., Clutter, W.E. and Cryer, P.E. (1981) Glossopharyngeal neuralgia with syncope secondary to tumor. *Am. J. Med.*, **71**, 165–168.

Griffiths, T.L., Warren, S.J., Chant, A.D.B. and Holgate, S.T. (1990) Ventilatory effects of hypoxia and adenosine infusion in patients after bilateral carotid endarterectomy. *Clin. Sci.*, **78**, 25–31.

Guz, A., Noble, M.I.M., Trenchard, D., Cochrane, H.L. and Makey, A.R. (1964) Studies on the vagus nerves in man: Their role in respiratory and circulatory control. *Clin. Sci.*, **27**, 293–304.

Guz, A., Noble, M.I.M., Widdicombe, J.G., Trenchard, D., Mushin, W.W. and Makey, A.R. (1966) The role of vagal and glossopharyngeal afferent nerves in respiratory sensation, control of breathing and arterial pressure regulation in conscious man. *Clin. Sci.*, **30**, 161–170.

Heath, D. (1991) The human carotid body in health and disease. *J. Pathol.*, **164**, 1–8.

Heath, D., Khan, Q., Nash, J. and Smith, P. (1989) Carotid body disease and the physician — chronic carotid glomitis. *Postgrad. Med. J.*, **65**, 353–357.

Hering, H.E. (1927) *Die Karotissinusreflexe auf Herz and Gefässe vom normal-physiologischen, pathologisch-physiologischen and Klinischen Standpunkt: Gleichzeitig Über die Bedeutung der Blutdrückzügler für den normalen und abnormen Kreislauf*. Dresden: Theodore Steinkopff.

Heymans, C., Bouckaert, I.J. and Regniers, P. (1933) *Le sinus carotidien*. Paris; Doin.

Holton, P. and Wood, J.B. (1965) The effects of bilateral removal of the carotid bodies and denervation of the carotid sinuses in two human subjects. *J. Physiol.*, **181**, 365–378.

Honda, Y., Watanabe, S., Hashizume, I., Satomura, Y., Hata, N., Sakakibara, Y. and Severinghaus, J.W. (1979) Hypoxic chemosensitivity in asthmatic patients two decades after carotid body resection. *J. Appl. Physiol. Respir. Environ. Exercise Physiol.*, **46**, 632–638.

Jannetta, P.J. and Gendell, H.M. (1979) Clinical observations on etiology of essential hypertension. *Surgical Forum*, **30**, 431–435.

Jannetta, P.J., Segal, R. and Wolfson, S.K., Jr. (1985) Neurogenic hypertension: Etiology and surgical treatment. I. Observations in 53 patients. *Ann. Surg.*, **201**, 391–398.

Jannetta, P.J., Segal, R., Wolfson, S.K., Jr., Dujovny, M., Semba, A. and Cook E.E. (1985) Neurogenic hypertension: Etiology and surgical treatment. II. Observations in an experimental nonhuman primate model. *Ann. Surg.*, **201**, 391–398.

Kezdi, P. (1954a) Sinoaortic regulatory system. *Arch. Intern. Med.*, **91**, 26–34.

Lampen, H. (1949) Über Entzügelungschockdruck bei Polyneuritis. *Deutsch. Med. Wochenshr.*, **74**, 536–540.

Lampen, H., Kezdi, P. and Kaufmann, L. (1949a) Entzügelungschockdruck am Menschen. *Klin Wchnschr*, **27**, 272–278.

Lampen, H., Kezdi, P., Kopperman, E. and Kaufmann, L. (1949b) Experimenteller Entzügelungschockdruck bei arterieller Hypertonie. *Ztschr Kreislaufforsch*, **38**, 577–592.

Lampen, H., Kezdi, P. and Kopperman, E. (1949c) Karotissinusblockade bei akuter und chronischerNephritis. *Ztschr Kreislaufforsch*, **38**, 726–737.

Lugliani, R., Whipp, B.J., Seard, C. and Wasserman, K. (1971) Effect of bilateral carotid-body resection on ventilatory control at rest and during exercise in man. *N. Engl. J. Med.*, **20**, 1105–1111.

McSwain, B. and Spencer, F.C. (1947) Carotid body tumor in association with carotid sinus syndrome. *Surgery*, **22**, 222–229.

Möller, E. (1942) Der Karotisdruckversuch beim Normalen, bei der Hypertonie und anderen veränderten Kreislaufeinstellungen im Zusammenhang mit der reflektorischen Selbststeuerung. *Arch Kreislaufforsch*, **10**, 185–209.

Onrot, J., Wiley, R.G., Fogo, A., Biaggioni, I., Robertson, D. and Hollister, A.S. (1987) Neck tumor with syncope due to paroxysmal sympathetic withdrawal. *J. Neurol. Neurosurg. Psych.*, **50**, 1063–1066.

Page, I. (1987) *Hypertension mechanisms*, pp. 707–719, Orlando: Grune and Stratton.

Page, I. and McCubbin, J.W. (1967) One facet of neural regulation. *N. Engl. J. Med.*, **276**, 335–338.

Parisi, R.A., Croce, S.A., Edelman, N.H. and Santiago, T.V. (1987) Obstructive sleep apnea following bilateral carotid body resection. *Chest.*, **91**, 922–927.

Reis, D.J. and Doba, N. (1972) Hypertension as a localizing sign of mass lesions of brainstem. *N. Engl. J. Med.*, **287**, 1354–1355.

Ripley, R.C., Hollifield, J.W. and Nies, A.S. (1977) Sustained hypertension after section of the glossopharyngeal nerve. *Am. J. Med.*, **62**, 297–302.

Robertson, D., Goldberg, M.R., Hollister, A.S., Wade, D. and Robertson, R.M. (1984) Baroreceptor dysfunction in man. *Am. J. Med.*, **76**, A49–A58.

Sanders, J.S., Ferguson, D.W. and Mark, A.L. (1988) Arterial baroreflex control of sympathetic nerve activity during elevation of blood pressure in normal man: Dominance of aortic reflexes. *Circulation*, **77**, 279–288.

Segal, R., Gendell, H.M., Canfield, D., Dujovny, M. and Jannetta, P.J. (1982) Hemodynamic changes induced by pulsatile compression of the ventrolateral medulla. *Angiology*, **33**, 161–172.

Sprenkle, J.M., Eckberg, D.L., Goble, R.L., Schelhorn, J.J. and Halliday, H.C. (1986) Device for rapid quantification of human carotid baroreceptor-cardiac reflex response. *J. Appl. Physiol.*, **60**, 727–732.

Swanson, G.D., Whipp, R.J., Kaufman, R.D., Aqleh, K.A., Winter, B. and Bellville, J.W. (1978) Effect of hypercapnia on hypoxic ventilatory drive in carotid body-resected man. *J. Appl. Physiol. Respir. Environ. Exercise Physiol.*, **45**, 971–977.

Tuckman, J., Slater, S.R. and Mendlowitz, M. (1965) The carotid sinus reflexes. *Am. Heart J.*, **70**, 119–135.

Volhard, F. (1948) Über die Pathogenese des roten (essentiellen) arteriellen Hochdrucks und der malignen Sklerose. *Schweiz. med. Wchnschr.*, **78**, 1189–1194.

Weiss, S. and Baker, J.P. (1933) The carotid sinus reflex in health and disease. Its role in the causation of fainting and convulsions. *Medicine*, **12**, 297–354.

Weisenburg, T.H. (1910) Cerebellopontine tumor diagnosed for six years as tic douloureux. The symptoms of irritation of ninth and twelfth cranial nerves. *JAMA*, **54**, 1600–1604.

Wezler, K. and Böger A. (1939) Die Dynamik des arteriellen Systems: Der arterielle Blutdruck und seine Komponenten. *Ergebn. Physiol.*, **41**, 292–606.

Winter, B. (1972) Bilateral carotid body resection for asthma and emphysema. *Int. Surg.*, **57**, 458–466.

12 Orthostatic Intolerance Syndrome, Vasoregulatory Asthenia and Other Hyperadrenergic States

Italo Biaggioni

Autonomic Dysfunction Center, Vanderbilt University,
Nashville, Tennessee 37232, USA

Orthostatic intolerance, characterized by symptoms of fatigue, palpitations and shortness of breath in the upright posture, usually results from autonomic cardiovascular deconditioning following bedrest. Orthostatic tachycardia is the most consistent finding in these cases, whereas actual orthostatic hypotension may not be apparent. A similar clinical picture is seen in patients in whom there is no history of immobilization or dehydration, and independent investigators have described a variety of syndromes characterized clinically by orthostatic intolerance. Different terms have been used to describe these patients, including sympathotonic or hyperadrenergic orthostatic hypotension, mitral valve prolapse-associated dysautonomia, neurocirculatory asthenia, hyperdynamic β-adrenergic state, primary hypovolemia, hyperbradykininism, and mast cell activation. It is not clear if these represent separate entities or form part of a spectrum of disease. In most cases an increase in sympathetic activity is found, but it is not certain whether sympathetic overactivity is the primary cause in these syndromes or if it is a secondary phenomenon. Likewise, decrease in plasma volume has been demonstrated in some, but not all cases. Chronic sympathetic activation with contraction of the vascular space may contribute to hypovolemia. There is also evidence of inappropriate venous pooling in the lower extremities. It has been suggested that this may be due to selective autonomic neuropathy in the lower limbs. It has also been proposed that the local or systemic release of vasodilators, such as dopamine, bradykinin, atrial natriuretic factor and histamine, contribute to this phenomenon. No single therapy is uniformly successful, and treatment of these patients can be a challenge. Patients may respond to fludrocortisone by improving central volume, central sympatholytic agents by reducing sympathetic activity, β-blockers, or ergotamine. These agents have been used alone or in combination. Patients may also benefit from a sensitive exercise conditioning program.

KEY WORDS: Tachycardia, hypotension, dysautonomia, hypovolemia, neuropathy, hyperbradykinism, mast cells

INTRODUCTION

Orthostatic intolerance usually results from autonomic cardiovascular deconditioning in patients subjected to prolonged periods — two to three weeks — of immobilization. Autonomic abnormalities, however, can be apparent even in previously

271

normal subjects after relatively short periods of 5° head-down tilt, as brief as 20 hours (Gaffney *et al.*, 1985). The clinical picture produced by prolonged bedrest or head-down tilt is dominated by an intolerance to maintenance of upright posture manifested by fatigue, lightheadedness, and fainting. Exaggerated orthostatic tachycardia is also apparent. The pathophysiology of the autonomic deconditioning produced by head-down tilt includes an initial shift of fluids into the central compartment, evidenced by an increase in central venous pressure. This is followed by compensatory sympathoinhibition and natriuresis, eventually resulting in a decrease in intravascular and interstitial volume. Because this clinical picture is similar to the physical deconditioning produced by weightlessness, head-down tilt has become a model to study weightlessness-induced deconditioning, a subject reviewed elsewhere in this book (Chapter 14).

On the other hand, a clinical picture similar to that produced by cardiovascular deconditioning can arise "spontaneously" in patients in whom there is no history of immobilization, dehydration or other illness that may explain their condition. These patients can be significantly disabled, with exertion limited by symptoms of fatigue, tachycardia or shortness of breath. The term "sympathotonic or hyperadrenergic orthostatic hypotension" has been coined to describe this syndrome because of evidence for increased sympathetic activity in these patients. However, as we will discuss below, it is uncertain whether the increased sympathetic activity is the cause or a consequence of the disorder. Furthermore, only a minority of patients actually present with significant orthostatic hypotension. We, therefore, prefer the term "orthostatic intolerance syndrome". On the other hand, heart rate is usually greatly increased on standing in these patients. Therefore, the terms "orthostatic tachycardic syndrome" or "postural tachycardia syndrome" are also used. More definite terminology will require a better understanding of the etiology and pathophysiology of these disorders. In this category we do not include patients in whom an etiology is apparent, such as diabetes mellitus, surreptitious use of drugs or diuretics, significant weight loss or chronic debilitating diseases.

HISTORICAL ASPECTS

Since the late 1800's physicians recognized the existence of a condition characterized by poor exercise tolerance in the absence of an obvious cause, such as prolonged inactivity. The first cases were reported among soldiers during the American War between the States and the terms "soldier's heart" or "irritable heart syndrome" were used to describe this condition (DaCosta, 1871). These terms also reflect the initial interpretation that this syndrome was due to a functional cardiac defect. This syndrome experienced a revival during World War I, and at that time it was postulated that this functional cardiac phenomenon was due to an inadequate neural adjustment of peripheral blood flow, hence the use of the terms "neurocirculatory asthenia" or "vasoregulatory asthenia". These patients were described as having large cardiac outputs (a cardiac "hyperkinetic" state) (Holmgren, 1957), although it was not clear whether this was a primary phenomenon or if it was secondary to inadequate peripheral vascular regulation. Similar phenomenology was reported by Frohlich and colleagues in the late 1960's to describe patients with limited exercise tolerance and increased cardiac output, measured using indocyanine green (Frohlich, Dustan and Page, 1966; Frohlich, Tarazi and Dustan, 1969). These investigators proposed that the underlying cause was an increase in β-receptor sensitivity, hence their use of the term "hyperdynamic beta-adrenergic state". Furthermore, they proposed that previous cases of "soldier's heart syndrome" and "vasoregulatory asthenia" could be explained by this condition (Frohlich, Tarazi and Dustan, 1969). It is of interest that, about the same time, prolapse of the mitral valve was recognized as a clinical entity (Barlow *et al.*, 1963, 1968). Mitral valve prolapse has also been claimed as an explanation for these disorders (Wooley, 1976). More recently, evidence of increased adrenergic activity has been found in these patients and the term "hyperadrenergic" or "sympathotonic orthostatic hypotension" are also used to describe them.

While these syndromes are commonly encountered in a general medicine practice and constitute one of the most frequent referrals in centers specialized in autonomic disorders, their incidence among the general population is not known and probably depends on the local bias of the medical community. In Europe, up to 5% of all general medicine outpatient visits are diagnosed as "constitutional hypotension" (Pemberton, 1989), a condition that can also be associated with symptoms of fatigue and limited exercise capacity. This syndrome is not recognized in the United States, where similar symptoms, because of their vagueness, are frequently interpreted as psychosomatic.

In summary, several syndromes which have been independently described by various investigators, are characterized by orthostatic intolerance and a "hyperadrenergic state", and share a striking clinical resemblance. Different terms have been coined to describe these syndromes, based on the underlying cause proposed by the investigators. It is unclear, however, if these syndromes represent discrete entities or if they represent different presentations within the spectrum of a disease. We postulate that these syndromes share overlapping pathophysiologic mechanisms. We will review their clinical and pathological characteristics, and will try to point out their similarities and differences.

MITRAL VALVE PROLAPSE-ASSOCIATED DYSAUTONOMIA (MVP-D)

Perhaps the best evaluated form of idiopathic orthostatic intolerance is that seen in a subset of patients with mitral valve prolapse (MVP), initially described indendently by Coghlan (Coghlan *et al.*, 1979) and Gaffney (Gaffney *et al.*, 1979). These patients are usually young females with symptomatic orthostatic intolerance and orthostatic tachycardia, but, in general, with little orthostatic hypotension. In addition to orthostatic dizziness, palpitation and exercise intolerance, these patients may also complain of atypical chest pain either at rest or during exertion. It is estimated that this syndrome is present in approximately 10% of patients with MVP. Unfortunately, it is difficult to determine its precise incidence, in part because these studies are done in specialized referral centers, introducing, therefore, a selection bias. Considering that MVP may be present in as much as 4-7% of the general population (Procacci et al., 1976; Taylor et al., 1989), it is possible that this type of "dysautonomia" affects a large number of subjects.

Exaggerated orthostatic tachycardia is found in most cases. Its absence should cast doubt on the diagnosis. On the other hand, most patients do not present with orthostatic hypotension, a fact not widely recognized that may account for delay in their diagnosis. While patients may have a significant narrowing of pulse pressure on standing, the average blood pressure obtained immediately on standing in most reported series is not significantly different from normals. Individual patients can present with significant orthostatic hypotension or, paradoxically, with an exaggerated increase in blood pressure on standing. In the vast majority of cases, therefore, the symptoms cannot be attributed to hypotension per se. It has been proposed that upright blood pressure is maintained by an increased peripheral vasoconstriction, at the expense of adequate blood flow to vital organs. While studies of regional blood flow are lacking, Gaffney and colleagues investigated the systemic hemodynamic responses to upright posture and to lower body negative pressure (LBNP) in these patients, a procedure that simulates the venous pooling produced by upright posture (Gaffney et al., 1979, 1983a). Hemodynamic parameters were not different from normal controls in the supine position. Stroke volume in the upright posture, however, was significantly lower in patients. Patients were divided into two groups depending on how they compensate for this decrease in stroke volume. Some patients maintained their upright blood pressure by an exaggerated increase in heart rate to normalize their cardiac output, while others maintained their upright blood pressure through an increased peripheral vasoconstriction. On the other hand, the decrease in stroke volume produced by upright posture in MVP-D patients could not be reproduced by LBNP. LBNP produced a similar decrease in stroke index and cardiac index in patients as in controls. These results, however, were confounded by a smaller degree of venous pooling (assessed indirectly by measurements of calf volume) produced in patients compared to controls, raising the possibility that LBNP produced a lesser stimuli in the patients. The authors emphasize that a "hyperdynamic state" with increased cardiac output was not observed in these patients, therefore differentiated them from patients with "vasoregulatory asthenia" (see below) (Gaffney et al., 1983a).

There has been great interest to determine whether these abnormalities are related to an abnormal autonomic cardiovascular regulation ("dysautonomia"). Unfortunately, only a handful of studies have investigated the autonomic function in these patients. There is agreement among reports that sympathetic activity is enhanced, based on increases in urinary noradrenaline (Boudoulas et al., 1980, 1983) and in plasma noradrenaline, especially in the upright posture (Pasternac et al., 1982; Boudoulas et al., 1983; Gaffney et al., 1983a). In agreement with the suggestion of increased sympathetic activity, Davies et al. found an exaggerated blood pressure overshoot during phase IV of the Valsalva manoeuvre (Davies *et al.*, 1987), and Gaffney *et al.* found an enhanced forearm vasoconstrictor response to LBNP (Gaffney *et al.*, 1979). On the other hand, there is less agreement about parasympathetic function in these patients. Coghlan *et al.* found a delay in the recovery from the bradycardia during the phase IV of the Valsalva manoeuvre, which was interpreted as an increase in vagal tone (Coghlan *et al.*, 1979). On the other hand, two groups of investigators found a decrease in the bradycardic response to the diving reflex (facial immersion in ice water), which was interpreted as decreased vagal responsiveness (Gaffney *et al.*, 1979; Taylor *et al.*, 1989). Gaffney *et al.* also investigated baroreceptor function and found a decrease in the bradycardic response to a continuous infusion of phenylephrine (Gaffney *et al.*, 1979).

Gaffney *et al.* proposed that the enhanced vasoconstriction observed during upright posture and LBNP was due to α-adrenoreceptor hypersensitivity (Gaffney *et al.*, 1983b). This hypothesis, to our knowledge, has not been tested directly; it remains possible that the enhanced vasoconstriction is simply due to increased noradrenaline release acting on normal α-adrenoreceptors. No differences in platelet α-adrenoreceptors are found in MVP patients without dysautonomia (Schatz *et al.*, 1990).

β-receptor function, on the other hand, has been studied in greater detail in MVP-D. It was initially suggested that these patients had hypersensitive β-receptors because intravenous infusion (Boudoulas *et al.*, 1983) or bolus injection (Davies *et al.*, 1987) of isoproterenol produces a greater increase in heart rate in them compared to normal controls. It is worthwhile noting that while isoproterenol produces a greater increase in the heart rate (β_1-receptor function), it does not consistently produce a greater decrease in blood pressure (β_2-receptor function) in these patients (Davies *et al.*, 1987). Isoproterenol administration can also reproduce the symptoms that spontaneously occur when these patients stand up, such as chest pain, fatigue, dyspnea and dizziness (Boudoulas *et al.*, 1983). Of interest, during World War I, both British (Fraser and Wilson, 1918) and American (Peabody *et al.*, 1918) army physicians reported increased sensitivity to epinephrine in patients with irritable "soldier's heart" and postulated that this condition was due to increased sympathetic sensitivity.

It must be noted that the finding of increased β-receptor responsiveness to isoproterenol is of special significance in the presence of increased plasma catecholamines. Just the opposite would be expected in this situation; chronic exposure to increased plasma catecholamines should result in β-receptor down-regulation and decreased responsiveness to isoproterenol (Feldman *et al.*, 1983). Recent studies have found a plausible explanation for this paradox. Davies *et al.* reported an apparent "supercoupling" of β-adrenoreceptors; while binding parameters of receptor number and affinity (β-max and kd) were not different from normal controls, activation of lymphocyte β-receptors from patients resulted in a greater cAMP accumulation than in lymphocytes from normal controls. This appears to be due to an increase in the number of receptors in the high affinity state (Davies *et al.*, 1987). These results have been independently confirmed by Anwar *et al.* (Anwar *et al.*, 1991) and suggest an abnormal coupling of the receptors to the guanine nucleotide regulatory complex (Davies *et al.*, 1991). Whether this is a structural alteration, or a functional modification secondary to an unidentified pathological process, is yet to be determined. While this alteration has been reported exclusively in MVP-D, it is not known whether it is present or absent in other syndromes of orthostatic intolerance unrelated to MVP.

To explain the role of the sympathetic nervous system in the pathophysiology of this disorder, it has been proposed that chronic sympathetic activation may eventually lead to a contraction of vascular space resulting in hypovolemia (Gaffney *et al.*, 1983b), analogous to the volume constriction observed in pheochromocytoma. In support of this hypothesis, a decrease in plasma volume has been found in these patients (Gaffney *et al.*, 1983b; Pasternac *et al.*, 1986) It is, of course, hard to determine whether sympathetic hyperactivity is the primary phenomenon leading to hypovolemia, or if the hypovolemia is the primary phenomenon leading to reflex sympathetic activation. Pasternac *et al.* have also reported that plasma atrial natriuretic factor (ANF) may be elevated in a subset of patients with MVP-D (Pasternac *et al.*, 1986). Approximately half of their patients, however, had normal plasma ANF and, therefore, its role in the regulation of plasma volume and in the pathophysiology of this disorder remains to be determined.

The role of MVP in the pathophysiology of this syndrome also remains unclear. Even though most publications stress the association between this syndrome and the presence of MVP, it is now evident that patients with normal mitral valves may present with symptomatology that is indistinguishable from patients with MVP-D. This issue is complicated further by preliminary reports indicating that normal subjects may develop echocardiographic signs of MVP when subjected to volume depletion. If this is true in patients with MVP-D, it opens the possibility that hypovolemia may be the primary phenomenon that produces a "functional" prolapse of the mitral valve. In our experience, more often than not, patients are referred with the diagnosis of MVP only to find equivocal signs of prolapse on re-examination of the echocardiogram. Taylor *et al.* studied 78 patients with MVP-D and 40 patients with identical symptomatology but without MVP, and compared them to a group of asymptomatic MVP patients and normal controls. No differences (in symptoms, hemodynamic responses to autonomic stimuli, or plasma catecholamines) were found between MVP-D and patients with orthostatic intolerance without MVP (Taylor *et al.*, 1989). As a group, patients with MVP-D were slightly more sensitive to the tachycardic response to isoproterenol than patients without MVP (Taylor *et al.*, 1989). It remains to be determined if patients with dysautonomia unrelated to MVP can be differentiated from MVP-D by the degree of β-adrenoreceptor hypersensitivity, by differences in upright cardiac output or by alterations in plasma volume.

There are a limited number of studies about treatment of these patients, and most of these studies are unblinded or uncontrolled. Because palpitations and anxiety are prominent in some patients, β-blockers have been tried and are useful in some patients (Boudoulas *et al.*, 1980). Clonidine has been given to reduce the "overactivity" of the sympathetic nervous system, and can also be helpful in some patients (Gaffney *et al.*, 1983b). Short term infusions of isoproterenol have paradoxically been used to "desensitize" the apparent supercoupled β-adrenoreceptors and improve symptoms (Taylor *et al.*, 1989).

In summary, abnormal cardiovascular responses to autonomic stimuli are found in a subset of patients with MVP. These autonomic alterations include an exaggerated heart rate response to standing and symptoms of orthostatic intolerance including fatigue, weakness, exercise intolerance, atypical chest pain, shortness of breath and presyncope. There is evidence of increased sympathetic activity and decreased plasma volume. These alterations may explain the decrease in upright stroke volume and the enhanced peripheral vasoconstriction observed in these patients. Upright cardiac output and blood pressure are maintained by orthostatic tachycardia and vasoconstriction, so that only a minority of patients have significant orthostatic hypotension.

ORTHOSTATIC INTOLERANCE UNRELATED TO MITRAL VALVE PROLAPSE

As mentioned previously, while patients with MVP may be particularly prone to develop this "dysautonomia", patients with normal mitral valves may have a clinical picture that is indistinguishable from MVP-D. It seems possible that these patients correspond to what was previously known as "vasoregulatory asthenia", "neurasthenia" and more recently "hyperadrenergic or sympathotonic orthostatic hypotension".

The term "hyperadrenergic orthostatic hypotension" was used by Cryer to describe diabetic patients with orthostatic hypotension and exaggerated heart rate and plasma noradrenaline responses to upright posture (Cryer, 1980). Kuchel *et al.* also used this term to describe 16 patients with symptoms of orthostatic intolerance (Kuchel *et al.*, 1985), including seven diabetics. The presence or absence of MVP was not reported. Blood volume was not measured, but urinary sodium excretion, measured while on sodium balance, was not different from normal controls. Free plasma noradrenaline tended to be higher in the patients as compared to normals, but the authors found the main defect to be an increase in upright total dopamine (free + sulfoconjugated dopamine). They postulated that the primary defect in these patients was an increase in dopamine release, especially in the upright posture, that would not be reflected in free dopamine measurements, because virtually all plasma dopamine is present in its conjugated form. The excessive dopamine release may produce natriuresis and vasodilatation, therefore contributing to the pathophysiology of this disorder.

A different pathophysiologic process has been proposed by Hoeldtke *et al.* (1989). They reported 5 patients with "sympathotonic orthostatic hypotension" with recent history of significant weight loss (2 cases related to neck cancer) or viral infections, and orthostatic intolerance (dizziness, headache and palpitations on standing). Age range was 21 to 61 years and two patients had a history of alcoholism. Heart rate responses to the Valsalva manoeuvre and sinus arrhythmia were normal. Plasma noradrenaline showed an exaggerated response on assuming the upright posture. Plasma volume and red cell volume were decreased in only one subject. The main finding in this study was a decrease in autonomic surface potentials and prolongation in the latency of the responses. These alterations were limited to the lower extremities, as palmar responses were normal. Although considerable overlap was found when compared to normal controls, these results were interpreted as a selective sympathetic neuropathy in lower extremities (Hoeldtke et al., 1989). In a subsequent study of six patients with idiopathic orthostatic tachycardia, autonomic neuropathy in lower extremities was present only in a minority of patients (Hoeldtke and Davis, 1991). In agreement with the hypothesis of selective autonomic neuropathy in lower extremities, Streeten found supersensitivity to the venoconstrictive effects of noradrenaline in the lower extremities of eight patients with "hyperadrenergic orthostatic hypotension" (seven idiopathic cases and one secondary to diabetes) (Streeten, 1990). Arterial blood pressure responses to intravenous infusion of phenylephrine were normal. Contractile venous responses to infused noradrenaline were measured with a linear variable differential transformer. The venous responses of hand veins were similar in patients compared to normal controls. On the other hand, the foot veins of patients with hyperadrenergic orthostatic hypotension were supersensitive to noradrenaline. This supersensitivity was similar to that observed in hand and foot veins of patients with true autonomic failure and, therefore, this finding is compatible with "denervation supersensitivity". These results were interpreted as anatomical or functional denervation of lower limb veins causing excessive gravitational blood pooling and orthostatic hypotension (Streeten, 1990). These findings are in agreement with previous studies showing increased orthostatic pooling of blood in the legs of patients with hyperadrenergic orthostatic hypotension using ^{99m}Tc-labelled erythrocytes (Streeten et al., 1988) Furthermore, symptomatic and hemodynamic improvement in the upright posture was obtained when this inappropriate venous pooling in the lower extremities was corrected with an inflated pressure suit (Streeten, 1990). Therefore, an exaggerated venous pooling in lower extremities is found in patients with idiopathic orthostatic intolerance. Autonomic neuropathy circumscribed to the lower extremities may be present in some patients, but the cause of this defect is not entirely clear. It has been suggested that some of these cases may represent an attenuated form of post-viral acute pandysautonomia (Schondorf and Low, 1993).

PRIMARY HYPOVOLEMIA

Fouad et al (1986) reported 11 patients with orthostatic tachycardia and orthostatic intolerance (palpitations, blurred vision, weakness, and dizziness). Five patients also had episodic flushing. Only in two cases was orthostatic hypotension present. Some patients were either hypertensive while supine or while upright. The presence of mitral valve prolapse was not reported. Cardiac output, measured by radionuclide methods, was in the low normal range, and total peripheral resistance was elevated. The authors emphasized that this is the major difference between this disorder and the hyperbeta-adrenergic syndrome, where cardiac output is increased (Frohlich, Tarazi and Dustan, 1969). Autonomic reflexes appeared to be normal, as were plasma catecholamines in the supine position. These patients were hypersensitive to isoproterenol administration, although this manoeuvre did not reproduce their symptoms. The major alteration found in these patients was a reduction of plasma volume as measured by iodinated albumin methods, hence the term "primary hypovolemia" used by the authors to describe their patients. These patients failed to respond to β-blockers, but responded partially with fludrocortisone, especially when combined with clonidine (Fouad et al., 1986). In addition to a decrease in plasma volume, a decrease in red cell mass has been found by Hoeldtke (Hoeldtke and Davis, 1991) and by us (unpublished observations) in patients similar to those described by Fouad et al. An explanation for this alteration is not apparent.

HYPERBRADYKININISM

Streeten et al. (1972) described five patients characterized by orthostatic lighthead-edness, facial flushing and tachycardia. The incidence of this syndrome appears to be familial, as there were a mother and daughter, and a brother and sister among the five patients. These patients had a narrowing of pulse pressure in the upright posture, but no significant orthostatic hypotension. These symptoms were episodic and aggravated by hot weather, meals, the luteal phase of the menstrual cycle and by caffeine. The legs assumed a purple-blue color in the upright posture, suggesting inappropriate venous pooling. These patients were found to have elevated bradykinin plasma levels, as determined by bioassay (Streeten et al., 1972). The authors postulated that the increased bradykinin would act as a vasodilator and play a role in the genesis of this syndrome. Patients were treated with propranolol, fludrocortisone, or cyproheptadine, with some improvement.

DYSAUTONOMIA ASSOCIATED WITH MAST CELL ACTIVATION

Among patients with idiopathic orthostatic intolerance followed at Vanderbilt's Autonomic Dysfunction Clinic, a significant proportion are characterized by episodes of flushing, palpitations, shortness of breath, chest discomfort, headache, lightheadedness and occasionally syncope. Hypotension or hypertension may be seen during these "attacks". After acute episodes, patients may complain of increased diuresis and/or diarrhoea, and invariably experience severe fatigue and sleepiness, which may last for several hours. Such episodes frequently are brought on by exposure to extremes of temperature, emotional stress, or exertion, and may occur premenstrually. Symptoms of orthostatic intolerance may be present between episodes and are typically worsened after an acute episode. These episodic symptoms are very similar to those seen in patients with mast-cell activation disorders (Roberts and Oates, 1991). In fact, an increase in urinary methylhistamine, a marker of mast cell activation, can be found in these patients (unpublished observations). Mast cell activation results in the release of vasoactive compounds such as histamine. This vasodilator, therefore, may contribute to the genesis of this disorder. Again, it is unclear if mast cell activation represents a primary or a secondary phenomenon. These patients have greater increased plasma noradrenaline in the upright posture and *in vitro* data suggests that neuropeptides such as Neuropeptide Y, which is co-released with noradrenaline, may be involved in mast cell activation (Arzubiaga *et al.*, 1991).

HYPERDYNAMIC BETA-ADRENERGIC CIRCULATORY STATE

This syndrome was reported in the late 1960's by Dr. Frohlich and colleagues (Frohlich, Dustan and Page, 1966; Frohlich, Tarazi and Dustan, 1969), approximately at the same time Barlow *et al.* described the prolapse of the mitral valve as a clinical entity (Barlow *et al.*, 1963; Barlow *et al.*, 1968). These patients have orthostatic intolerance characterized by fatigue and tachycardia, and may have a decrease, no change, or an increase in blood pressure on standing (Frohlich, Tarazi and Dustan, 1969). These patients were found to be hypersensitive to the tachycardic effects of isoproterenol, and were relieved by β-blockers. It is unclear whether these patients had the auscultatory signs of MVP. Of interest, the authors proposed that the "irritable heart" syndrome ("soldier's heart", "vasoregulatory asthenia") described during the War between the States could be explained by this condition (Frohlich, Tarazi and Dustan, 1969). As mentioned previously, a similar claim has been proposed for MVP. Moreover, in an attempt to explain the increased responsiveness to isoproterenol, the authors stated in 1969 that "only when the precise metabolic derangement, possibly in the adenyl cyclase system, is demonstrated in this clinical setting can this syndrome be definitely attributed to abnormal receptors". This statement presaged the recent findings by Davies *et al.* (1987) of "supercoupling" of β-receptors with increased stimulation of adenylate cyclase in patients with MVP-D. The main apparent difference between this syndrome and MVP-D, according to the few published studies, is that cardiac output is increased in this syndrome ("hyperkinetic cardiac state") but is normal in MVP-D and in "primary hypovolemia".

PSEUDOPHEOCHROMOCYTOMA

More than fifty years ago, Page described a syndrome of paroxysmal hypertension accompanied by flushing, palpitations, and orthostatic intolerance, with polyuria and/or diarrhoea following these episodes. Dr. Page postulated that the symptoms were the result of sympathetic hyperactivity from diencephalic origin (Page, 1935). A clinical clue in the differentiation with pheochromocytoma is the presence of flushing, which is rarely seen in pheochromocytoma. Since then, this condition has been called Page's syndrome, diencephalic hypertension, hyperadrenergic essential hypertension or pseudopheochromocytoma. These episodes can be precipitated by exertion, emotional stress, premenstrual syndrome, and other factors. While hypertension is a main feature of this syndrome, it otherwise shares most of the characteristics of the other syndromes of idiopathic orthostatic intolerance. The clinical picture is particularly similar to that observed in patients with mast cell-activation disorders, some of whom may have hypertensive crises during "attacks" (see above). Likewise, patients with idiopathic orthostatic intolerance may present with paradoxical orthostatic hypertension but usually of lesser magnitude than the one described by Page. It has been postulated that patients with pseudopheochromocytoma have surges of dopamine, probably from adrenal origin. Measurements of free dopamine may miss these surges because dopamine is almost entirely sulfonated, and, therefore, measurements of total dopamine (free + sulfonated) may be required to study this syndrome (Kuchel *et al.*, 1986).

SUMMARY OF PATHOPHYSIOLOGY

In summary, independent investigators have described a variety of syndromes characterized clinically by intolerance to the upright posture and evidence of sympathetic hyperactivity (Table 12.1). The clinical features of these patients resemble those seen after prolonged immobilization. These patients, however, do not have a history of prolonged bedrest or any precipitating factor that may explain the apparent cardiovascular deconditioning. While these syndromes have been described as discrete clinical entities, it is possible that the abnormalities described by the investigators reflect selection bias or their particular research interests. It is possible that these syndromes are part of a spectrum of diseases that share common pathophysiologic processes (Figure 12.1).

A common feature in all of these syndromes is an increased sympathetic activity. In some cases, but not all, a decrease in total plasma volume can be demonstrated. Patients selected for this abnormality, but otherwise indistinguishable from other syndromes, have been described ("primary hypovolemia"). The cause of this defect in the regulation of blood volume is not known. Chronic sympathetic activation with contraction of the vascular space may be a contributing factor. The role of the kidney and its handling of sodium has not, to our knowledge, been explored. Even patients with plasma volumes measured in the "normal range" may benefit acutely with fluid loading, or chronically with mineralocorticoids (unpublished observation), a finding that underscores the limitations in the current methodology used to estimate plasma volume. It is also possible that the decrease in plasma volume may be exaggerated in the central compartment in the upright posture due

TABLE 12.1

Clinical Features of Syndromes Characterized by Orthostatic Intolerance

	Flushing	Upright Hemodyn CO	SVR	Plasma Catechols	Isoproterenol HR Response	Plasma Volume	Venous Pooling
MVP-D	?	Nl, ↓	↑	↑	↑	↓	?
Non-MVP-D	?	?	?	↑	↑	↓, Nl	Y
Soldier's Heart/ Neurocirculatory Asthenia	?	↑	?	?	↑	?	?
Hyperdynamic β-adrenergic state	?	↑	?	?	↑	?	?
Primary Hypovolemia	Y	Nl	↑	Nl	↑	↓	?
Hyperbradykininism	Y	?	?	?	?	?	Y
Mast Cell Activation	Y	?	?	↑	?	Nl, ↓	?

The common features of these syndromes include orthostatic "intolerance" (symptoms of fatigue, palpitations, shortness of breath, dizziness, and syncope associated with the upright posture) and orthostatic tachycardia and extremely limited exercise capacity.

CO, cardiac output; SVR, systemic vascular resistance; HR, heart rate; Nl, normal; MVP, mitral valve prolapse; D, dysautonomia.

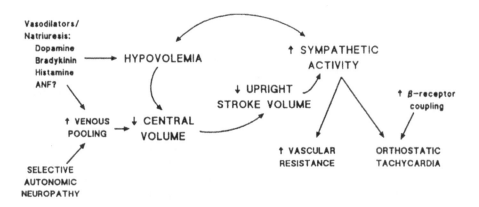

FIGURE 12.1   Proposed pathophysiology of orthostatic intolerance. See text for details.

to inappropriate venous pooling in the lower extremities, leading to a reduction in upright stroke volume. Although the cause of this excessive venous pooling is not entirely known, a selective autonomic neuropathy of the lower extremities has been found in some patients. The local or systemic release of vasodilators may also contribute to this phenomenon. Increase in plasma levels of dopamine, bradykinin, ANF and histamine is found in some patients, but could represent secondary events. The decrease in central volume will activate low pressure baroreceptors located in central veins and produce further sympathetic activation, closing a vicious circle that results in increased arterial vasoconstriction and orthostatic tachycardia (Figure 12.1). The latter may be exaggerated by the alteration in β-receptors described in patients with MVP-D. The role of MVP in the genesis of this entity remains controversial. Certainly its presence is not required in most patients for the full expression of this syndrome.

MANAGEMENT

As with any dysautonomia, the first step in the management of these patients is to rule out any correctable cause that may require specific treatment. Most of these, such as diabetes, significant weight loss, chronic debilitating diseases or prolonged immobilization will be obvious. Less obvious causes, such as surreptitious drug or diuretic abuse may require a urine drug screen. Treatment of these patients can be a challenge. No single therapy is uniformly successful. Medications that are useful in some patients may have no effect in others. In general, there are no reliable findings that predict the success of a particular therapeutic approach. We will attempt, however, to outline some general, although empiric, guidelines. Patients may show improvement to acute saline administration (e.g., 1 l intravenously over 1 hour). This response can be assessed objectively by measuring the hemodynamic and hormonal responses to head-up tilt before and after saline administration. Fludrocortisone (0.1 to 0.4 mg/dl orally) may be tried in these patients. We routinely administer sodium chloride tablets (2 g/d, with meals) together with fludrocortisone to assure adequate and stable sodium intake. Excessive fluid retention and hypokalemia should be avoided. Sodium chloride tablets may produce gastrointestinal symptoms that may improve with the use of coated preparations. In some patients, the increase in sympathetic activity is a predominant feature, with very high upright plasma noradrenaline levels. Central sympatholytic agents can be tried in these patients. It must be noted that they can be extremely sensitive to such agents and, therefore, very small doses should be tried first. Clonidine may be useful in some patients either alone (Gaffney *et al.*, 1983b) or in combination with fludrocortisone (Fouad *et al.*, 1986). We find α-methyldopa easier to use in these patients, at a starting dose of 125 mg bid.

Other drug therapies found to be useful include β-blockers and ergotamine (Hoeldtke and Davis, 1991). The latter is given as a preferential venoconstrictor, in an attempt to reduce the excessive venous pooling postulated to be present in these patients. It is probably best given by inhalation because of the erratic bioavailability of the oral preparations. Patients with evidence of "supercoupling" of β-adrenoreceptors appear to benefit from intravenous infusion of isoproterenol.

This paradoxical response is postulated to be mediated by desensitization of the supercoupled β-adrenoreceptors produced by the continuous infusion of isoproterenol (Davies *et al.*, 1987). Finally, anecdotal reports suggest that, as important as drug therapy, these patients may benefit from enrolling in a sensitive exercise conditioning program. It is probably prudent, however, to stop exertion whenever symptoms develop. Some patients express a preference for swimming, perhaps because of the added benefit of the hydrostatic pressure which may improve venous return and shift blood into the central compartment. Short-term exercise may correct some of the abnormalities found in β-adrenoreceptors in these patients (Anwar *et al.*, 1991). It must be reiterated that these recommendations are necessarily based on empirical results, and perhaps reflect the lack of a comprehensive understanding of the pathophysiology of these syndromes.

REFERENCES

Anwar, A., Kohn, S.R., Dunn, J.F., Hymer, T.K., Kennedy, G.T., Crawford, M.H., *et al.* (1991) Altered adrenergic receptor function in subjects with symptomatic mitral valve prolapse. *Am. J. Med. Sci.*, **302**, 89–97.

Arzubiaga, C., Morrow, J., Roberts, L.J.II and Biaggioni, I. (1991) Neuropeptide Y, a putative cotransmitter in noradrenergic neurons, induces mast cell degranulation but not prostaglandin D2 release. *J. Allergy Clin. Immunol.*, **87**, 88–93.

Barlow, J.B., Bosman, C.K., Pocock, W.A. and Marchand, P. (1968) Late systolic murmurs and nonejection (mid-late) systolic clicks: An analysis of 90 patients. *Br. Heart J.*, **30**, 203–218.

Barlow, J.P., Pocock, W.A., Marchand, P. and Denny, M. (1963) The significance of late systolic murmurs. *Am. Heart J.*, **66**, 443–452.

Boudoulas, H., Reynolds, J.C., Mazzaferri, E. and Wooley, C.F. (1980) Metabolic studies in mitral valve prolapse syndrome. *Circulation*, **61**, 1200–1205.

Boudoulas, H., Reynolds, J.C., Mazzaferri, E. and Wooley, C.F. (1983) Mitral valve prolapse syndrome: The effect of adrenergic stimulation. *J. Am. Coll. Cardiol.*, **2**, 638–644.

Coghlan, H.C., Phares, P., Cowley, M., Copley, D. and James, T.N. (1979) Dysautonomia in mitral valve prolapse. *Am. J. Med.*, **67**, 236–244.

Cryer, P.E. (1980) Disorders of sympathetic neural function in human diabetes mellitus: Hypoadrenergic and hyperadrenergic postural hypotension. *Metabolism*, **29**, 1186–1189.

DaCosta, J.M. (1871) An irritable heart: A clinical study of a form of functional cardiac disorder and its consequences. *Am. J. Med. Sci.*, **61**, 17–52.

Davies, A.O., Mares, A., Pool, J.L. and Taylor, A.A. (1987) Mitral valve prolapse with symptoms of beta-adrenergic hypersensitivity. *Am. J. Med.*, **82**, 193–201.

Davies, A.O., Su, C.J., Balasubramanyam, A., Codina, J. and Birnbaumer, L. (1991) Abnormal guanine nucleotide regulatory protein in MVP dysautonomia: Evidence from reconstitution of Gs. *J. Clin. Endocrinol. Metab.*, **72**, 867–875.

Feldman, R.D., Limbird, L.E., Nadeau, J., FitzGerald, G.A., Robertson, D. and Wood, A.J.J. (1983) Dynamic regulation of leukocyte beta adrenergic receptor agonist interactions by physiological changes in circulating catecholamines. *J. Clin. Invest.*, **72**, 164–170.

Fouad, F.M., Tadena-Thome, L., Bravo, E.L. and Tarazi, R.C. (1986) Idiopathic hypovolemia. *Ann. Intern. Med.*, **104**, 298–303.

Fraser, F. and Wilson, R.M. (1918) The sympathetic nervous system and the "irritable heart of soldiers". *Br. Med. J.*, **2**, 27–32.

Frohlich, E.D., Dustan, H.P. and Page, I.P. (1966) Hyperdynamic beta-adrenergic circulatory state. *Arch. Intern. Med.*, **117**, 612–619.

Frohlich, E.D., Tarazi, R.C. and Dustan, H.P. (1969) Hyperdynamic beta-adrenergic circulatory state. *Arch. Intern. Med.*, **123**, 1–7.

Gaffney, F.A., Karlsson, E.S., Campbell, W., Schutte, J.E., Nixon, J.V., Willerson, J.T., *et al.* (1979) Autonomic dysfunction in women with mitral valve prolapse syndrome. *Circulation*, **59**, 894–901.

Gaffney, F.A., Bastian, B.C., Lane, L.B., Taylor, W.F., Horton, J., Schutte, J.E., *et al.*, (1983a) Abnormal cardiovascular regulation in the mitral valve prolapse syndrome. *Am. J. Cardiol.*, **52**, 316–320.

Gaffney, F.A., Lane, L.B., Pettinger, W. and Blomqvist, C.G. (1983b) Effect of long-term clonidine administration on the hemodynamic and neuroendocrine postural responses of patients with dysautonomia. *Chest*, **83**, 436–438.

Gaffney, F.A., Nixon, J.V., Karlsson, E.S., Campbell, W., Dowdey, A.B.C. and Blomqvist, C.G. (1985) Cardiovascular deconditioning produced by 20 hours of bedrest of head-down tilt (-5°) in middle-aged healthy men. *Am. J. Cardiol.*, **56**, 634–638.

Hoeldtke, R.D. and Davis, K.M. (1991) The orthostatic tachycardia syndrome: Evaluation of autonomic function and treatment with octreotide and ergot alkaloids. *J. Clin. Endocrinol. Metab.*, **73**, 132–139.

Hoeldtke, R.D., Dworkin, G.E., Gaspar, S.E. and Israel, B.C. (1989) Sympathotonic orthostatic hypotension: A report of four cases. *Neurology*, **39**, 34–40.

Holmgren, A., Jonsson, B., Levander, M., Linderholm, H., Sjostrand, T. and Strom, G. (1957) Low physical working capacity in suspected heart cases due to inadequate adjustment of peripheral blood flow (vasoregulatory asthenia) *Acta. Med. Scand.*, **158**, 413.

Kuchel, O., Buu, N.T., Hamet, P., Larochelle, P., Gutkowska, J., Schiffrin, E.L., *et al.* (1985) Orthostatic hypotension: A posture-induced hyperdopaminergic state. *Am. J. Med. Sci.*, **289**, 3–11.

Kuchel, O., Buu, H.T., Larochelle, P., Hamet, P. and Genest, J., Jr. (1986) Episodic dopamine discharge in paroxysmal hypertension. *Arch. Intern. Med.*, **146**, 1315–1320.

Page, I.H. (1935) A syndrome simulating diencephalic stimulation occurring in patients with essential hypertension. *Am. J. Med. Sci.*, **190**, 9–14.

Pasternac, A., Tubau, J.F., Puddu, P.E., Krol, R.B. and DeChamplain, J. (1982) Increased plasma catecholamine levels in patients with symptomatic mitral valve prolapse. *Am. J. Med.*, **73**, 783–790.

Pasternac, A., Kouz, S., Gutkowska, J., Petitclerc, R., Taillefer, R., Cequier, A., *et al.* (1986) Atrial natriuretic factor: A possible link between left atrium, plasma volume, adrenergic control and renin-aldosterone in the mitral valve prolapse syndrome. *J. Hypertens.*, **4**, S76–S79.

Peabody, F., Clough, H., Sturgis, C., Wearn, J. and Tompkins, E. (1918) Effects of the injection of epinephrine in soldiers with "irritable heart". *JAMA*, **71**, 1912–1919.

Pemberton, J. (1989) Does constitutional hypotension exist? *Br. Med. J.*, **298**, 660–662.

Procacci, P.M., Savran, S.V., Schreiber, S.K. and Bryson, A.L. (1976) Prevalence of clinical mitral valve prolapse in 1169 young women. *New Engl. J. Med.*, **294**, 1086.

Roberts, L.J.II and Oates, J.A. (1991) Biochemical diagnosis of systemic mast cell disorders. *J. Invest. Dermatol.*, **96**, 19S–25S.

Schatz, I.J., Ramanathan, S., Villagomez, R. and MacLean, C. (1990) Orthostatic hypotension, catecholamines, and alpha-adrenergic receptors in mitral valve prolapse. *West J. Med.*, **152**, 37–40.

Schondorf, R. and Low, P.A. (1993) Idiopathic postural orthostatic tachycardia syndrome (POTS): An attenuated form of acute pandysautonomia? *Neurology*, **43**, 132–137.

Streeten, D.H.P., Kerr, L.P., Kerr, C.B., Prior, J.C. and Dalakos, T.G. (1972) Hyperbradykininism: A new orthostatic syndrome. *Lancet*, **2**, 1048–1053.

Streeten, D.H.P., Anderson, G.H., Jr., Richardson, R. and Thomas, F.D. (1988) Abnormal orthostatic changes in blood pressure and heart rate in subjects with intact sympathetic nervous function: Evidence of excessive venous pooling. *J. Lab. Clin. Med.*, **111**, 326–335.

Streeten, D.H.P. (1990) Pathogenesis of hyperadrenergic orthostatic hypotension. Evidence of disordered venous innervation exclusively in the lower limbs. *J. Clin. Invest.*, **86**, 1582–1588.

Taylor, A.A., Davies, A.O., Mares, A., Raschko, J., Pool, J.L., Nelson, E.B., *et al.* (1989) Spectrum of dysautonomia in mitral valve prolapse. *Am. J. Med.*, **86**, 267–274.

Wooley, C.F. (1976) Where are the diseases of yesteryear? DaCosta's syndrome, soldiers' heart, the effort syndrome, neurocirculatory asthenia, and the mitral valve prolapse syndrome. *Circulation*, **53**, 749–751.

13 Paroxysmal Autonomic Syncope

Ronald G. Victor, Charles M.T. Jost,

Richard L. Converse, Jr. and Tage N. Jacobsen

*Department of Internal Medicine, University of Texas Southwestern
Medical Center, Dallas, Texas 75235-9034, USA*

Activation of the sympathetic nervous system, with reflex vasoconstriction and tachycardia, is known to be a major compensatory adjustment to hypotensive states. In contrast, sympathoinhibition or sudden withdrawal of sympathetic reflex drive appears to be able to cause acute episodes of hypotension. Such episodes of sympathoinhibition can cause several clinical syndromes which collectively can be termed "paroxysmal autonomic syncope". It is well appreciated that such reactions occur during emotional fainting, but it is much less recognized that the same type of paroxysmal syncope may contribute importantly to the morbidity and mortality of cardiovascular emergencies such as myocardial infarction and haemorrhagic shock.

This chapter critically reviews what is known and what is not known about paroxysmal vasodepressor syncope, its determinants, mechanisms, and management.

KEY WORDS: Baroreflexes, sympathetic nervous system, C-fibers, vasodepressor syncope

INTRODUCTION

Activation of the sympathetic nervous system, with reflex vasoconstriction and tachycardia, is a major compensatory adjustment during many hypotensive states (Bedford and Jackson, 1916; Vatner, 1974; Abboud, 1985). In addition, the sudden withdrawal of sympathetic drive can be a primary cause of acute, episodic hypotension with reflex bradycardia and peripheral vasodilation, which are seemingly paradoxical autonomic responses in the setting of hypotension (Wallin and Sündlof, 1982; Scherrer *et al.*, 1990). Such sympathoinhibition appears to cause several clinical syndromes which collectively can be termed "paroxysmal autonomic syncope". Paroxysmal hypotension with inappropriately normal or frankly decreased heart rate and peripheral vascular resistance is well known to be the haemodynamic basis of emotional fainting (Lewis, 1932; Murray *et al.*, 1961;

Glick and Yu, 1963). However, it is much less well appreciated that the same type of vasovagal, or vasodepressor, reaction also may contribute importantly to the morbidity and mortality of cardiovascular emergencies such as acute myocardial infarction and haemorrhagic shock.

Most current medicine textbooks emphasize that sympathetically mediated peripheral vasoconstriction and tachycardia are the classic clinical signs of hypovolaemic shock and suggest that these compensatory adjustments may be lost only in the terminal, irreversible stage of shock (Hurst *et al.*, 1990; Braunwald, 1991; Guyton, 1991; West, 1991; Wilson *et al.*, 1991; Wyngaarden, Smith and Bennet, 1991). In contrast to this traditional teaching, there is increasing experimental evidence — but as yet little clinical recognition — that earlier stages of hypovolaemic hypotension also can be accompanied by bradycardia and vasodilation which are caused by the reversible, reflex withdrawal of sympathetic drive (Jansen, 1978; Secher, Jensen and Werner, 1984a; 1984b; Secher and Bie, 1985; Ludbrook, Secher, *et al.*, 1987; Lieshout *et al.*, 1991). In the past 20 years, our understanding of vasodepressor syncope has been improved greatly by the ability to record sympathetic nerve activity directly not only in experimental animals but also in conscious humans.

This chapter will critically review what is known and what is not known about vasodepressor syncope, its determinants, mechanisms and management. The major purpose of this review is to provide a conceptual framework to improve the scientific understanding and clinical recognition of vasodepressor syncope in the setting of hypovolaemic hypotension.

Three principal concepts will be presented.

1. In humans, withdrawal of sympathetic vasoconstrictor drive to skeletal muscle is the final neurophysiological pathway causing many forms of vasodepressor syncope.

2. Activation of inhibitory cardiac afferents is thought to be the primary peripheral reflex mechanism causing vasodepressor syncope, not only with acute myocardial infarction and with aortic stenosis, but also with many forms of hypovolaemic hypotension.

3. Activation of inhibitory central neural mechanisms may contribute to the development of vasovagal syncope, not only during emotional fainting, but also during hypovolaemia. The relative importance of these different mechanisms may vary greatly, depending upon the experimental model and the specific sympathetic outflow and species under study.

WITHDRAWAL OF SYMPATHETIC VASOCONSTRICTOR DRIVE AS THE PRIMARY CAUSE OF VASODEPRESSOR SYNCOPE

As early as 1793, the surgeon John Hunter first observed the effects of vasodilation during vasovagal syncope as a "scarlet appearance" of venous blood (Palmer,

1837). In 1895, Hill proposed that reduced activity in vasomotor nerves is the major cause of emotional fainting. In 1932, Sir Thomas Lewis coined the term "vasovagal" for this type of syncope to emphasize both the peripheral vascular and the cardiac components of this perplexing autonomic response. Because reversal of vagal bradycardia with atropine failed to restore blood pressure to normal during vasovagal syncope, he concluded that "the cause of syncope is mainly vasomotor and not vagal" (Lewis, 1932). Subsequent measurements of limb blood flow during vasovagal syncope have provided abundant support for this conclusion (Barcroft *et al.*, 1944; Barcroft and Edholm, 1945; Bridgen, Howarth and Sharpey-Shafer, 1950; Murray *et al.*, 1961; Glick and Yu, 1963; Epstein, Stampfer and Beiser, 1968; Wallin and Sündlof, 1982).

While activation of sympathetic cholinergic vasodilatation in skeletal muscle has been proposed as one mechanism of vasodepressor syncope in humans (Barcroft *et al.*, 1944; Barcroft and Edholm, 1945; Greenfield, 1951; Blair *et al.*, 1959; Greenfield, 1966; Roddie, 1977), the experimental data to support this theory are indirect and inconclusive and would appear to be inconsistent with the observation that atropine does not reverse the hypotension (Lewis, 1932; Blair *et al.*, 1959; Barcroft *et al.*, 1960). In contrast, microneurographic measurements of sympathetic nerve activity have provided direct neurophysiological evidence that withdrawal of sympathetic vasoconstrictor drive to skeletal muscle, and perhaps other vascular beds, is a primary mechanism causing vasovagal, or vasodepressor, syncope.

NORMAL INVERSE RELATION BETWEEN MUSCLE SYMPATHETIC NERVE ACTIVITY AND ARTERIAL PRESSURE

In 1969, the microneurographic technique for intraneural recordings of sympathetic nerve activity in humans was developed at the University of Uppsala, Sweden, by Hagbarth, Vallbo, Wallin, and colleagues; the details of this technique have been reviewed extensively (Vallbo *et al.*, 1979). Briefly, multiunit recordings of post-ganglionic sympathetic nerve activity are obtained in conscious, unanaesthetized humans by insertion of unipolar microelectrodes into human peripheral nerves. The sympathetic fibers travel in the motor and sensory fascicles, allowing separate recordings of sympathetic traffic targeted to skeletal muscle or to skin. In humans, muscle sympathetic nerve activity is postganglionic because the conduction velocity is 1 ms^{-1}, indicative of unmyelinated C-fibers, and is eliminated by ganglionic blockade. The muscle sympathetic nerve activity is predominantly vasoconstrictor activity because increased activity is accompanied by regional vasoconstriction and decreased activity by vasodilation (Vallbo *et al.*, 1979; Vissing, Scherrer and Victor, 1989). Under normal conditions, this sympathetic activity is under baroreflex regulation: it is locked to the cardiac cycle and is inversely related to blood pressure (Figure 13.1). The normal inverse relation between muscle sympathetic nerve activity and arterial pressure is also evident during pharmacological manipulations in arterial pressure (Figure 13.2).

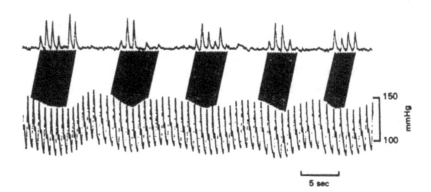

FIGURE 13.1   Relation between muscle sympathetic nerve activity (top panel) and arterial pressure (bottom panel) at rest. On this mean voltage display, the narrow peaks are spontaneous bursts of sympathetic discharge. Under baseline conditions, the sympathetic bursts occur mainly during spontaneous decreases in arterial pressure and are inhibited during increases in arterial pressure. (From Sündlof and Wallin, 1978. *Journal of Physiology (London)*, 274, 627. With permission.)

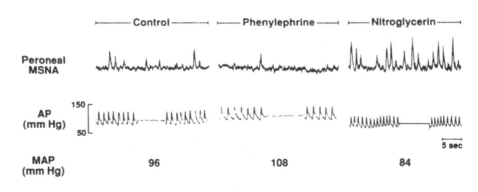

FIGURE 13.2   Segments of an original neurogram from one subject in our laboratory showing changes in muscle sympathetic nerve activity (MSNA), phasic arterial pressure (AP) and mean arterial pressure (MAP) during pharmacological alterations in arterial pressure. The sympathetic bursts were almost completely inhibited by an increase in mean arterial pressure of 12 mmHg (intravenous phenylephrine) and were greatly increased both in number and amplitude by a decrease in mean arterial pressure of 12 mmHg (intravenous nitroglycerine) (Unpublished data from our laboratory.)

RELATION BETWEEN MUSCLE SYMPATHETIC NERVE ACTIVITY AND ARTERIAL PRESSURE DURING VASOVAGAL SYNCOPE

In 1982, Wallin and Sündlof recorded muscle sympathetic nerve activity during two episodes of vasovagal syncope. As shown in Figure 13.3, during vasovagal syncope the normal inverse relation between arterial pressure and sympathetic nerve

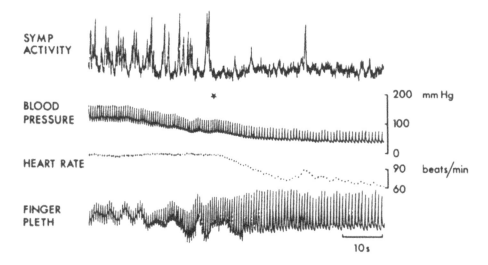

FIGURE 13.3 Decreases in muscle sympathetic nerve activity, blood pressure, heart rate and finger blood flow (pulse plethysmogram) during vasovagal syncope (asterisk) evoked by intravenous infusion of nitroprusside. (From Wallin and Sündlof, 1982. With permission.)

activity was lost: during the rapid decline in blood pressure, muscle sympathetic nerve activity did not increase but rather decreased markedly. This paradoxical suppression of sympathetic activity was accompanied by increased regional blood flow, providing direct evidence that such vasodilation is caused by the withdrawal of sympathetic vasoconstrictor drive.

However, the underlying mechanism triggering this sudden, paradoxical withdrawal of sympathetic drive is unknown and recently has been the subject of considerable research and debate. In general, vasovagal syncope has been attributed to two principal mechanisms: central inhibition of sympathetic outflow triggered by some emotional event (i.e. a simple emotional faint); and reflex inhibition of sympathetic outflow caused by mechanical activation of cardiac ventricular afferents during hypovolaemic hypotension.

ACTIVATION OF INHIBITORY CARDIAC AFFERENTS AS A POTENTIAL MECHANISM TRIGGERING VASODEPRESSOR SYNCOPE: THE "VENTRICULAR SYNCOPE HYPOTHESIS"

STUDIES IN EXPERIMENTAL ANIMALS

Studies in experimental animals have demonstrated that the heart is an important sensory organ containing polymodal afferent, or sensory, nerve endings that are located in the atria, venoatrial junctions and ventricles (Coleridge, Coleridge and Kidd, 1964; Öberg and Thorén, 1972b; Thorén, 1976, 1977; Coleridge and Coleridge, 1977; Thorén, 1979; Thorén, Noresson and Ricksten, 1979; Coleridge and Coleridge, 1980; Bishop, Malliani andd Thorén, 1983; Abboud, 1989). When activated by certain chemical and/or mechanical stimuli, these afferent nerve endings evoke a variety of excitatory and inhibitory cardiovascular reflexes (Öberg and White, 1970a; Bishop, Malliani and Thorén, 1983).

Many of the ventricular endings give rise to unmyelinated (C-fiber) vagal afferents that exert a predominantly inhibitory influence on sympathetic vasoconstrictor outflow (Coleridge, Coleridge and Kidd, 1964; Coleridge and Coleridge, 1977; Thorén, 1979; Bishop, Malliani and Thorén, 1983). Activation of these afferents with exogenous chemical agents such as veratridine or nicotine elicits the classic von Bezold–Jarisch reflex, characterized by hypotension, bradycardia, peripheral vasodilation and gastric relaxation (Bezold and Hirt, 1867; Jarisch and Zotterman, 1949; Abrahamsson and Thorén, 1973). Under physiological conditions ventricular C-fiber afferents are thought to function as mechanoreceptors, i.e. endings that are activated by deformation of their receptive fields (Bishop, Malliani and Thorén, 1983). Studies in anaesthetized cats have indicated that the two primary determinants of ventricular mechanoreceptor afferent discharge are ventricular filling pressure and contractility (Thorén, 1977). Thus, ventricular mechanoreceptor discharge normally increases during increases in left ventricular end-diastolic pressure produced by volume expansion and during infusion of positive inotropic agents. Furthermore, this activity decreases during infusion of negative inotropic agents and during mild and moderate reductions in left ventricular end-diastolic pressure produced by non-hypotensive haemorrhage.

In 1972, Öberg and Thorén (1972a) found that during more severe (and rapid) reduction in left ventricular end-diastolic pressure produced by hypotensive haemorrhage in cats, ventricular C-fiber discharge did not decrease further but showed a paradoxical burst of activity, which was followed promptly by the sudden onset of bradycardia. In contrast, haemorrhage-induced bradycardia was not preceded by a sudden change in activity of atrial receptor afferents, suggesting that the bradycardia is a reflex arising in the ventricles rather than the atria. The haemorrhage-induced bradycardia indeed was a reflex caused by the activation of C-fiber vagal afferents, because it was abolished by both selective cardiac vagal afferent section and by cervical vagal cooling, which selectively impairs conduction in afferent C-fibers while leaving intact conduction in myelinated efferent (motor) vagal fibers (Öberg and White, 1970b) (Figure 13.4).

These observations provided neurophysiological support for the theory, originally advanced in the 1950's by a number of investigators, that during severe hypovolaemic hypotension, inhibitory ventricular vagal C-fiber afferents are activated by excessive mechanical deformation of their receptive fields when the adrenergically stimulated heart contracts forcefully around an almost empty ventricular chamber (e.g. Jarisch, 1949; Henry, 1950; Sharpey-Shafer, 1956; Bishop, Malliani and Thorén, 1983; Mark and Mancia, 1983; Abboud, 1989) (Figure 13.5).

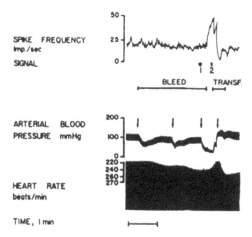

FIGURE 13.4 Segments of the original record from one cat showing effects of successive haemorrhages and transfusion on the spike frequency of a single ventricular vagal C-fiber afferent, arterial pressure and heart rate. The first two haemorrhages caused ventricular C-fiber afferent activity to decrease slightly and caused small and transient decreases in blood pressure which were partially offset by compensatory increases in heart rate. In contrast, the third haemorrhage caused a sudden explosive increase in C-fiber activity leading to bradycardia and a profound and sustained decrease in arterial pressure. These responses were reversed by transfusion. (Adapted from Öberg and Thorén, 1972a. *Acta Physiologica Scandinavica*, 85, 167. With permission.)

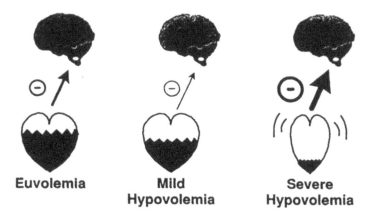

FIGURE 13.5 Cartoon illustrating the concept of paradoxical activation of ventricular afferents during hypovolaemic hypotension (the "ventricular syncope" hypothesis). The thickness of the arrows represent the inhibitory input from the ventricular vagal afferents during progressive reductions in cardiac filling. During euvolaemia, these afferents are thought to exert a tonic inhibitory influence on sympathetic vasomotor outflow. During mild hypovolaemia, as in non-hypotensive haemorrhage, the afferents are unloaded and their inhibitory input is reduced. During severe hypovolaemia, as with hypotensive haemorrhage, the ventricular afferent activity does not decrease further but rather increases paradoxically as the hypercontractile ventricle contracts forcefully around an almost empty ventricular chamber.

While there is substantial evidence that this reflex mechanism causes the sudden bradycardia during hypotensive haemorrhage in anaesthetized cats (Öberg and Thorén, 1970, 1972a), there has been much speculation but as yet no direct proof that the same mechanism causes the withdrawal of sympathetic vasoconstrictor drive to the peripheral circulation during haemorrhagic hypotension in either experimental animals or humans. In rats, haemorrhagic hypotension at first produces small and transient increases in heart rate and renal sympathetic nerve activity followed by sustained and parallel decreases in renal sympathetic activity as well as heart rate (and also splanchnic sympathetic activity) (Skoog, Mansson and Thorén, 1985; Victor et al., 1989) (Figure 13.6).

These investigators also showed that activation of vagal afferents plays a major role in causing this depressor response because the renal sympathetic inhibition was markedly attenuated by cervical vagotomy. However, these experiments: (1) indicate that mechanisms other than vagal afferent activation also contribute to haemorrhage-induced renal sympathoinhibition in the rat, because in some experiments the sympathoinhibition was only partially reversed by vagotomy; and (2) do not prove that the left ventricle indeed is the site of origin of this vagal afferent reflex, because in rats, unlike cats, mechanical probing has disclosed vagal afferents with receptive fields only in the atria and none in the ventricles (Thorén, 1979; Thorén, Noresson and Ricksten, 1979). In order to define precisely the receptive fields of the vagal fibers that form the afferent arm of this reflex, single fiber afferent recordings would need to be performed during haemorrhage in rats. Furthermore, there may be chemical, as well as mechanical, factors that trigger this vagal afferent reflex in the rat, thus this renal sympathoinhibitory response is abolished by administration of serotonergic antagonists in normal rats and is absent in rats that have a genetic deficiency in vasopressin (Morgan et al., 1988; Peuler et al., 1990a).

Although haemorrhage-induced bradycardia and renal sympathoinhibition now have been demonstrated in a number of mammalian species, there are major inter-species differences in the underlying mechanisms causing these vasodepressor reactions. In conscious dogs and rabbits, for example, progressive haemorrhage also has been shown to cause the same biphasic response in renal sympathetic nerve activity as that seen in rats: the sympathetic activity increases during mild and moderate haemorrhage, but then returns to or below the baseline level during severe hypotensive haemorrhage (Morita and Vatner, 1985; Morita, Manders and Vatner, 1985; Morita et al., 1988). Although activation of vagal afferents clearly is the primary mechanism causing the decreases in renal sympathetic nerve activity during haemorrhagic hypotension in the rat, vagal afferents do not play a primary role in causing this response in conscious dogs and rabbits, because in these species vagotomy has no detectable effect on haemorrhage-induced renal sympathoinhibition (Morita and Vatner, 1985).

Most of these studies have relied on renal sympathetic activity as the index of sympathetic nerve response to haemorrhage. However, there is recent evidence that regional sympathetic responses to hypotensive haemorrhage are regulated by highly specific reflex pathways that differ greatly, not only between the different species,

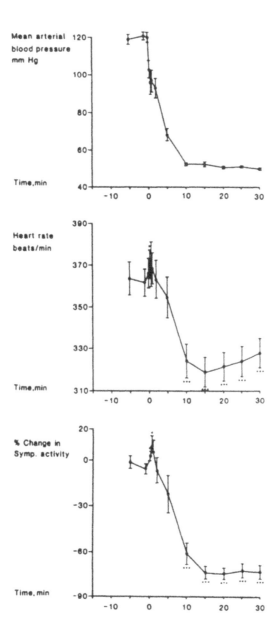

FIGURE 13.6   Decreases in mean arterial pressure, heart rate and renal sympathetic nerve activity during severe haemorrhage in chloralose-anaesthetized rats. (From Skoog, Mansson and Thorén, 1985. With permission.)

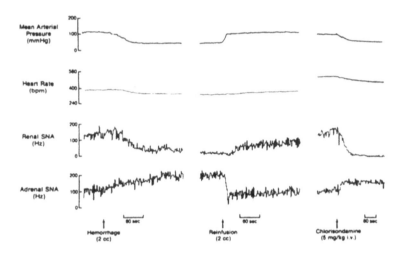

FIGURE 13.7   Segments of an illustrative record showing response of mean arterial pressure, heart rate and renal and adrenal sympathetic nerve activity (SNA) (displayed as a time–frequency histogram) during hypotensive haemorrhage, reinfusion of blood and ganglionic blockade with chlorisondamine. Haemorrhagic hypotension decreased renal SNA and heart rate, but simultaneously increased adrenal SNA in this rat. These variables returned to control after transfusion. Ganglionic blockade abolished SNA in the postganglionic renal nerve, but did not decrease and tended to increase SNA in the adrenal nerve, which contains some preganglionic fibers. (From Victor et al., 1989. With permission from the Editor.)

but even between sympathetic outflows to different organs that are in close anatomic proximity in the same animal. Thus, in rats, haemorrhagic hypotension triggers directionally opposite responses in sympathetic outflow to the kidney and adrenal glands, with renal sympathoinhibition but adrenal sympathoexcitation (Victor et al., 1989) (Figure 13.7).

Although a paradoxical increase in ventricular mechanoreceptor discharge is the most widely studied mechanism for the reflex withdrawal of sympathetic outflow during hypovolaemic hypotension (Öberg and Thorén, 1972a), there is also direct experimental evidence in anaesthetized cats that severe carotid sinus hypotenion triggers paradoxical increases in carotid baroreceptor discharge (Landgren, 1952) (Figure 13.8).

It is reasonable to speculate that during severe hypotension, paradoxical increases in carotid, as well as ventricular, mechanoreceptor discharge are caused by local deformation of the walls of cardiovascular structures in which the receptors are embedded (Landgren, 1952; Thorén, 1979). During hypotension, such paradoxical increases in carotid baroreceptor discharge might be further augmented by reflex sympathetic activation; increases in efferent sympathetic nerve activity and direct application of noradrenaline have both been shown to sensitize arterial baroreceptor

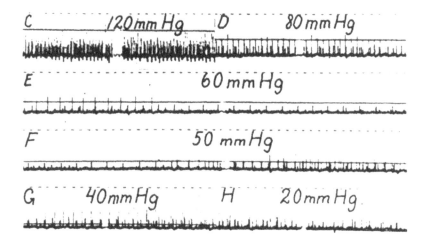

FIGURE 13.8 Segments of an original record showing effects of progressive reductions in carotid sinus pressure on single fiber carotid baroreceptor discharge in the cat. The discharge frequency of this fiber decreased progressively (almost to zero) as carotid sinus pressure was decreased progressively from 120 to 50 mmHg. As carotid sinus pressure was decreased below 50 mmHg, however, baroreceptor discharge did not decrease further but rather showed an abrupt, paradoxical increase in activity. (From Landgren, 1952. With permission.)

discharge (e.g. Felder, Heesch and Thames, 1983; Munch, Thorén and Brown, 1987). However, the functional importance of increased carotid baroreceptor discharge in causing reflex withdrawal of sympathetic vasoconstrictor outflow during hypovolaemic hypotension remains unknown, since sinoaortic denervation did not prevent renal sympathoinhibition during severe haemorrhage in either anaesthetized rats or conscious dogs (Morita and Vatner, 1985; Victor *et al.*, 1989).

Taken together, these observations from animal experiments suggest that multiple, complex, and perhaps some as yet undefined mechanisms are likely to be involved in the regulation of sympathetic outflow during severe hypotensive haemorrhage and that considerable caution should be taken in extrapolating these experimental data directly to humans.

STUDIES IN HUMANS

Most current textbooks of internal medicine and cardiology emphasize that activation of the sympathetic nervous system produces many of the classic clinical signs of hypovolaemic shock, including peripheral vasoconstriction and tachycardia. Although some textbooks suggest that acidaemia and hypoxaemia may attenuate the ability of the cardiovascular system to respond appropriately to increased sympathetic neural drive during the decompensated stage of hypovolaemic shock, none of these major medical textbooks discuss the possibility that hypovolaemic hypotension can trigger the abrupt reflex (and reversible) withdrawal of sympathetic drive, producing inappropriate bradycardia, vasodilation and worsening hypotension.

In the past 50 years, however, there have been a small number of reports indicating that hypovolaemic hypotension can trigger relative bradycardia in the clinical, as well as in the experimental, setting. Hypotension with an inappropriately "normal" heart rate has been reported in haemorrhagic shock in World War II air-raid casualties (Grant and Reeve, 1941; McMichael, 1944); in obstetric patients with acute intraperitoneal bleeding (Jansen, 1978); in patients presenting to the emergency room with severe but reversible hypovolaemic shock due to trauma and other causes (Secher *et al.*, 1984b; Secher and Bie, 1985; Rorsgaard and Secher, 1986; Sander-Jensen *et al.*, 1986) and in patients experiencing hypovolaemic hypotension during major abdominal surgery. In a study of 20 consecutive patients presenting to the Rigshospitalet in Copenhagen, Denmark with haemorrhagic shock, the average heart rate was 73 beats/min even though the average blood pressure was 81/55 mmHg (Sander-Jensen *et al.*, 1986). Furthermore, the heart rate increased to 102 beats/min during transfusion.

Thus, in contrast to the traditional teaching, these important studies strongly suggest that relative bradycardia: (1) is more common than previously recognized in patients with haemorrhagic hypotension; and (2) is rapidly reversed with volume expansion, suggesting that the bradycardia is caused mainly by a reflex mechanism rather than by irreversible damage to neural or cardiovascular tissues (i.e. that the shock has not yet reached the "irreversible phase").

Although sympathetic nerve activity has not yet been recorded in humans during haemorrhagic hypotension, direct measurements of muscle sympathetic nerve activity have documented acute sympathoinhibition (with bradycardia and classic vasovagal symptoms) in humans during vasodepressor reactions caused by upright tilt and infusion of nitroprusside (Wallin and Sündlof, 1982; Scherrer *et al.*, 1990). Because nitroprusside is a mixed vasodilator decreasing both preload and afterload, many authors have suggested that the vasodepressor reactions produced by upright tilt, nitroprusside and haemorrhage are all caused by the same mechanism: paradoxical withdrawal of sympathetic activity due to increased mechanical stimulation of ventricular afferents (i.e. "ventricular" or "neurocardiogenic" syncope) (Abboud, 1989; Almqvist *et al.*, 1989). While the data from animal experiments make this an attractive hypothesis, at present the experimental support for this hypothesis in humans is indirect and inconclusive.

Recently, ventricular mechanoreceptor-induced vasovagal syncope has been proposed as a common cause of otherwise unexplained syncope in patients with normal electrophysiological testing (Almqvist *et al.*, 1989; Chen *et al.*, 1989; Milstein *et al.*, 1989). These investigators based their initial conclusion on the finding that upright tilt, together with intravenous infusion of isoproterenol, provoked acute hypotension with bradycardia in 9 of 11 patients with a history of recurrent syncope, but did so in only 2 of 18 subjects with no such history. The authors hypothesized that the combination of decreased preload and ventricular dimension (produced by upright tilt) and increased cardiac contractility (produced by isoproterenol) increased the activity of ventricular mechanoreceptors, triggering reflex bradycardia and peripheral vasodilation. A subsequent study by Shalev *et al.* (1991) provided some echocardiographic support for the assumption that the isoproterenol–tilt test indeed produces a vigorous myocardial contraction, together with a decrease in left ventricular end-diastolic dimension, and further reported that prevention of the increased contractility with cardiac adrenergic blockade prevented vasavagal syncope in 3 of 5 patients during tilting alone and in 4 of 4 patients during tilting plus isoproterenol.

While provocative, these data should be interpreted cautiously because of small numbers of subjects, failure to control for emotional factors and other flaws in experimental design. A major oversight in these studies is that the test–retest reproducibility of the isoproterenol–tilt test to repeatedly provoke vasovagal syncope in a given patient has never been reported (Rea, 1989). Because emotional factors are likely to contribute greatly to the development of vasovagal syncope during tilt testing in habitual fainters, the metroprolol data cannot be interpreted meaningfully until further studies determine whether the vasovagal reaction does or does not habituate when a given patient is tested repeatedly. Furthermore, it is unclear whether the echocardiographic changes were the cause or the consequence of the syncope, since a small hypercontractile ventricle would be the inevitable consequence of sudden severe decreases in preload and afterload.

Thus, the heart rate and echocardiographic responses to the isoproterenol–tilt test have not yet provided a definitive test of the hypothesis that the left ventricle is the main reflexogenic area that triggers the paradoxical sympathetic inhibition during hypovolaemic hypotension.

We recently had an unexpected opportunity to test this hypothesis in a patient who had undergone cardiac transplantation, an operation that causes both efferent and afferent denervation of the ventricles. During this operation, the ventricular afferents are severed, but afferents from the remnant atria, venoatrial junctions, pulmonary veins and arterial baroreceptors are left intact. While we were recording muscle sympathetic nerve activity (microelectrodes in the peroneal nerve) in an attempt to assess arterial baroreflex function with intravenous infusion of nitroprusside, classic vasovagal symptoms suddenly developed in the patient (Scherrer *et al.*, 1990). These symptoms immediately followed the peak increases in sympathetic nerve activity and heart rate produced by the nitroprusside and were accompanied by the abrupt cessation of the sympathetic discharge and a slowing of the innervated remnant atrial heart rate (Figure 13.9).

This latter study documents that vasodilator-induced hypovolaemic hypotension can precipitate a classic vasodepressor syncope in a patient with transplantation–induced ventricular denervation. The failure of cardiac transplantation to prevent sympathetic inhibition and classic vasovagal syncope after the infusion of a vasodilator therefore indicates that ventricular mechanoreceptor activation is not the only mechanism that can trigger sympathetic inhibition during hypovolaemic hypotension and is consistent with animal data indicating that multiple, redundant, and perhaps as yet unidentified mechanisms may be involved. *It is important to emphasize, however, that these findings in a heart transplant recipient certainly do*

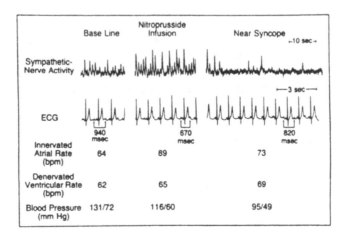

FIGURE 13.9 Segments of the original record in a 41 year-old heart transplant recipient 11 months after transplantation, showing the mean voltage neurogram of muscle sympathetic nerve activity and the electrocardiogram
at baseline, during the infusion of nitroprusside, and during a near syncope episode immediately following the
completion of the infusion. All measurements were performed with the patient supine. The numbers shown below
the electrocardiograph represent the P-P intervals of the innervated atrial remnant in milliseconds. In contrast to
the clear decrease in the rate of the innervated atrial remnant, the rate of the transplanted, denervated ventricle did
not decrease during the near syncopal episode. (From Scherrer *et al.*, 1990. With permission from the Editor.)

*not exclude the possibility that ventricular afferent activation is one of the
important mechanisms triggering hypovolaemia-induced vasovagal syncope in
normal humans.*

The finding of vasodilator-induced sympathetic inhibition in a patient with
cardiac denervation is consistent with previous findings that in conscious dogs
haemorrhagic hypotension-induced sympathoinhibition also is not prevented by
cardiac denevation (Morita and Vatner, 1985). The precise mechanism causing
the vasovagal reaction in this heart transplant recipient remains unknown, but
emotional factors are very unlikely to have been important. The patient had
no history of emotional fainting and experienced no pain or other emotional
event during the experimental protocol. However, these considerations do not
exclude the possibility that specific central neural mechanisms might play an
important role in causing this in other instances of hypovolaemia-induced vasovagal
syncope.

CENTRAL NEURAL MECHANISMS AS POTENTIAL CAUSES OF HYPOVOLAEMIA-INDUCED VASODEPRESSOR SYNCOPE

In experimental animals, stimulation of the limbic sympathoinhibitory centre in
the hypothalamus is well known to cause vasodilation in skeletal muscle with

bradycardia and hypotension (Smith, 1990). It is reasonable to speculate that emotional states might activate similar central neural pathways to elicit this pattern of autonomic circulatory response in conscious humans. Although very little is known currently about specific central autonomic pathways in humans, such information may be forthcoming with recent advances in technology, such as measurements of regional brain blood flow with radioisotopes and metabolic imaging with positron emission tomography.

Regardless of the precise central neural pathways involved, emotional and psychological factors must be considered carefully when interpreting studies of hypovolaemia-induced vasodepressor syncope in humans. Anxiety and fear elicited by the laboratory conditions and the influence of intravascular instrumentation, for example, may precipitate vasodepressor syncope in individuals prone to that type of reaction (Stevens, 1966; Abboud, 1989). Clearly the emotional and psychological context in which vasodepressor reactions occur are very different in: (1) a critically ill patient admitted to the emergency ward for severe haemorrhage; (2) a habitual fainter on a tilt table; and (3) a healthy volunteer subject resting comfortably (and supine) during experimental infusion of a vasodilator.

In addition to emotional activation of hypothalamic vasodilator pathways, two other specific central mechanisms have been implicated in triggering sympathetic inhibition during hypovolaemic hypotension. The first is that reduction in sympathetic vasomotor outflow occurs only in the terminal stage of decompensated, irreversible hemorrhagic shock when the severity and duration of brainstem ischaemia are sufficient to directly depress the firing of the central sympathetic neurons (Rothe, Schwendenmann and Selkurt, 1963; Lundgren, Lundwall and Mellander, 1964). This concept is stated concisely in the latest edition of the *Textbook of Medical Physiology*: "... there comes a point at which diminished blood flow to the vasomotor center itself is so depressed that the center becomes less active and finally totally inactive.... Fortunately, though the vasomotor center does not usually fail in the early stage of shock — only in the late stages" (Guyton, 1991).

In contrast to this traditional view, the wealth of experimental animal and human data reviewed in this chapter demonstrate that sympathoinhibition can occur at a much earlier stage of hypovolaemic hypotension than previously thought and that such sympathoinhibition, if diagnosed and treated with volume replacement, is often reversible.

A second theory is that activation of central opioid receptors contributes to the withdrawal of sympathetic outflow during hypotensive haemorrhage (Schadt *et al.*, 1984; Morita, Manders and Vatner, 1985; Schadt and Gaddis, 1986; Morita *et al.*, 1988). In conscious rabbits, progressive haemorrhage produces a biphasic pattern of autonomic response: initial increases in heart rate, renal sympathetic nerve activity, and total peripheral resistance during mild and moderate non-hypotensive haemorrhage, followed by a sudden precipitous fall in blood pressure accompanied by parallel decreases in heart rate, renal sympathetic nerve activity and total peripheral resistance to or below the pre-haemorrhage baseline levels. This secondary vasodepressor reaction was reversed by intravenous injection of the opiate receptor antagonist naloxone and was reproduced by administration of the

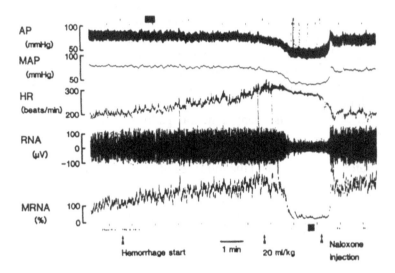

FIGURE 13.10   An original record from one experiment showing the effects of intravenous naloxone changes on phasic and mean arterial pressure (AP; MAP), heart rate (HR), renal sympathetic nerve activity displayed on the filtered neurogram (RNA) and the mean renal nerve activity (MRNA) during progressive haemorrhage in a conscious rabbit. During haemorrhage of less than 20 ml blood per kg body weight, arterial pressure was well maintained by compensatory increases in heart rate and renal sympathetic activity. During haemorrhage of greater than 20 ml/kg, however, arterial pressure fell precipitously as heart rate and renal sympathetic decreased activity to or below the baseline levels. The decreases in arterial pressure, heart rate and sympathetic activity were rapidly reversed by intravenous injection of naloxone. (From Morita *et al.*, 1988. With permission.)

opiate receptor agonist methionine-enkephalin (Morita, Manders and Vatner, 1985; Schadt and Gaddis, 1986; Morita *et al.*, 1988) (Figure 13.10).

OTHER CLINICAL SYNDROMES OF PAROXYSMAL AUTONOMIC SYNCOPE IN WHICH REFLEX WITHDRAWAL OF SYMPATHETIC DISCHARGE HAS BEEN IMPLICATED BUT UNPROVEN

The first suggestion that ventricular afferents might evoke a vasodepressor reaction in the clinical setting was the observation that bradyarrhythmias are much more common in patients during the initial phase of acute inferior–posterior than of the anterior myocardial infarction (Webb, Adgey and Pantridge, 1972). These clinical findings are consistent with data from experimental animals indicating that the vagal afferent endings that mediate bradycardia and sympathoinhibition during acute coronary occlusion are located preferentially in the inferior–posterior wall of the left ventricle (Thames *et al.*, 1978; Walker *et al.*, 1978).

As reviewed extensively elsewhere (Mark, 1983), activation of ventricular afferents has also been implicated in causing: (1) bradyarrhythmias and hypotension during coronary vasospasm (Perez-Gomez *et al.*, 1979) and thrombolytic therapy (Wei *et al.*, 1983) and balloon angioplasty with reperfusion involving the right coronary artery (Shakespeare *et al.*, 1991); and (2) exertional syncope in patients with severe aortic stenosis (Mark *et al.*, 1973). In the latter group of patients, bicycle exercise was found to produce vasodilation in the resting forearm (causing a "steel syndrome"), whereas the normal response is reflex forearm vasoconstriction (to decrease blood flow to non-working muscle). The abnormally forearm vasodilator responses in the patients was accompanied by an abnormally large exercise-induced rise in left ventricular end-diastolic pressure, presumably causing an abnormal large increase in the activity of left ventricular mechanoreceptors. Further evidence that the abnormal forearm vasodilator response was caused by a reflex arising in the left ventricle is that this response was not observed during exercise in patients with mitral, rather than aortic stenosis, and was normalized by successful aortic valve replacement. Without direct measurements of muscle sympathetic nerve activity, however, it is unknown whether the abnormal forearm vasodilator response to exercise in patients with aortic stenosis is caused by the reflex withdrawal of sympathetic vasoconstrictor drive, as occurs in vasovagal syncope, and/or by the engagement of an active (e.g. cholinergic) vasodilator pathway as described during isometric exercise in humans (Sanders, Mark and Ferguson, 1989).

Sympathetic withdrawal from activation of ventricular afferents might also contribute to orthostatic intolerance in highly trained athletes and after space flight (Levine, 1991; Levine *et al.*, 1991a,b). In addition, this reflex mechanism might offer a potential explanation for the development of supine hypotension in the last trimester of pregnancy (Kerr, 1965) and also a case report of unexplained recurrent, episodic hypotension and sympathetic insufficiency in a patient, who had undergone placement of an umbrella filter device in the inferior vena cava because of recurrent pulmonary embolism (Williams and Bashore, 1980). In these instances, intermittent mechanical reductions in venous return may have set the stage for activation of ventricular afferents.

There are many other forms of acute autonomic syncopes producing vasodepressor-type reactions. These have been reviewed extensively elsewhere (e.g. Manolis *et al.*, 1990) and include: (1) carotid sinus hypersensitivity (Walter, Crawley and Dorney, 1978; Coplan and Schweizter, 1984; Zee-Cheng and Gibbs, 1986; Holmes *et al.*, 1987; Kenny *et al.*, 1987; Tulchinsky and Krasnow, 1988); (2) glossopharyngeal neuralgia (Reddy *et al.*, 1987; Lagerlund *et al.*, 1988); (3) cough syncope (Sharpey-Shafer, 1953; MacIntosh, Estes and Warren 1956; Kerr and Eich, 1961; Hart *et al.*, 1982; Saito *et al.*, 1982; Strauss, Longstreth and Thiele, 1984); (4) micturition syncope (Proudfit and Forteza, 1959; Lyle, Monroe and Flinn, 1961; Kapoor, Peterson and Karpf, 1985; Fagius and Karhuvaara, 1989); (5) deglutition syncope (James, 1958; Lichstein and Chadda, 1972; Tomlinson and Fox, 1975; Wik and Hillestad, 1975; Palmer, 1976; Golf, 1977; Brick, Lowther and Deglin, 1978; Bortolotti, Cirignotta and Labo, 1982; Lipsitz *et al.*, 1983; Armstrong, MacMillan and Simon, 1985; Kadish, Wechsler and Marchlinski, 1986;

Lipsitz *et al.*, 1986); and (6) defecation syncope (Kapoor, Peterson and Karpf, 1986; Kollef and Schachter, 1990).

CONCLUSION

In conscious humans, however, the role played by opiate receptors in causing hypovolaemia-induced sympathoinhibition remains to be determined. In a recent preliminary report, intravenous naloxone did not prevent vasovagal syncope during venous pooling with lower body negative pressure in normal human subjects (Smith *et al.*, 1991). However, the dose of naloxone used in the human subjects, 0.1 mg/kg, was 30 times less than that used to reverse sympathoinhibition and vasodilation during haemorrhagic hypotension in conscious rabbits (Schadt and Gaddis, 1986; Morita *et al.*, 1988).

A final consideration is that there may be important interactions between central neural and peripheral reflex mechanisms in causing the integrated autonomic response that leads to vasodepressor syncope. Experiments in rats suggest that activation of the vagal afferents mediating sympathoinhibition during hypotensive haemorrhage engages a highly specific central neural pathway involving the locus coeruleus; this pathway is distinctly different from the engaged deactivation of the vagal afferents triggering sympathetic activation during non-hypotensive haemorrhage (Elam *et al.*, 1984; Elam, Thorén and Svensson, 1986). Vasopressin, and other humoral agents released in large amounts during haemorrhagic hypotension, might contribute to the development of sympathoinhibition by sensitizing mechanoreceptor afferent and/or by influencing the central integration of vagal afferent and arterial baroreceptor inputs via its actions in the area postrema (Undesser *et al.*, 1985; Hasser, Undesser and Bishop, 1987; Fitzpatrick *et al.*, 1990; Peuler *et al.*, 1990a,b; Brooks and Hatton, 1991).

REFERENCES

Abboud, F.M. (1985) Shock. In *Cecil Textbook of Medicine*, edited by J.B. Wyngaarden *et al.*, pp. 211–225. Philadelphia: WB Saunders.

Abboud, F.M. (1989) Ventricular syncope: is the heart a sensory organ? *New England Journal of Medicine*, **320**, 390–392.

Abrahamsson, H. and Thorén, P. (1973) Vomiting and reflex vagal relaxation of the stomach elicited from heart receptors in the cat. *Acta Physiologica Scandinavica*, **88**, 433–439.

Almquist, A., Goldenberg, I.F., Milstein, S., Chen, M-Y., Chen, X., Hansen, R., *et al.* (1989) Provocation of bradycardia and hypotension by isoproterenol and upright posture in patients with unexplained syncope. *New England Journal of Medicine*, **320**, 346–351.

Armstrong, P.W., McMillan, D.G. and Simon, J.B. (1985) Swallow syncope. *Canadian Medical Association Journal*, **132**, 1281–1284.

Barcroft, H. and Edholm, O.G. (1945) On the vasodilation in humans skeletal muscle during post-hemorrhagic fainting. *Journal of Physiology (London)*, **104**, 161–175.

Barcroft, H., Edholm, O.G., McMichael, J. and Sharpey-Schafer, E.P. (1944) Posthemorrhagic fainting, study by cardiac output and forearm flow. *Lancet*, **1**, 489–491.

Barcroft, H., Brod, J., Hejl, Z., Hirsjarvi, E.A. and Kitchin, A.H. (1960) The mechanism of the vasodilation in the forearm muscle during stress (mental arithmetic). *Clinical Sciences*, **19**, 577–586.

Bedford, E.A. and Jackson, H.C. (1916) The epinephrine content of the blood in conditions of low blood pressure and "shock". In *Proceedings of The Society for Experimental Biology and Medicine*, **13**, 85–87.

Bezold, A. von and Hirt, L. (1867) Uber die physiologischen Wirkungen des essigsauren Veratrins. *Unters Physiologische Laboratorium Wurzburg*, **1**, 73–156.

Bishop, V.S., Malliani, A. and Thorén, P. (1983) Cardiac mechanoreceptors. In *Handbook of Physiology*, Section 2: *The Cardiovascular System*, vol. 3, edited by J.T. Shepherd and F.M. Abboud. pp. 497–545. Bethesda MD: American Physiological Society.

Blair, D.A., Glover, W.E., Greenfield, A.D.M. and Roddie, I.C. (1959) Excitation of cholinergic vasodilator nerves to human skeletal muscles during emotional stress. *Journal of Physiology (London)*, **148**, 633–647.

Bortolotti, M., Cirignotta, F. and Labo, G. (1982) Atrioventricular block induced by swallowing in a patient with diffuse esophageal spasm. *Journal of the American Medical Association*, **248**, 2297–2299.

Braunwald, E. (1991) *Heart disease. A Textbook of Cardiovascular Medicine*, 4th edn. pp. 569–584. Philadelphia: W.B. Saunders.

Brick, J.E., Lowther, C.M. and Deglin, S.M. (1978) Cold water syncope. *Southern Medical Journal*, **71**, 1579–1580.

Brigden, W., Howarth, S. and Sharpey-Shafer, E.P. (1950) Postural changes in the peripheral blood-flow of normal subjects with observations on vasovagal fainting reactions as a result of tilting, the lordotic posture, pregnancy and spinal anesthesia. *Clinical Sciences*, **9**, 79–90.

Brooks, V.L. and Hatton, D.C. (1991) Hypotension during vasopressin receptor blockade: role of V2 receptors and sympathetic nervous system. *American Journal of Physiology*, **260**, H1878–H1887.

Chen, M-Y., Goldenberg, I.F., Milstein, S., Buetikofer, J., Almqvist, A., Lesser, J., et al. (1989) Cardiac electrophysiologic and hemodynamic correlates of neurally mediated syncope. *American Journal of Cardiology*, **63**, 66–72.

Coleridge, J.C.G. and Coleridge, H.M. (1977) Afferent C-fiber and cardiorespiratory chemoreflexes. *American Review of Respiratory Diseases*, **115**, 251–260.

Coleridge, H.M. and Coleridge, J.C.G. (1980) Cardiovascular afferents involved in regulation of peripheral vessels. *Annual Review of Physiology*, **42**, 413–427.

Coleridge, H.M., Coleridge, J.C.G. and Kidd, C. (1964) Cardiac receptors in the dog, with particular reference to two types of afferent ending in the ventricular wall. *Journal of Physiology (London)*, **174**, 323–329.

Coplan, N.L. and Schweizter, P. (1984) Carotid sinus hypersensitivity. Case report and review of the literature. *American Journal of Medicine*, **77**, 561–565.

Elam, M., Yao, T., Svensson, T.H. and Thorén, P. (1984) Regulation of locus coerulus neurons and splanchnic, sympathetic nerves by cardiovascular afferents. *Brain Research*, **290**, 281–287.

Elam, M., Thorén, P. and Svensson, T.H. (1986) Locus coerulus neurons and sympathetic nerves: activation by visceral afferents. *Brain Research*, **375**, 117–125.

Epstein, S.E., Stampfer, M. and Beiser, G.D. (1968) Role of the capacitance and resistance vessels in vasovagal syncope. *Circulation*, **37**, 524–533.

Fagius, J. and Karhuvaara, S. (1989) Sympathetic activity and blood pressure increase with bladder distention in humans. *Hypertension*, **14**, 511–517.

Felder, R.B., Heesch, C.M. and Thames, M.D. (1983) Reflex modulation of carotid sinus baroreceptor activity in the dog. *American Journal of Physiology*, **244**, H437–H443.

Fitzpatrick, A., Williams, T., Jeffery, C., Lightman, S. and Sutton, R. (1990) Pathogenic role for argenine vasopressin (AVP) and catecholamines (EP and NEP) in vasovagal reaction. *Journal of American College of Cardiology* (abstract), **15**, 98A.

Glick, G. and Yu, P.N. (1963) Hemodynamic changes during spontaneous vasovagal reactions. *American Journal of Medicine*, **34**, 42–51.

Golf, S. (1977) Swallowing syncope. *Acta Medica Scandinavica*, **201**, 585–586.

Grant, R.T. and Reeve, E.B. (1941) Clinical observations on air-raid casualties. *British Medical Journal*, ii, 293–297 and 329–332.

Greenfield, A.D.M. (1951) An emotional faint. *Lancet*, i, 1302–1303.

Greenfield, A.D.M. (1966) Survey of the evidence for active neurogenic vasodilatation in man. *Federation Proceedings*, **25**, 1607–1610.

Guyton, A.C. (1991) *Textbook of Medical Physiology*, 8th edn, pp. 263–271. Philadelphia: W.B. Saunders.

Hart, G., Oldershaw, P.J., Cull, R.E., Humphrey, P. and Ward, D. (1982) Syncope caused by a cough-induced complete atrioventricular block. *PACE*, **5**, 564–566.

Hasser, E.M., Undesser, K.P. and Bishop, V.S. (1987) Interaction of vasopressin with area postrema during volume expansion. *American Journal of Physiology*, **253**, R605–R610.

Henry, J.P. (1950) Studies of the physiology of negative acceleration. An approach to the problem of protection. *Air Force Technical Report H 5953*, Oct 1950, USAF Air Material Command WPAF Base, Ohio.

Hill, L. (1895) The influence of the force of gravity on the circulation of the blood. *Journal of Physiology (London)*, **18**, 15–33.

Holmes, F.A., Glass, J.P., Ewer, M.S., Terjanian, T. and Tetu, B. (1987) Syncope and hypotension due to carcinoma of the breast metastatic to the carotid sinus. *American Journal of Medicine*, **82**, 1238–1242.

Hurst, J.W., Schlant, R.C., Rackley, C.E., Sonnenblick, E.H. and Wenger, N.K. (1990) *The Heart*, 7th ed., pp. 442–461. New York: Mac Graw-Hill.

James, A.H. (1958) Cardiac syncope after swallowing. *Lancet*, **1**, 771–772.

Jansen, R.P.S. (1978) Relative bradycardia: a sign of acute intraperitoneal bleeding. *Australian and New Zealand Journal of Obstetrics and Gynaecology*, **18**, 206–208.

Jarisch, A. (1949) Detektorstoffe des Bezoldeffektes. *Wiener Klinische Wochenschrift*, **61**, 551–555.

Jarisch, A. and Zotterman, Y. (1949) Depressor reflexes from the heart. *Acta Physiologica Scandinavica*, **16**, 31–51.

Kadish, A.H., Wechsler, L. and Marchlinski, F.E. (1986) Swallowing syncope: observations in the absence of conduction system or esophageal disease. *American Journal of Medicine*, **81**, 1098–1100.

Kapoor, W.N., Peterson, J.R. and Karpf, M. (1985) Micturition syncope. *Journal of the American Medical Association*, **253**, 796–798.

Kapoor, W.N., Peterson, J.R. and Karpf, M. (1986) Defecation syncope. A symptom with multiples etiologies. *Archives of Internal Medicine*, **146**, 2377–2379.

Kenny, R.A., Lyon, C.C., Ingram, A.M., Bayliss, J., Lightman, S.L. and Sutton, R. (1987) Enhanced vagal activity and normal arginine vasopressin response in carotid sinus syndrome: implications for a central abnormality in carotid sinus hypersensitivity. *Cardiovascular Research*, **21**, 545–550.

Kerr, M.G. (1965) The mechanical effects of the gravid uterus in late pregnancy. *J. Obstet. Gynecol. Br. Comm.*, **72**, 513.

Kerr, A. and Eich, R.H. (1961) Cerebral concussion as a cause of syncope. *Archives of Internal Medicine*, **108**, 248–252.

Kollef, M.H. and Schachter, D.T. (1990) Defecation syncope caused by pulmonary embolism. *Annals of Internal Medicine*, **113**, 86.

Lagerlund, T.D., Harper, C.M. Jr., Sharbrough, F.W., Westmoreland, B.F. and Dale, A.J.D. (1988) Annals of electroencephalographic study of glossopharyngeal neuralgia with syncope. *Archives of Neurology*, **45**, 472–475.

Landgren, S. (1952) On the excitation mechanism of the carotid baroreceptors. *Acta Physiologica Scandinavica*, **26**, 1–34.

Lewis, T. (1932) Vasovagal syncope and the carotid sinus mechanism with comments on Gower's and Nothnagel's syndrome. *British Medical Journal*, **i**, 873–876.

Levine, B.D. (1991) Regulation of blood volume and cardiac filling in endurance athletes - the Frank-Starling mechanism as a determinant of orthostatic tolerance. *Medicine and Science in Sports and Exercise*, **24**, (in press).

Levine, B.D., Buckey, J.C., Fritsch, J.M., Yancy, C.W. Jr., Watenpaugh, D.E., Snell, P.G., *et al.* (1991a) Physical fitness and cardiovascular regulation: mechanisms of orthostatic intolerance. *Journal of Applied Physiology*, **70**, 112–122.

Levine, B.D., Lane, L.D., Buckey, J.C., Friedman, D.B., and Blomqvist, C.G. (1991b) Left ventricular pressure-volume and Frank-Starling relations in endurance athletes. Implications for orthostatic tolerance and exercise performance. *Circulation*, **84**, 1016–1023.

Lichstein, E. and Chadda, K.D. (1972) Atrioventricular block produced by swallowing. With documentation by His bundle recordings. *American Journal of Cardiology*, **29**, 561–563.

Lieshout, J.J. van, Wieling, W., Karemaker, J.M. and Eckberg, D.L. (1991) The vasovagal response. *Clinical Sciences*, **81**, 575–586.

Lipsitz, L.A., Nyquist, R.P. Jr., Wei, J.Y. and Rowe, J.W. (1983) Postprandial reduction in blood pressure in the elderly. *New England Journal of Medicine*, **309**, 81–83.

Lipsitz, L.A., Pluchino, F.C., Wei, J.Y., Minaker, K.L. and Rowe, J.W. (1986) Cardiovascular and norepinephrine responses after meal consumption in elderly (older than 75 years) persons with postprandial hypotension and syncope. *American Journal of Cardiology*, **58**, 810–815.

Ludbrook, J. (1987) Vasodilator responses to acute blood loss. *Australian and New Zealand Journal of Surgery*, **57**, 511–513.

Lundgren, O., Lundwall, J. and Mellander, S. (1964) Range of sympathetic discharge and reflex vascular adjustments in skeletal muscle during hemorrhagic hypotension. *Acta Physiologica Scandinavica*, **62**, 380–390.

Lyle, C.B. Jr., Monroe, J.T. Jr. and Flinn, D.E. (1961) Micturition syncope: report of 24 cases. *New England Journal of Medicine*, **26**, 982–986.

Manolis, A.S., Linzer, M., Salem, D. and Estes, M.N.A. (1990) Syncope: current diagnostic evaluation and management. *Annals of Internal Medicine*, **112**, 850–863.

Mark, A.L. (1983) The Bezold-Jarish reflex revisited: clinical implications of inhibitory reflexes originating in the heart. *Journal of American College of Cardiology*, **1**, 90–102.

Mark A.L. and Mancia, G. (1983) Cardiopulmonary baroreflexes in humans. In *Handbook of Physiology*, section 2: *The Cardiovascular System*, vol. 3, edited by J.T. Shepherd and F.M. Abboud. pp. 795–813. Bethesda MD: American Physiological Society.

Mark, A.L., Kioschos, J.M., Abboud, F.M., Heistad, D.D. and Schmid, P.G. (1973) Abnormal vascular responses to exercise in patients with aortic stenosis. *Journal of Clinical Investigation*, **52**, 1138–1146.

McIntosh, H.D., Estes, E.H. and Warren, J.V. (1956) The mechanism of cough syncope. *American Heart Journal*, **52**, 70–82.

McMichael, J. (1944) Clinical aspects of shock. *Journal of the American Medical Association*, **124**, 275–281.

Milstein, S., Buetikofer, J., Lesser, J., Goldenberg, I.F., Benditt, D.G., Gornick, C., *et al.* (1989) Cardiac asystole: a manifestation of neurally mediated hypotension bradycardia. *Journal of American College of Cardiology*, **14**, 1626–1632.

Morgan, D.A., Thorén, P., Wilczynski, E.A., Victor, R.G. and Mark, A.L. (1988) Serotonergic mechanisms mediate renal sympathoinhibition during severe hemorrhage in rats. *American Journal of Physiology*, **255**, H496–H502.

Morita, H. and Vatner, S. (1985) Effects of hemorrhage on renal nerve activity in conscious dogs. *Circulation Research*, **57**, 788–793.

Morita, H., Manders, W.T. and Vatner, S.F. (1985) Opiate receptor mediated decrease in renal activity during hemorrhage in conscious rabbits. *Circulation*, **72**, Suppl. 3, 244 (abstract).

Morita, H., Nishida, Y., Motochigawa, H., Uemura, N., Hosomi, H. and Vatner, S.F. (1988) Opiate receptor mediated decrease in renal nerve activity during hypotensive hemorrhage in conscious rabbits. *Circulation Research*, **63**, 165–172.

Munch, P.A., Thorén, P.N. and Brown, A.M. (1987) Dual effects of norepinephrine and mechanisms of baroreceptors stimulation. *Circulation Research*, **61**, 409–419.

Murray, A., Greene, M.D., Boltax, A.J. and Ulberg, R.J. (1961) Cardiovascular dynamics of vasovagal reactions in man. *Circulation Research*, **9**, 12–17.

Öberg, B. and Thorén, P. (1970) Increased activity in vagal cardiac afferents correlated to the appearance of reflex bradycardia during severe hemorrhage in cats. *Acta Physiologica Scandinavica*, **80**, 22A–23A.

Öberg, B. and White, S. (1970a) Circulatory effects of interruption and stimulation of cardiac vagal afferents. *Acta Physiologica Scandinavica*, **80**, 383–394.

Öberg, B. and White, S. (1970b) The role for vagal cardiac nerves and arterials baroreceptors in the circulatory adjustment to hemorrhage in the cat. *Acta Physiologica Scandinavica*, **80**, 395–403.

Öberg, B. and Thorén, P. (1972a) Increased activity in left ventricular receptors during hemorrhage or occlusion of caval veins in the cat. A possible cause of the vasovagal reaction. *Acta Physiologica Scandinavica*, **85**, 164–173.

Öberg, B. and Thorén, P. (1972b) Studies on left ventricular receptors, signalling in non-medullated vagal afferents. *Acta Physiologica Scandinavica*, **85**, 145–163.

Palmer, E.D. (1976) The abnormal upper gastrointestinal vasovagal reflexes that affect the heart. *American Journal of Gastroenterology*, **66**, 513–522.

Palmer, J.F. (1837) *Works of John Hunter*. Volume 3. London.

Perez-Gomez, F., Martin deDios, R., Rey, J. and Aguado, A.G. (1979) Prinzmetal's angina: reflex cardiovascular response during episode of pain. *British Heart Journal*, **42**, 81–87.

Peuler, J.D., Schmid, P.G., Morgan, D.A. and Mark, A.L. (1990a) Inhibition of renal sympathetic activity and heart rate by vasopressin in hemorrhaged diabetes insipidus rats. *American Journal of Physiology*, **258**, H706–H712.

Peuler, J.D., Edwards, G.L., Schmid, P.G. and Johnson, A.K. (1990b) Area postrema and differential reflex effects of vasopressin and phenylephrine in rats. *American Journal of Physiology*, **58**, H1255–H1259.

Proudfit, W.L. and Forteza, M.E. (1959) Micturition syncope. *New England Journal of Medicine*, **260**, 328–331.

Rea, R.F. (1989) Neurally mediated hypotension and bradycardia: which nerves? How mediated? *Journal of American College of Cardiology*, **14**, 1633–1634.

Reddy, K., Hobson, D.E., Gomori, A. and Sutherland, G.R. (1987) Painless glossopharyngeal neuralgia with syncope: a case report and literature review. *Neurosurgery*, **21**, 916–919.

Roddie, I.C. (1977) Human responses to emotional faint. *Irish Journal of Medical Sciences*, **146**, 395–417.

Rorsgaard, S. and Secher, N.H. (1986) Slowing of the heart during hypotension in major abdominal surgery. *Acta Anaesthesiologica Scandinavica*, **30**, 507–510.

Rothe, C.F., Schwendenmann, F.C. and Selkurt, E.E. (1963) Neurogenic control of skeletal muscle vascular resistance in hemorrhagic shock. *American Journal of Physiology*, **204**, 925–932.

Saito, D., Matsuno, S., Matsushita, K., Takeda, H., Hyodo, T. and Haraoka, S. (1982) Cough syncope due to atrioventricular conduction block. *Japanese Heart*, **23**, 1015–1020.

Sander-Jensen, K., Secher, N.H., Bie, P., Warberg, J. and Schwartz, T.W. (1986) Vagal slowing of the heart during hemorrhage: observations from 20 consecutive hypotensive patients. *British Medical Journal*, **292**, 364–366.

Sanders, J.S., Mark, A.L. and Ferguson, D.W. (1989) Evidence for cholinergically mediated vasodilation at the beginning of isometric exercise in the humans. *Circulation*, **79**, 815–824.

Schadt, J.C. and Gaddis, R.R. (1986) Cardiovascular responses to hemorrhage and naloxone in conscious barodenervated rabbits. *American Journal of Physiology*, **251**, R909–R915.

Schadt, J.C. Mc Known, M.D., Mc Known, D.P. and Franklin, D. (1984) Hemodynamic effects of hemorrhage and subsequent naloxone treatment in conscious rabbits. *American Journal of Physiology*, **247**, R497–R508.

Scherrer, U., Vissing, S.F., Morgan, B.J., Hanson, P. and Victor, R.G. (1990) Vasovagal syncope after infusion of a vasodilator in a heart-transplant recipient. *New England Journal of Medicine*, **322**, 602–604.

Secher, N.H. and Bie, P. (1985) Bradycardia during reversible hemorrhagic shock - a forgotten observation? *Clinical Physiology*, **5**, 315–323.

Secher, N.H., Jensen, K.S. and Werner, C. (1984a) Bradycardia during hypovolemic shock, clinical observations of heart rate and blood pressure. *Acta Physiologica Scandinavica*, **121**, 49A (abstract).

Secher, N.H., Sander-Jansen, K., Werner, C. and Warberg, J. (1984b) Bradycardia during severe but reversible hypovolemic shock in man. *Circulatory Shock*, **14**, 267–274.

Shakespeare, C.F., Crowther, A., Cooper, I.C., Katritsis, D., Coltart, J.D. and Webb-People, M.M. (1991) Autonomic reflexes initiated by coronary balloon inflation in man. *Circulation*, **84**, 268 (abstract).

Shalev, Y., Gal, R., Tchou, P.J., Anderson, A.A., Avitall, B., Akhtar, M., *et al.* (1991) Echocardiographic demonstration of decreased left ventricular dimensions and vigorous myocardial contraction during syncope induced by head-up tilt. *Journal of American College of Cardiology*, **18**, 746–751.

Sharpey-Schafer, E.P. (1953) The mechanism of syncope after coughing. *British Medical Journal*, **2**, 860–863.

Sharpey-Schafer, E.P. (1956) Emergencies in general practice. Syncope. *British Medical Journal*, **1**, 506–509.

Skoog, P., Mansson, J. and Thorén, P. (1985) Changes in renal sympathetic outflow during hypotensive haemorrhage in rats. *Acta Physiologica Scandinavica*, **125**, 655–660.

Smith, J.J. (1990) Circulatory response to the upright posture. *Critical Reviews in Clinical Laboratory Sciences*, pp. 30–33. Boca Raton: CRC Press.

Smith, M.L., Carlson, M.D., Sheehan, H.M. and Thames, M.D. (1991) Naloxone does not prevent vasovagal syncope during simulated orthostasis in humans. *Physiologist*, **34**, A27.2.

Strauss, M.J., Longstreth, W.T. and Thiele, B.L. (1984) Atypical cough syncope. *JAMA*, **251**, 1731.

Stevens, P.M. (1966) Cardiovascular dynamics during orthostasis and the influence of intravascular instrumentation. *American Journal of Cardiology*, **17**, 211–218.

Sündlof, G. and Wallin, B.G. (1978) Human muscle nerve sympathetic activity at rest. Relationship to blood pressure and age. *Journal of Physiology (London)*, **274**, 621–637.

Thames, M.D., Klopfenstein, H.S., Abboud, F.M., Mark, A.L. and Walker, J.L. (1978) Preferential distribution of inhibitory cardiac receptors with vagal afferents to the infero-posterior wall of the left ventricle activated during coronary occlusion in the dog. *Circulation Research*, **43**, 512–519.

Thorén, P.N. (1976) Activation of left ventricular receptors with nonmedulated vagal afferent fibers during occlusion of a coronary artery in the cat. *American Journal of Cardiology*, **37**, 1046–1051.

Thorén, P.N. (1977) Characteristics of left ventricular receptors with nonmedulated vagal afferents in cats. *Circulation Research*, **40**, 415–421.

Thorén, P. (1979) Role of cardiac vagal C-fibers in cardiovascular control. *Review of Physiology, Biochemistry and Pharmacology*, **86**, 1–94.

Thorén, P., Noresson, E. and Ricksten, S-E. (1979) Cardiac receptors with vagal non-medullated vagal afferents in the rat. *Acta Physiologica Scandinavica*, **105**, 295–303.

Tomlinson, I.W. and Fox, K.M. (1975) Carcinoma of the esophagus with "swallow" syncope. *British Medical Journal*, **2**, 315–316.

Tulchinsky, M. and Krasnow, S.H. (1988) Carotid sinus syndrome associated with an occult primary nasopharyngeal carcinoma. *Archives of Internal Medicine*, **148**, 1217–1219.

Undesser, K.P., Hasser, E.M., Haywood, J.R., Johnson, A.K. and Bishop, V.S. (1985) Interactions of vasopressin with the area postrema in arterial baroreflex function in conscious rabbits. *Circulation Research*, **56**, 410–417.

Vallbo, A.B., Hagbarth, K-E., Torebjork, H.E. and Wallin, B.G. (1979) Somatosensory, proprioceptive, and sympathetic activity in humans peripheral nerves. *Physiological Reviews*, **59**, 919–957.

Vatner, S.F. (1974) Effects of hemorrhage on regional blood flow distribution in dogs and primates. *Journal of Clinical Investigations*, **54**, 225–235.

Victor, R.G., Thorén, P., Morgan, D.A. and Mark, A.L. (1989) Differential control of adrenal and renal sympathetic nerve activity during hemorrhagic hypotension in rats. *Circulation Research*, **64**, 686–694.

Vissing, S.F., Scherrer, U. and Victor, R.G. (1989) Relation between sympathetic outflow and vascular resistance in the calf during perturbations in central venous pressure. Evidence for cardiopulmonary afferent regulation of calf vascular resistance in humans. *Circulation Research*, **65**, 1710–1717.

Walter, J.L., Thames, M.D., Abboud, F.M., Mark, A.L. and Klopfenstein, H.S. (1978) Preferential distribution of inhibitory cardiac receptors in the left ventricle of the dog. *American Journal of Physiology*, **235**, H188–H192.

Wallin, B.G. and Sündlof, G. (1982) Sympathetic outflow to muscles during vasovagal syncope. *Journal of the Autonomic Nervous System*, **6**, 287–291.

Walter, P.F., Crawley, I.S. and Dorney, E.R. (1978) Carotid sinus hypersensitivity and syncope: *American Journal of Cardiology*, **42**, 396–403.

Webb, S.W., Adgey, A.A.J. and Pantridge, J.F. (1972) Autonomic disturbances at onset of acute myocardial infarction. *British Medical Journal*, **3**, 89–92.

Wei, J.Y., Markis, J.E., Malagolg, M. and Braunwald, E. (1983) Cardiovascular reflexes stimulated by reperfusion of ischemic myocardium in acute myocardial infarction. *Circulation*, **67**, 796–801.

West, J.B. (1991) *Physiological Basis of Medical Practice*, 12th ed, pp. 318–319. Baltimore: Williams and Wilkins.

Wik, B. and Hillestad, L. (1975) Deglutition syncope. *British Medical Journal*, **3**, 747.

Williams, R.S. and Bashore, T.M. (1980) Paroxysmal hypotension associated with sympathetic withdrawal: A new disorder of autonomic vasomotor regulation. *Circulation*, **62**, 901–907.

Wilson, J.D., Braunwald, E., Isselbacher, K.J., Petersdorf, R.G., Martin, J.B., Fauci, A.S., *et al.* (1991) Shock. In *Harrison's Principles of Internal Medicine*, 12th edn, pp. 232–237. New York: Mac Graw-Hill.

Wyngaarden, J.B., Smith, L.H. and Bennet, J-C. (1991) Shock. In *Cecil Textbook of Medicine*, 19th edn, pp. 207–228. Philadelphia: W.B. Saunders.

Zee-Cheng, C.S. and Gibbs, H.R. (1986) Pure vasodepressor carotid sinus hypersensitivity. Unusual cause of recurrent syncope. *American Journal of Medicine*, **81**, 1095–1097.

14 Autonomic Responses to Microgravity and Bedrest: Dysfunction or Adaptation

Victor A. Convertino* and Rose Marie Robertson

From the Armstrong Laboratory, Brooks Air Force Base; and the Department of Medicine, Division of Cardiology, Vanderbilt University Medical Center

Experiments conducted during spaceflight and its groundbase analog, bedrest, have provided consistent data that demonstrate that numerous changes in cardiovascular function occur as part of the physiological adaptation process to the microgravity environment. These adaptations include elevated heart rate and venous compliance, lowered blood volume, central venous pressure and stroke volume, and attenuated autonomic reflex functions. Although most of these adaptations are not functionally apparent during microgravity and bedrest exposure, they manifest themselves during the return to the upright posture as orthostatic hypotension and instability. Successful development and application of effective and efficient therapeutic measures will require knowledge of physiological mechanisms underlying cardiovascular adaptation to microgravity and bedrest which can be obtained only through controlled, parallel groundbased research to complement carefully designed flight experiments. Continued research will provide benefits for both space and clinical application as well as enhance our basic understanding of cardiovascular homeostasis in humans.

KEY WORDS: Orthostatic intolerance, blood pressure, blood volume, cardiac function, hormones, baroreflex function

INTRODUCTION

In the early days of the U.S. and Soviet space programs, there was no certainty that man could survive the rigors of space travel. Now, after more than 30 years of human spaceflight, it is difficult to recall the concern surrounding even the early flights of

* Victor A. Convertino, Ph.D., AL/AOCIY, 2507 Kennedy Circle, Brooks AFB, TX 78235-5117. Telephone: (210) 536-3242, Telefax; (210) 536-2208

canine and primate substitutes for man. While predicted by some, the fact that man could not only survive in space but function very effectively, was contrary to the beliefs of a number of respected senior scientists. The pendulum has now swung in the other direction. More than 250 men and women have now flown in space. It has become commonplace to see and hear men and women living and working in a weightless (or more properly, microgravity) environment. With this familiarity, it has become attractive for some to assume that there are few significant medical problems associated with even long-term space travel. Certainly the inability of early cosmonauts and astronauts to ambulate was noted after the initial long flights, but this problem has received less public attention in the Shuttle era of short flights, after which the spacecraft is flown back to earth, appearing much like a commercial aircraft.

From the point of view of those engaging in spaceflight and those responsible for their care, the alterations in cardiovascular homeostasis that are seen even after brief exposure to microgravity are of great significance. Important orthostatic hypotension and excessive tachycardia with standing were consistently exhibited by early crewmembers (Blomqvist, 1983a; Sandler, 1988), and understanding and countering these concomitants of space flight has become a major focus of the manned space program.

Because of the common perception that a reduction in physiological capacity with exposure to microgravity is a form of 'deconditioning', it is relevant to ask whether changes in cardiovascular control that occur with space flight are truly dysfunctional or simply adaptive. During space flight itself, astronauts function well after an initial period of adjustment. As we will detail below, the cardiovascular system accustoms itself to the new environment, and crewmembers are able to undertake physically strenuous as well as intellectually demanding work. Only on return to the terrestrial gravity (1G) environment (or to the 1+G situation of reentry) do problems with cardiovascular homeostasis arise. Thus, one might well argue that man has in fact demonstrated an ability to adapt appropriately to two very different sets of conditions, and that only the speed of adaptation is problematic. In this chapter, our focus will be on the autonomic aspects of this cardiovascular adaptation, and we will utilize data primarily from the Soviet and U.S. space experience and related ground-based studies.

The majority of the U.S. physiologic data was obtained from the Apollo program, including its use of the first U.S. space station, Skylab, during three missions lasting 28, 59 and 84 days. Some data from shorter (2–10 day) missions utilizing the Space Shuttle are also included. The Soviet experience comes primarily from the Soyuz program, including its use of the Salyut space station, with some missions as long as 237 days. Some data from even longer (366 day) missions in the newer Mir space station are included. Much of both the U.S. and Soviet data are available only in technical documents, and the number of subjects studied has been small; on many flights, the missions were not primarily concerned with collecting biomedical data, and even when they were, not all crewmembers participated. An additional complicating factor is the fact that both individual astronauts and medical personnel responsible for their health have gradually initiated therapeutic approaches or "countermeasures" to avoid some of the troubling consequences of spaceflight noted during early missions. Thus, many questions remain that will only be answered by carefully controlled ground-based experiments or by missions such as the recent Spacelab Life Sciences 1, which approached these important questions in the most controlled manner during spaceflight to date. However, as even current countermeasures are not perfectly effective, there are useful data to be reviewed.

MICROGRAVITY AND BEDREST

As investigators began to address the altered cardiovascular responses that accompanied microgravity, they were limited by the fact that a comprehensive assessment of human physiology could not be accomplished in space. It became clear that ground-based models designed to induce quantitative as well as qualitative physiologic changes would be very useful. While water immersion was used at first, maintaining subjects in this situation for any length of time posed considerable logistical problems. Investigators noted that post-flight hemodynamic responses, referred to as "cardiovascular deconditioning", bore considerable similarity to those seen after prolonged bedrest. While not precisely analogous, bedrest does remove the head-to-foot gravitational gradient of the normal 1G environment, and thus mimics at least in part the physiological adaptation to spaceflight (Table 14.1). In addition, bedrest responses are of interest in and of themselves for their relevance to clinical issues of confinement. The difficulties inherent in rehabilitating otherwise healthy patients who have been forced to remain in bed by orthopedic problems, for example, made it clear that the bedrest-related changes in cardiovascular regulation merited further study.

While the first bedrest study was conducted as early as 1921 (Campbell and Webster, 1921), several studies were conducted during the 1960s launching an increased interest that has culminated in the completion of more than 160 clinical studies of the physiological responses to bedrest. These have varied in length from 2 to more than 200 days, with a minority employing a degree of head-down tilt, ranging from $-2°$ to $-15°$. Astronauts universally complained of a feeling of fullness in the head on the first day in space, and head-down tilt (specifically $-6°$) was found to better approximate this sensation and the resulting cardiovascular adaptations (Kakurin *et al.*, 1976). While there remains some controversy, horizontal and head-down tilt induce many of the same physiological effects, so both will be reported here.

TABLE 14.1

Comparison of changes reported during rest and exercise following spaceflight and 6° head-down bedrest (HDBR).

Physiological Variable	References	Microgravity Analog	Days of Exposure	N	Condition	% Δ
Body Weight	Thornton, Hoffler and Rummel (1977c)	Spaceflight	28	3	Rest	3%↓
Body Weight	Convertino et al. (1989)	6° HDBR	30	8	Rest	4 % ↓
Venous Pressure	Kirsch et al. (1984)	Spaceflight	7	2	Rest	58% ↓
Venous Pressure	Convertino et al. (1994)	6° HDBR	7	11	Rest	32% ↓
Baroreflex	Fritsch et al. (1992)	Spaceflight	5	12	Rest	40% ↓
Baroreflex	Convertino et al. (1990b)	6° HDBR	12	11	Rest	31% ↓
Plasma Volume	Fischer et al. (1967)	Spaceflight	4	2	Rest	9% ↓
Plasma Volume	Convertino et al. (1990b)	6° HDBR	3	11	Rest	12% ↓
Leg Volume	Thornton, Hoffler and Rummel (1977c)	Spaceflight	28	3	Rest	10% ↓
Leg Volume	Convertino et al. (1989)	6° HDBR	30	8	Rest	10% ↓
Oxygen Uptake	Kakurin et al. (1976)	Spaceflight	5	7	700 kgm/min	4% ↑
Oxygen Uptake	Convertino et al. (1981)	6° HDBR	7	5	600 kgm/min	5% ↑
Heart Rate	Kakurin et al. (1976)	Spaceflight	5	7	700 kgm/min	13% ↑
Heart Rate	Convertino et al. (1981)	6° HDBR	7	5	600 kgm/min	12% ↑
Oxygen Pulse	Kakurin et al. (1976)	Spaceflight	5	7	700 kgm/min	9% ↓
Oxygen Pulse	Convertino et al. (1981)	6° HDBR	7	5	600 kgm/min	8% ↓
Stroke Volume	Atkov et al. (1987)	Spaceflight	237	2	765 kgm/min	31% ↓
Stroke Volume	Hung et al. (1983)	0° HDBR	10	12	835 kgm/min	28% ↓
Ejection Fraction	Atkov et al. (1987)	Spaceflight	237	2	765 kgm/min	13% ↑
Ejection Fraction	Hung et al. (1983)	0° HDBR	10	12	825 kgmimin	20% ↑
Heart Rate	Bungo et al. (1985)	Spaceflight	7	9	Standing	47% ↑
Heart Rate	Conventino et al. (1990b)	6° HDBR	30	10	Standing	55% ↑

ORTHOSTATIC INTOLERANCE

MICROGRAVITY

Some degree of orthostatic instability has been seen in many individuals who have travelled in space. Early crewmembers were often carried away from their spacecraft on stretchers and gave no stand-up press conferences for days post-flight. It was clear to both U.S. and Soviet medical personnel that cardiovascular function was compromised on return to Earth, and that this compromise was most prominent in the upright posture. While blood pressure was usually maintained when determined in the upright position, it was accompanied by higher than normal heart rates; in some subjects this compensatory mechanism was not fully effective and syncope could occur during a "stand test". Significant "cardiovascular deconditioning" has been described in 30% of all Shuttle passengers and crew (Bungo and Johnson, 1983), and the Mercury, Gemini, Apollo and Skylab missions all produced orthostatic instability of varying degrees in most subjects (Hoffler, 1977; Blomqvist, 1983a). The Soviet experience was similar (Gazenko, Genin and Yegorov, 1981), although not publicly reported during the early phases of their program. Descriptions of 18 cosmonauts who worked aboard Mir between 1986 and 1990 clarified the fact that orthostatic instability was common post-flight, and was in fact so severe that cosmonauts wore anti-gravity or "G" suits (positive-pressure garments much like MAST trousers) for the first 2–3 days after landing (Grigoriev et al., 1991). The duration of symptoms was not dependent on the duration of exposure to microgravity, but all flights were relatively long, some extending for more than one year.

Decreased orthostatic stability, whether measured by simple changes in blood pressure and heart rate while standing quietly or by decreased tolerance to lower body negative pressure (LBNP) (Blomqvist, 1983a), is of particular interest to the U.S. space effort because of the space vehicle used. With the Space Shuttle, in which the final hours and minutes of landing are controlled by the pilot, there are actually increased head-to-foot G forces during the landing procedure. The possibility that the pilot might be significantly less able to tolerate these forces than prior to flight and might lose consciousness at this critical time is of great concern, and has sparked a serious interest in understanding and counteracting this problem. Pilots now wear G suits during landing and a number of other countermeasures are employed empirically. The rationale for their use and the experimental data supporting their choice will be reviewed below.

BEDREST

In general, bedrest studies have also demonstrated a reduction in tolerance of the upright posture after healthy subjects have been confined to bedrest. Baisch and colleagues (1984a) reported a 37% reduction in orthostatic tolerance after subjects had been exposed to 7 days of 6° head-down tilt. This reduced orthostatic stability after bedrest exposure, usually manifest as increased orthostatic tachycardia, has been corroborated in a number of similar studies (Chestukhin et al., 1979; Goldwater et al., 1980; Guell et al., 1980; Gaffney et al., 1985; Convertino et al., 1990b). In fact, the 40% incidence of syncope reported in one bedrest study (Convertino et al., 1990b) was remarkably similar to that reported in astronauts after spaceflight (Bungo et al., 1983). Goldwater et al. (1980) compared horizontal bedrest with 6° head-down tilt for 7 days and found that both produced significant orthostatic intolerance which was still evident on the 10th day of re-ambulation. While head-down tilt did accentuate the intolerance to LBNP, hemodynamic responses were sufficiently similar to allow the consideration of both types of studies. Guell and co-workers (1980) used 85° head-up tilt (HUT) rather than standing or LBNP as a post-bedrest trial and produced near-syncope in two of their four volunteers after 7 days at −4° bedrest. Generally, there has been great similarity in the hemodynamic responses to orthostasis in studies in which subjects have been challenged with passive standing, 70° HUT, or LBNP at 50–100 mmHg after bedrest confinement. Only a very few studies have reported no effect of bedrest, and these have tested "acceleration tolerance time," an aeronautical surrogate for the measure of orthostatic tolerance (Hunt *et al.*, 1984; Hollister *et al.*, 1986). In general, however, the ability to cope with an orthostatic challenge following prolonged confinement to bedrest is dramatically reduced.

While the duration of bedrest has varied considerably, with some lasting 4–6 months, as little as 20 hours of 5° head-down tilt led to essentially as much orthostatic tachycardia as 2–3 weeks of bedrest (Gaffney *et al.*, 1985). Clearly, the time course of the physiologic changes with both bedrest and microgravity is of interest, especially in view of the fact that some of the changes seen with spaceflight seem either to plateau or even to return to control levels over long-duration flights. If there was a more complete understanding of the detailed time course of cardiovascular adaptations induced by exposure to microgravity, physiologic mechanisms could be identified and appropriate countermeasures could be better designed. We will thus review the available data describing alterations in heart rate, arterial pressure, venous compliance, blood volume, cardiac function and neurohumoral factors, including baroreflex function, before considering the likely cause of "cardiovascular adaptation" in these microgravity environments.

HEART RATE

MICROGRAVITY

In general, the reported data from both American and Soviet experiences suggest that resting heart rate is altered only slightly during spaceflight. Carefully recorded electrocardiographic data from Skylab demonstrated modest elevations in the heart rate at rest (Smith *et al.*, 1976), but no change or a slight decrease were described in the reports from several Salyut missions, both at 23 days (Degtyarev *et al.*, 1974) and at 63 days (Degtyarev *et al.*, 1978b). Cosmonauts working in the Mir space station were also reported to have elevated heart rates during flight (Grigoriev *et al.*, 1991), and the resting heart rate post-flight remained elevated for at least 3 days after a 30-day Salyut mission. Following spaceflight, it seems most likely that the elevated heart rate is an appropriate response to a reduced stroke volume in an attempt to maintain cardiac output and arterial pressure.

BEDREST

At first glance, bedrest data regarding heart rate seems to be at variance with the determinations made in microgravity, as slower resting heart rates have been reported by a number of investigators (Taylor *et al.*, 1945; Baisch *et al.*, 1984a;

Bystrov, Zhernavkov and Savilov, 1986; Convertino *et al.*, 1990b). However, the slow rates have been seen in the first 24 hours of horizontal and head-down bedrest, and evaluations after this time period have shown a gradual return to baseline levels over the first week (Baisch *et al.*, 1984a; Bystrov, Zhernavkov and Savilov, 1986). Examined after the first week, an increase of 10–30 beats per minute is often seen between 10 and 120 days (Taylor *et al.*, 1945); some variability in the rate of return to normal and beyond likely accounts for the studies concluding only that bedrest elevates the resting heart rate (Birkhead, Haupt and Meyers, 1963; Miller, Johnson and Lamb, 1964; Beregovkin *et al.*, 1969; Melada *et al.*, 1975; Georgiyevskiy and Mikhaylov, 1978; Convertino *et al.*, 1990b).

ARTERIAL AND VENOUS PRESSURES

MICROGRAVITY

Arterial blood pressure at rest has not been altered importantly by the microgravity environment, with a rise from $113\pm7/73\pm3$ to $122\pm4/80\pm2$ observed during the 23-day Salyut mission (Degtyarev *et al.*, 1974), and similar findings during 16, 18, 49 and 63 day missions (Gurovskiy *et al.*, 1975; Vorobyev *et al.*, 1976; Rudnyy *et al.*, 1977). More recent Mir data (Grigoriev *et al.*, 1991) demonstrated no significant change in arterial blood pressure, although the diastolic pressure tended to decrease slightly and the pulse pressure to widen. The constancy of resting baseline arterial blood pressure in microgravity compared to terrestrial gravity supports the concept that changes in cardiovascular function associated with spaceflight represent appropriate homeostatic adaptations rather than dysfunctions.

Redistribution of body fluids toward the head and chest cavity during microgravity led many investigators to the reasonable expectation that central venous pressure would be increased during spaceflight. Surprisingly, Kirsch and co-workers (1984) observed nearly 60% reduction in CVP during 7 days of spaceflight. Although the use of a peripheral rather than central venous catheter raised some challenge to these findings, preliminary data from the recent SLS-1 mission confirms the reduced venous pressure in an astronaut instrumented with a central venous line. The reason for reduced CVP in microgravity is not clear, but may reflect an appropriate adaptation in as low gravity environment where the redistribution of fluid to the upper body may not require as large a filling pressure as that needed on earth.

BEDREST

Arterial pressure at bedrest for up to 6 months has generally remained unchanged, whether with head-down (Blomqvist *et al.*, 1980; Convertino *et al.*, 1990b) or horizontal (Sandler, 1988) bedrest (Figure 14.1A). While elevated blood pressure has occasionally been reported, it has generally been minor and transient. Central venous pressure, measured with the same technique used on astronauts (Kirsch *et al.*, 1984), is reduced during bedrest (Figure 14.1B). The CVP tends to parallel the

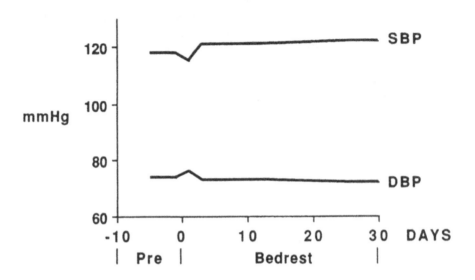

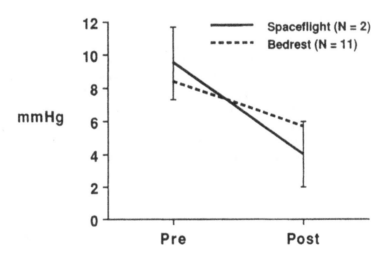

FIGURE 14.1   Mean systolic (SBP) and diastolic (DBP) arterial blood pressures before and during 30 days of bedrest (top panel) and mean central venous pressure before and after spaceflight (solid line) and bedrest (broken line) (bottom panel). Modified from Convertino *et al.* (1990b, 1994).

changes in vascular volume, with an initial increase followed by a fall to or below normal baseline levels (Katkov *et al.*, 1982; Sandler, 1988). The response of both arterial and venous blood pressures to bedrest is virtually identical to those reported in actual microgravity, supporting the argument that this analog provides a groundbased model to study cardiovascular adaptations to actual spaceflight.

CARDIAC FUNCTION

MICROGRAVITY

As mentioned above, man is clearly able to function well in space, engaging in challenging and difficult activities, both mentally and physically. Overall work performance, measured either as the ability to perform vigorous arm exercise (Butusov *et al.*, 1970) or leg exercise (Gazenko *et al.*, 1976; Degtyrev *et al.*, 1978a), was unimpaired during the Salyut missions. Early in these missions, in-flight heart rate, cardiac output and arterial pressure were higher during exercise than pre-flight, and these responses remained stable over 23 days (Degtyarev *et al.*, 1978a). Later in flight there was no change in the heart rate response to a standard exercise protocol (Vorobyev *et al.*, 1976) and with 3 carefully-studied Skylab astronauts, the same was true, with hemodynamic responses determined at 75% VO2max both before and after 84 days in space (Michel *et al.*, 1977). During a 140-day Salyut mission, the heart rate response to exercise rose only after day 41 at which time stroke volume had fallen. These changes tended to normalize the cardiac output during exercise, so that it remained essentially constant (Yegorov *et al.*, 1981). Thus, during space missions as long as 4–5 months, cardiac responses indicated no significant impairment of myocardial function.

Reduced intracardiac volume during and after spaceflight has been frequently demonstrated (Chekirda *et al.*, 1970; Blomqvist, 1983a; Atkov, 1985; Atkov, Bednenko and Fomina, 1987; Bungo *et al.*, 1987; Grigoriev *et al.*, 1991), with decreases in EDV from 8–50% over 96–175 day missions (Atkov, 1985). This decrease persisted for at least several days after a 30-day Salyut flight (Beregovkin *et al.*, 1976) and for as long as 14 days after flights in the U.S. program (Bungo *et al.*, 1987). The importance of this was demonstrated after Skylab 4, when the exercise SV was reduced by 24 ml with upright exercise, but only 5 ml supine (Michel *et al.*, 1977). Surprisingly, recent preliminary reports of data from Mir, while not presented in detail, suggested no change in intracardiac volumes over 7–8 months in flight. It is not clear why these results are at variance with previous data, unless the extensive countermeasures employed were exceptionally effective. In any event, the cause of lower cardiac volume is unclear but could represent something as simple as reduced filling or as serious as myocardial atrophy.

With space missions of duration longer than 5 months, there is evidence that cardiac output during physical work may be compromised. During the Mir missions, there was as much as a 15% fall in the cardiac output with exercise, primarily due to a decrease in stroke volume (Grigoriev *et al.*, 1991). This confirmed earlier Salyut 237-day data, which also documented a lower cardiac output during 125 and 175 watt exercises with echocardiographic studies showing decreased left ventricular volumes in diastole (Figure 14.2A) and systole, and a 30% decrease in stroke volume (Figure 14.2B). Even a higher heart rate during exercise could not compensate for the reduced stroke volume (Atkov, Bednenko and Fomina, 1987). One interpretation of the data from longer duration spaceflight is that the reduction in cardiac output represents a loss of cardiac function. However, evaluation of the Frank-Starling relationship and higher ejection fraction (Figure 14.2C) suggests

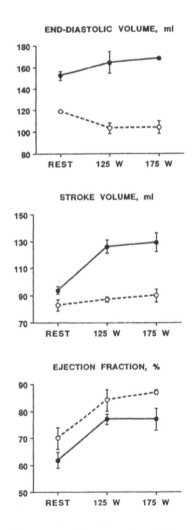

FIGURE 14.2   Mean (± SE) cardiac responses of two cosmonauts during rest and at 125 W and 175 W of exercise on a cycle ergometer before (closed circles and solid lines) and during (open circles and broken lines) a 237-day space mission. Modified from Atkov *et al.* (1987).

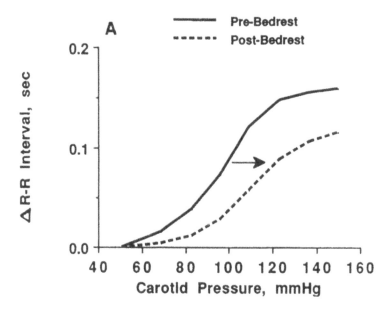

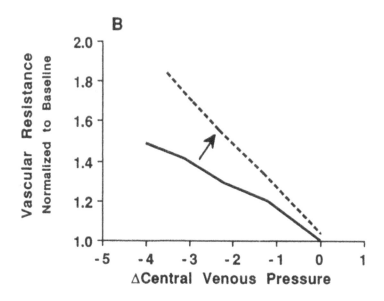

FIGURE 14.3   Mean stimulus-response relationships (N = 11) of the carotid-cardiac (top panel) and cardiopulmonary (bottom panel) baroreflexes before (solid lines) and after (broken lines) 6° head-down bedrest. Modified from Convertino *et al.* (1990b, 1994).

that cardiac function *per se* may not be adversely affected by spaceflight and that lower cardiac output simply reflects less myocardial filling due to hypovolemia (Atkov, Bednenko and Fomina, 1987).

BEDREST

Bedrest studies have confirmed data aquired in space. It was noted that EDV actually increased, as did SV, during the first hours of a 24-hour 5° head-down tilt study, but that SV had returned to baseline in 6 hours (Gaffney *et al.*, 1985; Bystrov, Zhernavkov and Savilov, 1986). A diuresis then follows and echocardiographic studies have shown a 6–12% fall in EDV over a period of several weeks (Saltin *et al.*, 1968; Sandler and Winter, 1978; Sandler, 1986; Sandler, Popp and Harrison, 1988b). Reduced EDV is associated with lower SV and cardiac output during exercise, even in the face of dramatically elevated heart rate (Saltin *et al.*, 1968; Hung *et al.*, 1983). However, increased ejection fraction during this exercise (Hung *et al.*, 1983) is consisient with spaceflight data in suggesting that the reduction in cardiac output reflects the effects of hypovolemia on cardiac filling without compromise of normal cardiac function.

VENOUS COMPLIANCE

MICROGRAVITY

As investigators began to address the issue of orthostatic instability following spaceflight, it was recognized that increased venous pooling with the upright posture might account for difficulty in maintaining the blood pressure while standing since LBNP tolerance was closely correlated with compliance (Klein, Wegmann and Kuklinski, 1977). In fact, both U.S. (Klein, Wegmann and Kuklinski, 1977; Thornton and Hoffler, 1977b) and Soviet (Yegorov and Itsekhovskiy, 1983; Thornton and Hoffler, 1977b) investigators found increased venous compliance in the legs early in spaceflight, as measured by occlusion plethysmography or by an increase in the fall in SV with LBNP (Grigoriev *et al.*, 1991). It was thought that this might be due to a number of factors, including a fluid shift of as much as 1–1.5 liters from the legs to the central compartment (Sandler, 1983), rendering the legs less fluid-replete, as well as a loss of muscle mass in the legs due to disuse in microgravity (Thornton and Hoffler, 1977).

BEDREST

Bedrest produces similar changes in lower extremity vascular compliance, and it has been postulated that the earliest changes are due to the central shift of fluid (Convertino *et al.*, 1989), estimated at 500–600 ml (Baisch and Klein, 1984b), with an additional later component added when the muscle compartment is reduced in mass (Convertino, Doerr and Stein, 1989b). This latter aspect can be prevented by electrically-stimulated muscle activity while bedrest continues, suggesting that muscle atrophy associated with inactivity can contribute to increased compliance.

BLOOD VOLUME

MICROGRAVITY

In addition to the increased venous compliance described above, volume home-ostasis is altered during exposure to microgravity. A fall in intravascular volume is observed during spaceflight (Convertino, 1990c), perhaps due in part to stimu-lation of pulmonary, atrial and carotid mechanoreceptors with a release of atrial natriuretic peptide (Graham and Zisfein, 1986) and a fall in ADH, renin and al-dosterone (Sandler, 1988a). As fluid intake may also be decreased by the nausea of the "space adaptation syndrome", there can be a rapid and significant negative fluid balance achieved. A significant rapid fall in plasma volume occurs (John-son, Kimzey and Driscoll, 1975; Burkovskaya et al., 1980; Dunn et al., 1984), with a more gradual reduction in red cell mass (Leach and Johnson, 1984; Johnson, Driscoll and LeBlanc, 1977; Burkuvskaya et al., 1980; Dunn et al., 1984), probably due to a fall in erythropoietin levels (Leach and Johnson, 1984). Reduction in total circulating blood volume is closely related to greater tachycardia during standing after spaceflight (Hoffler, 1977; Bungo, Charles and Johnson, 1985; Convertino, 1990), suggesting that vascular volume can be an important contributing factor to post-spaceflight orthostatic hypotension and instability.

BEDREST

Similar changes in intravascular volume are seen quite early in bedrest, with many studies having employed head-down tilt. Convertino and colleagues (1990b, 1992b) observed a 439-ml (12%) reduction in plasma volume after 3 days of 6° head-down bedrest which reached 15% by 30 days of exposure. The magnitude of change in this study was corroborated by reports of a 14% reduction in blood volume after 7 days of −6° tilt (Baisch et al., 1984a), and a 17% fall in plasma volume after 10 days (Blamick, Goldwater and Convertino, 1988). A 350 to 500 milliliter decrease in total blood volume can occur, however, as early as 24 hours (Blomqvist, Gaffney and Nixon, 1983b). A fall of 209 ml in plasma volume and of 533 ml in extracellular fluid was reported in 10 healthy volunteers after 48 hours (Fortney et al., 1991a). These reductions in circulating blood volume have been accompanied by a transient fall in ADH, plasma renin activity, and aldosterone (Leach, Altchuler and Cintron-Trevino, 1983a; Dallman et al., 1984; Convertino et al., 1990a). Longer bedrest studies (100–200 days) have resulted in as much as a 30% loss of plasma volume (Greenleaf, 1984). Like spaceflight, reductions in plasma and blood volume during bedrest are related to orthostatic hypotension and instability during re-ambulation, particularly when the duration of bedrest is relatively short (Hyatt and West, 1977; Sandler, 1986; Convertino et al., 1990b).

NEUROHUMORAL FACTORS

MICROGRAVITY

The fact that heart rate is changed very little and is often reduced with spaceflight suggests that there were not marked elevations of plasma catecholamines during microgravity exposure. This hypothesis is supported by the observations from a 25-day mission aboard the Mir station when measured plasma and urine levels of adrenaline, noradrenaline, and dopamine were similar to preflight levels (Kvetnansky et al., 1991) and that catecholamines were not altered from preflight levels in a 64-day Salyut mission (Tigranyan et al., 1977). However, after 84-days of spaceflight in the third U.S. Skylab mission, catecholamine excretion was reduced (Leach et al., 1983a). Following the return from microgravity to earth, catecholamine release is generally excessive compared to preflight levels (Kvetnansky et al., 1991; Leach and Rambaut, 1975). Together these data may indicate that the sympathetic nervous system has a slower time course of adaptation and that as the duration of microgravity exposure is increased, less sympathetic stimulation and discharge become apparent.

If plasma noradrenaline becomes low during extended spaceflight, the attenuated hormonal stimulation might lead to increased sensitivity of adrenoreceptors such that a given adrenergic discharge such as that observed postflight would be expected to lead to excessive tachycardia. If β_2 sensitivity were relatively greater than α_2 sensitivity, a normal vasoconstrictor response to upright posture might be transformed into a more vasodepressor and tachycardic response. Especially in the setting of hypovolemia, this adrenergic discharge might be ineffective in maintaining upright blood pressure and the baroreceptor-mediated increase in heart rate might be further enhanced. While this is largely speculative, it is an intriguing possibility that could have great practical importance for the development of possible treatments not yet tested. For example, since excess tachycardia is known to be a trigger for the Bezold-Jarisch response and vasovagal syncope, the use of a β-blocker with intrinsic sympathomimetic activity might diminish β-adrenoreceptor hypersensitivity. Combined with treatment to diminish volume losses, this might improve orthostatic stability at critical times of flight and landing. Clearly, important research remains to better define the sympathetic response to spaceflight and its effects on adrenoreceptor function.

BEDREST

Neurohumoral data from bedrest studies support the hypothesis addressed above. Davydova et al. (1986) found a decrease in noradrenaline early in 4.5° head-down tilt studies, although it had returned to control levels by the end of 120 days. This was supported by LaRochelle, Leach and Vernikos-Danellis (1982) who found a 5–27% decrease, and by Leach, Johnson and Suki (1983) who found a fall in noradrenaline especially during the early period of central fluid shifts. Investigators finding elevated levels have often not measured levels at this early time point (Krupina *et al.*, 1982; Natelson, De Roshia and Levin, 1982; Goldsmith, Francis and Cohn, 1985). Other investigators have made the point that head-down tilt reliably lowers noradrenaline well below seated levels (Guell *et al.*, 1986; Gharib *et al.*, 1988), and the early hours of spaceflight, i.e., the hours waiting for launch, are spent in precisely this position. However, over longer periods of bedrest (1–4 months), there is virtually no change in noradrenaline during bedrest compared to ambulatory baseline levels (Davydova *et al.*, 1986; Convertino *et al.*, 1990b).

The venoconstrictive response to exogenous noradrenaline following 2 weeks of bedrest was largely unaltered but was attenuated by administration of tyramine (Schmid *et al.*, 1968). Since tyramine releases noradrenaline from nerve endings, these results suggested that bedrest might inhibit the ability of the neuron to either release or synthesize noradrenaline while vascular responses remain intact. That adrenoreceptors might indeed become hypersensitive was suggested by data demonstrating a greater response of plasma renin activity to isoproterenol infusion after compared to before bedrest (Melada *et al.*, 1975). Thus, even if catecholamine levels are not dramatically altered, the lack of periodic bursts of sympathetic stimulation normally experienced during routine ambulatory activities in terrestrial gravity may act to hypersensitize the adrenergic receptor system during prolonged exposure to microgravity.

BAROREFLEX FUNCTION

MICROGRAVITY

While a decrease in intravascular volume certainly appears to play a major role in the adaptation to microgravity and in orthostatic instability following spaceflight, other factors must be involved. If the hypovolemia induced by spaceflight was the primary factor associated with postflight orthostatic hypotension, then treatments designed to restore or expand blood volume should be effective in promoting orthostatic stability. The attempt to restore vascular volume in astronauts by having them drink saline solutions just prior to re-entry has proven effective in reducing orthostatic instability after spaceflights of short duration (Bungo, Charles and Johnson, 1985; White *et al.*, 1991). However, as the duration of spaceflight lengthens, orthostatic instability persists despite fluid loading procedures (White *et al.*, 1991), suggesting that mechanisms other than vascular volume may be contributing to postflight orthostatic hypotension. The observation that standing heart rates begin to decline as the duration of spaceflight becomes longer than 10 to 12 days (Nicogossian *et al.*, 1991) raised the possibility that the attenuated cardioacceleratory baroreflexes may be part of the cardiovascular adapation to longer duration spaceflight and can contribute to orthostatic hypotension by failing to maintain an appropriate cardiac output response to postflight standing. Baroreceptor reflex dysfunction has subsequently been demonstrated after spaceflight (Nicogossian *et al.*, 1991; Fritsch *et al.*, 1992), but data are still sparse and the majority of the information about this important factor comes from ground-based bedrest studies.

BEDREST

Ground-based studies have provided strong evidence that baroreflex function is altered during and after bedrest (Shen *et al.*, 1988; Convertino *et al.*, 1990b; Convertino, 1992b). While we tend to think of baroreceptors in the carotid arteries and aortic arch as the primary sensors for regulating arterial blood pressure, low-pressure atrial and pulmonary receptors may produce a reflex vasoconstrictor response to minor changes in central blood volume, even in the absence of alterations in atrial pressure (Zoller *et al.*, 1972; Shepherd, 1982; Pool and Nicogossian, 1983). Low levels of LBNP (less than 20 mmHg) can activate low-pressure receptors, while higher levels (−40 mmHg) shift even more fluid to the legs, and reduce both CVP and arterial pressure, engaging both low- and high-pressure receptors (Mancia and Mark, 1983). Significant impairment of the vagally-mediated cardiac response to baroreflex stimulation was reported after 12 days of bedrest and was significantly correlated with the magnitude of orthostatic hypotension after bedrest (Convertino *et al.*, 1990b). Bedrest also induced resetting of the cardiopulmonary baroreflex control of vascular resistance so that the same reduction in central venous pressure caused greater vasoconstriction (Figure 14.3). We interpret this relationship as indicating a reduction in the reserve capacity for further vasoconstriction which could limit adequate elevation of peripheral vascular resistance and defense of arterial blood pressure during the orthostatic challenge of standing in 1G immediately after spaceflight or bedrest. It is clear that attenuation of autonomic baroreflex functions associated with control of cardiac and vascular resistance represent an appropriate adaptation to microgravity that can compromise orthostatic stability upon return to the upright gravity environment.

THERAPEUTIC APPROACHES

MICROGRAVITY

As noted initially, the countermeasures currently employed during spaceflight do not provide perfect protection against orthostatic challenge (Bungo and Johnson, 1983; Vorobyev *et al.*, 1986). However, it is now difficult to return to placebo-controlled trials given the sense among astronauts that there are several worthwhile options available. In addition, physical countermeasures such as exercise are difficult, if not impossible, to remove from this highly fit and motivated group of individuals. Nonetheless, we will attempt to examine and evaluate the best available data for each of the countermeasures that have been employed or considered. These primarily include exercise (2–2.5 hours/day in the Soviet space program), intermittent venous pooling with LBNP during flight, oral volume repletion with salt and water prior to landing, beta-blockade, clonidine, and the mineralcorticoid, fludrocortisone.

Because of the concept that orthostatic instability following spaceflight actually reflected cardiac dysfunction rather than hypovolemia, venous compliance and cardiovascular reflexes described above, exercise was a popular early countermeasure. It also had the potential to reduce muscle disuse atrophy. Skylab data suggested that the reduction in arm and leg strength seen when the exercise prescription was modest could be halved by working 1.5 hours/day on a treadmill rigged to allow "walking" or "jogging" in microgravity (Michel *et al.*, 1977; Thornton and Rummel, 1977). The Soviet program utilized both 2 hours/day of treadmill exercise against resistance, simulating a weight load of 50 kg (Vorobyev *et al.*, 1976, 1983), as well as spring-loaded "Penguin" or "Chibis" suits, which exerted a head-to-foot force and were worn 8 hours/day (Gurovskiy *et al.*, 1975). This extreme effort, with extensive aerobic exercise, did prevent much of the prior loss in muscle tone and strength, and a notably better heart rate and SV response to exercise, even over 63 days, but postflight orthostatic hypotension persisted. Even when just cardiovascular fitness itself is considered, there are data to suggest that those who are fittest at the beginning of flight tend to adapt more rapidly and lose a higher proportion of their VO2max (Convertina, Goldwater and Sandler, 1986; Klein, Wegmann and Kuklinski, 1977; Sandler, Popp and Harrison, 1988). Despite this, it is thought by many to be helpful, and there is insufficient data to consider the issue settled (Sandler, 1988).

Because of the increases in venous compliance mentioned above, it seemed reasonable to test the hypothesis that intermittent venous pooling could be effective as a treatment while in a microgravity environment. This procedure is currently part of the Soviet program for 5–7 days before re-entry. There are few supportive data from spaceflight experiences, however. Similarly, oral salt and water loading and the use of fludrocortisone to aid in fluid retention are empiric measures in the U.S. program, based largely on limited bedrest data.

BEDREST

Exercise again was an early measure utilized to mitigate the problems associated with bedrest. Even passive exercise, i.e., electrical stimulation of muscle activity, can have a positive effect on venous compliance. One bout of maximal exercise at the end of 10 days of bedrest restored exercise capacity (Convertino, 1987) and has acutely increased the responsiveness of the vagally-mediated cardiac baroreflex in ambulatory (Convertino and Adams, 1991) and bedrested subjects (Convertino *et al.*, 1992a). It has been shown to increase plasma volume by 12% (Gillen *et al.*, 1991) and improve orthostatic tolerance after water immersion (Stegemann *et al.*, 1975). These effects would seem to differ from those of more chronic exercise which, despite its ability to reduce leg muscle atrophy (Kas'yan *et al.*, 1980) and maintain plasma volume (Greenleaf, Wade and Leftheriotis, 1989), does not seem to prevent (Sandler, Popp and Harrison, 1988; Greenleaf, Wade and Leftheriotis, 1989) and may actually increase orthostatic intolerance (Bascands, *et al.*, 1984).

Intermittent venous pooling has been tested much more extensively during bedrest than in microgravity. Even modest exposures to LBNP (210 min/day) at levels of −30 to −60 mmHg or reversed gradient garments can nearly abolish the reduction in VO2max and endurance time seen with 15 days of bedrest (Convertino *et al.*, 1982; Fortney, 1991b). Plasma volume reduction has been reduced with a combination of LBNP at 30 mmHg for 4 hours/day and saline loading (Hyatt and West, 1977), and with venous pooling alone (Convertino *et al.*, 1982). However, there are a number of studies which demonstrate no benefit of these procedures with regard to improved orthostatic stability following bedrest (Vogt, 1965, 1966; Vogt and Johnson, 1967; Panferova, 1977; Sandler *et al.*, 1983).

Pharmacologic therapy has also been investigated in the bedrest setting. Presumably by decreasing peripheral β-adrenoreceptor-mediated vasodilation, beta-blockade with propranolol increased blood pressure stability at -70 mmHg of LBNP. Overall, however, it decreased the total time that LBNP was tolerated despite a raised systemic vascular resistance (Sandler *et al.*, 1985).

A number of investigators have explored the use of clonidine during bedrest and have found less fall in hematocrit and less change in renin activity and aldosterone with doses of 0.45 mg/day. Venous compliance was reduced and orthostatic tolerance seemed to be improved in at least two studies (Bonde-Petersen *et al.*, 1981; Norsk, Bonde-Petersen and Warberg, 1981). These results deserve further study, as do those with fludrocortisone, a mineralocorticoid that aids in the retention of salt and water. Although some studies seem to demonstrate a real benefit of volume expansion (Stevens and Lynch, 1965; Bohnn *et al.*, 1970; Hyatt, 1971), it is possible to restore plasma volume with persisting orthostatic instability. Some of the inconsistencies in the available data may reflect differences in the volume of fluid and salt loading and in the duration of bedrest exposure. Certainly one would not expect volume loading alone to immediately correct all aspects of the orthostatic instability of prolonged bedrest when other cardiovascular reflex control mechanisms may become impaired.

Despite the deficiencies noted above, real progress has been made in both understanding and beginning to treat the alterations in cardiovascular homeostasis associated with microgravity and with bedrest. Further investigation in this area will be essential for maintaining optimal health, safety and productivity of space travellers, especially if long-duration manned flights are planned. Further, such research should reap important benefits for the care of hospitalized patients confined to bed. Most importantly, continued research efforts in this area will enhance our understanding of cardiovascular control in man.

REFERENCES

Aĭkov, O.Yu. (1985) The state of cosmonauts' cardiovascular systems during long term orbital flights. *Kardiologicheskovo Nauchnovo Tsentra AMN SSSR*, **8**, 97–100.

Atkov, O.Yu., Bednenko, V.S. and Fomina, G.A. (1987) Ultrasound techniques in space medicine. *Aviat. Space Environ. Med.*, **58**(suppl 9), A69–A73.

Baisch, F., Beck, L., Muller, E.W. and Samel, A. (1984a) Cardiocirculatory adjustment during a 7 day microgravity simulation (6-degree head-down tilt, HDT) *Proceedings of the 35th Congress of the International Astronautical Federation*, Oct. 7–13, 1–5.

Baisch, F. and Klein, K.E. (1984b) Some physiologic consequences of G-induced body fluid shifts and muscle load reduction. *The Physiologist*, **27**, S103–S104.

Bascands, J.L., Gauquelin, G., Annat, G., Pequinot, J.P., Gharib, C. and Guell, A. (1984) Effect of muscular exercise during 4 days simulated weightlessness on orthostatic tolerance. *The Physiologist*, **27**, S63–S64.

Beregovkin, A.V., Buyanov, P.V., Galkin, A.V., Pisarenko, N.V. and Sheludyakov Y.Y. (1969) Results of investigation of the cardiovascular system during the after-effect of 70-day hypokinesis. *Problems in Space Biology*, **13**, 221–227.

Beregovkin, A.V., Vodolazov, A.S., Georgiyevskiy, V.S., Kalinichenko, V.V., Korelin, N.V., Mikhaylov, V.M., *et al.* (1976) Reactions of the cardiorespiratory system to a dosed physical load in cosmonauts after 30- and 63-day flights in the 'Salyut-4' orbital station. *Kosm. Biol. Aviakosm. Med.*, **10**(5), 24–29.

Birkhead, N.C., Haupt, C.J. and Meyers, R.N. (1963) Effects of prolonged bedrest on cardiodynamics. *Am. J. Med. Sci.*, **245**, 118–119.

Blamick, C.A., Goldwater, D.J. and Convertino, V.A. (1988) Leg vascular responsiveness during acute orthostasis following simulated weightlessness. *Aviat. Space Environ. Med.*, **59**, 40–43.

Blomqvist, C.G., Nixon, J.V., Johnson, R.L. and Mitchell, J.H. (1980) Early cardiovascular adaptation to zero gravity simulated by head-down tilt. *Acta Astronautica*, **7**, 543–553.

Blomqvist, C.G. (1983a) Cardiovascular adaptation to weightlessness. *Med. Sci. Sports Exerc.*, **15**, 428–431.

Blomqvist, C.G., Gaffney, F.A. and Nixon, J,V. (1983b) Cardiovascular responses to head-down tilt in young and middle-aged men. *The Physiologist*, **26**, S81–S82.

Bohnn, B.J., Hyatt, K.H., Kamenetsky, L.G., Calder, B.E. and Smith, W.B. (1970) Prevention of bedrest induced orthostatism with 9-alphafludrohydrocortisone. *Aerospace Med.*, **41**, 495–499.

Bonde-Petersen, F., Guell, A., Skagen, K. and Henriksen, O. (1981) The effect of clonidine on peripheral vasomotor reactions during simulated zero gravity. *The Physiologist*, **24**, S89–S99.

Bungo, M.W. and Johnson, P.C, Jr. (1983) Cardiovascular examinations and observations of deconditioning during the Space Shuttle orbital flight test program. *Aviat. Space Environ. Med.*, **54**, 1001–1004.

Bungo, M.W., Charles, C.B. and Johnson, P.C. (1985) Cardiovascular deconditioning during spaceflight and the use of saline as a countermeasure to orthostatic intolerance. *Aviat. Space Environ. Med.*, **56**, 985–990.

Bungo, M.W., Goldwater, D.J., Popp, R.L. and Sandler, H. (1987) Echocardiographic evaluation of space shuttle crewmembers. *J. Appl. Physiol.*, **62**, 278–283.

Burkovskaya, T.Y., Ilyukhin, A.V., Lobachik, V.l. and Zhidkov, V.V. (1980) Erythrocyte balance during 182-day hypokinesia. *Kosm. Biol. Aviakosm. Med.*, **14**, 50–54.

Butusov, A.A., Lyamin, V.R., Lebedev, A.A., Polyakova, A.P., Svistunov, I.B., Tishler, V.A. and Shulenin, A.P. (1970) Results of routine medical monitoring of cosmonauts during flight on the 'Soyuz-9' ship. *Kosm. Biol. Med.*, **4**(6), 35–39.

Bystrov, V.V., Zhernavkov, A.F. and Savilov, A.A. (1986) Human heart function in early hours and days of head-down tilt (Echocardiographic data) *Kosm. Biol. Aviakosm. Med.*, **20**, 42–46.

Campbell, J.A. and Webster, T.A. (1921) Day and night urine during complete rest, laboratory routine, light muscular work and oxygen administration. *Biochem. J.*, **15**, 660–664.

Chekirda, I.F., Bogdashevskiy, R.B., Yeremin, A.V. and Kolosov, I.A. (1970) Coordination structure of walking of Soyuz-9 crewmembers before and after flight. *Kosm. Biol. Med.*, **5**(6), 48–52.

Chestukhin, V.V., Katkov, V.Y., Seid-Guseynov, A.A., Shal'nev, B.I., Georgiyevskiy, V.S., Zybin, O.K., *et al.* (1979) Activity of the right cardiac ventricle and metabolism in healthy persons during an orthostatic test after short-term immobilization. *Patolog. Fiziol. Eksper. Terapiya.*, **2**, 36–40.

Convertino, V.A., Bisson, R., Bates, R., Goldwater, D. and Sandler, H. (1981) Effects of antiorthostatic bedrest on the cardiorespiratory responses to exercise. *Aviat. Space Environ. Med.*, **52**, 251–255.

Convertino, V.A., Sandler, H., Webb, P. and Annis, J.F. (1982) Induced venous pooling and cardiorespiratory responses to exercise after bedrect. *J. Appl. Physiol.*, **52**, 1343–1348.

Convertino, V.A., Goldwater, D.J. and Sandler, H. (1986) Bedrest-induced peak V02 reduction associated with age, gender and aerobic capacity. *Aviat. Space Environ. Med.*, **57**, 17–22

Convertino, V.A. (1987) Potential benefits of maximal exercise just prior to return from weightlessness. *Aviat. Space Environ. Med.*, **58**, 568–572.

Convertino, V.A., Doerr, D.F., Mathes, K.L., Stein, S.L. and Buchanan, P. (1989) Changes in volume, muscle compartment, and compliance of the lower extremities in man following 30 days of exposure to simulated microgravity. *Aviat. Space Environ. Med.*, **60**, 653–658.

Convertino, V.A., Doerr, D.F. and Stein, S.L. (1989) Changes in size and compliance of the calf following 30 days of simulated microgravity. *J. Appl. Physiol.*, **66**, 1509–1512

Convertino, V.A., Thompson, C.A., Benjamin, B.A., Keil, L.C., Savin W.M., Gordon, E.P., *et al.* (1990a) Hemodynamic and ADH responses to central blood volume shifts in cardiac-denervated humans. *Clin. Physiol.*, **10**, 55–67.

Convertino, V.A., Doerr, D.F., Eckberg, D.L., Fritsch, J.M. and Vernikos-Danellis, J. (1990b) Bedrest impairs human baroreflex responses and provokes orthostatic hypotension. *J. Appl. Physiol.*, **8**, 1458–1464.

Convertino V.A. (1990) Physiological adaptations to weightlessness: effects of exercise and work performance. *Exerc. Sports Sci. Rev.*, **18**, 119–165.

Convertino, V.A. and Adams, W.C. (1991) Enhanced vagal baroreflex response during 24 hours after acute exercise. *Am. J. Physiol.*, **260**, R570–R575.

Convertino, V.A., Doerr, D.F., Guell, A. and Marini, J.-F. (1992) Effects of acute exercise on attenuated vagal baroreflex function during bedrest. *Aviat. Space Environ. Med.*, **63**, 999–1003.

Convertino, V.A. (1992) Effects of exercise and inactivity on intravascular volume and cardiovascular control mechanisms. *Acta Astronautica*, **27**, 123–129.

Convertino, V.A., Doerr, D.F., Ludwig, D.A., and Vernikos, J. (1994) Effect of simulated microgravity on cardiopulmonary baroreflex control of forearm vascular resistance. *Am. J. Physiol. (Regulat. Integrative Comp. Physiol.)*, **266**, R1962–R1969.

Dallman, M.F., Vernikos, J., Keil, L.C., O'Hara, D. and Convertino, V.A. (1984) Hormonal, fluid and electrolyte response to 6° antiorthostatic bedrest in healthy male subjects. In: *Stress: Role of Catecholamines and Other Neurotransmitters*, edited by E. Usdin and R. Kvestnansky. pp. 1057–1077, New York: Gordon and Breach Sci. Publ. Inc..

Davydova, N.A., Shishkina, S.K., Korneyeva, N.V., Suprunova, Y.V. and Ushakov, A.S. (1986) Biochemical aspects of some neurohumoral system functions during long-term antiorthostatic hypokinesia. *Kosm. Biol. Aviakosm. Med.*, **20**, 91–95.

Degtyarev, V.A., Doroshev, V.G., Kalmykova, N.D., Kirillova, Z.A. and Lapshina, N.A. (1974) Dynamics of circulatory indices in the crew of the Salyut orbital station during an examination under rest conditions. *Kosm. Biol. Aviakosm. Med.*, **8**(2), 34–42.

Degtyarev, V.A., Doroshev, V.G., Kalmykova, N.D., Kukushkin, Y.A., Kirillova, Z.A., Lapshina, N.A., *et al.* (1978a) Dynamics of circulatory parameters of the crew of the Salut space station in functional test with physical load. *Kosm. Biol. Aviakosm. Med.*, **12**(3), 15–20.

Degtyarev, V.A. Doroshev, V.G., Kalmykova, N.D., Kirillova, Z.A., Lapshina, N.A., Lepskiy, A.A., *et al.* (1978b) Studies of hemodynamics and phase structure of cardiac cycle in the crew of Salyut-4. *Kosm. Biol. Aviakosm. Med.*, **12**(6), 9–14.

Dunn, C.D.R., Lange, R.D., Kimzey, S.L., Johnson, P.C. and Leach C.S. (1984) Serum erythropoietin titers during prolonged bedrest: relevance to the "anaemia" of spaceflight. *Eur J. Appl. Physiol.*, **52**, 178–182.

Fischer, C.L., Johnson, P.C. and Berry, C.A. (1967) Red blood cell and plasma volume changes in manned spaceflight. *J. Am. Med. Assoc.*, **200**, 579–583.

Fortney, S.M., Hyatt, K.H., Davis, J.E. and Vogel, J.M. (1991a) Changes in body fluid compartments during a 28-day bedrest. *Aviat. Space Environ. Med.*, **62**, 97–104.

Fortney, S.M. (1991) Development of lower body negative pressure as a countermeasure for orthostatic intolerance. *J. Clin. Pharmacol.*, **31**, 888–892.

Fritsch, J.M., Charles, J.B., Bennett, B.S., Jones, M.M. and Eckberg, D.L. (1992) Short-duration spaceflight impairs human carotid baroreceptor-cardiac reflex responses. *J. Appl. Physiol.*, **73**, 664–671.

Gaffney, F.A., Nixon, J.V., Karlsson, E.S., Campbell, W., Dowdey, A.B.C. and Blomqvist, C.G. (1985) Cardiovascular deconditioning produced by 20 hours of bedrest with head-down tilt (−5 degrees) in middle-aged healthy men. *Am. J. Cardiol.*, **56**, 634–638.

Gazenko, O.G., Gurovsky, N.N., Genin, A.M., Bryanov, I.I., Eryomin, A.V. and Egorov, A.D. (1976) Results of medical investigations carried out on board the Salyut orbital stations. *Life Sci. Space Res.*, **14**, 145–152.

Gazenko, O.G., Genin, A.M. and Yegorov, A.D. (1981) Summary of medical investigations in the USSR manned space missions. *Acta Astronautica*, **8**, 907–917.

Georgiyevskiy, V.A. and Mikhaylov, V.M. (1978) The effects of 60 days of bedrest on the circulatory system. *Fiziol Cheloveka*, **4**, 871–875.

Gharib, C., Gauquelin, G., Pequinot, J.M., Geelen, G., Bizollon, C. and Guell, A. (1988) Early hormonal effects of head-down tilt (−10 degrees) in humans. *Aviat. Space Environ. Med.*, **59**, 624–629.

Gillen, C.M., Lee, R., Mack, G.W., Tomaselli, C.M., Nishiyasu, T. and Nadel, E.R. (1991) Plasma volume expansion in humans after a single intense exercise protocol. *J. Appl. Physiol.*, **71**, 1914–1920.

Goldsmith, S.R., Francis, G.S. and Cohn, J.N. (1985) Effect of head-down tilt on basal plasma norepinephrine and renin activity in humans. *J. Appl. Physiol.*, **59**, 1068–1071.

Goldwater, D., Polese, A., Montgomery, L., London, L., Johnson, P., Yuster, D., *et al.* (1980) Comparison of orthostatic intolerance following horizontal or −6 degree head-down bedrest simulation of weightlessness. *Aerospace Med. Assoc. Preprints*, 28–29.

Graham, R.M. and Zisfein, J.B. (1986) Atrial Natriuretic Factor Regulation and Control in Circulatory Homeostasis. In *The Heart and Cardiovascular System, Scientific Foundation*, edited by Fozzard, H.A., Haber, E., Jennings, R.B., Katz. A.M. and Morgan, H.E., pp. 1559–1572. New York: Raven Press.

Greenleaf, J.E. (1984) Physiological responses to prolonged bedrest and fluid immersion in humans. *J. Appl. Physiol.*, **57**, 619–633

Greenleaf, J.E., Wade C.E. and Leftheriotis, G. (1989) Orthostatic responses following 30-day bedrest deconditioning with isotonic and isokinetic exercise training. *Aviat. Space Environ. Med.*, **60**, 537–542

Grigoriev, A.I, Bugrov, S.A., Bogomolov, V.V., Egorov, A.D., Polyakov, V.V., Tarasov, I.K., *et al.* (1991) Major medical results of extended flights on space station Mir in 1986–1990. *IAF/IAA-91-547*, 547 (Abstract).

Guell, A., Braak, L., Bousquet, J., Barrere, M. and Bes, A. (1980) Orthostatic tolerance and exercise response before and after 7 days simulated weightlessness. *The Physiologist*, **23**, S151–S152

Guell, A., Pequinot, J.M., Gauquelin, G., Bascands, J.L., Geelen, G., Allevard, A.M., *et al.* (1986) Volume regulating hormones during a 5-hour head-down tilt at −10 degrees: I-Epinephrine, norepinephrine and dopamine. *Proceedings of the 2nd International Conference on Space Physiology*, 177–179.

Gurovskiy, N.N., Yeremin, A.V., Gazenko, O.G., Yegorov, A.D., Bryanov, I.I. and Genin, A.M. (1975) Medical investigations during flights of the spaceships "Soyuz-13", "Soyuz-14" and the "Salyut-3" orbital station. *Kosm. Biol. Aviakosm. Med.*, **9**, 48–54.

Hoffler, G.W. (1977) Cardiovascular studies of U.S. space crews: an overview and perspective. In *Cardiovascular Flow Dynamics and Measurements*, edited by Hwang, N.H.C. and Normann, N.A. University Park Press, Baltimore, pp. 335–363.

Hollister, A.S., Tanaka, I., Imada, T., Onrot, J., Biaggioni, I., Robertson, D., *et al.* (1986) Modulation of plasma atrial natriuretic factor levels in human subjects by sodium loading and posture change. *Hypertension*, **8**, II106–111.

Hung, J., Goldwater, D., Convertino, V.A., McKillop, J.H., Goris, M.L. and DeBusk, R.F. (1983) Mechanisms for decreased exercise capacity following bedrest in normal middle-aged men. *Am. J. Cardiol.*, **51**, 344–348.

Hunt, S.C., McCarron, D.A., Smith, J.B., Ash, K.O., Bristow, M.R. and Willams, R.R. (1984) The relationship of plasma ionized calcium to cardiovascular disease endpoint and family history of hypertension. *Clin. Exp. Hypertens.*, **6**, 1397–1414.

Hyatt, K.H. (1971) Hemodynamic and Body Fluid Alterations Induced by Bedrest. In *Hypodynamic and Hypogravic Environments*, edited by Murray, R.H., *et al.* pp. 189–209. NASA SP-269.

Hyatt, K.H. and West, D.A. (1977) Reversal of bedrest-induced orthostatic intolerance by lower body negative pressure and saline. *Aviat. Space Environ. Med.*, **48**, 120–124.

Johnson, P.C., Kimzey, S.L. and Driscoll, T.B. (1975) Postmission plasma volume and red-cell mass changes in the crews of the first two Skylab missions. *Acta Astronautica*, **2**, 311–317.

Johnson, P.C., Driscoll, T.B. and LeBlanc, A.D. (1977) Blood volume changes. In: *Biomedical Results from Skylab*, edited by Johnston, R.S. and Dietlein, L.F. NASA SP-377. pp. 235–241.

Kakurin, L.I., Lobachik, V.I., Mikhailov, V.M. and Senkevich, Yu.A. (1976) Antiorthostatic hypokinesia as a method of weightlessness simulation. *Aviat. Space Environ. Med.*, **47**, 1083-1086.

Kas'yan, I.I., Talavrinov, V.A., Luk'yanchikov, V.I. and Kobzev, Y.A. (1980) Effect of antiorthostatic hypokinesia and spaceflight factors on change in leg volume. *Kosm. Biol. Aviakosm. Med.*, **14**, 51–55.

Katkov, V.Y., Chestukhin, V.V., Nikolayenko, E.M., Grozdev, S.V., Rumyantsev, V.V., Guseynova, T.M., *et al.* (1982) Central circulation in the healthy man during 7-day head down hypokinesia. *Space Biol. Aerospace Med.*, **16**, 64–72.

Kirsch, K.A. Rocker, L., Gauer, O.H. and Krause, R. (1984) Venous pressure in man during weightlessness. *Science*, **225**, 218–219.

Klein, K.E., Wegmann, H.M. and Kuklinski, P. (1977) Athletic endurance training — advantage for spaceflight? The significance of physical fitness for selection and training of Spacelab crews. *Aviat. Space Environ. Med.*, **48**, 215–222.

Krupina, T.N., Tizul, A.Y., Kuz'min, M.P. and Tsyganova, N.I. (1982) Clinicophysiological changes in man during long-term antiorthostatic hypokinesia. *Kosm. Biol. Aviakosm. Med.*, **16**, 29–34.

Kvetnansky, R., Noskov, V.B., Blazicek, P., Gharib, C., Popova, I.A., Gauquelin, G., *et al.* (1991) Activity of the sympathoadrenal system in cosmonauts during 25-day spaceflight on station Mir. *Acta Astronautica*, **23**, 109–116.

LaRochelle, F., Leach, C. and Vernikos-Danellis, J. (1982) Effects of age and sex on hormonal responses to weightlessness simulation. *The Physiologist*, **25**, S161–S162.

Leach, C.S. and Rambaut, P.C. (1975) Endocrine responses in long-duration manned spaceflight. *Acta Astronautica*, **2**, 115–127.

Leach, C.S., Altchuler, S.I. and Cintron-Trevino, N.M. (1983a) The endocrine and metabolic responses to spaceflight. *Med. Sci. Sports Exerc.*, **15**, 432–440.

Leach, C.S., Johnson, P.C. and Suki, W.N. (1983) Current concepts of spaceflight induced changes in hormonal control of fluid and electrolyte metabolism. *The Physiologist*, **26**, S24–S27.

Leach, C.S. and Johnson, P.C. (1984) Influence of spaceflight on erythrokinetics in man. *Science*, **225**, 216–218.

Mancia, G.A., and Mark, A.L. (1983) Arterial baroreflexes in humans. In *Handbook of Physiology. The Cardiovascular System. Peripheral Circulation and Organ Blood Flow*, edited by Shepherd, J.T., Abboud, F.M., and Geiger, S.R., pp. 755–793. Bethesda: American Physiological Society.

Melada, G.A., Goldman, R.H., Leutscher, J.A. and Zager, P..G. (1975) Hemodynamics, renal function, plasma renin and aldosterone in man after 5 to 14 days of bedrest. *Aviat. Space Environ. Med.*, **46**, 1049–1055.

Michel, E.L., Rummel, J.A., Sawin, C.F., Buderer, M.C. and Lem, J.D. (1977) Results of Skylab medical experiment M171 – metabolic activity. In *Biomedical Results from Skylab*, edited by Johnston, R.S. and Dietlein, L.F. NASA SP-377, pp. 372–287.

Miller, P.B., Johnson, R.L. and Lamb, L.E. (1964) Effects of four weeks of absolute bedrest on circulatory functions in man. *Aerospace Med.*, **35**, 1194–1200.

Natelson, B.H., DeRoshia, C. and Levin, B.E (1982) Physiological effects of bedrest. *The Lancet*, **1**, 51.

Nicogossian, A.E., Charles, C.B., Bungo, M.W. and Leach-Huntoon, C.S. (1991) Cardiovascular function in spaceflight. *Acta Astronautica*, **24**, 323–328.

Norsk, P., Bonde-Petersen, F. and Warberg, J. (1981) Cardiovascular effects of clonidine during 20 hr head-down tilt (−5 degrees) *The Physiologist*, **24**, S91–S92.

Panferova, N.Y. (1977) *Hemodynamics and the Cardiovascular System*, pp. 1–336. Moscow: Nauka.

Pool, S.L. and Nicogossian, A.E. (1983) Biomedical results of the Space Shuttle orbital flight test program. *Aviat. Space Environ. Med.*, **54**, S41–S49.

Rudnyy, N.M., Gazenko, O.G., Gozulov, S.A., Pestov, I.D., Vasilyev, P.V., Yeremin, A.V., *et al.* (1977) Main results of medical research conducted during the flight of two crews on the Salyut-5 orbital station. *Kosm. Biol. Aviakosm. Med.*, **11**(5), 33–41.

Saltin, B., Blomqvist, G., Mitchell, J.H., Johnson, R.L., Wildenthal, K. and Chapman, C.B. (1968) Response to exercise after bedrest and after training. *Circ.*, **38**(suppl 7), 1–78.

Sandler, H., and Winter, D.L. (1978) *Physiological responses of women to the first female bedrest study*, pp. 1–187. NASA SP-430.

Sandler, H. (1983) Cardiovascular adjustments to gravitational stress. In *Handbook of Physiology, Sec 2*, The Cardiovascular System, edited by Shepherd, J.T., Abboud, F.M. and Geiger, S.R., pp. 1025–1063. Bethesda: American Physiological Society.

Sandler, H., Webb, P., Annis, J., Pace, N., Grunbaum, B.W., Dolkas, D., *et al.* (1983) Evaluation of a reverse gradient garment for prevention of bedrest deconditioning. *Aviat. Space Environ. Med.*, **54**, 191–201.

Sandler, H., Goldwater, D.J., Popp, R.L., Spaccavento, L. and Harrison, D.C. (1985) Beta blockade in the compensation of bedrest cardiovascular deconditioning: Physiological and pharmacological observations. *Am. J. Cardiol.*, **55**, 114D–120D.

Sandler, H. (1986) Cardiovascular Effects of Inactivity. In *Inactivity — Physiological Effects*, edited by Sandler, H. and Vernikos, J. pp. 11–40. New York: Academic Press.

Sandler, H. (1988) *Cardiovascular effects of weightlessness and ground-based simulation*, pp. 1–43. NASA Technical Memorandum 8-8314.

Sandler, H., Popp, R.L. and Harrison, D.C. (1988) The hemodynamic effects of repeated bedrest exposure. *Aviat. Space Environ. Med.*, **59**, 1047–1054.

Schmid, P.G.. Shaver, J.A., McCally, M., Bensy, J.J. Pawlson, L.G. and Plenume, T.E. (1968) *Aerospace Med. Assoc. Preprints*, p. 104 (Abstract).

Shen, X.Y., Sun, Y.H., Xiang, Q.L., Meng, J.R., Xu, L.H., Yan, X.X., *et al.* (1988) The study of baroreceptor reflex function before and after bedrest. *The Physiologist*, **31**, S22–S23.

Shepherd, J.T. (1982) Reflex control of arterial blood pressure. *Cardiovasc. Res.*, **16**, 357–383

Smith, R.F., Stanton, K., Stoop, D., Janusz, W. and King, P. (1976) Quantitative electrocardiography during extended spaceflight: the second Skylab mission. *Aviat. Space Environ. Med.*, **47**, 353–359.

Stegemann, J., Meier, U., Skipka, W., Hartlieb, W., Hemme, B. and Tibes, U. (1975) Effects of multi-hour immersion with intermittent exercise on urinary excretion and tilt table tolerance in athletes and nonathletes. *Aviat. Space Environ. Med.*, **46**, 26–29.

Stevens, P.M. and Lynch, T.N. (1965) Effects of 9-alphafluorohydrocortisone on dehydration to prolonged bedrest. *Aerospace Med.*, **36**, 1151–1156.

Taylor, H.L., Erickson, L., Henschel, A. and Keys, A. (1945) The effect of bedrest on the blood volume of normal men. *Am. J. Physiol.*, **144**, 227–232.

Thornton, W.E. and Rummel, J.A. (1977) Muscular deconditioning and its prevention in spaceflight. In *Biomedical Results from Skylab*, edited by Johnson, R.S. and Dietlein, L.F. NASA SP-377, pp. 191–197.

Thornton, W.E. and Hoffler, G.W. (1977) Hemodynamic studies of the legs under weightlessness. In *Biomedical Results from Skylab*, edited by Johnson, R.S. and Dietlein, L.F. NASA SP-377, pp. 324–329.

Thornton, W.E., Hoffler, G.W. and Rummel, J.A. (1977) Anthropometric changes and fluid shifts. In *Biomedical Results from Skylab*, edited by Johnson, R.S. and Dietlein, L.F. NASA SP-377, pp. 330–338.

Tigranyan, R.A., Popova, I.A., Belyakova, M.I., Kalita, N.F., Tuzova, Y.G., Sochilina, L.B. and Davydova, N.A. (1977) Results of metabolic studies of the crew of the second expedition of the Salyut-4 orbital station. *Kosm. Biol. Aviakosm. Med.*, **11**(2), 48–53.

Vogt, F.B. (1965) Effect of extremity cuff-tourniquets on tilt table tolerance after water immersion. *Aerospace Med.*, **36**, 442–447.

Vogt, F.B. (1966) Effect of intermittent leg cuff inflation and intermittent exercise after 10 days bed recumbency. *Aerospace Med.*, **37**, 943–947.

Vogt, F.B. and Johnson, P.C. (1967) Plasma volume and extracellular fluid volume change associated with 10-day recumbency. *Aerospace Med.*, **38**, 21–25.

Vorobyev, Y.I., Gazenko, O.G., Gurovskiy, N.N., Nefedov, Y.G., Yegorov, B.B., Bayevskiy, R.M., *et al.* (1976) Preliminary results of medical investigations carried out during flight of the second expedition of the 'Salyut-4' orbital station. *Kosm. Biol. Aviakosm. Med.*, **10**(5), 3–18.

Vorobyov, E.I., Gazenko, O.G., Genin, A.M. and Egogrov, A.D. (1983) Medical results of Salyut-6 manned spaceflights. *Aviat. Space Environ. Med.*, **54**(Suppl 1), S31–S40.

Vorobyev, Y.I., Gazenko, O.G., Shulzhenko, Y.B., Grigoryev, A.I., Barer, A.S., Yegorov, A.D. and Skiba, A.I. (1986) Preliminary results of medical investigations during 5-month spaceflight aboard Salyut-7-Soyuz-T orbital complex. *Kosm. Biol. Aviakosm. Med.*, **20**(2), 27–34.

White, R.J., Leonard, J.I., Srinivasan, R.S. and Charles, J.B. (1991) Mathematical modeling of acute and chronic cardiovascular changes during extended duration orbiter (EDO) flights. *Acta Astronautica*, **23**, 41–51.

Yegorov, A.D., Itsekhovskiy, O.G., Polyakova, A.P., Turchaninova, V.F., Alferova, I.V., Savelyeva, V.G., *et al.* (1981) Results of studies of hemodynamics and phase structure of the cardiac cycle during functional test with graded exercise during 140-day flight aboard the Salyut-6 station. *Kosm. Biol. Aviakosm. Med.*, **15**(3), 18–22.

Yegorov, A.D. and Itsekhovskiy, O.G. (1983) Study of cardiovascular system during long-term spaceflights. *Kosm. Biol. Aviakosm. Med.*, **17**(5), 4–6.

Zoller, R.P., Mark, A.L, Abboud, F.M., Schmid, P.G. and Heisted, D.D. (1972) The role of low pressure baroreceptors in reflex vasoconstrictor responses in man. *J. Clin. Invest.*, **42**, 2967–2972.

15 Orthotopic Cardiac Transplantation: A Model of the Denervated Heart

Raymond Stainback

Adult Echocardiography Laboratory, 505 Parnassus Avenue, Moffitt Hospital, San Francisco, CA 94143-0214, USA

Human orthotopic cardiac transplantation is a unique model of complete cardiac denervation that has become pervasive at a number of medical centres in recent years. This growing population of patients presents an opportunity for expanding our knowledge of the complex array of counter-regulatory mechanisms responsible for blood pressure and heart rate control and volume homeostasis. Cardiac transplantation is distinct from pharmacological or disease models of cardiac denervation in that non-cardiac autonomic function is generally intact. Thus, with careful monitoring, relatively aggressive pharmacological and physiological manoeuvres may be performed in these patients without substantial danger of wide or prolonged blood pressure fluctuations. Many studies of both animal and human cardiac transplantation subjects have revealed the importance of multiple intrinsic mechanisms of cardiac adaptation that must be distinguished from extrinsic autonomic control mechanisms. Current knowledge of the mechanisms and clinical implications of cardiac denervation hypersensitivity are discussed in detail. Extensively documented is evidence for gradual cardiac reinnervation in various animal models. To date, the factors underlying functional cardiac reinnervation in animals are very poorly understood as are the reasons for a relative lack of cardiac reinnervation following human transplantation. In addition to outlining areas of future research in cardiac denervation physiology, this review should provide information useful in the clinical management of heart transplantation patients.

KEY WORDS: Cardiac, transplantation, denervation, hypersensitivity, reinnervation

THE HISTORY OF ORTHOTOPIC CARDIAC TRANSPLANTATION

The beginning of the colourful history of heart transplantation, reviewed by Griepp and Ergin (1984) and Reitz (1990), may be traced to 1905 when Alexis Carrel, then at the University of Chicago, described heterotopic transplantation of a puppy's heart into the neck of an adult dog (Carrel and Guthrie, 1905). Another milestone was reached during the mid-1940's by the Soviet surgeon V.P. Demikhov (1962), although his findings were not reported to the West until 1962. He demonstrated that intrathoracic heterotopic transplantation of the heart-lung block was technically feasible, with survivals of 2–6 days reported in several animals, notably, before the availability of cardiopulmonary bypass. During the 1950's, studies of animal models of heterotopic cardiac transplantation continued to address problems of cardiac graft rejection and organ preservation. Several attempts at direct orthotopic transplantation met with various technical difficulties and will not be reviewed here. A seminal report in 1960 by Richard Lower and Norman Shumway at Stanford University briefly described a successful technique for orthotopic transplantation of the canine heart (Lower and Shumway, 1960), which later became the basis for methods that continue to be employed in human cardiac transplantation. The recipient heart was removed by proximal division of the aorta and pulmonary artery and excision of the atria leaving a remnant common posterior atrial wall with intact pulmonary veins, vena cavae ostia and a ridge of interatrial septum, thus obviating the need for multiple venous anastomoses. Although these experiments involved excision and replacement of the dog's own heart, the Stanford group of surgeons recognized this procedure as a practical solution for otherwise inoperable heart disease if problems of homograft rejection could be overcome. Survival of one dog for 8 days indicated that complete maintenance of the circulation by a transplanted and denervated heart was possible (Lower, Stofer and Shumway, 1961). Long-term survival in dog autotransplants allowed careful evaluation of cardiac denervation physiology in the absence of rejection and paved the way for human transplantation by showing a normalization of cardiac output and venous pressures and, in fact, evidence for cardiac reinnervation (Dong et al., 1964).

Although cardiac transplantation results in complete extrinsic denervation of the donor heart with attendant depletion of myocardial catecholamines (Cooper et al., 1962), Daggett et al. (1967) observed no significant difference in left ventricular function curves over the same range of filling pressures when comparing autotrans- planted dogs with controls. Additional orthotopic transplantation studies in animals during the early and mid-1960's also supported the contention that denervation neither seriously impairs cardiac performance nor precludes adequate circulatory adaptation mechanisms (Willman et al., 1962; Lower, Dong and Shumway, 1965). Improvements in surgical and myocardial preservation techniques led to the first human cardiac transplantation at the University of Mississippi in 1964, although a chimpanzee xenograft was employed. The operation, performed by James Hardy et al. (1964) (Hardy and Chavez, 1968), was technically successful; but the small primate graft rapidly failed following removal of cardiopulmonary bypass. The first human cardiac allograft transplantation occurred in Cape Town, South Africa in 1967. Christiaan Barnard (1967, 1968) modified only slightly the canine tech- nique described by Lower, Stofer and Shumway (1961). The recipient heart was excised, leaving the right and left posterior atrial walls with intact vena cavae and pulmonary vein ostia, respectively. However, the donor interatrial septum was not excised in order to avoid damage to the atrioventricular (A-V) node (Figure 15.1). This method was later adapted in turn by the Stanford group (Stinson et al., 1969) in order to preserve internodal conducting pathways and to protect the donor sinus node by removing it from the region of the suture line, and this has remained the standard operative technique.

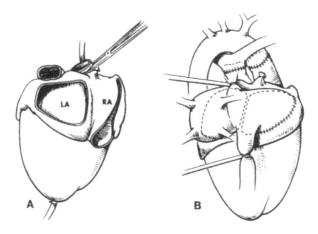

FIGURE 15.1 Drawing **A** shows the method by which the back of the donor left atrium (LA) was opened to avoid damage in the sinus node. Drawing **B** shows anastamosis of the donor left atrium (LA) and right atrium (RA) with the recipient atria. (Adapted from Barnard, 1968. With permission.)

In subsequent years combined heart-lung transplantation, as well as heterotopic heart transplantation, in humans have achieved a limited degree of clinical success. However, the number of patients undergoing these procedures has remained severely restricted in comparison with orthotopic transplantation because of pulmonary rejection and, with heterotopic transplantation, increased operative mortality among other difficulties. A small body of literature addressing the issue of cardiac denervation in humans after heart-lung or heterotopic transplantation exists (Banner *et al.*, 1989a; Yusuf *et al.*, 1989). This review is limited primarily to orthotopic transplantation as a model for cardiac denervation, however, since far greater and increasing numbers of patients exist. Furthermore, the variables of pulmonary and pulmonary vascular denervation (heart-lung transplantation) or a considerable mass of remnant, diseased, yet innervated myocardium (heterotopic transplantation) are not present.

The early experience in orthotopic cardiac transplantation resulted in a significant number of patients with relatively long survivals. These patients proved to be willing participants in many small studies which will be included here and immeasurably contributed to the understanding of cardiac denervation physiology. Unfortunately, the initial enthusiasm surrounding heart transplantation was dampened by generally poor long-term outcomes on the immunosuppressive regimens of the 1960's and 1970's (Pennock *et al.*, 1982). After several hundred operations at centres worldwide proved the technical feasibility of widespread application, a long hiatus in human cardiac transplantation prevailed from the mid 1970's until 1983 when the FDA approved cyclosporin-A for use as an immunosuppressive

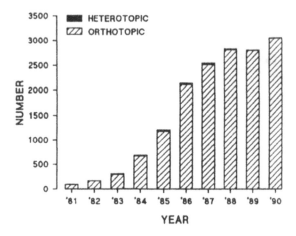

FIGURE 15.2   Number of orthotopic and heterotopic heart transplantations performed worldwide from 1982 through 1990. (From Kriett and Kaye, 1991. With permission.)

agent. The success of cyclosporin-A in the prevention of acute rejection (Oyer *et al.*, 1983) and other advances in management resulted in renewed interest in heart transplantation. A proliferation of centres, an expanding age range of recipients (currently newborn to 70 years) and improving 1, 5 and 10 year survival rates (80%, 70% and > 50% respectively) from the mid-1980's until the present have produced a substantial and growing post-cardiac transplantation population. The most current registry (Kaye, 1992) reported a 1991 total of 3000 heart transplantations from over 229 centres worldwide. In recent years less than 1% of transplants have been heterotopic with the large majority being orthotopic (Kriett and Kaye, 1991). Although a non-expanding donor supply has curtailed rapid growth in the procedure since 1988 (Figure 15.2), the sharply increased prevalence of cardiac transplant patients at many medical centres and new techniques for evaluating the autonomic nervous system represent new opportunities for exploring cardiac innervation. The growing body of data concerning cardiac denervation physiology, the exploration of intrinsic mechanisms of cardiac control in the grafted heart and new evidence for cardiac reinnervation in humans will be presented in this review.

NON-TRANSPLANTATION MODELS OF CARDIAC DENERVATION

Interestingly, development of non-transplantation models of selective cardiac denervation in intact animals occurred simultaneously with the development of successful orthotopic transplantation and provided information invaluable in the prediction of human performance after heart transplantation.   Since non-transplantation animal models of cardiac denervation continue to complement findings in cardiac transplantation, it is useful to compare the techniques. Cooper *et al.* (1961) introduced mediastinal neural ablation as a method for producing chronic cardiac denervation. Careful dissection of the entire heart and great vessels from surrounding mediastinal tissue, removal of the pericardium and stripping off the adventitial layers of the blood vessels left the paravertebral ganglia and the main vagal trunks intact while, theoretically, excising all extrinsic nerves entering and leaving the heart. Data from earlier studies, generally employing vagotomy and cervical and thoracic ganglionectomy, suffer from the confounding effects of non-specific denervation of lungs, viscera and systemic vascular beds and often a lack of total cardiac denervation (Peiss *et al.*, 1966). Approaches to cardiac denervation in animals, along with human cardiac denervation procedures aimed at treating intractable angina, vasospasm, ventricular arrhythmias and the congenital long QT syndrome are discussed in reviews by Kaye (1984) and Randall (1984).

Donald and Shepherd (1963) performed mediastinal neural ablation in mongrel dogs according to the method of Cooper *et al.* (1961) and observed subsequent cardiovascular adaptation in response to exercise.   Effective denervation was demonstrated by a lack of heart rate change with atropine, low-dose tyramine, startle, respiration, and direct electrical stimulation of vagal trunks and stellate ganglia. Normal animals increased cardiac output at low work loads primarily by increasing heart rate with little change in stroke volume. Denervated dogs, on the other hand, maintained adequate cardiac outputs in the face of exertional stress primarily by striking compensatory increases in stroke volume with little change in heart rate. At peak work loads augmentation of cardiac output was achieved by further increases in stroke volume in addition to a steady and more contributory increase in the heart rate, albeit to a considerably blunted peak. Exercise-induced stroke volume changes were attributed to the Frank-Starling mechanism, operative as a result of increased venous return produced by contracting skeletal muscle in conjunction with decreased skeletal muscle vascular resistance.

Heart rate changes were hypothesized to result from exercise-induced cate-cholamine release into the circulation along with a demonstrated hypersensitiv-ity to noradrenaline. However, two denervated dogs with adrenalectomy showed no difference in heart rate response with exercise. In additional exercise studies of greyhounds, Donald, Milburn and Shepherd (1964) demonstrated that competition-level racing times were nearly unaffected by cardiac denervation; although, in these brief, intense performances the percentage of work produced by anaerobic metabolism in denervated dogs could not be determined. Interestingly, the heart rate continued to rise after running was stopped with a slowing to baseline re-quiring 8–10 min compared with 2 min for normal dogs. Further observations in denervated greyhounds treated with β-blockers indicated that the cardiostimulatory effects of both sympathetic nerves and circulating catecholamines are necessary for maximal performance. Removal of either of these components only minimally af-fects performance while removal of both severely limits maximal exercise capacity.

Furthermore, intrinsic cardiac mechanisms in addition to circulating catecholamines appeared to account for some of the increase in heart rate after denervation, since heart rate increases during exercise were significant even in the face of β-blockade (Donald and Samueloff, 1966; Donald, Ferguson and Milburn, 1968).

Problems even with careful mediastinal neural ablation include gradual reinnervation (discussed below) and inadequate denervation in the majority of cases when rigorous testing is used (Peiss et al., 1966). Because of the high mortality and extensive postoperative care associated with animal autotransplantation and a need to perform invasive testing of cardiac function and innervation, more complex, often staged methods of non-transplantation total cardiac denervation surgery have been introduced (Geis et al., 1971; Randall et al., 1980; Kaye, 1984) in addition to a recently described method for selective parasympathetic ganglionectomy (Randall et al., 1986; 1987).

EVIDENCE FOR DEAFFERENTATION

The most widely recognized and clinically relevant consequence of cardiac denervation in human transplantation is the removal of afferent nociceptive fibres. The inability of patients to perceive myocardial ischemia (Gao et al., 1989; Schroeder and Hunt, 1992) necessitates frequent and costly coronary angiography to follow the course of graft coronary atherosclerosis. Because of C-fibre afferent pathway interruption, the Bezold- Jarisch cardiac inhibitory reflexes originating in cardiac sensory receptors (Mark, 1983) are absent in transplantation patients. Observations in transplant patients have confirmed a cardiac origin for various components of the Bezold-Jarisch reflex in man which were previously based mainly on animal model studies. Arrowood et al. (1989) showed that the afferent limb of the cardiac-depressor reflex could not be activated by intracoronary iodinated contrast in transplant patients, although this response is frequently seen in normally innervated subjects. Similarly, in autotransplanted dogs cryptenamine injection into the left ventricle does not activate the Bezold-Jarisch response (see Reinnervation-Animal Models, below). Scherrer et al. (1990a) surreptitiously but carefully recorded an episode of vasodilator-induced hypovolemic hypotension in an instrumented cardiac transplant patient. Depression of the patient's muscle sympathetic nerve activity and of the remnant atrial heart rate (see Atrial Activity, below) occurred in the setting of profound hypotension, indicating strong vagal activation despite ventricular denervation. That afferent ventricular mechanoreceptor activation is not essential for this clinical example of the Bezold-Jarisch reflex was a new observation.

Cox et al. (1989) demonstrated the presence of an adenosine-activated cardiac-sympathoexcitatory reflex arc by comparing normal subjects with transplant patients. He found that intracoronary adenosine stimulates cardiac afferent chemoreceptors, producing reflex hypertension and tachycardia in normally innervated but not in transplanted hearts.

Cardiac mechano- and chemoreceptors may play an important role in intravascular volume homeostasis. Reflex release of renin from the kidneys (Thames *et al.*, 1971; Thames, Jarecki and Donald, 1978) and atrial natriuretic peptide (ANP) from the heart itself may be profoundly affected by cardiac deafferentation. The independent effect of ventricular denervation upon intravascular volume status and renin release in humans is not clear (Held *et al.*, 1989) although cardiac transplantation offers a model with which to study the interaction. ANP is significantly elevated after cardiac transplantation (Singer *et al.*, 1986; Wilkins *et al.*, 1987; Deray *et al.*, 1990). Although neither the reason nor the clinical significance of this observation is understood, there is evidence for an impaired renal response to ANP in transplant patients (Wilkins *et al.*, 1988). Further investigation of the mechanisms of ANP release in cardiac transplant patients may shed light upon the modulatory role of the autonomic nervous system in normal subjects.

ATRIAL BEHAVIOUR

As described above, the donor heart is sutured at the atrial level to the recipient posterior and lateral atrial wall remnants which are left *in situ* along with the intact recipient sinus node and undisturbed neural connections (Stinson *et al.*, 1969). Although sinus node ischemia may result from division of the artery of the sinus node (Bieber, Stinson and Shumway, 1969), perfusion by bronchial collateral vessels (Moberg, 1967) is thought to explain preservation of remnant (recipient) sinus node activity in the majority of patients. The donor heart is excised so as to preserve sinoatrial and atrioventricular node function, and this is generally the case. However, a necropsy study reported a significant incidence of procurement-related donor and recipient A-V node ischaemic injury (Stovin and Hewitt, 1986), and chronotropic incompetence of the donor sinus node is not uncommon (Mackintosh *et al.*, 1982). A sufficient mass of recipient atrial tissue is present to produce a discrete P wave recognizable on the routine surface electrocardiogram in the majority of patients, although appropriately timed contractions contribute no more than 2-4 mmHg to ventricular filling (Stinson *et al.*, 1972b). The atrial and great vessel suture lines represent a boundary of denervation for the donor heart. Except for a single case report in which a functioning bridge of conduction tissue was thought to be present (Bexton *et al.*, 1983a), the donor heart conduction system is electrically isolated from the recipient atrial tissue in addition to being denervated. As demonstrated in animal transplant models, various physiological and pharmacological manoeuvres which reflexively stimulate or inhibit autonomic outflow affect the recipient atrial rate normally but have no effect upon the donor heart rate (Hallman *et al.*, 1969; Shaver *et al.*, 1969). While the atrial remnant generally exhibits normal sinus rhythm, fibrillation and flutter independent of the donor rhythm is occasionally observed (Leachman *et al.*, 1969; Stinson *et al.*, 1972b). At rest the remnant atrial rate is generally slower than the donor rate because of intact vagal tone (Higgins, Vatner and Braunwald, 1973). In a novel demonstration of recipient atrial activity, Kacet *et al.* (1991) produced "normal sinus node function" in a transplanted heart with sinus node incompetence requiring a pacemaker.   The implanted pacing system sensed the normally responsive recipient atrial rate which then triggered pacemaker stimulation of the donor atrium. Since the donor A-V node was functional, normal chronotropic responsiveness and A-V synchrony in the donor heart was established.

Atrial accrochage or entrainment of recipient and donor atria with a constant relationship between the recipient atrial rate and the donor ventricular rate has been observed (Leachman *et al.*, 1969).   By pacing the slower recipient atrium at rates slightly higher than the intrinsic donor rate, Mason and Harrison (1979) established entrainment that persisted despite slight increases or decreases in the recipient atrial pacing rate. Entrainment, or "phénomène d'accrochage" of the atria and ventricles has been observed in the setting of complete heart block (Segers, Lequime and Denolin, 1947).   Although a variety of theories were put forward to explain the phenomenon, observations in the transplant patient of a bi-directionality of this interatrial interaction (Mason and Harrison, 1979) support a mechanical tug theory (Carleton, 1976) by which one contracting chamber entrains another. By unknown mechanisms, stretching of conduction tissues may, then, be transducted into a change in the rate of depolarization. (see Heart Rate – Intrinsic Mechanisms, below)

RESPONSE TO EXERCISE

DYNAMIC EXERCISE

As predicted by early studies in the denervated dog (Donald, Milburn and Shepherd, 1964), the cardiac transplant patient is generally able to enjoy a normally active life and under ideal conditions, engage in substantial training regimens (Kavanagh *et al.*, 1988) and tolerate major surgical procedures under general anesthesia (Reitz *et al.*, 1977). Quite a number and variety of exercise studies in transplant patients have advanced our understanding of the important role of normal cardiac innervation in maintaining adequate circulation during stress. Equally intriguing has been an emerging knowledge of the various intrinsic mechanisms that provide for near-normal adaptation to varying circulatory demands in the absence of cardiac innervation.

The cardiac output increase required to meet tissue oxygen demands at low levels of exertion in normal subjects is accomplished by an increase in heart rate due to vagal withdrawal (Arai *et al.*, 1989) with little change in stroke volume or sympathetic outflow until peak work loads (Pope *et al.*, 1980; Quigg *et al.*, 1989). In cardiac transplant patients, on the other hand, the resting heart rate is elevated since no vagal tone is present. The resting cardiac output and intracardiac pressures are normal early (Lower *et al.*, 1968; Stinson *et al.*, 1975) and up to two years post transplantation (Stinson *et al.*, 1972a). At low work loads, circulatory demands are met almost entirely by an increase in the stroke volume and the Frank-Starling mechanism (Pope *et al.*, 1978; Schroeder, 1979; Pflugfelder *et al.*, 1987) with very little change in the donor heart rate occurring early in exercise. In contrast to normal subjects, increased venous return from skeletal muscle contraction and lowered systemic vascular resistance results in a prompt increase in central right-sided pressures (Pflugfelder, McKenzie and Kostuk, 1988) and in left ventricular end-diastolic pressure (Campeau *et al.*, 1970; Stinson *et al.*, 1972a), left ventricular end-diastolic volume (Pope *et al.*, 1980) and (via the Frank-Starling mechanism) an increased dV/dT. As the severity of exertion increases, a near-normal cardiac index is maintained by further increases in stroke volume along with a gradually increasing heart rate. The peak heart rate in cardiac transplant patients is dependent upon raised concentrations of circulating catecholamines (Bexton *et al.*, 1983c; Yusuf *et al.*, 1985). During recovery, the donor heart rate returns to normal in a considerably delayed fashion, reflective of gradual catecholamine clearance. Since noradrenaline and adrenaline levels during peak exercise are about the same in transplant patients as in normals, attenuation of the donor heart rate may theoretically be attributed to a lack of direct sympathetic innervation which in innervated hearts produces higher neural cleft noradrenaline concentrations than is measured in the circulation (Quigg *et al.*, 1989). In summary, augmentation of the cardiac output in transplant patients is heavily volume-dependent, with an abnormal heart rate-time relationship which is closely tied to the circulating catecholamine concentration.

Kavanagh *et al.* (1988) explored the cardiopulmonary response to long-term endurance training in a large number (n=23) of relatively young cardiac transplant patients. A significant increase in the maximum work capacity, a lowering of arterial blood pressure and a moderate decrease in the resting and exercise heart rate were exhibited by the most compliant patients. Peripheral adaptations such as increased muscular strength and metabolic efficiency are likely mechanisms, although down-regulation of cardiac and peripheral adrenergic receptors in relation to frequent, extreme work loads (and resulting high circulating catecholamine levels) must be considered as well. Functional reinnervation was not a factor.

While the transplanted heart may easily double its cardiac output during stress (Clark *et al.*, 1973), the adaptive mechanisms are not perfect. In heart transplant patients, the maximum oxygen consumption (Clark *et al.*, 1973) and maximum work load (Banner *et al.*, 1989b) are generally about half that of normals although they can be improved somewhat by training, and training bradycardia is not strictly innervation-dependent (Kavanagh *et al.*, 1988). For a given level of O_2 consumption, serum lactate levels and AVO_2 difference are greater in transplant patients than in normal controls (Schroeder, 1979; Mettauer *et al.*, 1991) and the cardiac output in general tends to be in the low or normal range at rest or with exercise (Stinson *et al.*, 1972a).

Ventricular performance – A number of recent observations in heart transplant patients have confirmed that myocardial contractility is an intrinsic property, independent of autonomic innervation. Aside from a hypertension-associated increase in stroke work index in patients on cyclosporin, the haemodynamic function of the transplanted heart appears normal as long as 5 years postoperatively (Frist *et al.*, 1987). In the absence of severe graft fibrosis, serial radioventriculograms (McGiffin *et al.*, 1984) showed normal left ventricular ejection fractions from 6 to 21 months postoperatively. Hosenpud *et al.* (1989) found abnormally elevated exercise filling pressures and low peak work loads during supine exercise early after transplantation to be related to undersizing of the donor heart. But, this association was not present at 12 months. Most investigators have demonstrated completely normal intracardiac pressures at rest and normal ventricular contractile reserve, even when differences in afterload were controlled (Borow *et al.*, 1989; von Scheidt *et al.*, 1991).

In the detailed comparisons of the exercise studies mentioned above, one must be aware of whether or not work was performed in the upright or supine position since ventricular performance may be affected by variable loading conditions. In supine bicycle ergometry, for example, an increase in venous return due to leg elevation may increase preload (Hosenpud *et al.*, 1989). Supine exercise may also be associated with higher afterload due to an increased component of isometric work from handgrip.

Aside from stimulation of β-receptors by circulating catecholamines, mechanisms of increased ventricular contractility in the transplant patient are, of necessity, non-neural. The intrinsic Frank-Starling effect, already mentioned, effectively increases heart output as the input increases. The Bowditch treppe effect is another non-neural mechanism in which (by poorly known mechanisms) an increase in heart rate, from whatever cause, produces an increase in contractile force. The Bowditch effect was more apparent in intact anaesthetized dogs than in conscious control dogs (Higgins *et al.*, 1973), and has probably been demonstrated to be clinically significant in transplant patients (Ricci *et al.*, 1979). Moreover, as noted by Traill (1990), the Bowditch effect may be a source of experimental error in studies of ventricular function when rate is not controlled. Another factor to be considered (Kent and Cooper, 1974) is the Anrep effect. This is the intrinsic property of the heart to increase its contractile force as the aortic pressure (afterload) increases, independent of ventricular end diastolic volume. However the Anrep effect was not observed in transplant patients performing isometric exercise (see below). A clear elevation arterial pressure with no change in heart rate was not associated with a change in left ventricular contractility or cardiac output as measured by Doppler echocardiography (Robson *et al.*, 1989).

Isometric Exercise

As in dynamic exercise, the arterial pressure in transplant patients increases appropriately during isometric exercise. The mechanism appears to be almost entirely due to a reflex increase in systemic vascular resistance rather than an increase in cardiac output via the Frank-Starling mechanism. In fact heart rate, cardiac geometry, cardiac output and ventricular contractility are unchanged during handgrip exercise in transplant patients. Both diastolic and systolic blood pressures rise due to an increase in systemic vascular resistance. (Savin *et al.*, 1980; Haskell *et al.*, 1981; Robson *et al.*, 1989). Transplant patients' response to handgrip differs significantly form that observed in normal innervated subjects. Normal volunteers also manifest an increase in arterial pressure, although the mechanism is via an increase in cardiac output due to reflex tachycardia with little change in stroke volume or peripheral vascular resistance (Martin *et al.*, 1974; Robson *et al.*, 1989).

ELECTROPHYSIOLOGY AND PHARMACOLOGY

CONDUCTION SYSTEM

Electrophysiological studies in human cardiac transplant patients have demonstrated that many activities of the sinus and A-V nodes and conduction system are more a function of their intrinsic properties than the results of direct autonomic innervation. For example, Cannom, Graham and Harrison (1973) showed that at rest and during donor atrial pacing, atrium to His bundle (A-H) and His bundle to ventricle (H-V) conduction times were normal. Infusions of noradrenaline, adrenaline and isoprenaline by the same investigators (Cannom *et al.*, 1975) demonstrated intact β-receptor function in denervated, transplanted hearts and that the decrease in A-H conduction time was normal for the degree of tachycardia produced by these agents. In other words, intrinsic properties of the sinus node, A-V node and the His bundle define their reactivity to adrenergic stimuli. The absence of dual sympathetic and parasympathetic innervation may subtly modify the functional characteristics of the sinus and A-V nodes, however. Mason, Stinson and Harrison (1976) found that the A-V node functional refractory period did not decrease as expected with increased heart rate suggesting a primary importance of autonomic innervation in this structure. Electrophysiological studies in a series of 14 transplant patients (Bexton *et al.*, 1983d) detected a variety of subtle sinus and A-V node conduction abnormalities, although they were of no demonstrated clinical significance. In an analysis of the sinus node postpacing recovery phenomenon in transplantation patients, Mason (1980) found a smooth pattern of increasing donor sinus node recovery times (SNRT) with increasing rates of overdrive atrial pacing up to a maximum of 200 beats per min. This was in contrast to SNRT displayed by the transplanted patients' innervated recipient sinus node and normal volunteer sinus nodes. In these controls greater fluctuations in SNRT with actual decrements in post-pacing pauses at pacing rates greater than 150 beats per min were seen. This disparity may be secondary to the absence of vagally mediated acetylcholine release at the sinus node and perinodal tissues in the denervated donor heart in response to rapid atrial pacing. Bexton *et al.* (1984) observed similar predictable "intrinsic responses" to pacing of the donor sinus node. Accordingly, variable autonomic activation during rapid atrial pacing in innervated hearts may partially explain the lack of specificity of SNRT in identifying sinus node dysfunction in nontransplant patients (Gupta *et al.*, 1974; Crook *et al.*, 1977).

Recently, Heinz *et al.* (1990) showed that sinus node abnormalities early after transplantation were present in 47.6% of patients, comparable to the 50% reported by Mackintosh *et al.* (1982). These abnormalities appeared to be transient, however, even when related to prolonged ischaemic times during preservation hypothermia. Reevaluation on postoperative day 21 demonstrated sinus node abnormalities in only 21.4% of patients.   The need for a permanent pacemaker was predicted best by a slow early postoperative donor heart rate, and the function of lower pacemakers was unreliable in cases of sinus node failure.   Prolonged monitoring in a larger transplantation series (n=25) supported the observation that clinically significant sinus arrhythmias early after transplantation frequently resolve, even when related to prolonged donor heart ischaemic time and sinus node dysfunction (Jacquet et al., 1990). That information on the long-term natural history of conduction abnormalities after transplantation remains scanty, however, must be considered when interpreting electrophysiological data in transplanted hearts.

Arrhythmias

In a review of the first 47 heart transplantation patients treated at Stanford University before the cyclosporin era, Berke et al. (1973) and Schroeder et al. (1974) noted that cardiac arrhythmias, both atrial and ventricular, are frequently found on follow up, particularly in the presence of rejection and possibly ischaemia. These reports verified that innervation is not required for the generation of ventricular arrhythmias, even though cardiac denervation in the dog model seemed to protect against ischaemia-induced ventricular tachycardia (Ebert, Vanderbeek and Sabiston, 1970). While virtually every type of atrial and ventricular arrhythmia has been observed in cardiac transplant patients (Mason and Harrison, 1979), Bexton et al. (1983d) found no significant rhythm disturbance during ambulatory electrocardiographic monitoring in asymptomatic transplant patients, even though a significant number of them had conduction abnormalities on formal electrophysiological testing. In 25 transplant patients continuously monitored postoperatively until discharge from the hospital, a high incidence of ventricular premature beats and non-sustained ventricular tachycardia was not related to ischaemic time, but did correlate significantly with rejection episodes (Jacquet et al., 1990). A growing body of data supports re-entry as the primary mechanism for ventricular tachycardia, and it is conceivable that rejection and myocardial inflammation may be a trigger. A unique form of graft atherosclerosis is the major cause of donor heart loss after one year (Jamieson, 1992), and in this setting, myocardial infarction is not uncommon (Gao et al., 1989). The arrangement of surviving myocardial fibres and connective tissue in an infarct zone serves as a substrate for reentry and ventricular tachycardia in innervated patients (de Bakker et al., 1990) and presumably in transplanted hearts as well. The modulatory effects of denervation on ventricular arrhythmias have not been explored.

Cardiovascular Drugs

The cardiac transplantation model enables separation of autonomic and direct effects of various agents upon myocardial tissues. Leachman et al. (1971) noted in one transplant patient that neither the donor heart rate nor ventricular dP/dT changed after administration of 1.25 mg of digoxin intravenously, suggesting that acute cardiac effects of this drug are due primarily to its vagotonic activity. That the acute effects of digoxin on the sinus and A-V nodes are dependent upon autonomic innervation was verified by Goodman et al. (1975a, 1975b) although a modest direct effect upon the A-V node of chronic digoxin administration is manifested by a lowered threshold for Wenckebach block during rapid atrial pacing (Ricci *et al.*, 1978).

Quinidine has both autonomically mediated vagolytic (atropine-like) and direct depressant effects upon the conduction system. When the effect of quinidine on donor hearts was compared with that of normally innervated hearts (Mason *et al.*, 1977) and remnant atrial activity in the same transplant patients (Josephson *et al.*, 1974) the vagolytic effect was predominant. That is, quinidine increased heart rate and decreased A-H conduction time in innervated hearts, while in the denervated hearts, a direct depressant effect upon both the sinus and A-V nodes slowed the heart rate and A-H conduction time. In toto, these findings argue against the utility of digitalization prior to quinidine in transplanted hearts with rapid atrial fibrillation, although this has not been tested. Disopyramide, another class IA antiarrhythmic drug, has also been studied in transplant patients (Bexton *et al.*, 1983b). As with quinidine, the vagolytic activity of disopyramide was predominant over direct sinus and A-V node depressant effects as demonstrated by acceleration in the innervated recipient sinus node with concomitant slowing of the donor sinus rate and lengthening of A-V conduction time. On electrocardiograms, QRS complex widening and QT interval prolongation was similar in transplanted hearts to that observed in controls.

Tricyclic antidepressant drugs are used cautiously in most cardiac patients because of their anticholinergic and quinidine-like properties. The use of nortriptyline (Shapiro, 1991) and desipramine (Kay *et al.*, 1991) for major depressive episodes in small numbers of cardiac transplant patients was not associated with significant electrocardiographic or haemodynamic changes, although comparison with the donor atrium or innervated control hearts was not made.

The hypotensive and direct A-V node slowing effects of verapamil are theoretically accentuated (particularly after intravenous administration) in the cardiac transplant patients since reflex sympathetic stimulation of the heart rate is not available to maintain cardiac output in the face of peripheral vasodilatation (Qi *et al.*, 1987). Verapamil has been used without reported difficulty in cardiac transplant patients, although detailed analysis of its haemodynamic effects in this population have not appeared in the literature.

The short-term haemodynamic effects of diltiazem and nifedipine were compared in cardiac transplant patients and renal transplant patients. Striking differences in response to the drugs between the two groups were felt to be due mainly to cardiac denervation (Roy *et al.*, 1989). The hypotensive effect of a single dose of either drug was more pronounced in heart transplant than in renal transplant patients. In heart transplant patients, diltiazem produced hypotension with small decreases in the heart rate and cardiac index. The hypotension after diltiazem in renal transplant patients was associated with a reflex increase in both the heart rate and cardiac index. Nifedipine, a more potent vasodilator than diltiazem, had neither negative inotropic nor negative chronotropic effects upon transplanted hearts. However, the hypotensive effect of nifedipine was equal to that of diltiazem and was due primarily to a blunted heart rate response to vasodilatation in comparison with the renal transplantation group. Thus, in heart transplant patients each calcium blocker produced the same magnitude of blood pressure lowering, but by differing effects on the cardiovascular system. Furthermore, one may surmise the potentially detrimental effects of the injudicious use of potent vasodilator agents (hydralazine, nitroprusside, nifedipine) in patients with cardiac sympathetic denervation, particularly in settings of volume depletion.

The human heart transplantation model was used by Epstein *et al.* (1990) to test the hypothesis that the bradycardiac effects of low doses of atropine are centrally mediated. In cardiac transplant patients the innervated, remnant sinus node exhibited slowing in response to low doses of atropine and acceleration after high doses of atropine. The denervated donor heart, however, was unaffected by either high or low therapeutic doses of atropine, supporting the theory that, regardless of dose, atropine's cardiac effects are predominantly centrally mediated.

The effects of adrenergic agonists and β-blockers upon transplanted hearts are discussed elsewhere in this review.

HEART RATE – INTRINSIC MECHANISMS

Although pharmacological methods have been used to study intrinsic cardiac function (Jose and Taylor, 1969), anatomically denervated heart models avoid the direct cardiac and systemic effects of β-blockers and atropine. Interest in intrinsic mechanisms of heart rate control has been heightened by observations of cardioacceleration during exercise despite β-blockade in denervated (Donald and Samueloff, 1966) and transplanted (Bexton *et al.*, 1983c; Yusuf *et al.*, 1985) hearts. It is important to distinguish this phenomenon from the Bainbridge reflex (Bainbridge, 1915) which is cardioacceleration resulting from right atrial stretch receptor activation of a neural reflex arc (Bishop, Malliani and Thoren, 1983). Since the donor heart is denervated, by definition, the Bainbridge reflex is not operative in transplant patients, as demonstrated by an absence of heart rate change after volume loading by passive leg raising sufficient to raise left ventricular end-diastolic volume and cardiac output (Pope *et al.*, 1978). Yet, in isolated mammalian hearts a significant chronotropic effect of increasing right atrial pressure can be demonstrated in the absence of innervation (Blinks, 1956). Furthermore, power spectral analysis of heart rate variability in transplant patients has provided additional evidence of an intrinsic, possibly stretch-related mechanism of heart rate variability in transplanted hearts (Bernardi *et al.*, 1989, 1990; Zeuzem *et al.*, 1991). The high-frequency component of beat-to-beat changes in the R-R interval significantly increases with deep respiration in transplanted and atropine-treated hearts. Each transplant patient (n=10) demonstrated acceleration of the heart rate in phase with inspiration and deceleration of the heart rate in phase with expiration (Zeuzem *et al.*, 1991). In a review Pathak (1973) proposed that autoregulation of the chronotropic response through pacemaker stretch via increased venous return is analogous to the well-established inotropic response of the heart to stretching of myocardial fibres (Frank-Starling mechanism).   Autoregulation of the heart rate by pacemaker stretch may represent a primitive adaptation mechanism with phylogenetic variability – assuming more importance in lower avian and reptilian species which lack sophisticated neurohumoral responses (Pathak, 1973). Changes in the donor and recipient atrial rates as manifested by atrial accrochage or entrainment (discussed above) also support the existence of an important intrinsic mechanism of heart rate variability. The mechanical tugging of one atrium may affect the rate of the atrium to which it is sutured.   Because of the difficulties in achieving complete cardiac denervation in *in situ* animal models (discussed elsewhere in this review), the human cardiac transplant patient may be an ideal model for further investigation of this putative neurally-independent phenomenon.

POWER SPECTRAL ANALYSIS OF HEART RATE VARIABILITY

Power spectral analysis of heart rate variability from the surface electrocardiogram is a relatively new non-invasive method of assessing the integrity of neural feedback mechanisms and the phasic and tonic components of efferent sympathovagal control of the heart (Malliani *et al.*, 1991a, 1991b). Orthotopic cardiac transplant patients are a unique model in which to study beat-to-beat variation in the anatomically denervated heart.   Transplant patients have been used to validate changes in rate variability observed in pharmacological studies and observations made in autonomic failure patients such as diabetics (Zeuzem *et al.*, 1991). Prominent, high-frequency respiratory oscillations in heart rate variability are not present in transplant patients and disappear in normals treated with atropine (Zeuzem *et al.*, 1991), but not in normals treated with α- or β-blockers (Pfeifer *et al.*, 1982). Thus, alterations in the high frequency component are reflective of cardiac vagal activity. Certain low-frequency components of R-R variability are associated with combined vagal and sympathetic activation. Observations of persistent respiratory oscillations in R-R variability in denervated patients (Bernardi *et al.*, 1989) may be due to intrinsic stretch-induced changes in heart rate (see Heart Rate – Intrinsic Mechanisms). Fallen *et al.* (1988) used spectral analysis of heart rate variability to demonstrate possible functional reinnervation in a transplanted human heart.

CORONARY VASCULAR REACTIVITY

Hodgson *et al.*   (1989) used quantitative angiography, subselective Doppler catheterization and regional α- and β-blockade in a large number of heart transplant patients and innervated controls to assess the role of adrenergic tone in the coronary circulation under resting conditions.   The methodological problems associated with measuring coronary blood flow in small resistance vessels is considerable. Nonetheless, the data acquired by Hodgson *et al.*   (1989) strongly suggests that in resting patients, α-mediated coronary vascular tone is negligible in both transplanted and in control hearts. No change in coronary vascular resistance or in coronary flow reserve was seen in either group after α-blockade or selective β1-blockade, although non-selective β-blockade did increase vascular resistance, presumably via β2-receptor inhibition. The effect of adrenergic innervation on coronary flow reserve under stress conditions was not studied, however.

Epicardial coronary vasospasm in the cardiac transplant patient has been reported by several catheterization laboratories (Buda *et al.*, 1981; Cattan *et al.*, 1988; Kushwaha, Mitchell and Yacoub, 1990), demonstrating conclusively that vascular reactivity is not dependent upon innervation. While direct sympathetic innervation contributes to coronary vasospasm (Yasue *et al.*, 1976), and circulating catecholamines may play a role in the denervated patient, in fact, the poorly understood mechanisms of coronary vasospasm probably involve a combination of neural, humeral and endothelial factors (Kaski *et al.*, 1986). Intracoronary infusion of acetylcholine in transplant patients produces vasospasm at sites of coronary stenosis as well as in angiographically normal segments (Fish *et al.*, 1988; Nellessen *et al.*, 1988; Rowe *et al.*, 1991). Since the normal vasodilatory effect of acetylcholine in coronary arteries is thought secondary to stimulated release of endothelial relaxation factor (Furchgott and Zawadzki, 1980; Furchgott and Vanhoutte, 1989), endothelial damage whether from rejection, atherosclerosis or ischaemia may explain paradoxical coronary vasoconstriction in transplant patients (Ludmer *et al.*, 1986).

DENERVATION HYPERSENSITIVITY – ADRENERGIC

The paravertebral location of cardiac sympathetic ganglia insures sectioning of all postganglionic sympathetic axons during procurement of the donor heart. Removal of the sympathetic nerve terminal catecholamine uptake-1 apparatus, up-regulation of β-receptors and enhancement of post-receptor processes such as adenylate cyclase activity are all plausible mechanisms for hypersensitivity to certain sympathomimetic agents observed in transplanted hearts. Elimination of presynaptic nerve terminals theoretically slows local clearance of endogenous or exogenously administered uptake-1 sensitive catecholamines such as adrenaline and noradrenaline from cardiac effector sites, thereby promoting hypersensitivity. Moreover, elimination of neuronal noradrenaline release could influence β-receptor density and post-receptor transduction events. Data from animal models aimed at defining modes of cardiac adrenergic hypersensitivity have been mixed. The inotropic response of isolated feline hearts to isoprenaline (no presynaptic uptake) and noradrenaline suggested a primarily presynaptic mechanism of noradrenaline hypersensitivity (Dempsey and Cooper, 1968). Inotropic and chronotropic responses to noradrenaline, isoprenaline and prenalteral (a β1-specific agonist without presynaptic uptake) in chronically denervated, awake dogs also supported a predominance of a presynaptic mechanism of noradrenaline hypersensitivity. However, an increase in β-receptor density and enhanced coupling to adenylate cyclase seemed to explain a slight hypersensitivity to isoprenaline and to prenalteral which was observed only when neural reflex mechanisms in control animals were blocked (Vatner *et al.*, 1985). In non-working, abdominally placed heterotopic cardiac grafts in rabbits and rats, both β-receptors and isoprenaline-stimulated adenylate cyclase levels were elevated (Lurie, Bristow and Reitz, 1983).

In human heterotopic and orthotopic cardiac grafts Yusuf *et al.* (1987) observed an exaggerated chronotropic response to isoprenaline, suggesting up-regulation of β-receptors in transplanted hearts. Importantly, however, normal volunteer controls and control remnant hearts in that investigation were not pretreated with atropine. Subsequent human studies have supported the concept that significant up-regulation of β-receptors does not occur and that the mechanism of adrenergic hypersensitivity in cardiac transplantation is purely presynaptic in origin (Gilbert *et al.*, 1989). Both the inotropic (von Scheidt *et al.*, 1992) and chronotropic (Gilbert *et al.*, 1989; Quigg *et al.*, 1989; Port *et al.*, 1990a; Horn *et al.*, 1991) responses of donor hearts to isoprenaline were unchanged from either atropine-treated controls or recipient atrial responses. The chronotropic response to adrenaline (uptake-1 sensitive) was exaggerated in the donor heart relative to the recipient atrial response in the face of unaltered β-receptor density (Gilbert *et al.*, 1989). von Scheidt *et al.* (1992) similarly observed supersensitivity to the inotropic effects of adrenaline but not isoprenaline and tested the dependence upon uptake-1 mechanisms by pretreating control subjects with the uptake-1 inhibitor desipramine. In this setting, the control subjects acquired adrenaline hypersensitivity equivalent to that of the transplant patients.

That normal β-receptor density and adenylate cyclase levels were found in donor right ventricular myocardial biopsies (Denniss *et al.*, 1989) also supports the hypothesis of a purely presynaptic mechanism of adrenergic hypersensitivity in the denervated human heart. Although the total myocardial β-receptor density post-transplantation may be normal, subtype populations appear to be altered. In whole human allografts with normal ventricular function at the time of retransplantation Port *et al.* (1990b) observed normal total β-receptor density, but a surprising increase in the ordinarily small percentage of myocardial β2-receptors. In serial right ventricular biopsy specimens obtained from eight donor hearts before transplantation and up to 18 months after transplantation, Brodde, Khamssi and Zerkowski (1991) also found no difference in β-receptor density but a progressive decrease in the β1:β2 receptor ratio. The mechanism of an increasing β2-receptor population is not known, although an association with hypertension or cyclosporin therapy — both generally present in transplant patients – has been suggested (Brodde, Khamssi and Zerkowski, 1991). The pathophysiological consequences of increased β2-receptors is not known. However, an increase in the importance of circulating adrenaline in modifying the rate and inotropic state of the transplanted heart might be expected, since the affinity of adrenaline for β2-receptors is approximately 100-fold greater than that of noradrenaline, while the two endogenous catecholamines have almost equal affinities for β1-receptors. (Lands *et al.*, 1967).

In congestive heart failure, partial cardiac denervation exists due to high autonomic drive with attendant myocardial catecholamine depletion (Chidsey, Braunwald and Morrow, 1965), decreased noradrenaline uptake (Sandoval *et al.*, 1989) and β2-receptor down-regulation. Descriptions of β1-receptor down-regulation out of proportion to the β2-receptor population (Bristow *et al.*, 1986) and altered coupling of β-receptors to adenylate cyclase in end-stage congestive heart failure (Bristow *et al.*, 1989) may assist in devising optimal strategies for inotropic support in these patients. Similarly, mechanisms of adrenergic denervation hypersensitivity and emerging knowledge of altered post-receptor events has clinical implications in the management of transplant patients. Because myocardial catecholamines are absent after cardiac transplantation (Regitz *et al.*, 1990), inotropic support of the donor heart is better effected by direct-acting β-adrenergic agents such as isoprenaline or adrenaline, rather than an indirect agent such as dopamine which relies partly on presynaptic release of noradrenaline for its effects on the myocardium (Port *et al.*, 1990a). Furthermore, absence of the cardiac presynaptic uptake-1 mechanism may augment the inotropic effect of adrenaline in heart transplant patients by endowing it with a higher β- to α-adrenergic (or inotropic to vasoconstrictor) effect ratio than in innervated patients (Bristow, 1990). In addition, the increased percentage of myocardial β2-receptors after transplantation means that the non-selective β-agonists adrenaline, isoprenaline and dobutamine would potentially occupy more β-receptors, providing greater inotropic support than relative β1-selective agents such as noradrenaline. The implications of increased myocardial β2-receptors after transplantation (Brodde, Khamssi and Zerkowski, 1991) are not fully understood, and the cardiac effects of β2-specific agonists in cardiac transplant patients have not been reported.

DENERVATION HYPERSENSITIVITY – CHOLINERGIC

After cardiac transplantation or local neural ablation experiments, postganglionic sympathetic effector sites are effectively denervated and demonstrate (above) adrenergic supersensitivity. Often not appreciated, however, is that transplantation models of cardiac denervation result in preganglionic sectioning of cardiac vagal efferents. It is, in fact, the intrinsic parasympathetic (nicotinic) ganglion cell which is denervated, while cardiac effector sites (muscarinic) are decentralized but not, in the absence of ganglionic cell dysfunction or degeneration, denervated. Unchanged levels of total myocardial acetylcholinesterase after parasympathectomy. (Willman *et al.*, 1964; Woods, 1970) suggests that ganglion cell degeneration does not occur allowing continued regulation of muscarinic receptors. Herein may lie the explanation for why the heart was until recently regarded as an exception to Cannon's "law of denervation" (Cannon and Haimovici, 1939) in regard to cholinergic responsivity, while skeletal muscle sensitivity to acetylcholine, for example, increases dramatically following denervation (Miledi, 1960). Very limited data concerning measurement of myocardial muscarinic receptor density in denervated animals has been contradictory, possibly because of inter-species differences or variance in method. Roeske *et al.* (1978) reported a 45% reduction in rat left ventricular acetylcholinesterase activity and a significant increase in muscarinic receptor density in non-working abdominal heterotopically transplanted hearts. In surgically denervated, intact (working) canine hearts Vatner *et al.* (1985) found up-regulation of β-receptors with evidence for presynaptic hypersensitivity to noradrenaline, but down-regulation of muscarinic receptors. This group pointed out, however, that absence of muscarinic cholinergic receptors located on presynaptic sympathetic nerves (Sharma and Banerjee, 1978) and not down-regulation could be the explanation for their findings. In pharmacological studies, Hageman, Urthaler and James (1977) failed to demonstrate acetylcholine hypersensitivity after extrinsic parasympathectomy in dogs. In a canine model of chronic cardiac denervation, supersensitivity to the negative inotropic effects (Priola and Spurgeon, 1977) and the negative dromotropic effects (Priola *et al.*, 1983) of intracoronary acetylcholine was not present. However, supersensitivity to intracoronary nicotine was demonstrated, implying that denervated intrinsic cardiac parasympathetic ganglia do obey Cannon's law of denervation supersensitivity.

Before responses to parasympathomimetic agents in cardiac denervation are understood, the degree of parasympathetic ganglia intactness in various models, including orthotopic transplantation, must be ascertained. Moreover, reflex and direct (Sharma and Banerjee, 1978) effects of acetylcholine upon sympathetic neuronal noradrenaline release in the innervated heart must be considered. Unlike animal models of denervation in which the atria are sometimes left intact, human cardiac transplantation removes portions of the donor posterior atrial wall, which may contain parasympathetic ganglia.   While preganglionic parasympathetic innervation of the heart has been described in great detail (Randall, 1984), descriptions of intrinsic cardiac parasympathetic ganglia have only recently become available. In the dog, parasympathetic innervation of the sinus and A-V nodes by anatomically distinct (Ardell and Randall, 1986) and functionally independent projections of the vagus (Lazzara *et al.*, 1973; O'Toole, Ardell and Randall, 1986) has been demonstrated. Blomquist, Priola and Romero (1987) observed a lack of significant ventricular parasympathetic ganglia in the canine model. Phenol disruption of neural fibres traversing the atrioventricular groove reduced the negative inotropic effect of intracoronary nicotine, while the response to acetylcholine was unchanged. Detailed structure-function experiments by Randall *et al.* (1986) localized preganglionic parasympathetic projections subserving canine sinus and A-V nodes to small epicardial fat pads containing nerve fibres and neural ganglia. Removal of a small fat pad at the junction of the inferior vena cava and inferior left atrium selectively eliminated A-V node (negative dromotropic) responsiveness to vagal stimulation without affecting the heart rate (sinus node) response. Dissection of a small fat pad overlying the right pulmonary vein-left atrial junction selectively abolished sinus node responsiveness to vagal stimulation, without affecting the A-V node-function (Randall *et al.*, 1987). A similar arrangement of epicardial fat pads containing parasympathetic ganglia and fibres selectively innervating the sinus node and A-V node were also observed in the non-human primate (Billman *et al.*, 1989).

Since significant interspecies variability in cardiac autonomic innervation occurs, it was not known whether comparable observations in humans existed until Carlson *et al.* (1992) described the presence of nerve fibres and ganglia in a fat pad about 1 cm in diameter located near the sinus node at the margin of the right superior pulmonary vein, the superior vena cava and the right atrium in human autopsies. Intraoperative stimulation of this fat pad in patients undergoing cardiac surgery (n=13) increased the sinus cycle length without affecting A-V node conduction. Presumably, reports of selective A-V node parasympathetic fibres and ganglia in the human will be forthcoming as well. Kaseda and Zipes (1988) provided pharmacological evidence in dogs that these small epicardial fat pads contain large numbers of intrinsic cardiac parasympathetic ganglia. Careful dissection of these structures produced complete cardiac parasympathetic denervation, presumably at the postganglionic level, and supersensitivity to intracoronary acetylcholine, whereas prior investigations of preganglionic parasympathectomy (Priola *et al.*, 1983; Smith, Priola and Anagnostelis, 1985) demonstrated supersensitivity to the affects of nicotine but not acetylcholine.

Although new information concerning human cardiac parasympathetic innervation is emerging, to date there have been no reported studies of direct-acting parasympathomimetic agents in cardiac transplant patients. If, as indicated by Carlson *et al.* (1992), the locations of human cardiac parasympathetic ganglia are similar to the dog and non-human primate, it is likely that an unknown portion of them would be excluded during donor procurement or injured during preservation or in surgery. It is, therefore, difficult to predict whether parasympathetic denervation hypersensitivity (if present) would predominate at the nicotinic (ganglionic) level or at the muscarinic level at given effector sites (myocardium, sinus node or A-V node). Furthermore, the functional status of chronically denervated parasympathetic ganglia, even if present, in transplanted heart is not known. Immunohistochemical studies of explanted human cardiac allografts found that intrinsic cardiac ganglia were present in the atrial subepicardium and the atrioventricular groove but not in the ventricles (Wharton *et al.*, 1990). Thus, denervated parasympathetic ganglia do not necessarily degenerate and may retain some degree of functional activity with a modulatory effect on muscarinic receptors. However, Stemple *et al.* (1978) showed that edrophonium, an acetylcholinesterase inhibitor, had no effect on the donor heart sinus or A-V node activity in transplant patients, implying functionally inactive postganglionic cholinergic fibres which do not accumulate or release acetylcholine. Ellenbogen *et al.* (1990) hypothesized that the exaggerated negative chronotropic and negative dromotropic responses to intravenous adenosine cardiac transplant patients could be due to up-regulation of post-muscarinic receptor transduction mechanisms which are coupled to adenosine receptors. On the other hand, the contribution of absent reflex and tonic sympathetic activity to this "hypersensitive response" is not known. Intracoronary acetylcholine has been administered to human cardiac transplant patients in studies of vascular reactivity (Fish *et al.*, 1988; Nellessen *et al.*, 1988; Rowe *et al.*, 1991). But, injection into the right coronary artery, which gives rise to the artery of the sinus node, was intentionally avoided; and negative dromotropic and negative chronotropic supersensitivity to acetylcholine were not specifically examined. There are currently no reports of nicotine administration in human cardiac transplantation patients.

CARDIAC REINNERVATION – ANIMAL MODELS

The technique of autograft transplantation (Lower, Stofer and Shumway, 1961) has been applied to a number of animal models so that the physiological effects of denervation and reinnervation may be studied in the absence of limited recipient survival and organ dysfunction due to rejection. Functional cardiac reinnervation occurs after autotransplantation across species and in allografted dogs within months of surgery and approaches that of control animals within 1-2 years.

Efferent Reinnervation

In four autotransplanted dogs surviving more than 2 years after surgery, Dong *et al.* (1964) reported respiratory sinus arrhythmia in two animals; and in all four, responses to intravenous tyramine, noradrenaline, acetylcholine and direct electrical stimulation of the vagus nerve were indicative of extrinsic cardiac innervation. However, the time course of reinnervation was not known. In a larger series, Willman, Cooper and Hanlon (1964) followed reinnervation in 25 autotransplanted dogs using pharmacological methods and direct electrical stimulation of the vagus nerve and stellate ganglion. Gradual return of sympathetic and parasympathetic neural connections were evident 12-24 months postoperatively, but reinnervation was not detected during the first 11 months. Subsequent investigators have found that consistent, if limited, autonomic reinnervation occurs much earlier – usually within 3-6 months after transplantation (Kontos, Thames and Lower, 1970). This difference may be due to more judicious use of anaesthetic agents during autonomic reflex testing. This time course of reinnervation is in agreement with observed regrowth rates for myelinated fibres (Guth, 1956). After autotransplantation, Peiss *et al.* (1966) demonstrated atropine-sensitive slowing of the heart rate with vagal nerve stimulation at 26 days postoperatively in one dog, with evidence of parasympathetic and sympathetic reinnervation in another dog at 74 days. Individual dogs studied at 1, 2 and 3 years exhibited near-normal functional cardiac innervation. Nonetheless, histological evidence for the return of probable adrenergic fibres in association with cardiac blood vessels was evident only by 406 days of autotransplantation in dogs (Norvell and Lower, 1973), suggesting that only a minimal number of regenerated fibres are required for functionally detectable reinnervation.

Kontos, Thames and Lower (1970) also confirmed the presence of early (3-6 months) return of cardiac autonomic responses in autotransplanted dogs. Significantly, their findings were extended to allotransplanted dogs treated with methylprednisolone and azathioprine, suggesting that host and recipient immunological differences do not absolutely preclude reinnervation. In using Holstein calves, Seki and Danielson (1971) provided new evidence that reinnervation after autotransplantation is not species specific. In Rhesus monkeys and in dogs, Kondo, Matheny and

Hardy (1972) showed that reinnervation occurred within 2-4 months of surgery, although Willman and Hanlon (1969) reported no signs of reinnervation in a single autotransplanted Baboon 1 year after surgery. Kaye, Wells and Tyce (1979) showed that reinnervation after surgical cardiac denervation in dogs was enhanced by nerve growth factor.

Afferent Reinnervation

As reviewed by Coleridge and Coleridge (1979), unmyelinated C-fibres travelling with and actually greatly outnumbering myelinated vagal efferents are variably responsive to mechanical and chemical stimuli and are purported to be the afferent limb of a variety of vagally mediated cardiac reflexes. In fact, an extensive system of afferent vagal C-fibres exists in the great vessels, lungs and abdominal viscera, as well as the heart (Coleridge and Coleridge, 1977). Intracoronary and epicardial veratrum alkaloid application has been used to identify specific sites of increased left ventricular chemoreceptor activity (Dawes, 1947). Mechanical and locally produced chemical stimuli such as bradykinin, prostaglandins and adenosine undoubtedly activate both sympathetic and parasympathetic afferent pathways.

Mechanoreceptors

While cardiac efferent reinnervation in animals occurs early after auto- and allotransplantation, significant afferent mechanoreceptor reinnervation has been difficult to demonstrate, even up to 2 years postoperatively (Thames et al., 1971). A putative reflex loop for maintaining circulating volume homeostasis has been described. Well loaded, volume-sensitive cardiac stretch receptors tonically inhibit renal sympathetic nerve activity via central mechanisms, which in turn suppress renin release. Even though baseline renin levels were within the normal range, Thames et al. (1971) confirmed earlier findings (Willman et al., 1967) of abnormally increased circulating blood volumes in autotransplanted dogs. Two years post-transplantation, the rise in plasma renin was significantly blunted relative to sham-operated animals after volume depletion by haemorrhage or diuretic administration, suggesting a lack of cardiac afferent reinnervation. Using a different approach, Mohanty et al. (1986b) concluded that cardiac "reafferentation" can occur long after transplantation. In three dogs 8-12 years post-autotransplantation, acute volume loading produced renal nerve activity inhibition. However, renal nerve suppression was not significantly different from short-term autotransplanted dogs, possibly due to compensatory afferent receptor activation in the remnant atria and great vessels. The response to ventricular chemoreceptor stimulation with veratrum alkaloid in the same animals (see below) more strongly supported late afferent reinnervation.

Chemoreceptors

The Bezold-Jarisch effect is a cardioinhibitory, vasodepressor response to activation of afferent ventricular chemoreceptors by veratrum alkaloids (Dawes and Comroe, 1954; Krayer, 1961). This loop of cardiac chemoreceptor stimulation of vagal efferent fibres has also been examined as an indicator of cardiac afferent reinnervation. That afferent reinnervation is possible was first suggested by Thames *et al.* (1971), who elicited bradycardia and hypotension in response to left ventricular injection of the cryptenamine (a mixture of veratrum alkaloids) in a single dog 2 years following autotransplantation. Mohanty *et al.* (1986b) found that left ventricular cryptenamine injection in a small number of recently autotransplanted dogs (n=2) did not produce a Bezold-Jarisch response. However, in a small number of long-term autotransplanted dogs, evidence for significant afferent reinnervation was obtained. In three dogs from 8-12 years post-autotransplantation, injection of cryptenamine into the left ventricular cavity produced marked suppression of renal nerve activity and significant hypotension.

Myocardial Catecholamines

The myocardium is richly supplied by postganglionic sympathetic fibres and contains high levels of catecholamines (Shore *et al.*, 1958). After autotransplantation in dogs, myocardial noradrenaline levels are rapidly and almost completely depleted (Cooper *et al.*, 1962). In the event of sympathetic reinnervation with adrenergic fibre regrowth, one would predict a longitudinal increase in myocardial noradrenaline. Willman *et al.* (1963) measured near-normal levels of catecholamines in left atrial and left ventricular myocardium in two reinnervated dogs 1 year after autotransplantation. However, in serial left ventricular biopsies post-autotransplantation in six dogs, Ebert and Sabiston (1970) found that myocardial catecholamine remained very depleted at 18 months, although somewhat higher than the negligible levels measured at 4 months.

By measuring myocardial noradrenaline content in all four cardiac chambers in dogs followed for 2 years, Kaye *et al.* (1977) provided evidence that reinnervation (as logic would dictate) occurs first in both atria, followed by a ventricular base-to-apex sequence. In agreement with Ebert and Sabiston (1970) the "reinnervated" myocardial noradrenaline levels were but a fraction of normal, prompting the speculation that up-regulation or hypersensitivity of adrenergic receptors must be in part responsible for "normalization" of cardiac autonomic reflexes with only partially regenerated adrenergic fibres. More recently, Mohanty *et al.* (1986a) measured myocardial noradrenaline, adrenaline and dopamine levels in early and late transplanted dogs. Their findings confirmed those of previous investigators (Peiss *et al.*, 1966; Kaye *et al.*, 1977; Pierpont, DeMaster and Cohn, 1984) showing baseline regional differences in myocardial catecholamine content with atrial levels significantly higher than ventricular levels and ventricular base levels modestly, but significantly higher than in the ventricular apex in sham-operated dogs. Yet, shortly after autotransplantation, noradrenaline levels became negligible in all four cardiac chambers.

To date there are no reports of myocardial acetylcholine or acetylcholinesterase levels or other histological markers of intrinsic cardiac parasympathetic fibres indicative of post-transplantation parasympathetic reinnervation.

CARDIAC REINNERVATION – HUMANS

FUNCTIONAL REINNERVATION?

While ample evidence exists for functional reinnervation in animals after cardiac transplantation, the issue of reinnervation in human subjects has become increasingly controversial and has centred mainly around cardiac sympathetic efferents (Bristow, 1990). Until recently it had been accepted that cardiac reinnervation does not occur in humans. Isolated reports suggestive of clinically detectable cardiac reinnervation are found in the literature. Fallen et al. (1988) reported a single patient with a near-normal power spectral analysis of heart rate variability, and responses to exercise consistent with intact sympathetic and vagal tone 33 months after orthotopic cardiac transplantation. Rudas et al. (1991) observed that patients studied 33-83 months after cardiac transplantation (n=18) had developed earlier heart rate increases with orthostasis and exercise and earlier heart rate decreases post-exercise when compared to early cardiac transplant patients. In this group, it is unclear whether improved heart rate responsiveness implies reinnervation or the development of intrinsic mechanisms of heart rate control (Bernardi et al., 1989). Although Wilson et al. (1991) found that resting donor heart rates are slightly slower long after cardiac transplantation, it is possible that progressive donor sinus node dysfunction (Bexton et al., 1984) is as plausible an explanation as parasympathetic reinnervation. In a single patient 3 years post-transplantation, Rudas, Pflugfelder and Kostuk (1992) reported heart rate deceleration during a vasodepressor syncope episode associated with instrumentation. Appropriate orthostatic cardioacceleration and deceleration in the same patient suggested parasympathetic reinnervation. Increasingly reported are anecdotal cases of angina-like chest pain in association with acute ischaemic events late after cardiac transplantation (see below). Not to be forgotten, however, are the observations of many investigators, following patients up to 61 months post-cardiac transplantation, who have failed to demonstrate physiological or pharmacological evidence for reinnervation (Stinson et al., 1972a; Pope et al., 1980). In fact, available data suggest that functionally apparent cardiac reinnervation rarely, if ever, occurs in humans after orthotopic cardiac transplantation. On the other hand, new and growing evidence for limited, heterogeneous autonomic reinnervation as detected by more sensitive probes (discussed below) is becoming increasingly convincing. Continued exploration of the mechanisms involved may indeed have clinical implications for both the growing numbers of long-term heart transplant survivors and for autonomic failure patients.

Myocardial Catecholamine Levels and Noradrenaline Release

Regitz et al. (1990) found no measurable noradrenaline in right ventricular septum biopsy specimens in a small number of long-term (>5 years) cardiac transplantation survivors. However, Wilson et al. (1991) recently described the first laboratory evidence of possible reefferentation of the heart in cardiac transplant patients using a potentially more sensitive probe of postganglionic sympathetic fibre intactness. Tyramine causes the release of noradrenaline from adrenergic nerve terminals by displacing it from large storage vesicles. Employing a technique developed by Forman *et al.* (1984) 12 patients studied within 5 months of cardiac transplantation had no elevation of coronary sinus noradrenaline in response to intracoronary tyramine administration, consistent with complete cardiac sympathetic denervation and post-transplantation myocardial catecholamine depletion (Cooper *et al.*, 1962; Regitz *et al.*, 1990). However, in 78% of 50 patients studied more than 1 year after cardiac transplantation, a significant, if subnormal, elevation in coronary sinus noradrenaline occurred in response to tyramine. Furthermore, coronary sinus noradrenaline increased after sustained handgrip manoeuvres in a smaller percentage of the same patients, implying reflex release of noradrenaline from centrally stimulated extracardiac sympathetic ganglia with reformed cardiac axons. The lower number of responses to handgrip manoeuvres is not surprising, since handgrip is a relatively poor stimulus for reflex sympathetic activation (Robertson *et al.*, 1979). Moreover, the degree of reinnervation probably varies widely between patients. The possibility of the intrinsic cardiac neural ganglion cell origin of noradrenaline has been raised (Gu and Muralidharan, 1992), although the likelihood of this has been disputed (Wilson, McGinn and Stark, 1992). It is unlikely that noradrenaline was released into the coronary sinus from remnant recipient atrial tissue, since the coronary sinus does not drain this region nor do donor coronary arteries perfuse the area.

While not supporting the hypothesis of cardiac sympathetic reinnervation in humans, the negative biopsy data of Regitz *et al.* (1990) do not rule out limited sympathetic reinnervation of uncertain functional significance. In denervated dogs, Kaye *et al.* (1977) found evidence for a logical atrial and then ventricular base-to-apex sequence of reinnervation. During routine myocardial biopsies with sampling primarily of the mid and apical right ventricular septum (Baughman, 1990), atrial and heterogeneous, basal ventricular innervation would not be detected.

Nuclear Medicine Techniques

Radioisotopes of pharmaceuticals which trace the neuronally dependent noradrenaline uptake-1 mechanism (Crout *et al.*, 1964; Tobes *et al.*, 1985) have been applied as quantitative, non-invasive methods of assessing regional cardiac adrenergic innervation. Iodine-123 metaiodobenzylguanidine ([^{123}I] MIBG) is a noradrenaline analogue developed as an imaging agent for pheochromocytomas (Sisson *et al.*, 1981). Using traditional imaging techniques and single photon emission tomography (SPECT), [^{123}I] MIBG scanning has shown reduced cardiac noradrenaline uptake in patients with idiopathic dilated cardiomyopathy (Schofer *et al.*, 1988) post myocardial infarction (Dae *et al.*, 1986) and with diabetic autonomic neuropathy (Sisson *et al.*, 1987). Glowniak *et al.* (1989), the first to apply this technique to cardiac transplant patients, showed absent active cardiac noradrenaline uptake in the early postoperative period.

Because reinnervation in humans is likely to be limited and heterogeneous, positron emission tomography (PET) is probably superior to conventional (SPECT) methods, due to improved spatial resolution. Wieland *et al.* (1990) validated neuronal mapping of the canine left ventricle using the false neurotransmitter

6-[^{18}F] fluorometaraminol and PET. [^{11}C] Hydroxyephedrine ([^{11}C] HED) also specifically traces the uptake and storage of noradrenaline in presynaptic adrenergic nerve terminals. Schwaiger et al., (1990) showed homogeneous uptake of [^{11}C] HED throughout the left ventricle in normal humans with an 82% global reduction in left ventricular uptake and storage in patients examined early after cardiac transplantation. [^{11}C] HED PET studies were extended to specifically address the question of adrenergic fibre reinnervation in a group of patients more than 2 years status post-cardiac transplantation (Schwaiger et al., 1991). In the late cardiac transplant patients, left ventricular apex and inferior wall tracer retention was not improved over that seen in early transplant patients. However, considerable tracer retention in the anterior basal and anterior septal walls approaching 60% of normal was present, strongly suggesting a heterogeneous base-to-apex progression of late reinnervation. Of note, however, is that tracer uptake in the thin-walled atria and right ventricle could not be quantified because of spatial resolution limitations and that no evidence of functional reinnervation could be detected by routine clinical testing in these patients.

Immunohistochemical and Ultrastructural Studies

As with investigations of myocardial catecholamines (above), studies seeking histo-logical evidence for cardiac reinnervation in man remain limited and controversial. Observations of sparse myocardial ganglion cells or nerves using conventional his-tological methods have been made in postmortem cardiac or biopsy specimens from long-term cardiac transplant patients (Milam et al., 1970; Rowan and Billingham, 1988). In a cross-sectional study of patients at variable times up to 12 years after cardiac transplantation, electron microscopic evaluation of neural elements in right ventricular biopsy specimens showed no change in axon or axon bundle density over time (Rowan and Billingham, 1988). In one patient previously studied with [^{11}C] HED PET (Schwaiger et al., 1991) and found to have increased uptake of tracer in the basal portion of the left ventricle, thorough histological evaluation of the entire heart was possible because of death 60 months after transplantation. Although an increased density of axonal tissue was found in the base compared with the apex, in keeping with the decreasing gradient of tracer uptake from base to apex, histologi-cal techniques in this and the other studies (above) could not distinguish between possibly regenerated adrenergic fibres and possible intrinsic parasympathetic fibres that had never degenerated. Wharton et al. (1990) applied sensitive immunohisto-chemical methods for neural marker proteins and enzymatic or peptide transmitter content in an attempt to characterize cardiac nerve type in non-transplanted and previously transplanted hearts. Heart-lung and heterotopic cardiac allografts ex-planted at the time of retransplantation appeared to contain only viable intrinsic parasympathetic nerves and ganglia with an atrial to ventricular gradient.

Afferent Reinnervation

The scant information that exists on reinnervation of the human heart pertains to afferent nociceptive pathways. Rare instances of chest pain in post-cardiac transplantation patients in which a cardiac source was highly suspected have appeared in the literature (Buda *et al.*, 1981; Uretsky, 1992). Two patients reported to have myocardial release of noradrenaline in response to intracoronary tyramine, suggesting efferent reinnervation (Wilson *et al.*, 1991), were later reported to have chest pain consistent with angina pectoris in association with well-documented acute coronary ischaemic events (Stark, McGinn and Wilson, 1991). In one patient, accelerating exertional and rest angina was associated with electrocardiographic changes and a new severe left anterior descending (LAD) coronary artery stenosis. Furthermore, chest pain and elevation of the electrocardiographic ST segment occurred after inflation of an angioplasty balloon in the vessel. In this patient angioplasty was successful and chest pain did not recur. Notably, three other patients included in the tyramine study (above) by Wilson *et al.* (1991) and not demonstrating coronary sinus noradrenaline release experienced dramatic acute ischaemic events without any associated chest pain (Stark, McGinn and Wilson, 1991). The perception of pain secondary to myocardial ischaemia implies the presence of afferent nociceptive pathways. Myocardial nociception is probably dependent upon small non-myelinated or thinly myelinated C-fibres which travel with larger myelinated sympathetic efferents (Brown, 1979). Moreover, sympathetic denervation in humans has successfully relieved angina (Leriche and Fontaine, 1928; Lindgren and Olivecrona, 1947). The mediators for afferent nociceptive receptor stimulation in response to ischaemia may be locally produced compounds such as bradykinin (Lombardi *et al.*, 1981) and adenosine (Sylvén *et al.*, 1986, 1988; Lagerqvist *et al.*, 1990). Adenosine administered in the left anterior descending coronary artery produces a reflex hypertensive response which is abolished in cardiac transplant patients (Cox *et al.*, 1989). However, return of the adenosine-mediated pressor reflex has not been used to test for reafferentation in long-term cardiac transplant patients. That the majority of clinical events in cardiac transplant patients are either silent or with atypical symptoms (Gao *et al.*, 1989) does not necessarily rule out some degree of reinnervation even in those patients, since reafferentation, in particular, may be absent or may not have occurred in the particular area of infarction. On the other hand, one could hypothesize that massive release of metabolites such as adenosine from large areas of ischaemia or infarction could excite nociceptive fibres in the innervated remnant right atrium after drainage from the coronary sinus.

NORMAL AUTONOMIC FUNCTION AFTER HEART TRANSPLANTATION?

The discussion of adrenergic receptor changes after transplantation has focused on the cardiac muscarinic and β-receptors. However, abnormalities in the afferent limbs of autonomic reflexes and peripheral adrenergic receptors resulting from congestive heart failure preceding transplantation and the post-cardiac transplantation milieu (cyclosporin, hypertension, neurohumoral changes) may independently alter cardiovascular responses. These factors may be of particular relevance in studies employing behaviour of the innervated recipient atrium as a control for the denervated donor heart. In general, studies to date support a return to normal of non-cardiac adrenergic function in heart failure patients who are treated with transplantation. In longitudinal studies pre- and post-transplantation, severely elevated noradrenaline levels quickly fell to within the normal range after surgery both at baseline (Levine, Olivari and Cohn, 1986; Olivari *et al.*, 1987) and in response to provocative manoeuvres (Levine, Olivari and Cohn, 1986), although neither of these studies included normal volunteers. Ellenbogen *et al.* (1989) found that arterial baroreflexes, as assessed by intravenous bolus injections or phenylephrine, return to normal early after transplantation and remain normal up to 2 years except in some patients who were elderly, hypertensive, or both. Thus, severe baroreflex abnormalities found in low output congestive heart failure are primarily due to the attendant neurohumoral and reversible neural changes, rather than permanent anatomical changes in arterial receptor or central baroreflex mechanisms. It is important to consider that hypertension and advanced age (discussed by Ryan and Lipsitz in Chapter 4 of this book) have profound influences upon baroreflex response and must be considered as independent variables in studies of autonomic reactivity in transplant patients.

Baseline mean arterial pressure and forearm vascular resistance were elevated in cardiac transplant patients evaluated by Mohanty *et al.* (1987). Impaired reflex increases in forearm vascular resistance and noradrenaline release during lower body negative pressure in these patients illustrates the importance of afferent ventricular mechanoreceptors in the baroreflex arc. These findings may support a significant tonic inhibitory effect of ventricular afferents upon sympathetic activity. However, in control renal transplant patients with hypertension and on a similar immunosuppressive regimen, similar baseline elevations in forearm vascular resistance were present. Nonetheless, net forearm vascular resistance, mean arterial pressure and noradrenaline level responses to the cold pressor test were unimpaired in the cardiac transplant patients, indicating no generalized impairment in cardiovascular reflexes. Furthermore, baroreflex responses to phenylephrine (Ellenbogen *et al.*, 1989) suggest that input from atrial, pulmonary and aorto-arterial receptive areas compensate for ventricular baroreceptor deafferentation.

Data from some investigators provide evidence for persistence or development of baseline autonomic dysfunction after transplantation. In a longitudinal study, Smith *et al.* (1990) found that parasympathetic activity, as measured by heart period variability in the innervated recipient atrium, improved substantially after transplantation but remained subnormal. Few studies have addressed the status of α-receptor function in the cardiac transplant patient. Borow *et al.* (1989) noted that 60% more methoxamine was required to produce a pressor effect in transplant patients, compared with normal controls, suggesting down-regulation of peripheral $\alpha 1$-receptors. Scherrer *et al.* (1990b) observed significantly increased muscle-sympathetic nerve activity in both cardiac transplant patients and myasthenia gravis patients on cyclosporin compared with similar patients not on cyclosporin. Although this apparent increase in sympathetic activity may explain possible $\alpha 1$-receptor down-regulation and a mechanism for cyclosporin-associated hypertension, circulating catecholamine levels in heart transplant patients on cyclosporin have been reported as normal (Olivari *et al.*, 1987). Furthermore, a variety of mechanisms other than sympathetic activation, cited by Sherrer *et al.* (1990b), have been proposed for the hypertensive effects of cyclosporin.

Since the development of sensitive assays for noradrenaline, plasma levels of this and other catecholamines have been used clinically to make inferences of sympathetic neural activity (Goldstein, 1981). Although heart rate responses in transplant patients are largely dependent upon circulating catecholamines, most reports have not found plasma noradrenaline levels (Levine, Olivari and Cohn, 1986; Olivari *et al.*, 1987) or the other catecholamines (Gilbert *et al.*, 1989) to be elevated outside normal ranges whether at rest, at peak dynamic exercise levels (Pope *et al.*, 1980; Banner *et al.*, 1989b; Quigg *et al.*, 1989) or during isometric exercise (Roca *et al.*, 1991). Although Mettauer *et al.* (1991) reported that catecholamines were markedly elevated above normal during maximal exercise, direct comparisons with normal controls undergoing a similar level of stress were not made. In fact, catecholamine levels in transplant patients have only rarely (Banner *et al.*, 1989b) been compared with normal volunteers in the same study.

DISCUSSION

The goal of this review has been to outline certain historical and most recent findings relating cardiac innervation and orthotopic cardiac transplantation. The transplanted heart should be viewed not only as an extraordinary means of extending the useful lives of many patients. Cardiac transplantation is a model through which we may extend our understanding of intrinsic cardiac adaptation mechanisms and ferret out the complex autonomic modulatory mechanisms existent in the normally innervated heart. It must be emphasized that many of the findings included herein are based upon investigations of small numbers of cardiac transplantation patients or animals. The time-course of events following transplantation, namely issues of reinnervation, are currently based upon cross-sectional studies of subjects with very few long-term longitudinal studies in cardiac transplant patients being available. This situation should improve. Clinical success has produced an ever-growing population of orthotopic transplantation patients who are generally altruistic with regards to aiding in the advancement of knowledge in cardiovascular physiology and pathophysiology. Accordingly, many of the ideas presented should be refined in the near future.

Orthotopic transplantation affords several advantages over other models of cardiac denervation, some of which will be reemphasized. Transplantation provides a model of isolated cardiac denervation. While autonomic failure patients may be employed in investigations of cardiac denervation, impairment of autonomic reflexes may affect all organ systems and denervation may be only partial and difficult to quantify. Surgical methods of cardiac denervation afford a specificity of effect not afforded by pharmacological denervation. The transplantation model is useful in identifying direct versus indirect cardiac effects of pharmacological agents or physiological manoeuvres upon the heart since the problem of "accentuated antagonism" by opposing limbs of the autonomic nervous system is not present. The human cardiac transplantation model excels over animal models when longitudinal studies are required since functional reinnervation, while a possibility that must be controlled for, is difficult to demonstrate. One convenience available in transplant patients (and used by several of the investigators mentioned above as a control for the donor heart in the same patient) is a functioning and fully innervated recipient sinus node adjoining remnant atrial tissue.

The neurohumeral and metabolic changes caused by immunosuppression, variable levels of rejection and hypertension are problematical when cardiac transplantation patients are compared with normal controls (see above: Normal Autonomic Function after Cardiac Transplantation?). Nonetheless, further research employing the human orthotopic transplant model may further our understanding of autonomic control mechanisms in the normally innervated heart. In addition we may learn to utilize post-denervation changes in adrenergic or cholinergic receptors and other cellular mechanisms in order to better treat those with autonomic failure and heart transplantation. Perhaps functional reinnervation of the heart and other organs will be possible in the future.

REFERENCES

Arai, Y., Saul, J.P., Albrecht, P., Hartley, L.H., Lilly, L.S., Cohen, R.J., et al. (1989) Modulation of cardiac autonomic activity during and immediately after exercise. Am. J. Physiol., 256, H132–H141.

Ardell, J.L and Randall, W.C. (1986) Selective vagal innervation of sinoatrial and atrioventricular nodes in canine heart. Am. J. Physiol., 251, H764–H773.

Arrowood, J.A., Mohanty, P.K., Hodgson, J.M., Dibner-Dunlap, M.E. and Thames, M.D. (1989) Ventricular sensory endings mediate reflex bradycardia during coronary arteriography in humans. Circulation, 80, 1293–1300.

Bainbridge, F.A. (1915) The influence of venous filling upon the rate of the heart. J. Phys., 64–84.

Banner, N.R. Lloyd, M.H., Hamilton, R.D., Innes, J.A., Guz, A. and Yacoub, M.H. (1989a) Cardiopulmonary response to dynamic exercise after heart and combined heart-lung transplantation. Br. Heart J., 61, 215–223.

Banner, N.R., Patel, N., Cox, A.P., Patton, H.E., Lachno, D.R. and Yacoub, M.H. (1989b) Altered sympathoadrenal response to dynamic exercise in cardiac transplant recipients. Cardiovasc. Res., 23, 965–972.

Barnard, C.N. (1967) A human cardiac transplant: an interim report of a successful operation performed at Groote Schuur Hospital. S. A. Med. J., 48, 1271–1274.

Barnard, C.N. (1968) What we have learned about heart transplants. J. Thorac. Cardiovasc. Surg., 56, 457–468.

Baughman, K.L. (1990) History and Current Techniques of Endomyocardial Biopsy. In: Heart and Heart-Lung Transplantation, edited by W.A. Baumgartner, B.A. Reitz and S.C. Achuff, pp. 165–182. Philadelphia: W.B. Saunders Company.

Berke, D.K., Graham, A.F., Schroeder, J.S. and Harrison, D.C. (1973) Arrhythmias in the denervated transplanted human heart. Circulation, 47(suppl 3), 113–115.

Bernardi, L., Keller, F., Sanders, M., Reddy, P.S., Griffith, B., Meno, F., et al. (1989) Respiratory sinus arrhythmia in the denervated human heart. J. of Applied Physiol., 67, 1447–1455.

Bernardi, L., Salvucci, F., Suardi, R., Soldá, P.L., Calciati, A., Perlini, S., et al. (1990) Evidence for an intrinsic mechanism regulating heart rate variability in the transplanted and the intact heart during submaximal dynamic exercise. Cardiovasc. Res., 24, 969–981.

Bexton, R.S., Hellestrand, K.J., Cory-Pearce, R., Spurrell, R.A.J., English, T.A.H. and Camm, A.J. (1983a) Unusual atrial potentials in a cardiac transplant recipient. Possible synchronization between donor and recipient atria. *J. Electrocardiol.*, **16**, 313–322.

Bexton, R.S., Hellestrand, K.J., Cory-Pearce, R., Spurrell. R.A.J., English T.A.H. and Camm, A.J. (1983b) The direct electrophysiologic effects of disopyramide phosphate in the transplanted human heart. *Circulation*, **67**, 38–45.

Bexton, R.S., Milne, J.R., Cory-Pearce, R., English, T.A.H. and Camm, A.J. (1983c) Effect of beta blockade on exercise response after cardiac transplantation. *Br. Heart J.*, **49**, 584–588.

Bexton, R.S., Nathan, A.W., Hellestrand, K.J., Cory-Pearce, R., Spurrell, R.A., English, T.A.H., *et al.* (1983d) Electrophysiological abnormalities in the transplanted human heart. *Br. Heart J.*, **50**, 555–563.

Bexton, R.S., Nathan, A.W., Hellestrand, K.J., Cory-Pearce, R., Spurrell, R.A., English, T.A.H., *et al.* (1984) Sinoatrial function after cardiac transplantation. *J. Am. Coll. Cardiol.*, **3**, 712–723.

Bieber, C.P., Stinson, E.B. and Shumway, N.E. (1969) Pathology of the conduction system in cardiac rejection. *Circulation*, **39**, 567–575.

Billman, G.E., Hoskins, R.S., Randall, D.C., Randall, W.C., Hamlin, R.L. and Lin, Y.C. (1989) Selective vagal postganglionic innervation of the sinoatrial and atrioventricular nodes in the non-human primate. *J. Auton. Nerv. Syst.*, **26**, 27–36.

Bishop, V.S., Malliani, A. and Thoren, P. (1983) Cardiac mechanoreceptors. In *Handbook of Physiology* edited by J.T. Shepherd and F.M. Abboun, pp. 497–498. Baltimore: The Williams & Wilkins Company.

Blinks, J.R. (1956) Positive chronotropic effect of increasing right atrial pressure in the isolated mammalian heart. *Am. J. Physiol.*, **186**, 299–303.

Blomquist, T.M., Priola, D.V. and Romero, A.M. (1987) Source of intrinsic innervation of canine ventricles: a functional study. *Am. J. Physiol.*, **252**, H638–H644.

Borow, K.M., Neumann, A., Arensman, F.W. and Yacoub, M.H. (1989) Cardiac and peripheral vascular responses to adrenoceptor stimulation and blockade after cardiac transplantation. *J. Am. Coll. Cardiol.*, **14**, 1229–1238.

Bristow, M.R. (1990) The surgically denervated, transplanted human heart. *Circulation*, **82**, 658–660.

Bristow, M.R., Ginsburg, R., Umans, V., Fowler, M., Minobe, W., Rasmussen, R., *et al.* (1986) Beta1- and beta2-adrenergic-receptor subpopulations in nonfailing and failing human ventricular myocardium: Coupling of both receptor subtypes to muscle contraction and selective beta1-receptor down-regulation in heart failure. *Circ. Res.*, **59**, 297–309.

Bristow, M.R., Hershberger, R.E., Port, J.D., Minobe, W. and Rasmussen, R. (1989) Beta1- and beta2-adrenergic receptor-mediated adenylate cyclase stimulation in nonfailing and failing human ventricular myocardium. *Mol. Pharmacol.*, **35**, 295–303.

Brodde, O.E., Khamssi, M and Zerkowski, H.R. (1991) Beta-adrenoceptors in the transplanted human heart: unaltered beta-adrenoceptor density, but increased proportion of beta2-adrenoceptors with increasing post transplant time. *Arch. Pharmacol.*, **344**, 430–436.

Brown, A. (1979) Cardiac reflexes. In: *Handbook of Physiology – The Cardiovascular System I*, edited by R.M. Berne, N. Sperelakis and S.R. Geiger, pp. 677–689. Baltimore: Williams and Wilkins Company.

Buda, A.J., Fowles, R.E., Schroeder, J.S., Hunt, S.A., Cipriano, P.R., Stinson, E.B. *et al.* (1981) Coronary artery spasm in the denervated transplanted human heart. *Am. J. Med.*, **70**, 1144–1149.

Campeau, L., Pospisil, L., Grondin, P., Dyrda, I. and Lepage, G. (1970) Cardiac catheterization findings at rest and after exercise in patients following cardiac transplantation. *Am. J. Cardiol.*, **25**, 523–528.

Cannom, D.S., Graham, A.F. and Harrison, D.C. (1973) Electrophysiological studies in the denervated transplanted human heart: Response to arterial pacing and atropine. *Circ. Res.*, **32**, 268–278.

Cannom, D.S., Rider, A.K., Stinson, E.B. and Harrison, D.C. (1975) Electrophysiologic studies in the denervated transplanted human heart. II. Response to norepinephrine, isoproterenol and propranolol. *Am. J. Cardiol.*, **36**, 859–866.

Cannon, W.B. and Haimovici, H. (1939) The sensitization of motoneurones by partial "denervation". *Am. J. Physiol.*, **126**, 731–740.

Carleton, R.A. (1976) An unusual mechanism of arrhythmia production. *J. Electrocardiol.*, **9**, 371–373.

Carlson, M.D., Geha, A.S., Hsu, J., Martin, P.J., Levy, M.N., Jacobs, G., *et al.* (1992) Selective stimulation of parasympathetic nerve fibers to the human sinoatrial node. *Circulation*, **85**, 1311–1317.

Carrel, A. and Guthrie, C.C. (1905) The transplantation of veins and organs. *Am. Med.*, **X; 27**, 1101–1102.

Cattan, S., Drobinski, G., Artigout, J-Y., Grogogea, Y. and Cabrol, C. (1988) Coronary artery spasm in a transplant patient. *Eur. Heart J.*, **9**, 557–560.

Chidsey, C.A., Braunwald, E. and Morrow, A.G. (1965) Catecholamine excretion and cardiac stores of norepinephrine in congestive heart failure. *Am. J. Med.*, **39**, 442–451.

Clark, D.A., Schroeder, J.S., Griepp, R.B., Stinson, E.B., Dong, E., Shumway, N.E., *et al.* (1973) Cardiac transplantation in man: review of first three years experience. *Am. J. Med.*, **54**, 563–576.

Coleridge, J.C.G. and Coleridge, H.M. (1977) Afferent C-fibers and cardiorespiratory chemoreflexes. *Am. Rev. Res. Dis.*, **155**, 251–260.

Coleridge, J.C.G. and Coleridge, H.M. (1979) Chemoreflex regulation of the heart. In: *Handbook of Physiology – The Cardiovascular System I*, edited by R.M. Berne, N. Sperelakis and S.R. Geiger, pp. 653–676. Baltimore: Williams & Wilkins Company.

Cooper, T., Gilbert, J.W.Jr., Bloodwell, R.D. and Crout, J.R. (1961) Chronic extrinsic cardiac denervation by regional neural ablation. *Circ. Res.*, **9**, 275–281.

Cooper, T., Willman, V.L., Jellinek, M. and Hanlon, C.R. (1962) Heart autotransplantation: Effect on myocardial catecholamine and histamine. *Science*, **138**, 40–41.

Cox, D.A., Vita, J.A., Treasure, C.B., Fish, R.D., Selwyn, A.P. and Ganz, P. (1989) Reflex increase in blood pressure during the intracoronary administration of adenosine in man. *J. Clin. Invest.*, **84**, 592–596.

Crook, B., Kitson, D., McComish, M.J. and Jewitt, D. (1977) Indirect measurement of sinoatrial conduction time in patients with sinoatrial disease and in controls. *Br. Heart J.*, **39**, 771–777.

Crout, J.R., Alpers, H.S., Tatum, E.L. and Shore, P.A. (1964) Release of metaramiol (Aramine) from the heart by sympathetic nerve stimulation. *Science*, **145**, 828–829.

Dae, M., Herre, J., Botvinick, E., Huberty, J., O'Connell, W., Davis, J., *et al.* (1986) Scintigraphic detection of denervated myocardium after infarction. *J. Nucl. Med.*, **27**, 949 (abstract).

Daggett, W.M., Willman, V.L., Cooper, T. and Hanlon, C.R. (1967) Work capacity and efficiency of the autotransplanted heart. *Circulation*, **35** (Suppl I), I96–I104.

Dawes, G.S. (1947) Studies on veratrum alkaloids. *J. Pharm. Exp. Ther.*, **89**, 325–342.

Dawes, G.S. Comroe, J.H. (1954) Chemoreflexes from the heart and lungs. *Phys. Rev.*, **34**, 167–201.

de Bakker, J.M.T., Coronel, R., Tasseron, S., Wilde, A.A.M., Opthof, T., Janse, M.J., *et al.* (1990) Ventricular tachycardia in the infarcted, Langendorff-perfused human heart: role of the arrangement of surviving cardiac fibers. *J. Am. Coll. Cardiol.*, **15**, 1594–1607.

Demikhov, V.P. (1962) *Experimental Transplantation of Vital Organs*, New York: Consultants Bureau.

Dempsey, P.J. and Cooper, T. (1968) Supersensitivity of the chronically denervated feline heart. *Am. J. Physiol.*, **215**, 1245–1249.

Denniss, A.R., Marsh, J.D., Quigg, R.J., Gordon, J.B. and Colucci, W.S. (1989) Beta-adrenergic receptor number and adenylate cyclase function in denervated transplanted and cardiomyopathic human hearts. *Circulation*, **79**, 1028–1034.

Deray, G., Maistre, G., Desruenne, M., Eurin, J., Barthelemy, C., Masson, F., *et al.* (1990) Atrial natriuretic peptide level and intracardiac pressure in cardiac transplant recipients. *Eur. J. Clin. Pharmacol.*, **38**, 219–221.

Donald, D.E. and Samueloff, S.L. (1966) Exercise tachycardia not due to blood-borne agents in canine cardiac denervation. *Am. J. Physiol.*, **211**, 703–711.

Donald, D.E. and Shepherd, J.T. (1963) Response to exercise in dogs with cardiac denervation. *Am. J. Physiol.*, **205**, 393–400.

Donald, D.E., Milburn, S.E. and Shepherd, J.T. (1964) Effect of cardiac denervation on the maximal capacity for exercise in the racing greyhound. *J. Appl. Physiol.*, **19**, 849–852.

Donald, D.E., Ferguson, D.A. and Milburn, S.E. (1968) Effect of beta-adrenergic receptor blockade on racing performance of greyhounds with normal and with denervated hearts. *Circ. Res.*, **22**, 127–134.

Dong, E.Jr., Hurley, E.J., Lower, R.R. and Shumway, N.E. (1964) Performance of the heart two years after autotransplantation. *Surgery*, **56**, 270–274.

Ebert, P.A. and Sabiston, D.C.Jr. (1970) Pharmacologic quantitation of cardiac sympathetic reinnervation. *Surgery*, **68**, 123–127.

Ebert, P.A., Vanderbeek, R.R. and Sabiston, D.C.Jr. (1970) Effect of chronic cardiac denervation on arrhythmias after coronary artery ligation. *Card. Res.*, **4**, 141–147.

Ellenbogen, K.A., Mohanty, P.K., Szentpetery, S. and Thames, M.D. (1989) Arterial baroreflex abnormalities in heart failure. *Circulation*, **79**, 51–58.

Ellenbogen, K.A., Thames, M.D., DiMarco, J.P., Sheehan, H. and Lerman, B.B. (1990) Electrophysiological effects of adenosine in the transplanted human heart. *Circulation*, **81**, 821–828.

Epstein, A.E., Hirschowitz, B.I., Kirklin, J.K., Kirk, K.A., Kay, G.N. and Plumb, V.J. (1990) Evidence for a central site of action to explain the negative chronotropic effect of atropine: studies on the human transplanted heart. *J. Am. Coll. Cardiol.*, **15**, 1610–1617.

Fallen, E.L., Kamath, M.V., Ghista, D.N. and Fitchett, D. (1988) Spectral analysis of heart rate variability following human heart transplantation; evidence for functional reinnervation. *J Auton. Nerv. Syst.*, **23**, 199–206.

Fish, R.D., Nabel, E.G., Selwyn, A.P., Ludmer, P.L., Mudge, G.H. and Kirshenbaum, J.M., *et al.* (1988) Responses of coronary arteries of cardiac transplant patients to acetylcholine. *J. Clin. Invest.*, **81**, 21–31.

Forman, M.B., Robertson, D., Goldberg, M., Bostick, D., Uderman, H., Perry, J.M., *et al.* (1984) Effect of tyramine on myocardial catecholamine release in coronary heart disease. *Am. J. Cardiol.*, **53**, 476–480.

Frist, W.H., Stinson, E.B., Oyer, P.E., Baldwin, J.C. and Shumway, N.E. (1987) Long-term hemodynamic results after cardiac transplantation. *J. Thorac. Cardiovasc. Surg.*, **94**, 85–693.

Furchgott, R.F. and Vanhoutte, P.M. (1989) Endothelium-derived relaxing and contracting factors. *FASEB*, **3**, 2007–2017.

Furchgott, R.F. and Zawadzki, J.V. (1980) The obligatory role of endothelial cells in the relaxation of arterial smooth muscle by acetylcholine. *Nature*, **288**, 373–377.

Gao, S.Z., Schroeder, J.S., Hunt, S.A., Billingham, M.E., Valantine, H.A. and Stinson, E.B. (1989) Acute myocardial infarction in cardiac transplant recipients. *Am. J. Cardiol.*, **64**, 1093–1097.

Geis, W.P., Tatooles, C.J., Kaye, M.P. and Randall, W.C. (1971) Complete cardiac denervation without transplantation: a simple and reliable technique. *J. Appl. Physiol.*, **30**, 289–293.

Gilbert, E.M., Eiswirth, C.C., Mealey, P.C., Larrabee, P., Herrick, C.M. and Bristow, M.R. (1989) Beta-adrenergic supersensitivity of the transplanted human heart is presynaptic in origin. *Circulation*, **79**, 344–349.

Glowniak, J.V., Turner, F.E., Gray, L.L., Palac, R.T., Lagunas-Solar, M.C. and Woodward, W.R. (1989) Iodine-123 metaiodobenzylguanidine imaging of the heart in idiopathic congestive cardiomyopathy and cardiac transplant. *J. Nuclear. Med.*, **30**, 1182–1191.

Goldstein, D.S. (1981) Plasma norepinephrine as an indicator of sympathetic neural activity in clinical cardiology. *Am. J. Cardiol.*, **48**, 1147–1154.

Goodman, D.J., Rossen, R.M., Ingham, R., Rider, A.K. and Harrison, D.C. (1975a) Sinus node function in the denervated human heart: effect of digitalis. *Br. Heart J.*, **37**, 612–618.

Goodman, D.J., Rossen, R.M., Cannom, D.S., Rider, A.K. and Harrison, D.C. (1975b) Effect of digoxin on atrioventricular conduction: studies in patients with and without cardiac autonomic innervation. *Circulation*, **51**, 251–256.

Griepp, R.B. and Ergin, M.A. (1984) The history of experimental heart transplantation. *Heart Transplant.*, **3**, 145–150.

Gupta, P.K., Lichstein, E., Chadda, K.D. and Badui, E. (1974) Appraisal of sinus nodal recovery time in patients with sick sinus syndrome. *Am. J. Cardiol.*, **34**, 265–270.

Gu, J. and Muralidharan, S. (1992) Sensory reinnervation of the heart after cardiac transplantation (letter). *New Engl. J. Med.*, **326**, 67.

Guth, L. (1856) Regeneration in the mammalian peripheral nervous system. *Phys. Rev.*, **36**, 441–478.

Hageman, G.R., Urthaler, F. and James, T.N. (1977) Differential sensitivity to neurotransmitters in denervated canine sinus node. *Am. J. Physiol.*, **233**, H211–H216.

Hallman, G.L., Leatherman, L.L., Leachman, R.D., Rochelle, D.G., Bricker, D.L., Bloodwell, R.D., *et al.* (1969) Function of the transplanted human heart. *J. Thorac. Cardiovasc. Surg.*, **58**, 318–325.

Hardy, J.D., Chavez, C.M., Kurrus, F.D., Neely, W.A., Eraslan, S., Turner, D., *et al.* (1964) Heart transplantation in man. *JAMA*, **13**, 1133–1141.

Hardy, J.D. and Chavez, C.M. (1968) The first heart transplant in man. *Am. J. Card.*, **22**, 772–781.

Haskell, W.L., Savin, W.M., Schroeder, J.S., Alderman, E.A., Ingles, N.B., Daughters, G.T., *et al.* (1981) Cardiovascular responses to handgrip isometric exercise in patients following cardiac transplantation. *Circ. Res.*, **48**(suppl. I), I156–I161.

Heinz, G., Ohner, T., Laufer, G., Gasic, S. and Laczkovics, A. (1990) Clinical and electrophysiologic correlates of sinus node dysfunction after orthotopic heart transplantation. *Chest*, **97**, 890–895.

Held, P., Yusuf, S., Mathias, C., Dhalla, N., Theodoropoulos, S. and Yacoub, M. (1989) Renin response to sympathetic stimulation in cyclosporin-treated heart-transplant patients. *Am. J. Cardiol.*, **63**, 1142–1144.

Higgins, C.B., Vatner, S.F. and Braunwald, E. (1973) Parasympathetic control of the heart. *Pharmacol. Rev.*, **25**, 119–149.

Higgins, C.B., Vatner, S.F., Franklin, D. and Braunwald, E. (1973) Extent of regulation of the heart's contractile state in the conscious dog by alteration in the frequency of contraction. *J. Clin. Invest.*, **52**, 1187–1194.

Hodgson, J.M., Cohen, M.D., Szentpetery, S. and Thames, M.D. (1989) Effects of regional alpha- and beta-blockade on resting and hyperemic coronary blood flow in conscious, unstressed humans. *Circulation*, **79**, 797–809.

Horn, E.M., Danilo, P., Apfelbaum, M.A., Barr, M.L., Pipino, P., Powers, E.R., *et al.* (1991) Beta-adrenergic receptor sensitivity and guanine nucleotide regulatory proteins in transplanted human hearts and autotransplanted baboons. *Transplantation*, **52**, 960–966.

Hosenpud, J.D., Morton, M.J., Wilson, R.A., Pantely, G.A., Norman, K.J., Cobanoglu, M.A., *et al.* (1989) Abnormal exercise hemodynamics in cardiac allograft recipients 1 year after cardiac transplantation. *Circulation*, **80**, 525–532.

Jacquet, L., Ziady, G., Stein, K., Griffith, B., Armitage, J., Hardesty, R., *et al.* (1990) Cardiac rhythm disturbances early after orthotopic heart transplantation: prevalence and clinical importance of the observed abnormalities. *J. Am. Coll. Cardiol.*, **16**, 832–837.

Jamieson, S.W. (1992) Investigation of heart transplant coronary atherosclerosis. *Circulation*, **85,3**, 1211–1213.

Jose, A.D. and Taylor, R.R. (1969) Autonomic blockade by propranolol and atropine to study intrinsic myocardial function in man. *J. Clin. Invest.*, **48**, 2019–2031.

Josephson, M.E., Seides, S.F., Batsford, W.P., Weisfogel, G.M., Akhtar, M., Caracta, A.R., *et al.* (1974) The electrophysiological effects of intramuscular quinidine on the atrioventricular conducting system in man. *Am. Heart J.*, **87**, 55–64.

Kacet, S., Molin, F., Lacroix, D., Prat, A,. Pol, A., Warembourg, H., *et al.* (1991) Bipolar atrial triggered pacing to restore normal chronotropic responsiveness in an orthotopic cardiac transplant patient. *PACE*, **14**, 1444–1447.

Kaseda, S. and Zipes, D. (1988) Supersensitivity to acetylcholine of canine sinus and AV nodes after parasympathetic denervation. *Am. J. Physiol.*, **255**, H534–H539.

Kaski, J.C., Crea, F., Meran, D., Rodriguez, L., Araujo, L., Chierchia, S., *et al.* (1986) Local coronary supersensitivity to diverse vasoconstrictive stimuli in patients with variant angina. *Circulation*, **74**, 1255–1265.

Kavanagh, T., Yacoub, M.H., Mertens, D.J., Kennedy, J., Campbell, R.B. and Sawyer, P. (1988) Cardiorespiratory responses to exercise training after orthotopic cardiac transplantation. *Circulation*, **77**, 162–171.

Kay, J., Bienenfeld, D., Slomowitz, M., Burk, J., Zimmer L., Nadolny, G., *et al.* (1991) Use of tricyclic antidepressants in recipients of heart transplants. *Psychosomatics*, **32**, 165–170.

Kaye, M.P. (1984) Denervation and reinnervation of the heart. In: *Nervous Control of Cardiovascular System*, edited by W.C. Randall, pp. 278–306. New York: Oxford University Press.

Kaye, M.P. (1992) The Registry of the International Society for Heart and Lung Transplantation: Ninth Official Report – 1992. *J. Heart Lung. Trans.*, **11**, 599–606.

Kaye, M.P., Randall, W.C., Hageman, G.R., Geis, W.P. and Priola, D.V. (1977) Chronology and mode of reinnervation of the surgically denervated canine heart: functional and chemical correlates. *Am. J. Physiol.*, **233**, H431–H437.

Kaye, M.P., Wells, D.J. and Tyce, G.M. (1979) Nerve growth factor-enhanced reinnervation of surgically denervated canine heart. *Am. J. Physiol.*, **236**, H624–H628.

Kent, K.M. and Cooper, T. (1974) The denervated heart: a model for studying autonomic control of the heart. *New Engl. J. Med.*, **291**, 1017–1021.

Kondo, Y., Matheny, J.L. and Hardy, J.D. (1972) Autonomic reinnervation of cardiac transplants: Further observations in dogs and rhesus monkeys. *Ann. Surg.*, **176**, 42–48.

Kontos, H.A., Thames, M.D. and Lower, R.R. (1970) Responses to electrical and reflex autonomic stimulation in dogs with cardiac transplantation before and after reinnervation. *J. Thorac. Cardiovasc. Surg.*, **59**, 382–392.

Krayer, O. (1961) The History of the Bezold-Jarisch effect. *Arch. Exp. Pathol. Pharmacol.*, **240**, 361–368.

Kriett, J.M. and Kaye, M.P. (1991) The Registry of the International Society for Heart and Lung Transplantation: Eighth official report – 1991. *J. Heart Lung. Trans.*, **10**, 491–498.

Kushwaha, S., Mitchell, A.G. and Yacoub, M.H. (1990) Coronary artery spasm after cardiac transplantation. *Am. J. Cardiol.*, **66**, 1515–1519.

Lagerqvist, B., Sylven, C., Helmius, G. and Waldenstrom, A. (1990) Effects of exogenous adenosine in a patient with transplanted heart. Evidence for adenosine as a messenger in angina pectoris. *Upsala J. Med. Science*, **95**, 137–145.

Lands, A.M., Arnold, A., McAuliff, J.P., Luduena, F.P. and Brown, T.G. (1967) Differentiation of receptor systems activated by sympathomimetic amines. *Nature*, **214**, 597–598.

Lazzara, R., Scherlag, B.J., Robinson, M.J. and Samet, P. (1973) Selective *in situ* parasympathetic control of the canine sinoatrial and atrioventricular nodes. *Circ. Res.*, **32**, 393–401.

Leachman, R.D., Cokkinos, D.V.P., Zamalloa, O. and Alvarez, A. (1969) Electrocardiographic behavior of recipient and donor atria after human heart transplantation. *Am. J. Cardiol.*, **24**, 49–53.

Leachman, R.D., Cokkinos, D.V., Cabrera, R., Leatherman, L.L. and Rochelle, D.G. (1971) Response to the transplantated, denervated human heart to cadiovascular drugs. *Am. J. Cardiol.*, **27**, 272–276.

Leriche, R. and Fontaine, R. (1928) The surgical treatment of angina pectoris. *Am. Heart J.*, **3**, 649–671.

Levine, T.B., Olivari, M.T. and Cohn, J.N. (1986) Effects of orthotopic heart transplantation on sympathetic control mechanisms in congestive heart failure. *Am. J. Cardiol.*, **58**, 1035–1040.

Lindgren, I. and Olivecrona, H. (1947) Surgical treatment of angina pectoris. *J. Neurosurgery*, **4**, 19–39.

Lombardi, F., Bella, P.D., Casati, R. and Malliani, A. (1981) Effects of intracoronary administration of bradykinin on the impulse activity of afferent sympathetic unmyelinated fibers with left ventricular endings in the cat. *Circ. Res.*, **48**, 69–75.

Lower, R.R. and Shumway, N.E. (1960) Studies on orthotopic homotransplantation of the canine heart. *Surg. Forum*, **11**, 18–19.

Lower, R.R., Stofer, R.C. and Shumway, N.E. (1961) Homovital transplantation of the heart. *J. Thorac. Cardiovasc. Surg.*, **41**, 196–204.

Lower, R.R, Dong, E. and Shumway, N.E. (1965) Long-term survival of cardiac homografts. *Surgery*, **58**, 111–119.

Lower, R.R., Kontos, H.A., Kosek, J.C., Sewell, D.H. and Graham, W.H. (1968) Experiences in heart transplantation. *Am. J. Cardiol.*, **22**, 766–771.

Ludmer, P.L., Selwyn, A.P., Shook, T.L., Wayne, R.R., Mudge, G.H., Alexander, W., *et al.* (1986) Paradoxical vasoconstriction induced by acetylcholine in atherosclerotic coronary arteries. *New Engl. J. Med.*, **315**, 1046–1051.

Lurie, K.G., Bristow, M.R. and Reitz, B.A. (1983) Increased β-adrenergic receptor density in an experimental model of cardiac transplantation. *J. Thorac. Cardiovasc. Surg.*, **86**, 195–201.

Mackintosh, A.F., Carmichael, D.J., Wren, C., Cory-Pearce, R. and English, A.H. (1982) Sinus node function in first three weeks after cardiac transplantation. *Br. Heart J.*, **48**, 584–588.

Malliani, A., Pagani, M., Lombardi, F. and Cerutti, S. (1991a) Cardiovascular neural regulation explored in the frequency domain. *Circulation*, **84**, 482–492.

Malliani, A., Pagani, M., Lombardi, F. and Cerutti, S. (1991b) Clinical and experimental evaluation of sympathovagal interaction: power spectral analysis of heart rate and arterial pressure variabilities. In: *Reflex Control of the Circulation*, edited by I.H. Zucker and J.P. Gilmore, pp. 938–964. Boca Raton: CRC Press.

Mark, A.L. (1983) The Bezold-Jarisch reflex revisited: clinical implications of inhibitory reflexes originating in the heart. *J. Am. Coll. Cardiol.*, **1**, 90–102.

Martin, C.E., Shaver, J.A., Leon, D.F., Thompson, M.E., Reddy, P.S. and Leonard, J.J. (1974) Autonomic mechanisms in hemodynamic responses to isometric exercise. *J. Clin. Invest.*, **54**, 104–115.

Mason, J.W. (1980) Overdrive suppression in the transplanted heart: effect of the autonomic nervous system on human sinus node recovery. *Circulation*, **62**, 688–696.

Mason, J.W. and Harrison, D.C. (1979) Electrophysiology and electropharmacology of the transplanted human heart. In: *Cardiac Arrhythmias: Electrophysiology, Diagnosis and Management*, edited by O.S. Narula, pp. 66–81. Baltimore: Williams and Wilkins Company.

Mason, J.W., Stinson, E.B. and Harrison, D.C. (1976) Autonomic nervous system and arrhythmias: Studies in the transplanted denervated human heart. *Cardiology*, **61**, 75–87.

Mason, J.W., Winkle, R.A., Rider, A.K., Stinson, E.B. and Harrison. D.C. (1977) The electrophysiologic effects of quinidine in the transplanted human heart. *J. Clin. Invest.*, **59**, 481–489.

McGiffin, D.C., Karp, R.B., Logic, J.R., Tauxe, W.N. and Ceballos, R. (1984) Results of radionuclide assessment of cardiac function following transplantation of the heart. *Ann. Thorac. Surg.*, **37**, 382–386.

Mettauer, B., Lampert, E., Lonsdorfer, J., Levy, F., Geny, B., Kretz, J.G., *et al.* (1991) Cardiorespiratory and neurohormonal response to incremental maximal exercise in patients with denervated transplanted hearts. *Transplantation Proc.*, **23**, 1178–1181.

Milam, J.D., Shipkey, F.H., Lind, C.J., Nora, J.J., Leachman, R.D., Rochelle, D.G., *et al.* (1970) Morphologic findings in human cardiac allografts. *Circulation*, **XLI**, 519–535.

Miledi, R. (1960) The acetylcholine sensitivity of frog muscle fibres after complete or partial denervation. *J. Physiol.*, **151**, 1–23.

Moberg, A. (1967) Anastomoses between extracardiac vessels and coronary arteries via bronchial arteries. *Acta Radiologica Diagnosis*, **6**, 167–192.

Mohanty, P.K., Thames, M.D., Capehart, J.R., Kawaguchir A., Ballon, B. and Lower, R.R. (1986a) Afferent reinnervation of the autotransplanted heart in dogs. *J. Am. Coll. Cardiol.*, **7**, 414–414.

Mohanty, P.K., Sowers, J.R., Thames, M.D., Beck, F.W.J., Kawaguchi, A. and Lower, R.R. (1986b) Myocardial norepinephrine, epinephrine and dopamine concentrations after cardiac autotransplantation in dogs. *J. Am. Coll. Cardiol.*, **7**, 419–424.

Mohanty, P.K., Thames, M.D., Arrowood, J.A., Sowers, J.R., McNamara, C. and Szentpetery, S. (1987) Impairment of cardiopulmonary baroreflex after cardiac transplantation in humans. *Circulation*, **75**, 915–921.

Nellessen, U., Lee, T.C., Fischell, T.A., Ginsburg, R., Masuyama, T., Alderman, E.L., *et al.* (1988) Effects of acetylcholine on epicardial coronary arteries after cardiac transplantation without angiographic evidence of fixed graft narrowing. *Am. J. Cardiol.*, **62**, 1093–1097.

Norvell, J.E. and Lower, R.R. (1973) Degeneration and regeneration of the nerves of the heart after transplantation. *Transplantation*, **15**, 337–344.

O'Toole, M.F., Ardell, J.L. and Randall, W.C. (1986) Functional interdependence of discrete vagal projections to SA and AV nodes. *Am. J. Physiol.*, **251**, H398–H404.

Olivari, M.T., Levine, T.B., Ring, W.S., Simon, A. and Cohn, J.N. (1987) Normalization of sympathetic nervous system function after orthotopic cardiac transplantation man. *Circulation*, **76**, V62–V64.

Oyer, P.E., Stinson, E.B., Jamieson, S.W., Hunt, S.A., Perlroth, M., Billingham, M., *et al.* (1983) Cyclosporine in cardiac transplantation: a 2 1/2 year follow-up. *Transplant Proc.*, **XV**(Suppl. 1) 2546–2552.

Pathak, C.L. (1973) Autoregulation of chronotropic response of the heart through pacemaker stretch. *Cardiology*, **58**, 45–64.

Peiss, C.N., Cooper, T., Willman, V.L. and Randall, W.C. (1966) *Circulation*, ulatory responses to electrical and reflex activation on the nervous system after cardiac denervation. *Circ. Res.*, **19**, 153–165.

Pennock, J.L., Oyer, P.E., Reitz, B.A., Jamieson, S.W., Bieber, C.P., Wallwork, J., *et al.* (1982) Cardiac transplantation in perspective for the future. *J. Thorac. Cardiovasc. Surg.*, **83**, 168–177.

Pfeifer, M.A., Cook, D., Brodsky, J., Tice, D., Reenan, A. Swedine, S., *et al.* (1982) Quantitative evaluation of cardiac parasympathetic activity in normal and diabetic man. *Diabetes*, **31**, 339–345.

Pflugfelder, P.W., Purves, P.D., McKenzie, F.N. and Kostuk, W.J. (1987) Cardiac dynamics during supine exercise in cyclosporine-treated orthotopic heart transplant recipients: assessment by radionuclide angiography. *J. Am. Coll. Cardiol.*, **10**, 336–341.

Pflugfelder, P.W., McKenzie, F.N. and Kostuk, W.J. (1988) Hemodynamic profiles at rest and during supine exercise after orthotopic cardiac transplantation. *Am. J. Cardiol.*, **61**, 1328–1333.

Pierpont, G.L., DeMaster, G. and Cohn, J.N. (1984) Regional differences in adrenergic function within the left ventricle. *Am. J. Physiol.*, **246**, H824–H829.

Pope, S.E., Ingels, N.B.Jr., Daughters, G.T. and Schroeder, J.S. (1978) *In vivo* demonstration of the Frank-Starling mechanism in the human denervated heart. *Am. J. Cardiol.*, **41**, 432.

Pope, S.E., Stinson, E.B., Daughters, G.T., Schroeder, J.S., Ingels, N.B.Jr. and Alderman, E.L. (1980) Exercise response of the denervated heart in long-term cardiac transplant recipients. *Am. J. Cardiol.*, **46**, 213–218.

Port, J.D., Gilbert, E.M., Larrabee, P., Mealey P, Volkman, K., Ginsburg, R., *et al.*, (1990a) Neurotransmitter depletion compromises the ability of indirect-acting amines to provide inotropic support in the failing human heart. *Circulation*, **81**, 929–938.

Port, J.D., Skerl, L., O'Connell, J.B., Renlund, D.G., Larrabee, P. and Bristow, M.R. (1990b) Increased expression of β2-adrenergic receptors in surgically denervated, previously transplanted human ventricular myocardium. *J. Am. Coll. Cardiol.*, **15**, 84A.

Priola, D.V. and Spurgeon, E.A. (1977) Cholinergic sensitivity of the denervated canine heart. *Circ. Res.*, **41**, 600–606.

Priola, D.V., Curtis, M.B., Anagnostelis, C. and Martinez, E. (1983) Altered nicotinic sensitivity of AV node in surgically denervated canine hearts. *Am. J. Physiol.*, **245**, H27–H32.

Qi, A., Tuna, I.C., Gornick, C.C., Barragry, T.P., Blatchford, J.W., Ring, W.S., *et al.* (1987) Potentiation of cardiac electrophysiologic effects of verapamil after autonomic blockade or cardiac transplantation. *Circulation*, **75**, 888–893.

Quigg, R.J., Rocco, M.B., Gauthier, D F. and Creager, M.A., Hartley, L.H. and Colucci, W.S. (1989) Mechanism of the attenuated peak heart rate response to exercise after orthotopic cardiac transplantation. *J. Am. Coll. Cardiol.*, **14**, 338–344.

Randall, W.C. (1984) Selective autonomic innervation of the heart. In: *Nervous Control of Cardiovascular Function*, edited by W.C. Randall, p. 46. New York: Oxford University Press.

Randall, W.C., Kaye, M.P., Thomas, J.X. and Barber, M.J. (1980) Intrapericardial denervation of the heart. *J. Surg. Res.*, **29**, 101–109.

Randall, W.C., Ardell, J.L., Calderwood, D., Milosavljevic, M. and Goyal, S.C. (1986) Parasympathetic ganglia innervating the canine atrioventricular nodal region. *J. Auton. Nerv. Syst.*, **16**, 311–323.

Randall, W.C., Ardell, J.L., Wurster, R.D. and Molosavljevic, M. (1987) Vagal postganglionic innervation of the canine sinoatrial node. *J. Auton. Nerv. Syst.*, **20**, 12–23.

Regitz, V., Bossaler, C., Strasser, R., Schuler, S., Hetzer, R. and Fleck, E. (1990) Myocardial catecholamine content after heart transplantation. *Circulation*, **82**, 620–623.

Reitz, B.A. (1990) The history of heart and heart-lung transplantation. In: *Heart and Heart-Lung Transplantation*, edited by W.A. Baumgartner, B.A. Reitz and S.C. Achuff, pp. 1–14. Philadelphia: W.B. Saunders Company.

Reitz, B.A., Baumgartner, W.A., Oyer, P.E. and Stinson, E.B. (1977) Abdominal aortic aneurysmectomy in long-term cardiac transplant survivors. *Arch. Surg.*, **112**, 1057–1059.

Ricci, D.R., Orlick, A.E., Reitz, B.A., Mason, J.W., Stinson, E.B. and Harrison, D.C. (1978) Depressant effect of digoxin on atrioventricular conduction in man. *Circulation*, **57**, 898–903.

Ricci, D.R., Orlick, A.E., Alderman, E.L., Ingels, N.B., Daughters, G.T., Kusnick, C.A., *et al.* (1979) Role of tachycardia as an inotropic stimulus in man. *J. Clin. Invest.*, **63**, 695–703.

Robertson D., Johnson, G.A., Robertson, R.M., Nies, A.S., Shand, D.G. and Oates, J.A. (1979) Comparative assessment of stimuli that release neuronal and adrenomedullary catecholamines in man. *Circulation*, **59**, 637–642.

Robson, S.C., Furniss, S.S., Heads, A., Boys, R.J., McGregor, C. and Bexton, R.S. (1989) Isometric exercise in the denervated heart: a Doppler echocardiographic study. *Br. Heart J.*, **61**, 224–230.

Roca, J., Caturla, M.C., Hjemdahl, P., Masotti, M., Augé, J.M. and Crexells, O.C. (1991) Left ventricular dynamics and plasma catecholamines during isometric exercise in patients following cardiac transplantation. *Eur. Heart J.*, **12**, 928–936.

Roeske, W.R., Lund, D., Schmid, P., Chen, M., Kelly, S., Roskoski, R., *et al.* (1978) Alterations in muscarinic cholinergic receptors in transplanted rat hearts (Abstract). *Circulation*, **57 and 58**(suppl II), II-20.

Rowan, R.A. and Billingham, M.E. (1988) Myocardial innervation in long-term heart transplant survivors: a quantitative ultrastructural survey. *J. Heart Transplant.*, **7**, 448–452.

Rowe, S.K., Kleiman, N.S., Cocanougher, B., Smart, F.W., Minor, S.T., Raizner, A.E., *et al.* (1991) Effects of intracoronary acetylcholine infusion early versus late after heart transplant. *Transplant. Proc.*, **23**, 1193–1197.

Roy, L.F., East, D.S., Browning, F.M., Shaw, D., Ogilvie, R.I., Cardella, C., *et al.* (1989) Short-term effects of calcium antagonists on hemodynamics and cyclosporine pharmacokinetics in heart-transplant and kidney-transplant patients. *Clin. Pharmacol. Ther.*, **46**, 657–667.

Rudas, L., Pflugfelder, P.W., Menkis, A.H., Novick, R.J., McKenzie F.N. and Kostuk, W.J. (1991) Evolution of heart rate responsiveness after orthotopic cardiac transplantation. *Cardiol.*, **68**, 232–236.

Rudas, L., Pflugfelder, P.W. and Kostuk, W.J. (1992) Vasodepressor syncope in a cardiac transplant recipient: A case of vagal re-innervation. *Can. J. Cardiol.*, **8**, 403–405.

Sandoval, A.B., Gilbert, E.M., Rose, C.P. and Bristow, M.R. (1989) Cardiac norepinephrine uptake and release is decreased in dilated cardiomyopathy. *Circulation*, **80**, II-393.

Savin, W.M., Alderman, E.L., Haskell, W.L., Schroeder, J.S., Ingels, N.B., Daughters, G.T., *et al.* (1980) Left ventricular response to isometric exercise in patients with denervated and innervated hearts. *Circulation*, **61**, 897–901.

Scherrer, U., Vissing, S., Morgan, J., Hanson. P. and Victor, R.G. (1990a) Vasovagal syncope after infusion of a vasodilator in a heart-transplant recipient. *New Engl. J. Med.*, **322**, 602–604.

Scherrer, U., Vissing, S.F., Morgan, B.J., Rollins, F.A., Tindall, R.S.A., Ring, S., *et al.* (1990b) Cyclosporine-induced sympathetic activation and hypertension after heart transplantation. *New Engl. J. Med.*, **323**, 693–699.

Schofer, J., Spielmann, R., Schuchert, A., Weber, K. and Schulüter, M. (1988) Iodine-123 meta-iodobenzylguanidine scintigraphy: a noninvasive method to demonstrate myocardial adrenergic nervous system disintegrity in patients with idiopathic dilated cardiomyopathy. *J. Am. Coll. Cardiol.*, **12**, 1252–1258.

Schroeder, J.S. (1979) Hemodynamic performance of the human transplanted heart. *Transplant Proc.*, **11**, 304–308.

Schroeder. J.S. and Hunt, S.A. (1992) Chest pain in heart-transplant recipients. *New Engl. J. Med.*, **324**, 1805–1806.

Schroeder, J.S., Berke, D.K., Graham, A.F., Rider, A.K. and Harrison, D.C. (1974) Arrhythmias after cardiac transplantation. *Am. J. Cardiol.*, **33**, 604–607.

Schwaiger, M., Kalff, V., Rosenspire, K., Haka, M.S., Molina, E., Hutchins, G.D., *et al.* (1990) Noninvasive evaluation of sympathetic nervous system in human heart by positron emission tomography. *Circulation*, **82**, 457–464.

Schwaiger, M., Hutchins, G.D., Kalff, V., Rosenspire, K., Haka, M.S., Mallett, S., *et al.* (1991) Evidence for regional catecholamine uptake and storage sites in the transplanted human heart by positron emission tomography. *J. Clin. Invest.*, **87**, 1681–1690.

Segers, M., Lequime, J. and Denolin, H. (1947) Synchronization of auricular and ventricular beats during complete heart block. *Am. Heart J.*, **33**, 685–691.

Seki, S. and Danielson, G.K. (1971) Characteristics of cardiac reinnervation in growing calves. *J. Thorac. Cardiovasc. Surg.*, **62**, 602–607.

Shapiro, P.A. (1991) Nortriptyline treatment of depressed cardiac transplant recipients. *Am. J. Psychiatry*, **148,3**, 371–373.

Sharma, V.K. and Banerjee, S.P. (1978) Presynaptic muscarinic cholinergic receptors. *Nature*, **272**, 276–278.

Shaver, J.A., Leon, D.F., Gray, S., Leonard, J.J. and Bahnson, H.T. (1969) Hemodynamic observations after cardiac transplantation. *New Engl. J. Med.*, **281**, 822–827.

Shore, P.A., Cohn, V.H. Highman, B. and Maling, H.M. (1958) Distribution of norepinephrine in the heart. *Nature*, **181**, 848–849.

Singer, D.R.J., Buckley, M.G., MacGregor, G.A., Khaghani, A., Banner. N.R. and Yacoub, M.H. (1986) Raised concentrations of plasma atrial natriuretic peptides in cardiac transplant recipients. *Br. Med. J.*, **293**, 1391–1392.

Sisson, J.C., Frager, M.S., Valk, T.W., Gross, M.D., Swanson, D.P., Wieland, D.M., *et al.* (1981) Scintigraphic localization of pheochromocytoma. *New Engl. J. Med.*, **305;1**, 12–17.

Sisson, J.C., Shapiro, B., Meyers, L., Mallett, S., Mangner, T.J., Wieland, D.M., *et al.* (1987) Metaiodobenzylguanidine to map scintigraphically the adrenergic nervous system in man. *J. Nucl. Med.*, **28**, 1625–1636.

Smith, D.C., Priola, D.V. and Anagnostelis, C. (1985) Comparison of *in vivo* and *in vitro* cholinergic responses of normal and denervated canine hearts. *J. Pharmacol. Exp. Ther.*, **235**, 37–44.

Smith, M.L., Ellenbogen, K.A., Eckberg, D.L., Sheehan, H.M. and Thames, M.D. (1990) Subnormal parasympathetic activity after cardiac transplantation. *Am. J. Cardiol.*, **66**, 1243–1246.

Stark, R.P., McGinn, A.L. and Wilson, R.F. (1991) Chest pain in cardiac-transplant recipients. *New Engl. J. Med.*, **324**, 1791–1794.

Stemple, D.R., Hall, R.J.C., Mason, J.W. and Harrison, D.C. (1978) Electrophysiological effects of edrophonium in the innervated and the transplanted denervated human heart. *Br. Heart J.*, **40**, 644–649.

Stinson, E.B., Dong, E.Jr., Iben, A.B. and Shumway, N.E. (1969) Cardiac transplantation in man. *Am. J. Surg.*, **118**, 182–187.

Stinson, E.B., Griepp, R.B., Schroeder, J.S., Dong, E.Jr. and Shumway, N.E. (1972a) Hemodynamic observations one and two years after cardiac transplantation in man. *Circulation*, **45**, 1183–1194.

Stinson, E.B., Schroeder, J.S., Griepp, R.B., Shumway, N.E. and Dong. E.Jr. (1972b) Observations on the behavior of recipient atria after cardiac transplantation in man. *Am. J. Cardiol.*, **30**, 615–622.

Stinson, E.B., Caves, P.K., Griepp, R.B., Oyer, P.E., Rider, A.K. and Shumway, N.E. (1975) Hemodynamic observations in the early period after human heart transplantation. *J. Thorac. Cardiovasc. Surg.*, **69**, 264–270.

Stovin, P.G.I. and Hewitt, S. (1986) Conduction tissue in the transplanted human heart. *J. Pathol.*, **149**, 183–189.

Sylvén, C., Beermann, B., Jonzon, B. and Brandt, R. (1986) Angina pectoris-like pain provoked by intravenous adenosine in healthy volunteers. *Br. Med. J.*, **293**, 227–230.

Sylvén, C., Borg, G., Brandt, R., Beermann, B. and Jonzon, B. (1988) Dose-effect relationship of adenosine provoked angina pectoris-like pain – a study of the psychophysical power function. *Eur. Heart J.*, **9**, 87–91.

Thames, M.D., Ul-Hassan, Z., Brackett, N.C.Jr., Lower, R.R. and Kontos, H.A. (1971) Plasma renin responses to hemorrhage after cardiac autotransplantation. *Am. J. Physiol.*, **221**, 1115–1119.

Thames, M.D., Jarecki, M. and Donald, D.E. (1978) Neural control of renin secretion in anesthetized dogs. *Circ. Res.*, **42**, 237–245.

Tobes, M.C., Jaques, S., Wieland, D.M. and Sisson, J.C. (1985) Effect of uptake-one inhibitors on the uptake of norepinephrine and metaiodobenzylguanidine. *J. Nucl. Med.*, **26**, 897–907.

Traill, T.A. (1990) Physiology and function of the transplant allograft. In: *Heart and Heart-Lung Transplantation*, edited by W.A. Baumgartner., B.A. Reitz and S.C. Achuff, pp. 266–278. Philadelphia: W.B. Saunders Company.

Uretsky, B.F. (1992) Sensory reinnervation of the heart after cardiac transplantation (letter). *New Engl. J. Med.*, **326**, 66–67.

Vatner, D.E., Lavallee, M., Amano, J., Finizola, A., Homcy, C.J. and Vatner, S.F. (1985) Mechanisms of supersensitivity to sympathomimetic amines in the chronically denervated heart of the conscious dog. *Circ. Res.*, **57**, 55–64.

von Scheidt, W., Neudert, J., Erdmann, E., Kemkes, B.M., Gokel, J.M. and Autenrieth, G. (1991) Contractility of the transplanted, denervated human heart. *Am. Heart J.*, **121**, 1480–1488.

von Scheidt, W., Bohm, M., Schneider, B., Reichat, B., Erdmann, E. and Autenrieth, G. (1992) Isolated presynaptic inotropic β-adrenergic supersensitivity of the transplanted denervated human heart *in vivo Circulation*, **85**, 1056–1062.

Wharton, J., Polak, J.M., Gordon, L., Banner, N.R., Springall, D.R., Rose, M., *et al.* (1990) Immunohistochemical demonstration of human cardiac innervation before and after transplantation. *Circ. Res.*, **66**, 900–912.

Wieland, D.M., Rosenspire, K.C., Hutchins, G.D., Van Dort, M., Rothley, J.M., Mislankar, S.G., *et al.*, (1990) Neuronal mapping of the heart with 6-[^{18}F] fluorometaraminol. *J. Med. Chem.*, **33**(3), 956–964.

Wilkins, M.R., Gammage, M.D., Lewis, H.M. and Tan, L.B. (1987) Raised concentrations of plasma atrial natriuretic peptides in cardiac transplant recipients. *Br. Med. J.*, **294**, 122.

Wilkins, M.R., Gammage, M.D., Lewis, H.M., Tan, L.B. and Weissber. P.L. (1988) Effect of lower body positive pressure on blood pressure, plasma atrial natriuretic factor concentration, and sodium and water excretion in healthy volunteers and cardiac transplant recipients. *Cardiovasc. Res.*, **22**, 231–235.

Willman, V.L. and Hanlon C.R. (1969) Structural and functional changes in cardiac transplants. *Transplantation Proc.*, **1**, 713–715.

Willman, V.L., Cooper, T., Cian, L.G. and Hanlon, C.R. (1962) Autotransplantation of the canine heart. *Surgery, Gynecol. Obstet.*, **115**, 299–302.

Willman, V.L., Cooper, T., Cian, L.G. and Hanlon, C.R. (1963) Neural responses following autotransplantation of the canine heart. *Circulation*, **27**, 713–716.

Willman, V.L., Cooper, T. and Hanlon, C.R. (1964) Return of neural responses after autotransplantation of the heart. *Am. J. Physiol.*, **297**, 187–189.

Willman, V.L., Jellinek, M., Cooper, T., Tsunekawa, T., Kaiser, G.C. and Hanlon, C.R. (1964) Metabolism of the transplanted heart: effect of excision and reimplantation on myocardial glycogen, hexokinase and acetylcholine esterase. *Surgery*, **56**, 266–269.

Willman, V.L., Nerhavy, J.P., Pennell, R. and Hanlon, C.B. (1967) Response of the autotransplanted heart to blood volume expansion. *Ann. Surg.*, **166**, 513–517.

Wilson, R.F., Christensen, B.V., Olivari, M.T., Simon, A., White, C.W. and Laxson, D.D. (1991) Evidence for structural sympathetic reinnervation after orthotopic cardiac transplantation in humans. *Circulation*, **83**, 1210–1220.

Wilson, R.F., McGinn, A.L. and Stark, R.P. (1992) Sensory reinnervation of the heart after cardiac transplantation (letter). *New Engl. J. Med.*, **326**, 67–68.

Woods, R.I. (1970) The innervation of the frog's heart. *Proc. Roy. Soc. Lond.*, **176**, 43–54.

Yasue, H., Touyama, M., Kato, H., Tanaka, S. and Akiyama, F. (1976) Prinzmetal's variant form of angina as a manifestation or alpha-adrenergic receptor-mediated coronary artery spasm: documentation by coronary arteriography. *Am. Heart J.*, **91**, 148–155.

Yusuf, S., Theodoropoulos, S., Dhalla, N., Mathias, C. and Yacoub, M. (1985) Effect of beta blockade on dynamic exercise in human heart transplant recipients. *Heart Transplant.*, **4**, 312–314.

Yusuf, S., Theodoropoulas, S., Mathias, C.J., Dhalla, N., Wittes, J., Mitchell, A., *et al.* (1987) Increased sensitivity of the denervated transplanted human heart to isoprenaline both before and after beta-adrenergic blockade. *Circulation*, **75**, 696–704.

Yusuf, S., Theodoropoulos, S., Dhalla, N., Mathias, C.J., Teo, K.K., Wittes, J., *et al.* (1989) Influence of beta blockade on exercise capacity and heart rate response after human orthotopic and heterotopic cardiac transplantation. *Am. J. Cardiol.*, **64**, 636–641.

Zeuzem, S., Olbrich, H.G., Seeger, C., Kober, G., Schoffling, K. and Caspary, W.F. (1991) Beat-to-beat variation of heart rate in diabetic patients with autonomic neuropathy and in completely cardiac denervated patients following orthotopic heart transplantation. *Intl. J. Cardiol.*, **33**, 105–114.

16 Orthostatic Hypotension Induced by Drugs and Toxins

Neal Benowitz

San Francisco General Hospital Medical Center,
University of California, San Francisco, USA

The role of drugs or toxins should be considered in any diagnostic evaluation of a patient with orthostatic hypotension. Orthostatic hypotension induced by drugs occurs in a variety of settings. It may be seen as a manifestation of a desired therapeutic effect (such as with the use of guanethidine in the treatment of hypertension), as an acute adverse reaction, to a drug (as with the use of tricyclic antidepressants), as a chronic manifestation of drug toxicity (as in orthostatic hypotension secondary to autonomic neuropathy caused by therapy with vincristine), or as in a drug interaction (as in quinidine-induced orthostatic hypotension in patients taking β-blockers). Drug-induced orthostatic hypotension is of particular concern in the elderly who often take combinations of medications and whose ability to compensate for drug-induced perturbations of cardiovascular function is deficient. Toxin-induced orthostatic hypotension may present with transient symptoms (as with marijuana), a prolonged but reversible condition (ciguatera fish poisoning) or as a permanent and disabling disorder (Vacor poisoning). In this chapter I will review the pathophysiology of orthostatic hypotension induced by a variety of drugs and toxins and, where relevant, discuss the management of chemical-induced orthostatic hypotension.

MECHANISMS OF CHEMICAL-INDUCED ORTHOSTATIC HYPOTENSION

GENERAL MECHANISMS

General mechanisms of orthostatic hypotension have been discussed elsewhere in this volume. Drugs and chemicals induce orthostatic hypotension by interfering with the normal mechanisms of the regulation of blood pressure. Major actions include: (1) depletion of the blood volume; (2) interference with the function of the

TABLE 16.1

Mechanisms of chemical-induced orthostatic hypotension

Hypovolaemia

Diuretics

Adrenal insufficiency

    secondary to discontinuation of

    chronic glucocorticoid therapy

Reduced central nervous system
    sympathetic activity

Barbiturates and other sedative drugs

Methyldopa

L-Dopa

Bromocriptine

Δ^9-THC (marijuana)

Autonomic ganglionic blockade

Trimethaphan

α-Adrenergic blockers

Phentolamine

Phenoxybenzamine

Prazosin

Terazosin

Labetolol

Verapamil

Quinidine

Imipramine and other tricyclic

    antidepressants

Trazodone

Chlorpromazine

Postganglionic sympathetic blockers

Guanethidine

Debrisoquine

Bretylium

Phenelzine and other MAOIs

Direct vasodilators

Nitrates

Morphine and other opiates

Insulin

Captopril and other ACE inhibitors

Autonomic Neuropathy

Alcohol (Wernicke's)

Amiodarone

Vincristine

Cisplatinum

Ciguatera toxin (fish poisoning)

Vacor (rat poisoning)

Release of vasoactive
    substances from microbes

Ivermectin

Diethylcarbamazine

Choroquine and pyrimethamine

Mefloquine and pyrimethamine

Vasovagal

Any hypotensive drug reaction

with an intense sympathetic response

may trigger a vasovagal episode

sympathetic nervous system at a variety of sites, resulting in failure to maintain vascular resistance, venous tone or cardiac output with upright posture; (3) direct vasodilation, which lowers vascular resistances and/or venous tone. Vasodilators may act directly on blood vessels or via involvement with the renin–angiotensin system. Table 16.1 presents a classification according to the primary mechanism by which drugs and toxins may induce orthostatic hypotension.

In general, when drugs induce orthostatic hypotension by depleting the blood volume or by vasodilation alone, there will be evidence of a compensatory sympathetic neural-reflex response to upright posture, i.e. tachycardia, palpitations, sweating, increased circulating catecholamines, etc. This sort of problem may be termed hyperadrenergic orthostatic hypotension. When drugs that impair sympathetic function are involved in the pathogenesis of orthostatic hypotension, the expected sympathetic reflex response will be blunted or absent (hypoadrenergic orthostatic hypotension).

Vasovagal syncope may be triggered by drugs — primarily those that produce hyperadrenergic orthostatic hypotension by vasodilation. The typical response in such patients is a period of reflex tachycardia after assuming an upright posture, followed after several minutes by sudden bradycardia and hypotension with signs and symptoms of hypoperfusion as well as increased vagal discharge. The mechanism is believed to be activation of afferent C-fibres in ventricular mechanoreceptors, owing to the intense myocardial contractile state. Activation of these receptors, similar to the Bezold–Jarisch reflex, produces parasympathetic discharge and inhibits sympathetic responses, resulting in bradycardia and vasodilation with no increase in plasma catecholamines. This type of vagal discharge can be produced in the laboratory by administering isoproterenol to people during upright tilting (Almquist *et al.*, 1989).

IMPORTANCE OF AGING

The elderly are more susceptible to drug-induced orthostatic hypotension than are younger people; and drugs are an important cause of syncope and falls in the elderly. Falls in the elderly are a serious cause of morbidity and mortality due to fractures, particularly hip fractures, owing to the presence of osteoporosis in this population. Typically, drug-induced orthostatic hypotension in the elderly results from therapeutic use of drugs such as sedative-hypnotics, diuretics, antihypertensive drugs, cardiac nitrates and/or antidepressants. The elderly are less able to compensate for drug-induced disturbances in cardiovascular function than are younger people. In particular, the elderly may have abnormal baroreceptor function, with less acceleration of the heart rate and cardiac contractility, as well as a diminished ability to conserve sodium and water, compared to younger people (Docherty, 1990). Figure 16.1 illustrates the susceptibility of the elderly but not the young to orthostatic hypotension after a moderate depletion of sodium by diuretic treatment (Shannon *et al.*, 1986). Another possible contributor to the high incidence of drug-induced orthostatic hypotension is that the elderly often take multiple drugs, and are more exposed to adverse drug interactions.

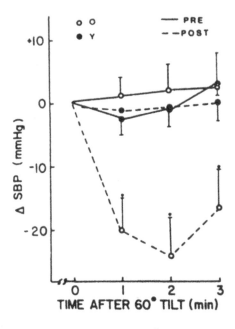

FIGURE 16.1   Change in systolic blood pressure (SBP) during 60 degree upright tilting in six young (Y) and old (O) subjects before (PRE) and after (POST) diuresis with 100 mg/day hydrochlorothiazide for 2 days and a diet containing 3g sodium/day. Despite similar urinary losses of sodium significant onthostatic hypotension was seen in the older but not the young subjects. (From Shannon *et al.* (1986) with permission.)

Because of the likelihood and serious consequences of orthostatic hypotension in the elderly, the physicians' focus should be prevention of orthostatic hypotension in this population. Orthostatic hypotension may be prevented or minimized by the use of vasoactive drugs in low doses and with a gradual escalation of doses, and a careful consideration of the risk–benefit issues in deciding to employ pharmacotherapy with vasoactive drugs in the elderly.

DRUG INTERACTIONS

Orthostatic hypotension is more likely to occur when the mechanisms of postural blood pressure adjustments are impaired at multiple sites. The most common circumstance predisposing to orthostatic hypotension is probably diuretic-induced hypovolaemia. In the presence of hypovolaemia, treatment with vasodilators such as angiotensin coverting enzyme (ACE)-inhibitors, α-blockers or cardiac nitrates may induce supine as well as orthostatic hypotension. Drugs such as β-blockers or postganglionic sympathetic blockers that inhibit sympathetic reflex responses to orthostatic stress may similarly predispose to orthostatic hypotension when vasodilators, including antiarrhythmic drugs such as quinidine, antidepressants or antipsychotic drugs, are administered. These interactions may be minimized by taking patients off diuretics for a few days to allow volume repletion before initiation of vasodilator therapy and/or by giving potentially interacting drugs in low doses with a gradual escalation of the dose as tolerated.

ORTHOSTATIC HYPOTENSION PRODUCED BY SPECIFIC CHEMICALS AND DRUGS

HYPOVOLAEMIC AGENTS

Diuretics

Diureties may produce orthostatic hypotension by volume depletion, particularly in susceptible populations. Susceptible populations include the elderly, patients taking vasodilators or sympathetic blocking drugs or those with underlying autonomic insufficiency. The latter may be exquisitely sensitive to blood volume and may become severely orthostatic with relatively minor reductions in the blood volume (Wagner, 1957).

Adrenal insufficiency

Adrenal insufficiency may follow the sudden withdrawal of corticosteroid therapy following prolonged treatment with high doses. It is associated with the inability to conserve sodium and hence hypovolaemia and orthostatic hypotension.

DRUGS THAT IMPAIR SYMPATHETIC FUNCTION

Drugs acting in the central nervous system

Barbiturates and *other sedative and anaesthetic drugs* reduce the central sympathetic neural output and may also have effects on venous tone. The result is predominantly venodilation with inadequate reflex responses to upright posture, resulting in orthostatic hypotension.

Methyldopa, L-dopa and *bromocriptine* are believed to act on central α-adrenergic neurons that modulate autonomic tone. The resultant effect is enhanced parasympathetic and reduced sympathetic output. Methyldopa does not typically produce orthostatic hypotension, but may do so in the presence of severe cardiac disease or when administered together with other vasoactive drugs. L-dopa and bromocriptine commonly produce some degree of orthostatic hypotension, which may complicate the management of parkinsonism. Parkinsonism is, in itself, associated with an element of autonomic dysfunction in some patients. Especially difficult in this regard is the management of patients with Shy-Drager syndrome (multiple systems atrophy), who have both severe Parkinsonian symptoms and severe orthostatic hypotension. In such cases L-dopa or bromocriptine therapy may aggravate the orthostatic hypotension from the autonomic failure, making it difficult or impossible to treat the motor system symptoms. Anticholinergic drugs and amantadine may be used without aggravating the orthostatic hypotension.

Δ^9-tetrahydrocannabinol (THC) is the psychoactive component of marijuana. THC produces orthostatic hypotension in everyone in high doses and in susceptible people in low doses (Benowitz and Jones, 1975). I am aware of people who developed severe symptomatic orthostatic hypotension after smoking a few puffs of marijuana or even after exposure to marijuana smoke in the environment. The mechanisms include venodilation and impairment of sympathetic vascular reflexes, although tachycardia is usually present. The site of action is thought to be the central nervous system, although the precise mechanisms are not well understood.

Drugs acting on autonomic ganglia

Trimethaphan camsylate is an antihypertensive drug that blocks autonomic ganglia. Its hypotensive effects depend on upright posture (i.e. reverse Trendelenberg position in the bed) to produce venous pooling. Excessive orthostatic hypotension can be reversed by placing the patient in the horizontal position and administering fluids.

α-Adrenoreceptor blockers

Drugs that block α-adrenoreceptors are prescribed for hypertension, benign prostatic hypertrophy, cardiac arrhythmias and depression. Orthostatic hypotension is a common occurrence in these patients. The non-selective α-blockers *phentolamine* and *phenoxybenzamine* are used acutely and chronically, respectively, to control blood pressures in patients with pheochromocytoma. Orthostatic hypotension is often the limiting factor in titration of the dose of phenoxybenzamine. This can be ameliorated to some degree by ensuring that patients are taking adequate sodium and fluids to expand their blood volume. Phenoxybenzamine is also used to treat obstruction of the bladder neck in patients with benign prostatic hyperplasia. Many of these patients are elderly and there is as high as a 30% incidence of side effects, primarily orthostatic hypotension and impotence (Abrams *et al.*, 1982).

The selective α-blockers *prazosin* and *terazosin* are used primarily to treat hypertension, although they are also used to treat benign prostatic hyperplasia. Selective antagonists of α_1 receptors allow noradrenaline that is released as a sympathetic reflex response to hypotension to feedback on α_2 receptors that are located presynaptically to further inhibit noradrenaline release. This minimizes the extent of reflex tachycardia as compared to that observed after treatment with non-selective α-blockers. Orthostatic hypotension is particularly a problem with the first dose of prazosin or terazosin, after which patients frequently become symptomatic of orthostatic hypotension (Figure 16.2) (Graham *et al.*, 1976). Tolerance develops rapidly to the orthostatic hypotension, in part due to a compensatory expansion of the blood volume. First dose orthostatic hypotension may be minimized by starting therapy with low doses and administering the first dose at bed time. Patients should be warned about the first effect of the first dose and instructed to get up out of bed slowly and to lie down if they begin to feel dizzy or light-headed.

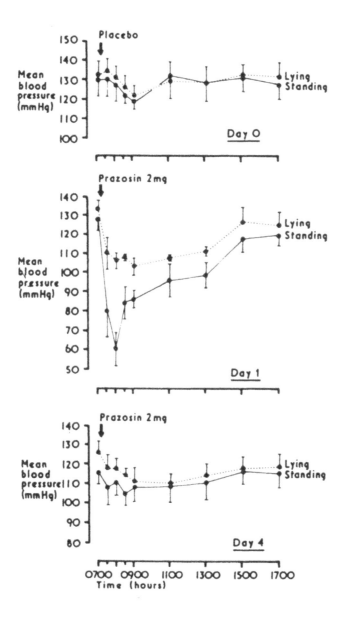

FIGURE 16.2   Mean arterial blood pressure in six hypertensive patients treated with a placebo tablet (day 0); a 2 mg prazosin tablet as the first dose (day 1); a 2 mg prazosin tablet as a test dose (day 4). On days 3 and 4 the patients received 2 mg prazosin three times a day. Prazosin produced severe orthostatic hypotension on day 1, but by day 4 the orthostatic response was similar to the placebo response. (From Graham *et al.* (1976) with permission.)

The antihypertensive drug labetolol is a combined non-selective α- and β-adrenoreceptor blocker. Symptomatic orthostatic hypotension is an occasional side effect of labetolol therapy (Vlachakis *et al.*, 1984).

Verapamil, used both in treating hypertension and arrhythmias, and *quinidine*, occasionally produce orthostatic hypotension, believed to be a consequence of α-adrenoreceptor blockade. Symptomatic orthostatic hypotension has occurred primarily when a combination of drugs that impair postural reflexes, such as quinidine plus verapamil, or quinidine plus β-blockers, is employed. In one case a patient was in atrial fibrillation and verapamil presumably blunted an increase in the ventricular response rate that would normally compensate for the quinidine-induced hypotension (Maisel, Motulsky and Insel, 1985). In a second case β-blockers blunted the reflex sinus tachycardia response to quinidine (Loon, Wilcox and Folger, 1986).

One of the more common causes of drug-induced orthostatic hypotension is the use of antidepressant and neuroleptic drugs which have α-blocking activity and may also blunt cardiovascular reflexes mediated by the central nervous system. Orthostatic hypotension is the most common cardiovascular side effect of *tricyclic antidepressants*, occuring in one series in 37% (7% symptomatic) of patients under 60 years old and 78% (42% symptomatic) of patients 60 years or older with atherosclerotic vascular disease receiving the drug (Muller, Goodman and Bellet, 1960). Two patients with pre-existing angina pectoris suffered acute myocardial infarction. In another series, 20% of 148 patients who were treated with *imipramine* (average dose 225 mg/day) had severe orthostatic hypotension that interfered with therapy and 4% experienced fall-related physical injuries (Glassman *et al.*, 1979).

Orthostatic hypotension from tricyclic antidepressants appears to be dose related, to occur early in the course of treatment, and to occur more commonly in the elderly (Middleton, Maisey and Millis, 1987). Some studies report a correlation between the extent of the orthostatic hypotension before treatment and the extent of the increase in orthostatic hypotension induced by drugs, suggesting that tricyclic antidepressants are interacting with some underlying disorder of cardiovascular regulation (Roose *et al.*, 1981). However other studies have not found such an association (Georgotas *et al.*, 1987).

Orthostatic hypotension has been studied most extensively with imipramine, but has also been described during treatment with amitriptyline, desipramine and clomipramine. The degree of orthostatic hypotension seems to be less with nortriptyline and possibly doxepin (Roose *et al.*, 1981; Neskes *et al.*, 1985). One study with the latter however, used doses below those commonly used to treat depression in clinical practice, making extrapolation of the results problematic.

The mechanisms of the production of orthostatic hypotension by tricyclic antidepressants (TCAs) are complex in part due to the many pharmacological actions of these drugs. TCAs, for example, block reuptake of noradrenaline, dopamine and serotonin. TCAs also block α-adrenoreceptors; and this is presumed to be the primary mechanism of orthostatic hypotension. Against this hypothesis is the observation that the potency of α-blockade *in vitro* does not correspond with the extent of orthostatic hypotension *in vivo*. For example, imipramine and nortriptyline are equipotent α-blockers (Sakalis et al., 1972), but the former is associated with more orthostatic hypotension than the latter. Desmethyldesipramine has less α-blocking activity, but produces a similar degree of orthostatic hypotension to imipramine.

Another mechanism of orthostatic hypotension appears to be impairment of cardiovascular reflexes. In seven women with orthostatic hypotension while taking tricyclic antidepressants (and other drugs as well), the resting heart rate was higher, the magnitude of sinus arrhythmia was lower, heart rate variation during and after release of Valsalva diminished, and the pressor response to hand-grip exercise was reduced compared to healthy controls (Middleton, Maisey and Millis, 1987). These findings are consistent with a combination of peripheral α-adrenergic blockade plus disturbed regulation of the heart rate due to anticholinergic actions and/or impaired central sympathetic reflex responses.

The non-tricyclic antidepressant trazodone and neuroleptics such as chlorpromazine occasionally produce orthostatic hypotension by peripheral α-blockade (Sakalis et al., 1972; Spivak, Radvan and Shine, 1987). Orthostatic hypotension due to these drugs is often associated with substantial reflex tachycardia.

The management of orthostatic hypotension due to TCA therapy has held considerable interest, because adequate doses of TCA are needed to optimize the antidepressant response. Limitation of therapy by orthostatic hypotension may result in therapeutic failure. Suggestions for reducing orthostatic hypotension have included selection of TCA's such as nortriptyline that are less likely to produce orthostatic hypotension, increasing the dietary intake of salt, and the use of support hose or abdominal binders (Pollack and Rosenbaum, 1987). A variety of pharmacological approaches to treating TCA-induced orthostatic hypotension have also been reported. These include salt tablets, fludrocortisone and non-steroidal antiflammatory drugs, which may ameliorate orthostatic hypotension but are often associated with increased body weight and the development of oedema, presumably related to the expansion in blood volume. Dihydroergotamine, a vasoconstricting drug, in doses of 10 mg per day, reduced the extent of orthostatic hypotension and related symptoms in one controlled study of 58 patients with orthostatic hypotension due to neuroleptics or TCAs (Thulesius and Berlin, 1986).

The use of yohimbine to treat TCA-induced orthostatic hypotension has recently produced considerable interest. Yohimbine is an α_2-adrenergic blocker that acts in the brain to enhance sympathetic outflow. Yohimbine treatment reduces the extent of orthostatic hypotension in patients receiving antidepressant drugs, in some cases with substantial symptomatic improvement (Lecrubier, Puech and Des Lauriers, 1981; Charney, Price and Heninger, 1986). The doses of yohimbine have ranged from 10–30 mg/day, given in three-four divided doses. Yohimbine may cause supine hypertension, particularly in patients with underlying hypertensive disorders, so the blood pressure must be monitored. At high doses, yohimbine may produce anxiety and/or aggravate underlying psychiatric disorders.

β-Adrenoreceptor blockers

β-Blockers when administered alone rarely, if ever, produce orthostatic hypotension. However, when combined with drugs with α-blocking or other vasodilating activities, orthostatic hypotension may occur.

Paradoxically, non-selective β-blockers may increase blood pressure in some patients with autonomic failure, as discussed elsewhere in this volume. The mechanism is believed to be the blockade of β_2-adrenoreceptors that mediate vasodilation, which may contribute to a decline in systemic vascular resistance when moving from the supine to the upright position. In many patients with autonomic failure, the heart rate increases minimally or not at all with standing due to the autonomic neuropathy, so loss of reflex tachycardia due to β-blocker therapy is not a problem. β-Blockers are not indicated in most patients with hyperadrenergic orthostatic hypotension, i.e. where there is a substantial response of the heart rate to orthostatic hypotension. An exception may be orthostatic hypotension secondary to prolapse of the mitral valve (the mitral valve prolapse syndrome).

Drugs acting on postganglionic sympathetic neurons

The antihypertensive drugs *guanethidine, debrisoquine* (and bretylium) act on presynaptic neurons to inhibit the release of noradrenaline. Clinical use of these agents is commonly associated with orthostatic hypotension that may limit the tolerability of the drug. Guanethidine therapy is associated with sodium and water retention, which compensates in part for the drug-induced hypotension (Gill, Mason and Bartter, 1964). Thus, in patients treated with guanethidine, the use of diuretics tends to aggravate orthostatic hypotension. Guanethidine has been administered intravenously to produce neural blockade for treatment of causalgia, reflex sympathetic dystrophies and Raynaud's disease. Severe and persistent orthostatic hypotension has been observed in a patient who received repetitive doses of intravenous guanethidine into the arm (Sharpe, Mibszkiewicz and Carli, 1987). *Monoamine oxidase inhibitors* (MAOIs), currently used as antidepressants, were at one time used as antihypertensive drugs. In one large retrospective case series, severe orthostatic hypotension, defined as cases where the patient reported passing out or having repeated episodes of falling, was found in 11% of people receiving phenelzine and 14% receiving tranylcypromine, as compared to 9% on imipramine (Rabkin *et al.*, 1985); In most cases orthostatic hypotension occurred within the first 2 months of drug treatment.

The mechanism of MAOI-induced orthostatic hypotension is believed to involve the accumulation of "false neurotransmitters". Tyramine is normally oxidized by monoamine oxidase in the intestinal mucosa and the liver. During MAOI therapy, tyramine reaches the systemic circulation in large quantities, where it is hydroxylated to octopamine, which is then taken up by noradrenergic nerve terminals. Octopamine is a very weak pressor agent and when released instead of noradrenaline results in orthostatic hypotension. The extent of the lowering of supine and upright blood pressure correlated with the percent inhibition of monoamine oxidase in platelets after 6 weeks of phenelzine therapy in one study (Robinson *et al.*, 1982), but no such correlation was found in another study (Georgotas *et al.*, 1987). Treatment considerations for MAOI-induced orthostatic hypotension are similar to those described previously for tricyclic antidepressants.

In these patients, however, yohimbine should be used with caution and starting at low doses, because MAOI can potentiate the pressor effects of this drug (Killian *et al.*, 1990).

VASODILATORS

As with sympathetic blocking drugs, direct vasodilators may produce or aggravate orthostatic hypotension. *Nitrates* commonly produce orthostatic hypertension, particularly after prolonged standing and/or when combined with other hypotensive drugs. With continuous or frequent dosing of cardiac nitrates, substantial tolerance to orthostatic hypotension develops within 48 h (Parker *et al.*, 1983). *Morphine* and other *narcotics* may produce orthostatic hypotension. The sites of action of narcotics may be both the peripheral and central nervous systems.

Insulin decreases blood pressure in diabetics with autonomic failure (Page and Watkins, 1976). The mechanism is not well characterized, but appears to involve venodilation and/or acute hypovolaemia, possibly due to the movement of fluid out of the vascular space. This effect is normally compensated for by sympathetic reflexes, as evidenced by an increase in the concentration of noradrenaline in the plasma (Miles and Hayter, 1968). The cardiovascular effects of insulin do not depend on induction of hypoglycaemia (Christensen *et al.*, 1980). In diabetics with autonomic insufficiency, compensating sympathetic responses are inadequate and orthostatic hypotension is aggravated immediately following dosing with regular insulin (Figure 16.3).

ACE inhibitors such as *captopril* are primarily arterial vasodilators and as such do not usually produce orthostatic hypotension. However, captopril may aggravate orthostatic hypotension in patients taking other hypotensive agents (White, 1986).

AUTONOMIC NEUROPATHY AND OTHER TYPES OF ORTHOSTATIC HYPOTENSION CAUSED BY OTHER DRUGS AND TOXINS

Alcohol (Wernicke's syndrome)

Orthostatic hypotension, defined as a fall in the diastolic blood pressure of at least 10 mmHg and a fall in the mean blood pressure of 20 mmHg, was found in 75% of 40 patients with Wernicke's encephalopathy (Birchfield, 1964). The mechanism of orthostatic hypotension was studied in a subset of nine of these patients. Autonomic dysfunction, as evidenced by hypotension without an adequate compensatory tachycardia during tilting, an abnormal heart rate response to Valsalva, a subnormal cold pressor response and hypersensitivity to exogenously infused catecholamines was described. Hypovolaemia was present in half of the eight subjects in which it was measured. However, the degree of hypovolaemia did not correlate with the extent of the orthostatic hypotension. Expansion of the blood volume by rapid infusion of sodium chloride solution diminished the degree of orthostatic hypotension, but the benefit was transient, disappearing in 24–36 h. The patients were very sensitive to the blood pressure lowering effects of hypovolaemia or of vasodilators such as nitroglycerin, consistent with a failure of

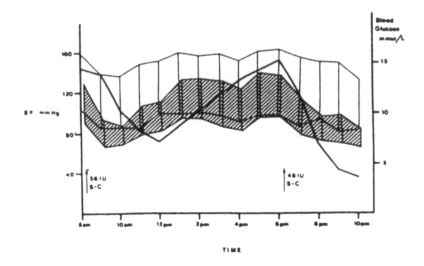

FIGURE 16.3 Supine (unhatched area) and upright (hatched area) blood pressure (BP) from 8:00 a.m. to 10:00 p.m. in a 48 year old diabetic man with severe autonomic neuropathy. The solid line shows the concentration of glucose in the blood. The arrows indicate the subcutaneous (S-C) administration of regular insulin. The upright blood pressure is noted to fall substantially after administration of insulin. (From Page and Watkins (1976) with permission.)

compensatory sympathetic reflexes. Based on pathological findings, the site of the autonomic lesion may be peripheral (peripheral nerves, preganglionic neurons of the intermediolateral cell column of the spinal cord) and/or central (hypothalamus, midbrain tegmentum, fourth ventrical roof nucleus).

Chronic abnormalities in circulatory reflexes (as assessed by response to Valsalva manoeuvre) and postural syncope have also been reported in patients with alcohol-induced peripheral neuropathy, in the absence of encephalopathy. In several cases circulatory reflexes return to normal after 2–3 months of nutrition therapy (Barraclough, Lond and Sharpey-Schafer, 1963).

Amiodarone

Amiodarone is known to occasionally produce neurotoxicity, primarily a peripheral neuropathy, associated with proximal muscle weakness. One case of severe incapacitating orthostatic hypotension was reported after 6 months of amiodarone therapy (Manolis *et al.*, 1987). Amiodarone was stopped because of suspected peripheral neuropathy; 2 weeks later after initiation of quindine and mexiletine, orthostatic syncope occurred. Orthostatic hypotension was discovered, without a compensatory increase in heart rate. Severe orthostatic hypotension persisted for an additional month after the other antiarrhythmic drugs were stopped. It was hypothesized that amiodarone-induced neurotoxicity produced autonomic dysfunction in this patient.

Antineoplastic agents

Vincristine is neurotoxic, and is well known to cause bowel disturbances (severe constipation) bladder dysfunction (urinary retention) as well as peripheral neuropathy. Vincristine therapy has also been associated with orthostatic hypotension, presumably due to autonomic neuropathy (Hancock and Naysmith, 1975). The affected patients have had evidence of peripheral neuropathy. Recovery from orthostatic hypotension occurred over several months after vincristine was discontinued.

In a prospective study of autonomic function, 33 patients who received vinca alkaloids were compared to 30 patients who received chemotherapy without vinca alkaloids (Roca *et al.*, 1985). Abnormal blood pressure responses to standing and abnormal heart rate responses to deep breathing and standing were more common in patients treated with vinca alkaloids (48–82%, depending on the test), than in patients not receiving vinca alkaloids (3–30%). Nearly total loss of urinary excretion of noradrenaline with intact afferent and preganglionic responses were found in one patient who developed orthostatic hypotension on vincristine, suggesting that the lesion has localised to the postganglionic adrenergic nerves (Carmichael *et al.*, 1970).

Cisplatinum occasionally produces neurotoxicity, including retrobulbar neuritis, seizures and peripheral neuropathy. Two patients with autonomic neuropathy including severe orthostatic hypotension with a subnormal reflex heart rate response, and, in one case, gastric paresis and recurrent vomiting, have been reported (Rosenfeld and Broder, 1984; Cohen and Mollman, 1987). Both patients had evidence of peripheral neuropathy as well. Cisplatinum can also cause orthostatic hypotension by inducing a salt-wasting nephropathy and hypovolaemia. Seven patients with this syndrome which includes hyporeninaemic hypoaldosteronism have been described following 2–4 months of cisplatinum therapy (total dose 200–600 mg/m^2) (Hutchinson *et al.*, 1988). Orthostatic hypotension was aggravated by fluid restriction and could be ameliorated with fludrocortisone and salt tablets.

Although not reported to produce symptomatic orthostatic hypotension, the anthracycline antineoplastic drugs *doxorubicin* and *epirubicin* have been found to be associated more frequently with an abnormal heart rate response to standing and a blunted blood pressure response to isometric hand grip compared with patients treated with other antineoplastic drugs (Viniegra *et al.*, 1990).

Antimicrobials

Zidovudine, an antiviral drug used for infection with human immunodeficiency virus (HIV), was reported to produce acute orthostatic hypotension in one patient 30–40 min after a dose, corresponding to the time of the peak level of zidovudine and its metabolite, zidovudine 5-o-glucuronide, in the blood (Loke, Murray-Lyon and Carter, 1990). The orthostatic hypotension resolved in 10 min. This patient had a subnormal plasma cortisol response to tetracosactrin (an adrenocorticotrophin hormone-like drug), indicating mild adrenal insufficiency. It was speculated that vasodilating effect of zidovudine combined with adrenal insufficiency was responsible for this patient's symptoms. As adrenal insufficiency is common in patients with HIV infection, this diagnosis needs to be considered as aetiologic or contributory in evaluating orthostatic hypotension due to any drug in patients with acquired immunodeficiency syndrome.

Ivermectin is a microfilaricide that is widely used in underdeveloped countries. In a community trial in Ghana, 14,911 people received a single dose of ivermectin, 130–200 mg/kg to control onchoceriasis (De Sole *et al.*, 1989). A syndrome described as "severe symptomatic postural hypotension" (SSPH) was observed in about 0.5% (34 people); in most cases this occurred within the first day of treatment, but occasionally occurred as late as 72 h after the dose. The reaction was defined as an inability of the patient to stand for 2 min due to severe dizziness or weakness attributable to a marked drop in blood pressure. The syndrome was often associated with excessive yawning, profuse sweating, tachycardia, fever and occasionally confusion. Upon recumbency, drowsiness and bradycardia were observed. The symptoms persisted for an average of 5 h (range 0.5–22 h).

No treatment other than the recumbency was needed for the majority of the patients with SSPH. Four patients were treated with hydrocortisone for persistent SSPH and appeared to recover within 2–4 h. The incidence of SSPH was highest in patients from villages in which the extent of microfilarial infection was greatest. This suggests that SSPH may be related to the killing of the microfilarial organisms rather than to ivermectin *per se*. A similar postural hypotension reaction has been observed after treatment of onchoceriasis with *diethyl-carbamazine* (Awadzi *et al.*, 1982).

Severe orthostatic hypotension has also been observed after initiation of antimicrobial therapy for falciparum malaria (Ekue *et al.*, 1988). *Choroquine, sulfadoxine* plus *pyrimethamine*, and *mefloquine* plus sulfadoxine plus pyrimethamine treatment regimens have been implicated. The time of onset after treatment and the duration of the orthostatic hypotension were 2–48 h and 3–72 h, respectively. It was speculated that falciparum malaria infection causes release of histamine, while pyrimethamine and chloroquine inhibit the metabolism of histamine by N-methyl transferase. The reaction appears to be dose-related in that few patients receiving a low dose of sulfadoxine and pyrimethamine developed orthostatic hypotension.

Ciguatera toxin (fish poisoning)

Ciguatera toxin is produced by dinoflagellates consumed by reef fish (Hughes and Merson, 1976). Intoxication is most common after ingestion of barracuda, red snapper or grouper and typically presents with vomiting, abdominal pain, myalgias, weakness, paresthesias of the mouth, face and extremities, and pruritus. A peculiar "hot and cold reversal", in which cold objects feel hot and vice versa is reported.

Cardiovascular features include bradycardia, hypotension and in some cases severe orthostatic hypotension (Bagnis, Kuberski and Laugier, 1979; Morris *et al.*, 1982). The nature of the autonomic disturbance was studied in one patient with severe orthostatic hypotension (Geller and Benowitz, 1992), This patient had evidence of both increase of vagal tone (the bradycardia and in part the orthostatic hypotension was reversed by atropine) and sympathetic insufficiency (low plasma catecholamines and hypersensitivity to the infused noradrenaline).

The pharmacological basis for this peculiar syndrome is unknown, but may involve both the central and peripheral nervous systems. The orthostatic hypotension seems to be reversible in most cases, resolving within 4–6 weeks in a few cases where recovery has been documented.

Vacor (N-3-pyridylmethyl-N'-p-nitrophenyl urea)

Vacor is a rodenticide that was sold in packets of cornmeal-like material containing 2% N-3-pyridylmethyl-N'-p-nitrophenyl urea. Vacor is no longer marketed in the United States, but is still available in some homes and for use by professional exterminators. Vacor ingestion (usually suicidal) produced an unusual syndrome including irreversible and severe autonomic dysfunction and insulin dependent diabetes mellitus (LeWitt, 1980). Severe orthostatic hypotension developed from 6 h to 2 days following injection of vacor. The autonomic disturbance in cardiovascular regulation has been characterized as having very low circulating catecholamines at rest without a response to upright posture and hypersensitivity to exogenous catecholamines, consistent with destruction of vascular adrenergic neurons (Benowitz et al., 1980; Osterman et al., 1981). In some patients the response of the heart rate to upright posture is impaired, but in others there is a substantial tachycardia, suggesting selective sparing of cardiac autonomic nerves. Other features of intoxication include a peripheral sensory and motor neuropathy and encephalopathy. Other features of the autonomic neuropathy include dysphagia, urinary retention, constipation, loss of sweating and impotence. These findings suggest generalized peripheral autonomic dysfunction. In many cases, the autonomic disturbances and diabetes have been irreversible, but in some cases at least partial recovery occurs, although such recovery may take many months (Osterman et al., 1981). Treatment of patients with orthostatic hypotension from ingestion of Vacor has often been difficult, but there has been some success with the use of a diet high in salt, fludrocortisone and pressor drugs, including injections of dihydroergotamine.

REFERENCES

Abrams, P.H., Shah, P.J.R. Stone, R. and Choa, R.G. (1982) Bladder outflow obstruction treated with phenoxy-benzamine. Br. J. Urol., 54, 527–530.

Almquist, A., Goldenberg, I.F., Milstein, S., Chen, M.Y., Chen, X., Gornick, C., et al. (1989) Provocation of bradycardia and hypotension by isoproterenol and upright posture in patients with unexplained syncope. N. Engl. J. Med., 320, 346–351.

Awadzi, K., Orme, M., Breckenridge, A.M. and Giles, H.M. (1982) The chemotherapy of onchocerciasis. VII: The effect of prednisone on the Mazzotti reaction. Ann. Trop. Med. Parasitol., 76, 331–338.

Bagnis, R., Kuberski, T. and Laugier, S. (1979) Clinical observations on 3,009 cases of ciguatera (fish poisoning) in the South Pacific. Am. J. Trop. Med. Hyg., 28, 1067–1073.

Barraclough, M.A., Lond, M.B. and Sharpey-Schafer, E.P. (1963) Hypotension by absent circulatory reflexes: effects of alcohol, barbiturates, psychotherapeutic drugs, and other mechanisms. Lancet, i, 1121–1126.

Benowitz, N.L. and Jones, R.T. (1975) Cardiovascular effects of prolonged delta-9-tetrahydrocannabinol ingestion. Clin. Pharmacol. Ther., 18, 287–297.

Benowitz, N.L., Byrd, R., Schambelan, M., *et al.* (1980) Dihydroergotamine treatment for orthostatic hypotension from Vacor rodenticide. *Ann. Intern. Med.*, **92**, 387–388.

Birchfield, R.I. (1964) Postural hypotension in Wernicke's disease: a manifestation of autonomic nervous system involvement. *Am. J. Med.*, **36**, 404–414.

Carmichael, S.M., Eagleton, L., Ayers, C.R. and Mohler, D. (1970) Orthostatic hypotension during vincristine therapy. *Arch. Intern. Med.*, **126**, 290–293.

Charney, D.S., Price, L.H. and Heninger, G.R. (1986) Desipramine-yohimbine combination treatment of refractory depression: implications for the β-adrenergic receptor hypothesis of antidepressant action. *Arch. Gen. Psychiatry*, **43**, 1155–1161.

Christensen, N.J., Gundersen, H.J.G., Hegedus, L., Jacobsen, F., Mogensen, C.E., Osterby, R., *et al.* (1980) Acute effects of insulin on plasma noradrenaline and the cardiovascular system. *Metabolism*, **29**, 1113–1145.

Cohen, S.C. and Mollman, J.E. (1987) Cisplatin-induced gastric paresis. *J. Neuroncol.*, **5**, 237–240.

De Sole, G., Awadzi, K., Remme, J., Dadzie, K.Y., Ba, O., Giese, J., *et al.* (1989) A community trial of ivermectin in the onchocerciasis focus of Asubende, Ghana. II. Adverse reaction. *Trop. Med. Parasit.*, **40**, 375–382.

Docherty, J.R. (1990) Cardiovascular responses in aging: a review. *Pharmacol. Rev.*, **42**, 104–125.

Ekue, K., Phiri, E.D., Mukunyadela, M., Sheth, U.K. and Wernsdorfer, W.H. (1988) Severe orthostatic hypotension during treatment of falciparum malaria. *Br. Med. J.*, **296**, 396.

Geller, R.J. and Benowitz, N.L. (1992) Orthostatic hypotension in ciguatera fish poisoning. *Arch. Int. Med.*, **152**, 2131–2133.

Georgotas, A., McCue, R.E., Friedman, E. and Cooper, T.B. (1987) A placebo-controlled comparison of the effect of nortriptyline and phenelzine on orthostatic hypotension in elderly depressed patients. *J. Clin. Pharmacol.*, **7**, 413–416.

Gill, J.R. Jr., Mason, D.T. and Bartter, F.C. (1964) Adrenergic nervous system in sodium metabolism: effects of guanethidine and sodium-retaining steroids in normal man. *J. Clin. Invest.*, **43**, 177–184.

Glassman, A.H., Giardina, E.V., Perel, J.M., Bigger, T.J. Jr., Kantor S.J. and Davies, M. (1979) *Lancet*, **i**, 468–472

Graham, R.M., Thronell, I.R., Gain, J.M., Bagnoli, C., Oates, H.F. and Stokes, G.S. (1976) Prazosin: the first-dose response phenomenon. *Br. Med. J.*, **ii**, 1293–1294.

Hancock, B.W. and Naysmith, A. (1975) Vincristine-induced autonomic neuropathy. *Br. Med. J.*, **3**, 207.

Hughes, J.M. and Merson, M.H. (1976) Fish and shellfish poisoning. *N. Engl. J. Med.*, **295**, 1117–1112.

Hutchinson, F.N., Perez, E.A., Gandara, Dr., Lawrence, J.H. and Kaysen G.A. (1988) Renal salt wasting in patients treated with cisplatin. *Ann. Intern. Med.*, **108**, 21–25.

Killian, T.J., Robertson, D., Biaggioni, I. Haile, V., Biscaia, I. and Robertson, R.M. (1990) Enhanced pressor effect of yohimbine with MAO inhibition in idiopathic orthostatic hypotension (abstract) *Circulation*, **82**, III–637.

Lecrubier, Y., Puech, A.J. and Des Lauriers, A. (1981) Favourable effects of yohimbine on clomipramine-induced orthostatic hypotension: a double-blind study. *Br. J. Clin. Pharmacol.*, **12**, 90–93.

LeWitt, P.A. (1980) The neurotoxicity of the rat poison Vacor: a clinical study of 12 cases. *N. Engl. J. Med.*, **302**, 73–77.

Loke, R.H., Murray-Lyon, I.M. and Carter, G.D. (1990) Postural hypotension related to zidovudine in a patient infected with HIV. *Br. Med. J.*, **300**, 163–164.

Loon, N.R., Wilcox, C.S. and Folger, W.A. (1986) Orthostatic hypotension due to quinidine and propranolol. *Am. J. Med.*, **81**, 1101–1104.

Maisel, A.S., Motulsky, H.J. and Insel, P.A. (1985) Hypotension after quinidine plus verapamil. *N. Engl. J. Med.*, **312**, 167–170.

Manolis, A.S., Tordjman, T., Mack, K.D. and Estes, N.A.M. (1987) Atypical pulmonary and neurological complications of amiodarone in the same patient; report of a case and review of the literature. *Arch. Intern. Med.*, 1805–1809.

Middleton, H.C., Maisey, D.N. and Millis, I.H. (1987) Do antidepressants cause postural hypotension by blocking cardiovascular reflexes? *Eur. J. Clin. Pharmacol.*, **31**, 647–653.

Miles, D.W. and Hayter, C.J. (1968) The effect of intravenous insulin on the circulatory responses to tilting in normal and diabetic subjects with special reference to baroreceptor reflex block and atypical hypoglycaemic reactions. *Clin. Sci.*, **34**, 419–430.

Morris, G.J., Lewin, P., Hargrett, N.T., Smith, W.C., Blake, P.A. and Schneider, R. (1982) Clinical features of ciguatera fish poisoning: a study of the disease in the US Virgin Islands. *Arch. Intern. Med.*, **142**, 1090–1092.

Muller, O.F., Goodman, N. and Bellet, S. (1960) The hypotensive effect of imipramine hydrochloride in patients with cardiovascular disease. *Clin. Pharmacol. Ther.*, **2**, 300–307.

Neshkes, R.E., Gerner, R., Jarvik, L.F., Mintz, J., Joseph, J., Linde, S., *et al.* (1985) Orthostatic effect of imipramine and doxepinin in depressed geriatric outpatients. *J. Clin. Psychopharmacol.*, **5**, 102–106.

Osterman, J., Zymslinski, R.W., Hopkins, C.B., Cartee, W., Lin, T. and Nankin, H.R. (1981) Full recovery from severe orthostatic hypotension after Vacor rodenticide ingestion. *Ann. Intern. Med.*, **141**, 1505–1507.

Page, M.M. and Watkins, P.J. (1976) Provocation of postural hypotension by insulin in diabetic autonomic neuropathy. *Diabetes*, **25**, 90–95.

Parker, J.O., Fung, H.L., Ruggirello, D. and Stone, J.A. (1983) Tolerance to isosorbide dinitrate; rate of development and reversal. *Circulation*, **68**, 1074–1080.

Pollack, M.H. and Rosenbaum, J.F. (1987) Management of antidepressant-induced side effects: a practical guide for the clinician. *J. Clin. Psychiatry*, **48**, 3–8.

Rabkin, J.G., Quitkin, F.M., McGrath, P., Harrison, W. and Tricamo, E. (1985) Adverse reactions to monoamine oxidase inhibitors. Part II. Treatment correlates and clinical management. *J. Clin. Psychopharmacol.*, **5**, 2–9.

Robinson, D.S., Nies, A., Corcella, J., Cooper, T.B., Spencer, C. and Keefover, R. (1982) Cardiovascular effects of phenelzine and amitriptyline in depressed outpatients. *J. Clin. Psychiatry*, **43**, 8–15.

Roca, E., Bruera, E., Politi, P.M., Barugel, M., Cedaro, L., Carraro, S., *et al.* (1985) Vinca alkaloid-induced cardiovascular autonomic neuropathy. *Cancer Treat. Rep.*, **69**, 149–151.

Roose, S.P., Alexander, H., Glassman, Siris, G.S., Walsh, T.B., Bruno, R.L., *et al.* (1981) Comparison of imipramine- and nortriptyline-induced orthostatic hypotension: a meaningful difference. *J. Clin. Pharmacol.*, **1**, 316–321.

Rosenfeld, C.S. and Broder, L.E. (1984) Cisplatin-induced autonomic neuropathy. *Cancer Treat. Rep.*, **68**, 659–660.

Sakalis, G., Curry, S.H., Mould, G.P. and Lader, M.H. (1972) Physiologic and clinical effects of chlorpromazine and their relationship to plasma level. *Clin. Pharmacol. Ther.*, **13**, 931–946.

Shannon, R.P., Wei, J.Y., Rosa, R.M., Epstein, F.H. and Rowe, J.W. (1986) The effect of age and sodium depletion on cardiovascular responses to orthostasis. *Hypertension*, **8**, 438–443.

Sharpe, E., Milaszkiewicz, R. and Carli, F. (1987) A case of prolonged hypertension following intravenous guanethidine block. *Anaesthesia*, **42**, 1081–1084.

Spivak, B., Radvan, M. and Shine, M. (1987) Postural hypotension with syncope possibly precipitated by trazodone. *Am. J. Psychiatry*, **144**, 1512–1513

Thuselsius, O. and Berlin, E. (1986) Dihydroergotamine therapy in orthostatic hypotension due to psychotropic drugs. *Int. J. Clin. Pharmacol. Ther. Toxicol.*, **24**, 465–467.

Viniegra, M., Marchetti, M., Losso, M., Navigante, A., Litovska, S., Senderowicz, A., *et al.* (1990) Cardiovascular autonomic function in anthracycline-treated breast cancer patients. *Cancer Chemother. Pharmacol.*, **26**, 227–231.

Vlachakis, N.D., Barr, J., Velasquez, M., Alecander, N. and Maronde, R. (1984) Acute effect of labetalol on blood pressure in relation to the sympathetic nervous system and plasma renin activity. *Clin. Pharmacol. Ther.*, **35**, 782–786.

Wagner, H.N. Jr. (1957) The influence of autonomic vasoregulatory reflexes on the rate of sodium and water excretion in man. *J. Clin. Invest.*, **36**, 1319–1327.

White, W.B. (1986) Hypotension with postural syncope secondary to the combination of chlorpromazine and captopril. *Arch. Intern. Med.*, **146**, 1833–1834.

17 Pheochromocytoma — Clinical Manifestations, Diagnosis and Management

William M. Manger and Ray W. Gifford, Jr.

*Department of Medicine, New York,
and Cleveland Clinic Foundation, Cleveland, Ohio, USA*

Pheochromocytoma is a rare tumor arising mainly from chromaffin cells in the adrenal medullae but also from those in the organ of Zuckerkandl and those associated with sympathetic nerves in extra-adrenal sites in the abdomen, chest and neck. Most tumors secrete adrenaline and noradrenaline continuously or episodically and cause sustained or paroxysmal hypertension. Left untreated, pheochromocytoma will almost invariably be lethal. Mortality and morbidity result from complications of severe hypertension and/or excess circulating catecholamines; 10% of tumors are malignant. Clinical and laboratory manifestations result from hypercatecholaminemia and hypertension and may mimic many diseases. Hypertension associated with severe headaches, generalized sweating and palpitations is particularly common; medullary thyroid carcinoma, hyperparathyroidism and neurocutaneous lesions may coexist with familial pheochromocytoma. Symptomatic patients with sustained or paroxysmal hypertension should be screened for pheochromocytoma by quantitating 24-hour urinary metanephrines and/or plasma catecholamines. Preoperative tumor location may be established by computed tomography, magnetic resonance imaging and radioisotope techniques. Expertise in drug management, surgical removal and postoperative care is essential. Radiotherapy and chemotherapy may be valuable in treating malignant pheochromocytoma.

KEY WORDS: Tumors, chromaffin cells, catecholamines, hypertension, carcinoma, sweating, palpitations, hyperparathyroidism, radiotherapy, chemotherapy

INTRODUCTION

A high index of suspicion and awareness of the vagaries of clinical and laboratory manifestations of pheochromocytoma are essential in detecting this tumor. There is no more important and treacherous case of hypertension to recognize since if left untreated, this "pharmacologic bomb" will almost invariably be lethal. Mortality and morbidity result mainly from complications of severe hypertension and/or excess circulating catecholamines (e.g., cerebrovascular and cardiovascular accidents, cardiomyopathy, cardiac decompensation, arrhythmias); only about 10% of tumors are malignant.

393

Clinical expressions of pheochromocytoma, often explosive and dramatic, are so variable that it can mimic a large variety of diseases. With the diagnostic modalities available today, the diagnosis and location of pheochromocytoma are almost always possible.

PREVALENCE

Fewer than 0.05% of Americans with sustained hypertension have pheochromocytoma; however, in estimating prevalence it must be appreciated that approximately 45% of patients with this tumor have only paroxysmal hypertension. Pheochromocytomas may occur at any age but are most common in the fourth and fifth decades.

ORIGIN

Pheochromocytomas arise from chromaffin cells in the adrenal medullae (90% of tumors), the organ of Zuckerkandl, and in chromaffin cells associated with sympathetic nerves and plexuses in extra-adrenal sites in the abdomen (including the urinary bladder), chest (< 2.0% of tumors), neck (< 0.1% of tumors); rarely tumors arise at the base of the skull, in the heart or spermatic cord. Multiple and extra-adrenal tumors occur more frequently in children (30% of cases) than adults (8% of cases). About 10% of pheochromocytoma cases are familial and at least 70% of these tumors are bilateral.

BIOSYNTHESIS AND CATABOLISM

Catecholamines (dopamine, noradrenaline and adrenaline) are synthesized by a series of enzymatic reactions in the following sequence: tyrosine → dopa → dopamine → noradrenaline → adrenaline. Some dopamine is catabolized to homovanillic acid (HVA) whereas some noradrenaline and adrenaline are converted to the metanephrines and vanillylmandelic acid (VMA); however, the major portion of catecholamines is conjugated. Biosynthesis occurs in chromaffin cells, sympathetic nerves and parts of the brain. Only dopamine and noradrenaline are synthesized in postganglionic sympathetic nerves. Some chromaffin cells in the adrenal medulla and certain cells in the brain can convert noradrenaline to adrenaline through the action of the enzyme phenylethanolamine-N-methyltransferase (PNMT). Adrenaline is more potent than noradrenaline in increasing metabolism and augmenting the rate and force of myocardial contractions whereas noradrenaline causes more intense vasoconstriction. Dopamine exerts a diuretic and natriuretic action; it can also augment myocardial contractions. Noradrenaline accounts for about 73% of free catecholamines in plasma whereas concentrations of adrenaline (14%) and dopamine (13%) are much less. Only small amounts of free noradrenaline and adrenaline are eliminated in the urine; the conversion of dopa to dopamine in the kidney results in relatively large concentrations of urinary dopamine. Catecholamines are mainly excreted in the urine as metabolites (metanephrine, VMA, HVA) and conjugates.

Most pheochromocytomas secrete both adrenaline and noradrenaline, but noradrenaline is usually the predominant amine. Some tumors secrete only noradrenaline or rarely only adrenaline. Very rarely, dopamine and even dopa may be secreted. Catecholamines released into the circulation exert cardiovascular and metabolic effects by stimulating specific protein cellular receptors (adrenoceptors) and thereby account for the physiologic and pharmacologic effects caused by these tumors.

PATHOPHYSIOLOGY

Pheochromocytomas average 5 cm in diameter and 70% of them weigh less than 70 g, although they may be microscopic or weigh 4000 g. Generally they are very vascular but some are avascular or cystic and may rarely contain calcium. Usually they are benign and encapsulated but roughly 10% are malignant as evidenced by invasion of adjacent structures or metastases. It is impossible to differentiate benign from malignant pheochromocytomas histologically; however, determination of DNA patterns by flow cytometry appears quite reliable in identifying whether tumors are malignant or benign (Klein et al., 1985; Hoska et al., 1986). Extra-adrenal pheochromocytomas are more often malignant than those in the adrenals. Furthermore, tumors containing dopamine are more apt to be malignant; however, the presence of dopamine or its precursor does not establish malignancy. Severity of symptoms depends primarily on the amount of catecholamines released into the circulation and whether tumor secretion is sustained or episodic. There is no good correlation between the size of tumors and clinical manifestations; small tumors freqently have a more rapid turnover of catecholamines and secrete more catecholamines than large tumors. Why some tumors cause sustained hypertension whereas others cause paroxysmal hypertension is unclear. Secretion from pheochromocytomas (none of which have any nervous innervation) seems to occur mainly by passive diffusion from tumor cells; however, some tumors actively secrete catecholamines simultaneously with other contents (eg, dopamine beta hydroxylase (DBH), ATP and chromogranins) in storage vesicles by exocytosis (O'Connor and Berstein, 1984; Manger and Gifford, 1990).

Some pheochromocytomas contain serotonin, vasoactive intestinal peptide (VIP), opioids (enkephalins and beta endorphin), alpha-melanocyte-stimulating hormone (α-MSH), adrenocorticotropic hormone (ACTH), calcitonin, somatostatin and others; secretion of these substances may play a role in symptomatology of some patients (Giraud et al., 1981; Wilson et al., 1981; Eiden et al., 1982; Sano et al., 1983; Voale et al., 1985; Garbini et al., 1986; Allen et al., 1987; Bostwick et al., 1987).

CLINICAL MANIFESTATIONS

Manifestations of pheochromocytoma, often diverse and numerous, can suggest a variety of diagnostic possibilities. Occurrence of hypertension is of great diagnostic importance since absence of sustained or paroxysmal hypertension almost eliminates pheochromocytoma as the cause of symptomatology suggesting excess circulating catecholamines; an exception is the rare patient with familial pheochromocytoma where hypertension may be absent and plasma and urinary catecholamines and their metabolites may be normal or slightly elevated.

One or more spontaneous attacks occur weekly in about 75% of patients; some experience one or more attacks daily whereas in others attacks may occur every few months. Attacks may sometimes be precipitated by palpation in the region of the tumor, postural changes, exertion, anxiety, trauma, pain, ingestion of certain foods or beverages containing tyramine (certain cheeses, beer or wine), or synephrine (citrus fruits), administration of certain drugs (histamine, glucagon, tyramine, phenothiazines, naloxone, metoclopramide, ACTH), arteriography in the region of the tumor, intubation, anesthesia, operative manipulation, and micturition or bladder distension (if pheochromocytoma occurs in the urinary bladder).

Patients complain of manifestations resulting mainly from hypercatecholaminemia and hypertension: headaches, excess sweating (generalized), palpitations, tachycardia (sometimes bradycardia), anxiety, tremulousness, chest and/or abdominal pain, nausea, vomiting, weakness, weight loss, dyspnea, warmth, visual disturbances, faintness, constipation, arm paresthesias, grand mal seizures and painless hematuria, frequency, nocturia and tenesmus (when the tumor is in the urinary bladder).

Hypertension may be sustained (with or without wide fluctuations) or paroxysmal or rarely hypertension may alternate with hypotension. Orthostatic hypotension in the untreated hypertensive or a paradoxic blood pressure response to certain antihypertensive drugs (beta or ganglionic blockers or guanethidine) or resistance to antihypertensive therapy or marked pressor responses to any of the conditions (mentioned above) which may precipitate attacks should suggest pheochromocytoma. Attacks may subside or worsen during pregnancy. Sudden hypercatecholaminemia can cause pronounced pallor (rarely flushing appears) and may cause patients to become fearful of impending doom. Sustained hypertension can cause retinopathy. Occasionally a slight fever develops. Children may display atypical manifestations, e.g., polydipsia, polyuria, puffy, red or cyanotic hands, and convulsive seizures.

Headache, the most common symptom, usually occurs during a paroxysm and is often severe, bilateral and throbbing. Sweating (generalized and sometimes drenching) and palpitations (with tachycardia or reflex bradycardia) are the next most common manifestations. Weight loss can be pronounced but may occur even if pheochromocytomas only cause paroxysmal hypertension. Severe constipation occasionally occurs in patients with sustained hypertension; rarely, secretion of VIP or serotonin from pheochromocytoma can cause severe diarrhoea.

ASSOCIATED PATHOLOGIC CONDITIONS

Coexistence of familial pheochromocytoma and medullary thyroid carcinoma (Sipple's syndrome) and frequently neoplasm or hyperplasia of the parathyroid glands characterizes multiple endocrine neoplasia (MEN) Type 2. Combination of medullary thyroid carcinoma (80% of cases) and pheochromocytoma (30% of cases) with mucosal neuromas, thickened corneal nerves, alimentary tract ganglioneuromatosis, and frequently a marfanoid habitus constitutes another familial entity, MEN Type 3; hyperparathyroidism rarely coexists in the latter. Medullary thyroid carcinomas may secrete calcitonin, serotonin and prostaglandin which may cause diarrhoea. All patients with pheochromocytomas should be screened for evidence of medullary thyroid carcinoma (or premalignant C-cell hyperplasia) and hyperparathyroidism following pheochromocytoma resection, since some pheochromocytomas can cause hypercalcitonemia and/or hypercalcemia. Rarely, pheochromocytomas or medullary thyroid carcinomas may secrete ACTH-like substance and cause Cushing's syndrome.

Since neurofibromatosis (von Recklinghausen's disease), often with the café-au-lait spots, afflicts 5% of persons with pheochromocytoma, screening of patients with hypertension and neurofibromatosis for pheochromocytoma is indicated.

The common embryonic origin of pheochromocytomas and associated thyroid and neurocutaneous lesions suggests a maldevelopment of the neural crest in these familial syndromes (Bolande, 1974).

DIAGNOSIS

Table 17.1 enumerates conditions whose manifestations may simulate pheochromocytoma; some can yield increased urinary catecholamines and their metabolites.

Consumption of some medications (amphetamines, monoamine oxidase inhibitors, decongestants and anorectics containing phenylpropanolamine) and illicit drugs (cocaine, phencyclidine, lysergic acid diethylamide) may cause hypertensive crises and symptoms mimicking pheochromocytoma. Factitious production of symptoms (pseudopheochromocytoma) should always be considered a possibility in emotionally disturbed persons having access to drugs.

Hemorrhagic necrosis of a pheochromocytoma may present as an acute abdomen or cardiovascular crisis. This rare complication requires prompt stabilizing treatment and tumor removal, otherwise the patient will almost certainly die.

A detailed history and physical examination are essential in deciding which patients to screen since 95% of those with pheochromocytoma complain of headaches and/or excess sweating and/or palpitations (Manger and Gifford, 1977).

All symptomatic patients with sustained or paroxysmal hypertension should be investigated for pheochromocytoma unless the cause of their hypertension is known. Even asymptomatic patients should be screened if (a) they have laboratory or electrocardiographic findings suggesting hypercatecholaminemia,

TABLE 17.1

Differential diagnosis of pheochromocytoma*+

All hypertension (paroxysmal and sustained)

Anxiety, tension states, neurosis, psychosis

Hyperthyroidism

Paroxysmal tachycardia

Hyperdynamic β-adrenergic circulatory state

Menopause

Vasodilating (migraine or cluster) headache

Coronary insufficiency syndrome

Acute hypertensive encephalopathy

Diabetes mellitus

Renal parenchymal or renal arterial disease with hypertension

Focal arterial insufficiency of the brain

Intracranial lesions (with or without increased intracranial pressure)

Autonomic hyperreflexia

Diencephalic seizure or syndrome

Preeclampsia or *eclampsia with convulsions*

Hypertensive crises associated with drug and food interactions with *monoamine oxidase inhibitors, some illicit drugs*, and drugs containing phenylpropanolamine

Carcinoid

Hypoglycemia

Mastocytosis

Familial dysautonomia

Acrodynia

Neuroblastoma, ganglioneuroblastoma, ganglioneuroma, neurofibromatosis (with or without renal arterial disease)

Unexplained shock

Acute infectious disease

Rare causes of paroxysmal hypertension (*acute medullary hyperplasia, acute porphyria, lead poisoning*, tabetic crisis, encephalitis, *clonidine withdrawal*, hypovolaemia with inappropriate vasoconstriction, pulmonary artery fibrosarcoma, pork hypersensitivity, dysregulation of hypothalmus, *tetanus, Guillain-Barre syndrome, factitious hypertension, baroreflex failure*)

* Modified from Manger, W.M., and Gifford, R.W. Jr. Current concepts of pheochromocytoma. Cardiovascular Medicine, 3: 289, 1978.

+ *Underlined* entries may have increased excretion of catecholamines and/or their metabolites.

(b) radiologic evidence of pheochromocytoma, (c) conditions known to coexist with pheochromocytoma or (d) a family history of pheochromocytoma. Screening pregnant patients with hypertension of unknown cause is mandatory since pregnancy and childbirth in the presence of pheochromocytoma carry a high fetal/maternal mortality.

Screening is most accurately performed by quantitating 24 hour urinary meta-nephrines and/or plasma catecholamines.

LABORATORY AND ELECTROCARDIOGRAPHIC ABNORMALITIES

Hypercatecholaminemia can cause hyperglycemia, hypermetabolism and increased plasma free fatty acids. Rarely, Cushing's syndrome (due to secretion of an ACTH-like substance by pheochromocytoma or medullary thyroid carcinoma) or polycythemia (due to secretion of erythropoietin by pheochromocytomas) may occur. Hypovolaemia afflicts some patients, particularly those with sustained hypertension.

Arrhythmias and ECG changes consistent with myocardial ischemia, damage or strain may appear; their transient appearance during a paroxysm of hypertension and symptoms suggesting pheochromocytoma support the diagnosis, especially in the absence of other causes. Hypertension, coronary atherosclerosis, myocardial infarction and catecholamine cardiomyopathy may produce ECG changes.

BIOCHEMICAL TESTS

When sustained hypertension results from pheochromocytoma, plasma cate-cholamines are invariably increased; elevations almost always occur in 24 hour urinary catecholamines and their metabolites but total metanephrine determinations are the most reliable (Manger and Gifford, 1990). With paroxysmal hypertension, plasma and urinary catecholamines and their metabolites may not be substantially elevated while the blood pressure remains normal. To establish the diagnosis in these individuals, one should obtain blood during a spontaneous or provoked hypertensive period or collect urine shortly after a hypertensive episode.

Occasionally persons with essential hypertension or hyperlabile blood pressures have borderline or modest elevations of plasma (500–2000 pg/ml) or urinary cat-echolamines or their metabolites. The clonidine suppression test (suppressing the sympathetic nervous system) can differentiate neurogenic from pheochromocytic hypertension by reducing plasma noradrenaline concentrations (by 50% and to nor-mal levels) in patients with neurogenic hypertension but not in those with pheochro-mocytoma (Bravo et al., 1981; Karlberg and Hedman, 1987).

Table 17.2 gives the upper limits of normal concentrations (these vary among laboratories) for catecholamines and metabolites and lists substances that can interfere with their determinations. Only a few drugs can lower urinary cate-cholamines or their metabolites to normal in patients with pheochromocytoma; however, radiopaque contrast media containing methylglucamine can cause false-negative metanephrine assays. In addition to conditions listed in Table 17.1 which can cause elevated catecholamines and their metabolites, severe hypotension or

TABLE 17.2

Effect of drugs and other interfering substances on plasma and urinary catecholamines and their metabolites*

Urine concentration rate		Effects	
Adult upper limit of normal (mg per 24th)		Increased apparent value	Decreased apparent value
Catecholamines		*Catecholamines*	Fenfluramine
Adrenaline	0.02	*Drugs containing catecholamines*	Mendalamine**
Noradrenaline	0.08	Isoproteronol**	Bromocriptine
		Levodopa	Some antihypertensives which
Total	0.10	Labetalol **	suppress adrenergic activity
		Phenothiazines	
Dopamine	0.20	Tricyclic antidepressants	
		Other fluorescent substances**	
		(eg, quinine, quinidine, bile,	
		B complex vitamins)	
		Rapid clonidine withdrawal	
		Ethanol	
		Ether	
Metanephrines		*Catecholamines*	*Meglumine* (x-ray contrast media)
Metanephrine	0.4	*Drugs containing catecholamines*	
Normetanephrine	0.9	*Labetalol****	
		Monoamine oxidase inhibitors	Fenfluramine
Total	1.3	*Phenothiazines****	Bromocriptine
		Benzodiazepines	
		Tricyclic antidepressants	Some antihypertensives which
		Rapid clonidine withdrawal	suppress adrenergic activity
		Methyldopa**	
		Ethanol	
		Ether	
Vanillylmandelic acid	6.5	Catecholamines	*Clofibrate*
		Drugs containing catecholamines	*Disulfiran*
		*Glycerol guaiacolate***	*Ethanol*
		Levodopa	*Monoamine oxide inhibitors*
		Labetalol**	Methyldopa
		*Nalidixic acid***	Fenfluramine
		Robaxin**	Phenothiazines
		Rapid clonidine withdrawal	Tricyclic antidepressants
		Ether	Some antihypertensives
			Bromocriptine
Plasma concentrations		*Catecholamines*	Fenfluramine
Adult upper limit of normal		*Drugs containing catecholamines*	Bromocriptine
(pg per ml) +		*Rapid clonidine withdrawal*	Some antihypertensives which
Adrenaline	275	Isoproterenol	suppress adrenergic activity
Noradrenaline	500	L-dopa	
Dopamine	120	Methyldopa	
+ (radioenzymatic assay)		Labetalol**	
		Ether	

* Many drugs interfere with fluorometric, colorimetric and less specific assays, making them unreliable; with most specific assays (e.g. spectrophotometric, radioenzymatic and HPLC) only *underlined* drugs may falsely indicate the presence or absence of pheochromocytoma. *Vasodilators (eg. Minoxidil, hydralazine) or hypoglycemic drugs may cause marked elevations of catecholamines and their metabolites.*

** Probably spurious interference.

*** Probably spurious interference with spectrophotometric but not HPLC assays. (From Pheochromocytoma — Method of William Muir Manger in Conn's Current Therapy, 612–618, W.B. Saunders, 1991)

stress (e.g., shock, heart failure, myocardial infarction, anoxia, acidosis, certain anesthetics, CNS stimulation, strenuous physical activity) can cause significantly elevelated levels of catecholamines and their metabolites.

Rarely, a provocative test (using glucagon) combined with quantitating plasma catecholamines may prove indispensable in detecting a paroxysmally secreting pheochromocytoma (Manger and Gifford, 1977). The test should not be performed in hypertensives (with BPs $\geq$ 170 mm Hg) or in patients with cerebrovascular or cardiovascular disease who would be at risk with a sudden blood pressure elevation. Provocative tests are relatively safe if performed in patients without contraindications and with precautions to counteract hypertensive crises, arrhythmias and hypotension. If desired, the hypertensive response can be avoided by administering prazosin (Minipress) for several days before the test or by giving 10 mg of nifedipine sublingually shortly before the test.

PREOPERATIVE LOCATION OF PHEOCHROMOCYTOMA

The presence of a relatively large amount of plasma or urinary adrenaline (or its metabolite, metanephrine) strongly suggests a pheochromocytoma in the adrenal gland; other tests can establish the location.

Computed tomography (CT) is extremely accurate in identifying lesions 1 cm or larger in the adrenals or 2 cm or larger in extra-adrenal abdominal locations; it can detect about 95% of abdominal pheochromocytomas. Small nonfunctional adrenal adenomas are sometimes visualised and should be considered in the differential diagnosis during localization procedures (Manger, Gifford and Hoffmann, 1985). CT should also prove valuable in revealing pheochromocytomas in the thorax and neck but experience is limited. CT is usually initially performed without contrast; if no tumor is visible the scan should be repeated with intravenous and oral contrast for optimal interpretation. CT is superior to and safer than angiography (now rarely indicated).

Use of the radioisotope ^{131}I-metaiodobenzylguanidine (^{131}I MIBG) provides a very specific and fairly sensitive technique for localizing pheochromocytomas since this isotope usually concentrates in these tumors. Yet, scanning with ^{131}I-MIBG fails to identify 15% of all pheochromocytomas (Sisson *et al.*, 1981).

Magnetic resonance imaging provides an especially valuable means for localizing and identifying a pheochromocytoma. Although not affording the same resolution as the CT scan, MRI imparts a particular pattern of high signal intensity which is characteristic for pheochromocytoma and is almost never seen in other adrenal tumors except primary or metastatic malignancy (Fink *et al.*, 1985; Baker *et al.*, 1987). MRI requires no contrast administration and causes no radiation, and it is superior to CT in locating extra-adrenal pheochromocytomas.

Hematuria or the occurrence of hypertensive episodes with micturition or urinary bladder distension is an indication for cystoscopy under α-adrenergic blockade.

If all attempts to localize a pheochromocytoma fail, sampling blood from various levels of the vena cava is sometimes helpful in localizing these tumors; samples with the highest catecholamine concentration can establish tumor location (Manger and Gifford, 1977).

PREOPERATIVE MANAGEMENT

Expertise and teamwork are essential for successful management of pheochromocytoma. Preoperative α-adrenergic blockade with prazosin, starting with 1 mg and increasing to 1 or 2 mg two to three times daily or phenoxybenzamine (Dibenzyline) 10 to 20 mg twice daily for a week or more and continued to the time of surgery usually prevents hypertensive crises and serious clinical manifestations, reverses hypovolaemia and promotes smooth anesthetic induction and a relatively stable blood pressure during surgery.

Total blockade, which causes orthostatic hypotension, should be avoided since it prevents the surgeon from (a) using blood pressure increases caused by tumor palpation during surgical exploration as a guide to tumor location and (b) immediately recognizing, by persistence of hypertension after tumor removal, presence of additional tumor(s).

For hazardous supraventricular arrhythmias or tachycardia or angina, β-adrenergic blockade is indicated, if there are no contraindications. *Since β-blockers can markedly increase blood pressure, they should not be used without first inducing α-adrenergic blockade.* Ventricular arrhythmias should be treated with lidocaine (Xylocaine). Although labetalol (Normodyne), an α- and β-adrenergic blocker may control blood pressure, it can occasionally cause hypertensive crisis in some patients with pheochromocytoma.

Surgical removal of the tumor, the curative procedure, should be performed expeditiously unless surgery is contraindicated. Coexistent medullary thyroid carcinoma and hyperparathyroidism should be excluded in all patients with pheochromocytoma (the relatives of those with familial pheochromocytoma should be screened for evidence of the MEN syndromes). Diagnosis and treatment of thyroid and parathyroid abnormalities should wait until after pheochromocytoma removal.

The presence and degree of bilateral renal function should be established before surgery in case the surgeon considers sacrificing a kidney during tumor removal.

Hypertensive crisis occurring spontaneously, or induced by angiography or provocative tests or resulting from anesthesia or operation can usually be controlled by a rapid intravenous bolus of phentolamine (Regitine) 3–5 mg; if there is little or no response in a minute or two, an additional 5 mg can be given and repeated as needed.   Since the effect of phentolamine is transient, repeated hypertensive crises (especially during surgery) may best be controlled with infusion of phentolamine or preferably sodium nitroprusside (Nipride of Nitropress), at a rate sufficient to maintain a relatively normal blood pressure (with impaired renal function or prolonged nitroprusside infusion, thiocyanate levels should be monitored since concentrations over 10 mg/dl can cause thiocyanate toxicity and psychosis).

If immediate operation is indicated, a significant blood volume deficit should be corrected with appropriate fluid or blood during the day before surgery to prevent severe postoperative hypotension. Normalization of blood pressure with α-adrenergic blocking drugs for one to two weeks prior to surgery will usually correct any deficit and avoid risks of blood transfusion.

Diagnostic procedures that involve any trauma or stress and abdominal palpation should be performed cautiously and with drugs available to treat hypertensive crises, arrhythmias and hypotension. The use of phenothiazines or morphine is contraindicated since they may precipitate hypertensive crisis or severe hypotension. If bilateral adrenalectomy is contemplated, steroid replacement is mandatory preoperatively.

OPERATIVE AND POSTOPERATIVE MANAGEMENT

Preoperative administration of diazepam (Valium) or meperidine (Demerol) will allay anxiety which might trigger catecholamine secretion; fentanyl (Innovar) and droperidol (Inapsine) should be avoided since they can cause the tumor to release catecholamines. Atropine is contraindicated because it can cause severe tachycardia in the presence of hypercatecholaminemia.

Before endotracheal intubation, the anesthesiologist must administer muscle relaxants and begin monitoring arterial and central venous pressure and ECG. It is critical during intubation and surgery to control hypertensive crises promptly with phentolamine or nitroprusside, and arrhythmias with intravenous propranolol (Inderal) or esmolol (Brevibloc) and/or lidocaine.   Intraoperative correction of blood volume deficits is essential. Isoflurane is the most ideal anesthetic, although enflurane and halothane are suitable.

An anterior transperitoneal incision is essential to remove intra-abdominal pheochromocytomas and to permit detection and removal of multiple and extra-adrenal tumors. If cholelithiasis, intra-abdominal neurofibromatosis, or vascular abnormalities are encountered they may require additional surgery. Operative mortality is 3.3% or less in medical centers with wide experience in treating pheochromocytoma.

Although controversial, bilateral adrenalectomy has been recommended in patients with pheochromocytoma in MEN type 2 or 3 syndromes because of the high probability that both adrenals harbor or will eventually develop pheochromocytomas.

Pheochromocytomas in the chest, neck and urinary bladder require special surgical removal, otherwise management is similar to that described above. Pheochromocytomas discovered during pregnancy are usually removed and the pregnancy continued if possible. If pregnancy is carried to term, caesarean section and tumor extirpation at delivery are advisable to avoid the stress of labor and vaginal delivery.

Close postoperative monitoring is essential until the patient's condition stabilizes. Hypotension may result from a blood volume deficit or hemorrhage from operative sites. Hypertension may result from fluid overload, pain, urinary retention, hypoxia, hypercarbia, or residual pheochromocytoma. Inadvertent renal artery ligation (resulting in renal ischemia and hyper-reninemia) can cause hypertension but not until a few days or weeks after surgery.

Hypoglycemia can occur within two hours after tumor removal and produce CNS manifestations, including coma; although a transient phenomenon, prompt infusion of a dextrose solution is mandatory. Starting an infusion of 5% dextrose immediately after tumor removal and continuing it for about 24 hours will prevent hypoglycemia and its complications (Pullerits and Reynolds, 1982).

About 75% of patients become normotensive following tumor removal; it is unclear why the rest remain hypertensive. Five-year survival for patients with benign tumors is 95%, whereas it is 50% or less when tumors are malignant.

To determine whether patients remain free of pheochromocytoma, 24 hour urinary metanephrines or plasma catecholamines are determined every 6 to 12 months for 5 years and then periodically thereafter, especially if any manifestations suggesting pheochromocytoma appear.

CHRONIC MEDICAL MANAGEMENT

If pheochromocytomas cannot be totally removed, as much as possible should be resected to minimize functioning tissue.

Prolonged treatment with α- and β-adrenergic blockade and metyrosine (Demser), an inhibitor of catecholamine synthesis, can effectively control blood pressure and symptoms for many years. Radiotherapy may occasionally prove highly effective, especially in treating bone metastases. Irradiation with large doses of ^{131}I-MIBG has sometimes reduced tumor size, catecholamine secretion, and symptomatology; however, long-term results have been disappointing and this therapy has been discontinued.

Combination chemotherapy with cyclophosphamide (Cytoxan), vincristine (Oncovin), and dacarbazine (DTIC-Dome) may prove valuable if malignant pheochromocytomas are causing signs and symptoms from metastases. A complete or partial response (i.e., reduced tumor mass, catecholamine secretion and symptomatology) may occur in over 50% of patients for up to 2 years.

Note: Many of the views expressed above have been reported in detail in: Manger, W.M., Gifford, R.W. Jr., Pheochromocytoma New York: Springer-Verlag, (1977) 1–398, and Manger, W.M., Gifford, R.W. Jr., Pheochromocytoma: A Clinical and Experimental Overview. Curr. Probl. Cancer (1985) 9, 1–89.

REFERENCES

Allen, J.M., Yeats, J.C., Causon, R., Brown, M.J. and Loom, S.R. (1987) Neuropeptide Y and its flanking peptide in human endocrine tumors and plasma. *J. Clin. Endocrinol. Metab.*, **64**, 1199–1204.

Baker, E.M., Spritzer, C., Blinder, R., Herfkens, R.J., Leight, G.S. and Dunnick, N.R. (1987) Benign adrenal lesions mimicking malignancy on MR imaging: report of two cases. *Radiology*, **163**, 669–671.

Bolande, R.P. (1974) The Neurocrestopathies: a unifying concept of disease arising in neural crest maldevelopment. *Hum. Pathol.*, **5**, 409–429.

Bostwick, D.G., Null, W.E., Holmes, D., Weber, E., Barchas, J.D. and Bensch, K.G. (1987) Expression of opioid peptides in tumors. *N. Engl. J. Med.*, **317**, 1439–1443.

Bravo, E.L., Tarazi, R.C. Fouad, F.M., Vidt, D.G. and Gifford, R.W. Jr. (1981) Clonidine-suppression test: a useful aid in the diagnosis of pheochromocytoma. *N. Engl. J. Med.*, **305**, 623–626.

Eiden, L.E., Giraud, P., Hotchkiss, A. and Brownstein, M.J. (1982) Enkephalins and VIP in human pheochromocytomas and bovine and adrenal chromaffin cells. In *Regulatory peptides from molecular biology to function: advances in biochemical psychopharmacology.* edited by M. Trabucci and E. Costa, vol 33, New York: Raven Press, pp. 387–395.

Fink, I.J., Reinig, J.W., Dwyer, A.J., Doppman, J.L. Linehan, W.M. and Keiser, H.R. (1985) MR imaging of pheochromocytoma. *J. Comput. Assist. Tomogr.*, **9**, 454–458.

Garbini, A., Mainardi, M., Grimi, M., Repaci, G., Nanni, G. and Bragherio, G. (1986) Pheochromocytoma and hypercalcemia due to ectopic production of parathyroid hormone. *NY. State J. Med.*, **86**, 25–27.

Giraud, P., Eiden, L.E., Audegier, Y., Gillioz, P., Conte-Devolx, B., Bouderesque, F., Eskay, R. and Oliver, C. (1981) ACTH, α-MSH and β-endorphin in human pheochromocytoma. *Neuropeptides*, **1**, 236–252.

Hoska, Y., Rainwater, L.M., Grant, C.S., Fareow, G.M., van Heerden, J.A. and Lieber, M.N. (1986) Pheochromocytoma: Nuclear deoxyribonucleic acid patterns studied by flow cytometry. *Surgery*, **100**, 1003–1010.

Karlberg, B.E. and Hedman, L. (1987) Value of clonidine suppression test in the diagnosis of pheochromocytoma. *Acta Med. Scand. [suppl.]*, **714**, 15–21.

Klein, A.F., Kay, S., Ratliff, J.E. White, F.K.H. and Newsome, H.H. (1985) Flow cytometric determinations of ploidy and proliferation patterns of adrenal neoplasms: an adjunct to histological classification. *J. Urol.*, **134**, 862–866.

Manger, W.M. and Gifford, R.W. Jr. (1977) *Pheochromocytoma*, pp. 1–398, New York: Springer-Verlag.

Manger, W.M. and Gifford, R.W. Jr. (1990) Pheochromocytoma. In *Hypertension: Pathophysiology, Diagnosis and Management*, edited by J.H. Laragh and B.M. Brenner, New York: Raven Press, pp. 1639–1659.

Manger, W.M., Gifford, R.W. Jr. and Hoffman, B.B. (1985) Pheochromocytoma: a clinical and experimental overview. *Curr. Probl. Cancer*, **9**, 46–47.

O'Connor, D.T. and Berstein, K.N. (1984) Radioimmunoassay of chromogranin A in plasma as a measure of exocytotic sympathoadrenal activity in normal subjects and patients with pheochromocytoma. *N. Engl. J. Med.*, **311**, 764–770.

Pullerits, J. and Reynolds, C. (1982) Pheochromocytoma: a clinical review with emphasis on pharmacological aspects. *Clin. Invest. Med.*, **5**, 258–265.

Sano, T., Saito., H., Inaba, H., Hizawa, K., Saito, S., Yamanoi, A., Mizunuma, Y., Matsumura, M., Yeasa, M. and Hiraishi, K. (1983) Immunoreactive somatostatin and vasoactive intestinal polypeptide in adrenal pheochromocytoma: an immunochemical and ultrastructural study. *Cancer*, **52**, 282–289.

Sisson, J.C., Frager, M.S., Valk, T.W., Gross, M.D., Sivanson, D.P., Wieland, D.M., Tobes, M.C., Beierwaltes, W.H. and Thompson, N.W. (1981) Scintigraphic localization of pheochromocytoma. *N. Engl. J. Med.*, **305**, 12–17.

Voale, F., Dellorto P., Moro, E., Gozzaglio, L. and Goggi, G. (1985) Vasoactive intestinal polypeptide-somatostatin- and calcitonin-producing adrenal pheochromocytoma associated with the watery diarrhea (WDHH) syndrome. *Cancer*, **55**, 1099–1106.

Wilson, S.P., Cubeddu, L.X., Chang, K.J. and Viveros, O.H. (1981) Metenkephalin, leuenkephalin and other opiate-like peptides in human pheochromocytoma tumors. *Neuropeptides*, **1**, 273–281

18 The Non-Pharmacological Management of Autonomic Dysfunction

Irwin J. Schatz

Department of Medicine, Johns A. Burns School of Medicine,
University of Hawaii, Honolulu, Hawaii 96813, USA

Autonomic dysfunction induces a wide range of responses. The physician must be familiar with the natural history of the causative disorder and the patient should understand his/her condition and participate as much as possible in chronic therapy. Blood pressure and bladder control, sexual function, gastrointestinal motility, and sleep all may be altered. The most common clinical problem is postural hypotension; specific treatment should be aimed at reducing or eliminating debilitating symptoms of inadequate organ perfusion. Conditions which favor the pooling of blood and impede venous return must be treated, and increasing total blood volume, interdicting all diuretic and sedative drugs, improving left ventricular filling pressure and reducing post-prandial hypotension are essential. Each of these areas may lend themselves to non-pharmacologic therapy. The less common autonomic dysfunctions affecting other organs often require the advice and help of urologists, gastroenterologists and sleep specialists.

KEY WORDS: Orthostatic hypotension, therapy, impotence, postprandial hypotension, bladder dysfunction, atrial pacing, sleep, diabetes

INTRODUCTION

Disability from disorders of the autonomic nervous system will depend, in great part, upon the nature of the disease process and the compensatory physiological mechanisms that are available to the patient.

As a number of different clinical conditions may be responsible, management requires that the physician understands the natural history of the illness. The clinician should be able to transmit that knowledge so that the intelligent and motivated patient can become a partner in therapy. Ultimately, much of the responsibility for day to day care may devolve, as it should, upon the patient. The principal goal of the physician is to make the patient as independent as possible, within the limitations of the disability induced by the disease.

The clinician must be able to assess the degree to which the patient will be successful in working through those natural psychological and intellectual barriers that impede successful treatment. Nowhere is this more important than when dealing with a chronic, debilitating and usually progressive disorder, such as occurs in many patients with disturbances of autonomic function. That patient who depends excessively upon family and physician for day to day care and decision making will have a much less satisfying existence than the individual who can take over much of the ongoing management.

Furthermore, the physician must be able to anticipate how well the individual patient will ultimately deal with progressive disability. Will family members provide mature and meaningful support for the patient? Are there others, suffering from the same or similar disorders, who can create a sense of community and sharing? Have the psychological defence mechanisms of anger, dependency and guilt been dealt with successfully? Clearly, each patient responds differently. The physician's responsibility is to predict the patient's response and develop management strategies that will utilize the patient's own strengths and minimize weaknesses (see Table 18.1).

ORTHOSTATIC HYPOTENSION

The goal of treatment is to reduce the *symptoms* of inadequate organ perfusion. As cerebrovascular autoregulation permits adequate blood flow to the brain even with systolic blood pressures of 70-80 mmHg, treatment should not be aimed at the level of blood pressure, but is driven by symptoms and signs of cerebral hypoxia. When definitive therapy is judged necessary, non-pharmalogical methods should be instituted in all patients before treating with drugs. In some, judicious use of these measures may obviate the need for pharmacological agents.

Furthermore, the necessary participation by the patient in such non-pharmacological manoeuvres creates a positive mood in the patient and adds to the potential for successful management.

ELIMINATE OR DECREASE ANY CONDITION THAT FAVOURS THE POOLING OF BLOOD AND THAT IMPEDES VENOUS RETURN

When the patient is upright, he or she should be instructed to repeatedly flex the calf muscles in order to enhance the venous pump and, therefore, increase venous return to the heart. Standing motionless is deleterious.

The use of so-called "stadium chairs or derby chairs" (a small seat which unfolds but may be carried as a cane) is useful when patients develop symptoms upon walking.

Patients should be encouraged to obtain recliner chairs so that they may rest without going to bed.

Slow and gradual ambulation upon arising from bed in the morning, and after meals, with adequate hand support nearby, is essential.

TABLE 18.1
Non-pharmacological management of the dysautonomias

Organ System	Therapeutic Measures
Cardiovascular	
Postural hypotension	Eliminate venous pooling
	Increase blood volume (nocturnal head-up tilt), increased salt intake
	Increase venous return and left ventricular filling pressure
	(pressure support garment)
	Atrial pacing in appropriately selected patients
Postprandial hypotension	Take frequent, small meals
	Use coffee
Bladder	Triple-voiding technique
	Condom drainage
	Catheter drainage
	Sphincterotomy
Genital (impotence)	Penile prostheses (external or internal)
Gastrointestinal	Frequent small meals
	Avoid gastroesophageal reflux
	High fibre diet in selected patients
Respiratory	Avoid sedatives, alcohol
(sleep disturbances)	Use low flow oxygen at night
	Use nasal continuous positive airway pressure
	Tracheostomy and ventilators in worst cases

Activities that decrease the effective volume of the circulating blood must be avoided. These include manoeuvres that induce a Valsalva response, including straining at stool or lifting heavy objects. Such efforts substantially decrease venous return. Working for any period of time with the arms above the shoulder level or at other forms of strenuous isometric exercise also may impede venous return.

Exposure to hot baths or showers may also reduce the effective volume of the circulating blood and tip the patient over into a symptomatic phase of hypotension. Some patients may not tolerate even the vasodilatation induced by fever; accordingly, vigorous treatment of pyrexia is warranted in most subjects with postural hypotension.

Hot and humid weather not only permits excessive loss of salt, but the vasodilatation decreases peripheral resistance and venous return. If personal circumstances permit, patients with symptomatic orthostatic hypotension will discover that they are much more comfortable in a temperate or even cold climate than when in the tropics or semi-tropics.

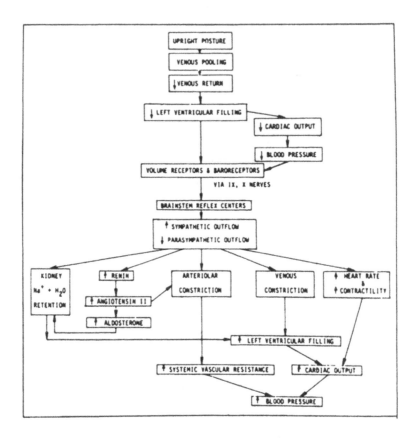

FIGURE 18.1   Changes in early morning blood pressure (lying and standing) in a patient with autonomic failure secondary to multiple system atrophy, studied when he slept sitting for 10 days with one interruption. Changes in body weight are also shown. (Reprinted with permission from Bannister and Ardill, 1969.)

INCREASE TOTAL BLOOD VOLUME

Head-up tilt at night

MacLean and Allen (1940) first demonstrated the value of this measure. Later it was shown conclusively that a significant increase in body weight occurs, probably because of a progressive augmentation of the volume of the extracellular fluid with the head-up position at night (Bannister and Ardill, 1969). Much of this volume was diuresed on the one night when their experimental subject slept flat. Furthermore, there was a substantially greater loss of nocturnal salt and water until head-up tilt at night was introduced (Figure 18.1). It is likely, therefore, that such head-up tilt reduces renal artery pressure and stimulates the release of renin, the formation of angiotensin II, the production of aldosterone and a consequent increase in blood volume. Indeed, some patients with previously incapacitating postural hypotension have been maintained satisfactorily for years by this form of therapy alone (Bannister and Mathias, 1988). Clearly, this important treatment modality will work only in those patients who can still release renin; fortunately most patients with neurogenic forms of orthostatic hypotension maintain this capability (Schatz, 1986).

Increased salt intake

This measure will increase the total blood volume. There is a surprising amount of resistance from patients when first approached with this advice; the physiological basis for an increase in salt intake must be explained and emphasized repeatedly to patients. Obviously, in those with severe supine hypertension, or any signs of congestive heart failure, an increased ingestion of salt is contraindicated.

Interdict all diuretic drugs

These agents are contraindicated. There is no place for their use either to control supine hypertension or to treat congestive heart failure if present. Patients must understand that any drug that causes a diuresis will make their orthostatic hypotension worse.

INCREASE THE LEFT VENTRICULAR FILLING PRESSURE

Lower body external pressure

Applying graduated pressure to the lower half of the body (Figure 18.2) will reduce venous pooling upon standing and increase the central blood volume, left ventricular filling pressure, cardiac output and cerebral perfusion.

The physiological basis for counter pressure support was first provided in 1941 (Stead and Ebert, 1941). These investigators demonstrated that pressure applied to the lower part of the body by immersion in water prevents the decrease in blood pressure normally seen when upright. Clearly, pressure is greatest at the base of the ankle and gradually decreases in a cephalad direction. This "gradient" may occur even when the venous muscle pump no longer exists. The principle of applying counter pressure has evolved coincident with the development of G-suits in aviation. The support garments now available may be custom fitted and will provide graduated support from the toes to the waist (Sheps, 1976). It is made of elastic mesh and forms a single piece. The material is porous and allows for adequate ventilation, but some patients in warmer climates will have difficulty in wearing it. Patients should be told to put it on while they are in bed, before rising in the morning, and must not remove it except to shower and when retiring at night. This requires considerable motivation for most patients, but the rewards are usually worth the discomfort. In those who will not tolerate the entire garment, partial relief from orthostatic symptoms may be provided by garments which reach mid-thigh. Unfortunately, calf-length support hose usually are of no value. For those patients who succeed in using a gradient support garment, such as described, substantial

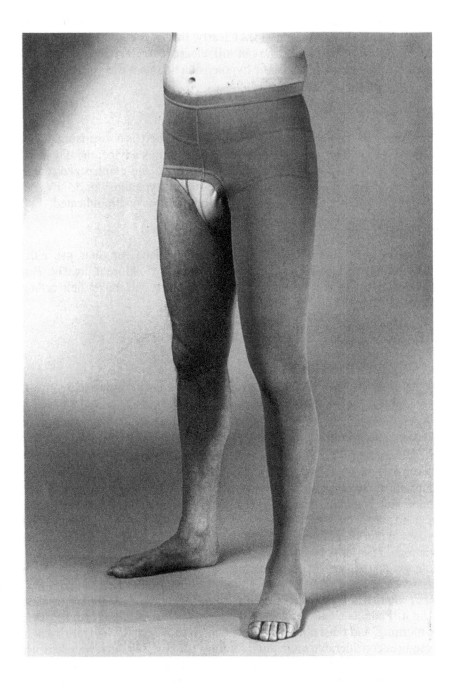

FIGURE 18.2   Pressure gradient support garment used to improve venous return to the heart. Pressure is greatest at ankles and decreases gradually in a cephalad direction. (Photo courtesy of Jobst Institute, Toledo, Ohio.)

relief may be anticipated (Levin, Ravenna and Weiss, 1964; Thomas and Schirger, 1968).

Occasional patients who are anhidrotic and wear this garment may develop dermatitis (Fox, 1971). Meticulous attention to proper nutrition and care of the skin is essential in these individuals.

There is a theoretical objection that such external pressure will reduce the myogenic reaction to stretch, which may be present even in partially denervated vessels; this is probably outweighed by the value of increasing the central blood volume and left ventricular filling pressure (Schatz, 1984).

REDUCE POSTPRANDIAL HYPOTENSION

The ingestion of food results in a complex release of hormones (principally pancreatic and gastrointestinal peptides), which have substantial effects on the cardiovascular system. Much of this is mediated through the activity of the autonomic nervous system. In spite of a marked increase in intestinal blood flow, the systemic blood pressure will remain unchanged in normal individuals, probably because of compensatory activation of the sympathetic nervous system and the release of vasoactive hormones (Quamar and Read, 1986; Mathias, Da Costa and Bannister, 1989). Cardiac output, stroke volume and heart rate will increase, but overall peripheral resistance falls. Catecholamines and plasma renin are stimulated but there are no changes in the cutaneous circulation.

Autonomic dysfunction, however, results in significant orthostatic hypotension in the postcibal state. This was first noted in hypertensive patients who had been given ganglionic blockers for treatment of their high blood pressure (Smirk, 1953). Subsequently, postprandial hypotension was confirmed as a clinical problem particularly in the elderly, and in those who ingested large carbohydrate loads (Seyer-Hansen, 1977; Robertson, Wade and Robertson, 1981). Fortunately, attenuation of these responses is possible by simple manipulation of the diet, as follows.

1. Patients should be advised to eat four to six equally spaced small meals during the day.

2. Excessive amounts of carbohydrates at one meal should be avoided.

3. Vigorous upright activity after eating is poorly tolerated; patients should also be cautioned not to attempt standing immediately after eating.

4. The ingestion of caffeine shortly before meals significantly improves the postprandial hypotension in patients with autonomic failure (Onrot et al., 1985). The amount of caffeine required probably amounts to no more than one or two strong cups of coffee per meal. The mechanism is uncertain, but it is not related to specific cardiovascular stimulation. Onrot and colleagues speculate that caffeine may attenuate the postprandial depressor response by blocking adenosine-mediated splanchnic vasodilatation. This would have the effect of minimizing the decrease in central blood volume. It is appropriate to advise patients with symptomatic postprandial hypotension to take regular coffee

either shortly before or with their meals as a non-pharmacological means of treating their blood pressure. Those who develop insomnia should be told to take the caffeine only with breakfast and midday meals and avoid it after 5 p.m. In the author's experience, the manipulation of the intake of caffeine during the day may be of substantial benefit to patients with postprandial hypotension.

ATRIAL PACING

There are isolated reports of the successful use of artificial electronic atrial pacing to treat postural hypotension (Moss, Galser and Topole, 1980; Kristinsson, 1983). The increased cardiac output that results is due to both a more rapid heart rate and the positive inotropic effect of the induced tachycardia (Ricci et al., 1979). Documentation of an increase in stroke volume secondary to temporary atrial pacing in the catheterization laboratory while the patient is standing should be available before permanent insertion of a pacemaker.

BLADDER DYSFUNCTION

Disorders due to neurogenic disease result in insufficient emptying of the bladder because of the loss of reflex detrusor contraction. Sympathetic and parasympathetic control of bladder activity is complex; the degree of involvement of either and/or both of these systems will vary from patient to patient. However, it has been suggested that sympathetic rather than parasympathetic efferent activity is necessary for the maintenance of proximal urethral competence (Kirby et al., 1985).

Frequency, urgency and a reduced force of stream are initial complaints. Obviously, this will be difficult to differentiate from the effects of benign prostatic hypertrophy. As the disease progresses, however, stress incontinence will appear, and investigation will disclose a large residual volume of urine. Lastly, as the process progresses, the bladder may become atonic and completely incompetent (Kirby and Fowler, 1989).

The first principal of effective management is to make sure that one is not dealing with prostatic obstruction. A grossly overdistended bladder should undergo initial drainage by catheter; this will improve bladder sensation and restore some intrinsic contractility. The so-called triple voiding technique is useful in preventing excessive urinary retention; patients will benefit by voiding intentionally at certain times rather than waiting for a conscious feeling of distention of the bladder. Chronic condom drainage may be necessary in some. Finally, sphincterotomy will reduce urethral resistance, but this should be implemented with caution. Obviously, destruction of the proximal urethral sphincter, which may already be involved by the neuropathic process, should be avoided. In fact, substantial urinary incontinence may be the outcome of this procedure (Zinke et al., 1974).

GENITAL DYSFUNCTION

Impotence is a common and major problem in the man with diseases of the autonomic nervous system. Sophisticated methods of differentiating between that small group of patients with psychogenic impotence from the majority of those men who cannot develop or maintain an effective erection because of organic disease are available. Clearly, expert urologic advice is necessary. Once established, however, difficulties with impotence may now be successfully managed in the majority of patients.

The normal erectile response depends upon a complex interaction of a number of different functions. These include adequate arterial supply, normal neurological function, and sufficient hormonal stimulation. The greatest regulatory role is served by both the central and peripheral nervous systems (Gordon and Weinstein, 1989). Motor and sensory fibres, carried principally along the sacral parasympathetic nerves, are influenced by excitatory and inhibitory stimuli from higher central nervous system sources.

PENILE PROSTHESES

Several different forms of external appliances have been devised. They incorporate several characteristics: an external splint, negative pressure or vacuum, and constriction. Such external appliances have proved successful and relatively easy to use for most patients (Witherington, 1990).

The surgical insertion of a permanent prosthesis should be reserved for those patients in whom medications (the intracavernous injection of vasoactive drugs) or the use of external appliances fail, after adequate trial. The surgical implantation of permanent penile prostheses does require careful discussion and consideration by the patient and his physician. Obviously, a urologist experienced in the advantages and drawbacks of such therapy must be involved in the management of the patient.

GASTROINTESTINAL SYSTEM

The most common cause of neurogenic motility disorders of the gastrointestinal system is diabetes mellitus. This may present either as a disorder of one part of the gut, or as a diffuse alteration of gastrointestinal motility. Gastroparetic complaints consist of recurrent nausea, vomiting and epigastric pain. Symptoms may wax and wane, and may ultimately lead to malnutrition or serious bleeding from Mallory-Weiss tears. In those with dilatation of the colon, diarrhoea may alternate with constipation (Malagelada, 1984).

Management of problems of gut motility in patients with autonomic dysfunction is most difficult. Obviously, patients will learn which foods are best digested; in general, it is prudent to advise frequent small meals. Because gastroesophageal reflux commonly occurs, patients should sit for periods of time after eating; this becomes difficult in those in whom postprandial hypotension is a complication of their autonomic dysfunction.   Efforts should be made to exclude possible malabsorbtion and consequent malnutrition. High fibre diets, in some way, improve emptying of the bowel.

SLEEP DISTURBANCES

Autonomic dysfunction may cause a variety of respiratory disturbances both in wakefulness and sleep. These are often intermingled with dysrhythmic breathing, which may become worse during sleep. This is due to the physiological vulnerability of brainstem centres to autonomic dysfunction (Chokroverty, 1986). Manifestations may consist of breathlessness, daytime sleepiness, marked fatigue and disturbed nocturnal sleep. Sleep studies will disclose several different abnormalities, including severe apnoea of both the central and peripheral (upper airway obstructive) types. Periodic breathing may also be apparent, with episodes of inspiratory gasps and transient respiratory arrests (Chokroverty, 1989).

Non-pharmacological management consists of avoiding medications or foodstuffs, including alcohol, which may depress the respiratory centre. In addition, the use of continuous positive airway pressure may cause a marked improvement in many patients with obstructive forms of sleep apnoea (Issa and Sullivan, 1986). If it is demonstrated that important and prolonged episodes of apnoea occur during sleep, accompanied either by cardiac arrhythmias or marked oxygen desaturation, mechanical ventilation during sleep may be necessary. This is accomplished most efficiently by a tracheostomy and the use of a mechanical respirator, by diaphragmatic pacing or by a negative-pressure ventilator. Clearly, these modalities would be reserved for the most severely affected patients (Godfrey et al., 1979). Such individuals must demonstrate severe hypoxia and hypercapnea at night. Fortunately, in the majority of patients with autonomic dysfunction and sleep disturbance, symptoms will be mild to moderate. It is important to exclude nasal obstruction or mild forms of congestive heart failure which, of course, will impede effective management of these conditions.

SUMMARY

There is a wide diversity of responses to autonomic dysfunction, ranging from a minor irritant on the one hand to a catastrophic impediment to daily living on the other. All affected patients will benefit from a sympathetic physician who understands the natural history of the disease process that is present. The relationship between the physician and the patient is paramount, for the practitioner will need to promote full and active participation by the patient in his or her care. Common sense measures that improve daily functions will often be of great use;

these may make the difference between a patient who is depressed and almost bedridden and one who can still lead a useful and productive life.

REFERENCES

Bannister, R. and Ardill, F. (1969) An assessment of various methods of treatment of idiopathic orthostatic hypotension. *Quarterly Journal of Medicine*, **38**, 377–395.

Bannister, R. and Mathias, C. (1988) Management of postural hypotension. In *Autonomic Failure*, 2nd edn, edited by R. Bannister, pp. 576. Oxford: Oxford University Press.

Chokroverty, S. (1986) Sleep and breathing in neurological disorders. In *Breathing Disorders of Sleep*, edited by N.H. Edelman and T.V. Santiago, pp. 225–264. New York: Churchill Livingstone.

Chokroverty, S. (1989) Sleep apnea and respiratory disturbances in multiple system atrophy with autonomic failure. In *Autonomic Failure*, 2nd edn, edited by R. Bannister, pp. 432–450. Oxford: Oxford University Press.

Fox, J.L. (1971) Orthostatic hypotension following bilateral percutaneous cordotomy: a case report of treatment with anti-G suit. *Acta Neurochirgerie*, **24**, 219–224.

Godfrey, C., Man, M., Jones, R., *et al.* (1979) Primary alveolar hypoventilation managed by negative-pressure ventilators. *Chest*, **76**, 219–221.

Gordon, B.E. and Weinstein, S.H. (1989) Impotence: its evaluation and treatment. *Missouri Medicine*, **86**, 632–636.

Issa, F. and Sullivan, C. (1986) Reversal of central sleep apnea using nasal CPAP. *Chest*, **90**, 165–171.

Kirby, R.S. and Fowler, C.J. (1989) Bladder and sexual dysfunction in diseases affecting the autonomic nervous system. In *Autonomic Failure*, 2nd edn, edited by R. Bannister, p. 431. Oxford: Oxford University Press.

Kirby, R.S., Fowler, C.J., Gosling, J.A. and Bannister, R. (1985) Bladder dysfunction in distal autonomic neuropathy of acute onset. *Journal of Neurology, Neurosurgery and Psychiatry*, **8**, 762–767.

Kristinsson, A. (1983) Programmed atrial pacing for orthostatic hypotension. *Acta Med. Scandinavica*, **214**, 79–83.

Levin, J.M., Ravenna, P. and Weiss, M. (1964) Idiopathic orthostatic hypotension: treatment with a commercially available counterpressure suit. *Archives of Internal Medicine*, **114**, 145–148.

MacLean, A.R. and Allen, E.V. (1940) Orthostatic hypotension and orthostatic tachycardia. Treatment with the "head-up" bed. *Journal of the American Medical Association*, **115**, 2162–2167.

Malagelada, J.R. (1984) Potential pharmacological approaches to management of gut motility disorders. *Scandinavian Journal of Gastroenterology (Suppl)*, **96**, 111–126.

Mathias, C., Da Costa, D. and Bannister, R. (1988) Postcibal hypotension in autonomic disorders. In *Autonomic Failure*, 2nd edn, edited by R. Bannister, pp. 367–380. Oxford: Oxford University Press.

Moss, A.J., Galser, W. and Topole (1980) Atrial tachypacing in the treatment of a patient with primary orthostatic hypotension. *New England Journal of Medicine*, **302**, 1456–1457.

Onrot, J., Goldberg, M.R., Biaggioni, I., Hollister, A.S., Kincaid, D. and Robertson, D. (1985) Hemodynamic and humoral effects of caffeine in autonomic failure, *New England Journal of Medicine*, **313**, 549–554.

Quamar, M.I. and Read, A.E. (1986) The effect of feeding and sham feeding on the superior mesenteric blood flow in man. *Journal of Physiology (London)*, **377**, 59.

Ricci, D.R., Orlick, A.E. and Alderman, E.L. (1979) Role of tachycardia as an inotropic stimulus in man. *Journal of Clinical Investigations*, **63**, 695–703.

Robertson, D., Wade, D. and Robertson, R.M. (1981) Postprandial alterations in cardiovascular hemodynamics in autonomic dysfunctional states. *American Journal of Cardiology*, **48**, 1048–1052.

Schatz, I.J. (1984) Orthostatic hypotension II: clinical diagnosis, testing and treatment. *Archives of Internal Medicine*, **144**, 1037–1041.

Schatz, I.J. (1986) *Orthostatic Hypotension*, 1st edn, pp. 93–95. Philadelphia: F.A. Davis.

Seyer-Hansen, K. (1977) Postprandial hypotension. *British Medical Journal*, **2**, 1262.

Sheps, S.G. (1976) Elastic garment in orthostatic hypotension. In *International Symposium on Neural Control of the Cardiovascular System and Orthostatic Regulation*, edited by H. Denolin and J.C. Demanet, *Cardiology*, **61**, 271–278.

Smirk, F.M. (1953) Action of a new methonium compound (MNB 2050A) in arterial hypotension. *Lancet*, i, 457.

Stead, E.A. Jr. and Ebert, R.R.V. (1941) Postural hypotension: a disease of the sympathetic nervous system. *Archives of Internal Medicine*, **67**, 546–562.

Thomas, J.E. and Schirger, A. (1968) Orthostatic hypotension: etiologic considerations, diagnosis and treatment. *Medical Clinics of North America*, **52**, 809–816.

Witherington, R. (1990) External penile appliances for management of impotence. *Seminars in Urology*, **8**, 124–128.

Zinke, H., Campbell, J.T., Palumbo, P.J. and Furlow, W.L. (1974) Neurogenic vesical dysfunction in diabetes: another look at vesical neck resection. *Journal of Urology*, **111**, 488–490

19 Pharmacological Treatment of Orthostatic Hypotension

Jack Onrot

University of British Columbia, Vancouver, Canada

Symptomatic orthostatic hypotension is a fall in blood pressure on standing that leads to symptoms of cerebral hypoperfusion. Drug therapy of this entity is designed to alleviate symptoms with minimal side effects, and not just to treat the pressure falls. Correctable problems such as volume depletion or hypotensive medications must be dealt with initially. In autonomic failure, approaches (both drug and non-drug) are then tailored to reverse the operative pathophysiologic factors. Supine hypertension is often the major limiting factor to drug therapy. The assortment of drugs employed is testimony to the difficulty in finding effective agents in this disease.

Mineralocorticoids (9-alpha fludrocortisone) are the cornerstone of therapy. These drugs improve volume status and sensitize vessels to endogenous pressors. Other agents enhance sympathetic nervous system (SNS) activity. Yohimbine activates the SNS centrally, monoamine oxidase inhibitors prevent catecholamine breakdown, and dihydroxyphenylserine is a catecholamine precursor. Some drugs mimic sympathetic activity. Alpha agonists such as phenylpropanolamine, phenylephrine, ephedrine, midodrine, amphetamines and clonidine work as exogenous veno- and arteriolo-pressor agents. Other non-adrenergic vasopressors such as dihydroergotamine, caffeine, and cyclooxygenase inhibitors can also be tried. Miscellaneous agents also used include beta blockers, dopamine antagonists, and somatostatin and vasopressin analogues.

Drug usage is often additive and effectiveness is assessed empirically, using symptoms as the major monitored parameter. Pharmacologic therapy should go hand in hand with patient education and non-drug manoeuvres.

KEY WORDS: Autonomic failure, mineralocorticoid drug therapy, sympathetic nerves, catacholamines, β-blockers, yohimbine

INTRODUCTION

Orthostatic hypotension is a disabling symptom in patients with autonomic failure. Unfortunately, it is very difficult to treat as evidenced by the number of drugs tested and available for therapy. Supine hypertension is the major limiting factor in treatment. A rational understanding of the pathophysiology of orthostatic hypotension in autonomic failure helps to devise strategies for treatment. 9 α-Fludrocortisone remains the mainstay of therapy. Shorter-acting agents are useful, as they can be pressor when needed and have disappeared from the circulation at times when pressor effects are undesirable. Drugs to treat this entity are often used empirically. As with all therapies, their success must be evaluated periodically and the most useful end-point is the relief of symptoms. Standing times can also be useful to assess drug therapy. Drug therapy should go hand in hand with patient education and non-pharmacological adjunctive manoeuvres.

INTRODUCTION AND PATHOPHYSIOLOGICAL RATIONALE FOR THERAPY

The preceding chapters in this volume detail disorders of the autonomic nervous system. Although there are many aspects of autonomic dysfunction that require treatment (e.g. urinary or gastrointestinal symptomatology), this chapter will deal primarily with the treatment of one specific disabling symptom of autonomic failure: orthostatic hypotension.

Normally, systolic pressure will fall by approximately 10 mmHg and diastolic pressure will increase by about 5 mmHg on assuming the upright posture. Symptomatic orthostatic hypotension is defined as any fall in blood pressure upon standing that produces symptoms of cerebral hypoperfusion (Onrot et al., 1986). Although acute exaggerated declines in standing blood pressure are important diagnostically (e.g. volume depletion, haemorrhage, etc.), the above operational definition implies that only patients with accompanying symptoms can be treated for their hypotension. Symptoms of orthostatic hypotension range from mild transient lightheadedness to syncope. Often there may be associated visual disturbances ("brownouts", tunnel vision), dyspnoea, profound weakness, or discomfort of the neck and head.

With assumption of the upright posture there is gravity-induced blood pooling in the venous system below the heart. This in turn leads to a fall in blood pressure and activation of reflex responses designed to maintain upright pressure. Orthostatic hypotension will result from a breakdown anywhere in these reflex responses. A review of the haemodynamic derangements present in autonomic failure will help in understanding the rationale behind the therapeutic options presented below. Patients with autonomic failure are characterized by an exaggerated fall in venous return on standing (MacLean, Allen and Magath, 1944), often assessed by left ventricular end-diastolic volumes (Ibrahim et al., 1974; Kronenberg et al., 1990; Belzberg et al., 1993). This fall in venous return is the primary cause of the observed reduction in cardiac output, because contractility is not affected in autonomic failure (Kronenberg et al., 1990). In addition, there is a relative inability to increase the heart rate on standing (Onrot et al., 1985a; Kronenberg et al., 1990; Belzberg et al., 1993).

How does autonomic failure lead to an exaggerated fall in venous return? With upright posture there is an initial fall in venous return, which leads to reflex sympathetic activation, splanchnic bed vasoconstriction and the mobilization of blood from the capacitance bed to the central circulation (Johnson et al., 1974; Greenway, 1983; Onrot et al., 1985b; Belzberg et al., 1993), thus maintaining preload. However, in autonomic failure this splanchnic venoconstrictor response is blunted (Onrot et al., 1985b; Belzberg et al., 1993), presumably secondary to decreased splanchnic autonomic outflow (Low, Thomas and Dyck, 1978). Venoconstriction elsewhere is also diminished. For instance, forearm reflex venoconstriction is decreased in autonomic failure (Page et al., 1955; Mason, Kopin and Braunwald, 1966). However, these reflex changes in limb venomotor tone are less important in maintaining upright blood pressure (Samueloff, Browse and Shepherd, 1966).

Autonomic failure is also characterized by a normal or elevated systemic vascular resistance in the supine position, with no change or a slight fall in resistance with upright posture (Bickelmann, Lippschutz and Brunjes, 1961; Niarchos et al., 1966; Ibrahim et al., 1974; Magrini, Ibrahim and Tarazi, 1976; Rohan-Chabot, Said and Ziza, 1980; Kronenberg et al., 1990; Belzberg et al., 1993). This is in contrast to a normal 20-35% increase in peripheral resistance with standing. Studies of forearm vascular resistance confirm the inadequate vasoconstrictor responses to various stimuli in patients with autonomic failure (Stead and Ebert, 1941; Mason, Kopin and Braunwald, 1966).

Thus, failure to increase sympathetic outflow may lead to orthostatic hyoptension at various levels: (1) failure to maintain venous return secondary to inadequate systemic and splanchnic venoconstrictor responses; (2) failure to adequately increase heart rate; and (3) inability to appropriately increase peripheral resistance due to inadequate arteriolar vasoconstriction. This constellation of defects will lead to a fall in cardiac output and, therefore, blood pressure, which cannot be overcome by increasing peripheral resistance. A rational pharmacological approach must address these underlying haemodynamic abnormalities.

In addition to the haemodynamic alterations, there are other known abnormalities in patients with autonomic failure that have important therapeutic implications, such as altered control of salt and volume, supine hypertension, "denervation hypersensitivity" and postprandial hypotension.

Patients with autonomic failure not only have lower total blood volumes (Ibrahim et al., 1974), but, in addition, reductions in the central blood volume are even more marked (Ibrahim et al., 1974; Magrini, Ibrahim and Tarazi, 1976), suggesting peripheral translocation of intravascular volume. Cardiac filling pressures are in the low-normal range, even while supine (Magrini, Ibrahim and Tarazi, 1976). Decreased plasma volume, especially central volume, will exaggerate any tendency to orthostatic hypotension, and conversely, optimizing total and/or central blood volume will improve venous return and relieve symptoms in these patients. Often, salt supplements and/or discontinuation of the use of diuretics is helpful for these reasons (Onrot et al., 1986).

This decrease in blood volume is due to inadequate conservation of sodium by the kidney (Wilcox, Aminoff and Slater, 1977; Wilcox et al., 1984). In normal subjects, sympathetic stimulation leads to reabsorption of sodium (Gill, 1979), and promotes secretion of renin with subsequent aldosterone-induced salt retention by the distal tubule (Gordon et al., 1967). These responses are decreased in autonomic failure (Slaton and Biglieri, 1967; Bliddal and Nielsen, 1970; Onrot et al., 1985a).

Furthermore, these patients have excessive nocturnal diuresis (Davidson, Smith and Morgan, 1976), presumably due to the above factors as well as a pressure natriuresis secondary to the elevated supine pressure commonly encountered. This leads to relative hypovolaemia and worsening of symptoms early in the day, especially upon arising (Mann *et al.*, 1983).

Approximately one-half of the patients with autonomic failure and chronic orthostatic hypotension also have supine hypertension, due to an elevated resting total peripheral resistance (Mason, Kopin and Braunwald, 1966; Niarchos *et al.*, 1966; Ibrahim *et al.*, 1974; Magrini, Ibrahim and Tarazi, 1976; Mann *et al.*, 1983; Onrot *et al.*, 1985b; Kronenberg *et al.*, 1990). The reasons for this are unclear. Supine hypertension may be severe and put the patient at risk for acute or chronic hypertensive complications. Unfortunately, treatment of symptomatic orthostatic hypotension with pressor agents may aggravate supine hypertension. This consideration all too often limits drug therapy. Because orthostatic hypertension is mostly a daytime problem when patients try to remain upright, and supine hypertension is mostly a problem at night (Mann *et al.*, 1983), one tries to administer short-acting pressor agents as early as possible during the day. An initial dose 30 min before arising will often help patients to function when getting out of bed in the morning. Patients should be cautioned not to lie down during the day, especially at times when pressor drugs are being administered. The effects of these short-acting pressor agents will hopefully be gone at night when the patient is supine in bed. "Head-up tilt" at night is helpful in lowering supine pressure, improves upright haemodynamics, and is a mainstay of therapy (MacLean and Allen, 1940; Mathias *et al.*, 1969; Onrot *et al.*, 1986). (see Chapter 18). Similarly, a small dose of a short-acting antihypertensive agent at bedtime can counteract severe supine hypertension (Jones and Reid, 1980). Calcium-channel blockers and nitroglycerin skin patches that can be divided for dose titration (e.g. Nitropatch) appear to be especially helpful (Biaggioni, I. and Robertson, D., personal communication). Treating supine hypertension during the night can theoretically improve daytime orthostatic hypotension by decreasing pressure diuresis and nocturia. On the other hand, patients should be cautioned that orthostatic hypotension may worsen during the night while on treatment with vasodilators.

Patients with autonomic failure exhibit supersensitivity to exogenous sympathomimetic amines (Robertson *et al.*, 1984). This is due to "denervation-induced" upregulation of catecholamine receptors (Hui and Connolly, 1981; Kafka *et al.*, 1984), combined with an inability to buffer haemodynamic changes because of the hypofunctioning autonomic nervous system. Thus, even small doses of pressor agents may lead to exaggerated rises in blood pressure and dangerous hypertension, especially while supine. This can be exploited for therapy by using very low doses of pressor agents and even using agents that are unlikely to be pressor in normal subjects (Onrot *et al.*, 1986).

Postprandial worsening of orthostatic symptomatology is commonly encountered, and even supine and sitting pressures fall after meals (Robertson, Wade and Robertson, 1981; Onrot *et al.*, 1985c; Hoeldtke, Boden and O'Dorisio, 1986; Hoeldtke *et al.*, 1986; Hoeldtke and Israel, 1989; Mathias *et al.*, 1989). In normal subjects, the ingestion of food, especially carbohydrates (Bannister *et al.*, 1984), leads to increased mesenteric blood flow (Qamar and Read, 1986), but blood pressure is maintained by compensatory venous and arteriolar constriction (Mathias *et al.*, 1989). Patients with autonomic failure, lacking these reflex responses, exhibit falls in postprandial pressure even while supine. The increase in mesenteric flow is attributed to the meal-induced release of gut hormones such as insulin (Brown, Polinsky and Baucom, 1989), other gut or pancreatic polypeptides (Mathias *et al.*, 1989), or adenosine (Onrot *et al.*, 1985c). These observations have led to the treatment of postprandial hypotension with agents that can reduce or oppose these hormonal responses, such as somatostatin (Hoeldtke, Boden and O'Dorisio, 1986) or caffeine (Onrot *et al.*, 1985c). Pressor agents can be given so that they have peak effects when needed most in the postprandial period.

GENERAL PRINCIPLES OF DRUG THERAPY

The goal of drug therapy is to reduce or eliminate symptoms of orthostatic hypotension, and not just to treat falls in blood pressure. In addition, this goal hopefully will be accomplished with minimal adverse effects (especially exacerbation of supine hypertension). Empirical interventions are often necessary, and periodic reassessment of efficacy are required. Monitoring supine and upright pressure is, of course, very useful, but improvement of symptoms by history is more important. Similarly, the "standing time" is often more useful than the absolute fall in blood pressure. Standing time is defined as the length of time a patient can stand motionless before the appearance of hypotensive symptoms. A standing time of 5 min usually indicates adequate symptomatic response of therapy.

Based on our knowledge of autonomic physiology, the hemodynamics of autonomic failure, and the special considerations outlined previously, it is clear that there can be many basic strategies in therapy (Table 19.1). One can try to stimulate what is left of the patient's own sympathetic nervous system, prevent the breakdown of catecholamines or enhance the pressor response to any given level of sympathetic activity. Alternatively, exogenous sympathomimetics can be given to replace endogenous catecholamines. Other pressor agents can similarly be used in an attempt to maintain upright pressures. As the orthostatic hypotension is most clearly related to falls in venous return, optimizing this factor is crucial to therapy. Special attention must be paid to the early morning and postprandial periods when symptoms are worst.

Drugs to be discussed are listed in Table 19.2. A number of reviews on this subject have been published (Mathias *et al.*, 1969; Thomas *et al.*, 1981; Onrot *et al.*, 1986; Blomqvist, 1987; Lipsitz, 1989; Ahmad and Watson, 1990).

MINERALOCORTICOIDS

9-α-Fludrocortisone is currently the mainstay of drug therapy for orthostatic hypotension in autonomic failure. Fludrocortisone raises supine and standing

TABLE 19.1
Strategies in Drug Therapy

- Increase catecholamine production
  - centrally
  - peripherally
- Decrease catecholamine breakdown/reuptake
- Improve end-organ response to catecholamines
- Use sympathomimetic pressor agents
- Use other pressor agents
- Optimize venous return
  - optimize blood volumes
  - use venoconstriction
  - shift blood volume to centrally-circulating compartment
  - reduce fall in venous return with meals
  - reduce blood pooling in response to gravity

TABLE 2
Drugs Used in Autonomic Failure

Type of Drug	Examples
Mineralocorticoids	9-α-fludrocortisone
Sympathetic "stimulators"	Yohimbine
	Tyramine ± monoamine oxidase inhibitors
	Dihydroxyphenylserine
α vasoconstrictor agonists	Phenylpropanolamine
	Ephedrine
	Phenylephrine
	Midodrine
	Clonidine
Other vasopressors	Dihydroergotamine
	Caffeine
	Prostaglandin inhibitors
β blockers	
Dopamine antagonists	
Somatostatin analogues	
Vasopressin analogues	

pressures and reduces symptoms (Hickler *et al.*, 1959; Mathias *et al.*, 1969; Schatz, Miller and Frame, 1976; Chobanian *et al.*, 1979). It may be especially useful in diabetic autonomic failure (Campbell, Ewing and Clarke, 1976) and has been used in orthostatic hypotension associated with the mitral valve prolapse syndrome (Onrot, J. and Robertson, D., unpublished data).

Mineralocorticoids act on the distal renal tubule to promote the reabsorption of sodium in exchange for the loss of K^+ and H^+. This counteracts the tendency to lose sodium seen in autonomic failure (Wilcox, Aminoff and Slater, 1977). The end result is an increase in the total blood volume, which, as noted, is often reduced in autonomic failure. As these patients are exquisitely sensitive to changes in volume, the blood pressures increase due to improved venous return in the standing position (Belzberg *et al.*, 1993). In addition, mineralocorticoids increase vascular responsiveness to endogenous catecholamines by an unknown mechanism (Raab, Humphreys and Lepeschkin, 1950; Schmid, Eckstein and Abboud, 1966; Davies *et al.*, 1979).

The starting dose is 0.1 mg and can be titrated up to about 1 mg/day. It is unclear, however, if doses greater than 0.4-0.6 mg provide additional benefit. The effectiveness of fludrocortisone may improve with dietary supplements of salt. Hypokalaemia and hypomagnesaemia may be encountered and patients may need supplementation. Oedema can be expected, is generally well-tolerated and implies optimal blood volumes. Treatment is limited by supine hypertension, hypokalaemia, and signs, symptoms or x-ray evidence of cardiac failure. As with other drugs used, higher supine pressures may have to be accepted to keep the patient functional. Most patients will have mild to moderate supine hypertension on therapy (160-200/90-110 mmHg), but pressures above 200/110 mmHg are cause for concern and require a decrease in dosage.

SYMPATHETIC "STIMULATORS"

These agents augment the activity of the patient's own sympathetic nervous system to increase the concentration of endogenous catecholamines as vascular receptors. As receptors are supersensitive to stimulation by agonists, small increases in endogenous catecholamines may have marked pressor effects.

Yohimbine

Yohimbine is an α_2-receptor antagonist. Central α_2 stimulation inhibits the activity of the sympathetic nervous system, and yohimbine, by interfering with this, enhances the sympathetic outflow (Goldberg and Robertson, 1983; Goldberg, Hollister and Robertson, 1983). In patients with autonomic failure, 2.5-5 mg yohimbine induced a small but significant rise in plasma noradrenaline and sustained elevations in seated blood pressure (Onrot *et al.*, 1987a; Biaggioni, Robertson and Robertson, 1993). Side effects were related to sympathetic stimulation, including anxiety, diarrhoea and tremulousness, and may limit therapy in some patients. Supine hypertension was encountered. Patients who can tolerate chronic therapy may have long-term improvements in standing times and symptoms (Onrot et al., 1987a). Yohimbine has also been used with some success in orthostatic hypotension induced by tricyclic antidepressants (LeCrubier, Puech and Des Lauriers, 1981; Seibyl et al., 1989) and in Parkinsonian patients on levodopa (Montastruc et al., 1981). Yohimbine can be used to treat impotence (Morales et al., 1983) and this added indication can theoretically be exploited in males with milder forms of autonomic failure.

Tyramine + monoamine oxidase inhibitors

Tyramine, an indirectly acting amine, will cause release of noradrenaline from storage pools in nerve terminals (Smith, 1973). Tyramine is often combined with a monoamine oxidase inhibitor to prevent breakdown of noradrenaline and further increase the levels available to act at receptors. A combination of 5-10 mg tyramine and 30-60 mg tranylcypromine (Diamond, Murray and Schmid, 1970; Lewis et al., 1972; Nanda, Johnson and Keogh, 1976) has been used with some success, but response is unpredictable (Ibrahim et al., 1979) and supine hypertension may be severe, as noted in one trial of tyramine in combination with phenelzine (Davies, Bannister and Sever, 1978). Adverse effects on the central nervous system may also be encountered.

Dihydroxyphenylserine

Dihydroxyphenylserine is an analogue of noradrenaline with a carboxyl group. As this agent is already β-hydroxylated, it can be converted into noradrenaline by dopa decarboxylase and does not require dopamine β-hydroxylase. Thus, patients with the rare syndrome of a deficiency of dopamine β-hydroxylase may be treated effectively with 500-1500 mg dihydroxyphenylserine daily in divided doses (Biaggioni and Robertson, 1987; Man in't Veld et al., 1987). Dihydroxyphenylserine may not be useful in other causes of autonomic failure.

α-AGONISTS

These agents act directly or indirectly to stimulate α-adrenergic receptors which vasoconstrict veins and arterioles. This mechanism improves venous return and systemic vascular resistance in order to maintain upright blood pressure.

Because of receptor hypersensitivity to pressor agents in autonomic failure and the inability to buffer pressor stimuli, patients are especially sensitive to α-agonists. A dose which has no pressor effect in normal subjects may be pressor in autonomic failure (Biaggioni et al., 1987). However, these agents are limited by their short duration of action (which, however, can also be an advantage) and by unacceptable supine hypertension. In fact, symptomatic improvement often only occurs at the expense of severe supine hypertension, suggesting a narrow therapeutic index (Davies, Bannister and Sever, 1978). This side effect may be reduced by: (1) administering the larger doses earlier in the day when symptoms are most severe;

(2) avoiding administration in the late afternoon and evening to limit residual pressor effects upon retiring for the night; (3) instructing the patient to avoid the supine posture for 3 h after each dose; and (4) elevating the head of the bed. Other side effects common to this group include piloerection, scalp tingling and stimulation of the central nervous system.

Phenylpropanolamine

Phenylpropanolamine hydrochloride (widely used in over-the-counter appetite suppressants and cold remedies) is a direct α-agonist and acts directly to release noradrenaline (Trendelenburg *et al.*, 1962). Doses of 12.5-25 mg produced an average rise greater than 30/15 mmHg for 90 min in 12 seated patients (Biaggioni *et al.*, 1987). These doses have little or no pressor effects in normal subjects. Little information is available on its chronic use.

Ephedrine

Ephedrine is also a direct agonist with some ability to release noradrenaline. It has been widely used (at doses of about 25 mg T.I.D.), but the pressor response may vary greatly and supine hypertension most often limits its use (Ghrist and Brown, 1928; Barnett and Wagner, 1958; Parks *et al.*, 1961; Davies, Bannister and Sever, 1978).

Phenylephrine

Phenylephrine, a direct-acting α-agonist, has similar effects. Supine hypertension may even be provoked by ocular administration of this agent (Robertson, 1979). Intranasal phenylephrine may be used in patients just before activity because it is short-acting (Robertson, D., unpublished data). This approach may be dangerous because of unpredictable hypertension, especially if the patient lies down.

Midodrine

Midodrine, a longer-acting α-agonist (duration of action 4-6 h), has been used with success at 7.5-30 mg/day in divided doses (Schirger *et al.*, 1981). One double-blind study suggested a pressor role for midodrine in mildly impaired patients, but this agent was depressor in the more severely impaired patients, presumably by producing volume depletion (Kaufmann *et al.*, 1988). It does not cross the blood–brain barrier and this lessens the incidence of central stimulating side effects.

Clonidine

Paradoxically, although the α_2-antagonist, yohimbine, is effective in autonomic failure, the α_2-agonist, clonidine, is also effective in some patients with this disorder. Postjunctional vascular α_2 receptors exist and play a role in the regulation of vascular tone (Elliott and Reid, 1983; Onrot *et al.*, 1987b). They may be more numerous in veins than arterioles (De Mey and Vanhoutte, 1981). Patients with autonomic failure have hypersensitive α-receptors and exogenous peripheral α_2-receptor activation may lead to enhanced vasoconstriction in this setting. Central α_2-receptor activation, on the contrary, inhibits sympathetic outflow and tends to be depressor. The latter effect is minimal in patients with autonomic failure who have little or no remaining sympathetic activity. Indeed, clonidine is pressor in patients with severe degrees of autonomic failure (Robertson *et al.*, 1983; Onrot *et al.*, 1987b). This pressor effect is almost certainly due to peripheral α_2 vasoconstriction because it is not blocked by prazosin, an α_1-antagonist (Onrot *et al.*, 1987b). The use of clonidine has been limited by sedation and dry mouth.

In patients with milder disease, clonidine's central sympathoinhibitory response predominates over the vasoconstrictor peripheral α_2 stimulation and blood pressure may fall, as seen in normal subjects. However, patients with milder degrees of autonomic failure may benefit from their central sympathetic outflow being stimulated. Thus, the α_2-antagonist yohimbine may be useful in mild autonomic failure because of its ability to augment sympathetic outflow, and the α_2-agonist clonidine may be useful in severe autonomic failure because of its ability to vasoconstrict via hypersensitive peripheral α_2-receptors.

Amphetamine

Amphetamine and methylphenidate may act by reducing the reuptake of noradrenaline as well as by direct α-receptor activation. The effectiveness of these drugs in the treatment of orthostatic hypotension has not been well-studied, although they have been used occasionally (Korns and Randall, 1939; Sellers, 1969; Thomas *et al.*, 1981).

OTHER VASODEPRESSOR AGENTS

Dihydroergotamine

This ergot alkaloid is intriguing because it has relatively selective venoconstrictor effects (Mellander and Nordenfelt, 1970) and can, therefore, "mobilize" blood volume from the capacitance vessels towards the central compartment, increasing venous return (Nordenfelt and Mellander, 1972). Its major metabolites also have active venoconstrictor effects (Aellig, 1984). Low oral bioavailability has limited the effectiveness of this agent (Bobik *et al.*, 1981).

Dihydroergotamine has proven useful in several trials at doses of 10-40 mg daily (Nordenfelt and Mellander, 1972; Bevegard, Castenfors and Lindblad, 1974; Tikholov, 1976; Jennings, Ester and Holmes, 1979; Benowitz *et al.*, 1980). A trial of inhaled dihydroergotamine, in order to improve bioavailability, showed promise for this agent (Biaggioni *et al.*, 1990). Caffeine 250 mg p.o. in combination with subcutaneous dihydroergotamine led to improved symptoms and prevented postprandial hypotension (Hoeldtke *et al.*, 1986). Side effects such as angina or myocardial infarction secondary to the raised pressures and vasoconstriction actions may occur (Benedict and Robertson, 1979). Ergotamine, useful in treating migraines, has better bioavailability than dihydroergotamine and has been tried (Chobanian et al., 1983).

Caffeine

This commonly encountered methylxanthine raises blood pressure in normal subjects (Robertson et al., 1978). In autonomic failure, a 250 mg dose has been used alone (Onrot et al., 1985c; Lenders et al., 1988) or in combination with dihydroergotamine (Hoeldtke et al., 1986) in an attempt to prevent postprandial hypotension. One or two cups of coffee (100-200 mg), given once a day with breakfast, may be the most effective regimen in order to avoid tolerance and to deliver the drug when it is most needed, i.e. after the morning meal when the symptoms are worst.

Prostaglandin inhibitors

Indomethacin can raise the mean pressure by about 20 mm Hg in most (Kochar, Itskowitz and Albers, 1979; Goldberg, Robertson and FitzGerald, 1985), but not all (Crook, Robertson and Whorton, 1981), patients with severe autonomic failure. Indomethacin can increase the sensivity to endogenous pressor hormones such as noradrenaline and angiotensin II (Davies et al., 1980). Alternative non-steroidal anti-inflammatory drugs have been used to reduce side effects.

Further side effects of indomethacin and other non-steroidal anti-inflammatory drugs can be encountered in addition to supine hypertension and most elderly people should receive concomitant anti-ulcer drugs to prevent gastritis. Response to these agents is unpredictable.

β-BLOCKERS

β-blockers have been shown to elevate peripheral resistance, presumably by blockade of β_2 vasodilator receptors. β_2 hypersensitivity exceeds β_1 hypersensitivity in these patients (Robertson et al., 1984). Thus, in these patients, the pressor effects resulting from blockade of vasodilator β_2-receptors predominate over the depressor effects resulting from cardiac β_1 blockade. The vasoconstrictor response to tilt can be augmented by β blockade (Skagen, 1983).

Clinical studies have shown an increased peripheral resistance and attenuation of postural hypotension (Chobanian et al., 1977; Brevetti et al., 1981). Negative inotropic and chronotropic actions can limit their usefulness, and side effects of β-blockers are common in the elderly population. Pindolol, the partial agonist β-blocker has been used (Man in't Veld and Schalekamp, 1981) and may have β-agonist effects in the setting of low basal sympathetic tone. Other studies have found this agent less useful (Davies et al., 1981). In general, as with non-steroidal anti-inflammatory drugs, the response is unpredictable in individual patients.

DOPAMINE ANTAGONISTS

The occasional patient may have excessive levels of dopamine that contribute to vasodilatation. In these patients, dopamine antagonists such as metoclopramide may be of benefit (Kuchel *et al.*, 1980, 1985). Domperidone has been used with success in diabetic autonomic neuropathy (Lopes de Faria *et al.*, 1988). Many patients, however, do not respond to these agents.

SOMATOSTATIN ANALOGUES

Food leads to stimulation of many gastrointestinal hormones that can be vasodilatory and act on the splanchnic system to decrease venous return or on the peripheral circulation to reduce arteriolar resistance (Mathias *et al.*, 1989). Somatostatin and somatostatin analogues can attenuate the secretion of these vasodilator peptides (Hoeldtke, Boden and O'Dorisio, 1986; Mathias *et al.*, 1988).

Octreotide (SMS-201-995) is a somatostatin analogue that reduces falls in postprandial pressure and attenuates symptoms when given subcutaneously to patients with autonomic failure (Hoeldtke, Boden and O'Dorisio, 1986; Jansen *et al.*, 1989). Octreotide can be infused subcutaneously using a pump controlled by the patient. Thus, this agent can be used when needed postprandially and turned off when the patient is supine. Surprisingly, octreotide may be of use in the fasting state as well (Hoeldtke and Israel, 1989). Its pressor effect is independent of the sympathetic nervous system in multiple system atrophy, but octreotide may suppress the adrenergic response to ingestion of glucose in idiopathic orthostatic hypotension (Hoeldtke *et al.*, 1989). Side effects such as diarrhoea and malabsorption are commonly encountered.

VASOPRESSIN ANALOGUES

Occasionally vasopressin analogues have been administered nasally. Patients can experience long-term improvements with a nasal spray of desmopressin (Mathias *et al.*, 1986).

REFERENCES

Aellig, W.H. (1984) Investigation of the venoconstrictor effect of 8'hydroxydiergotamine, the main metabolite of dihydroergotamine in man. *Eur J. Clin. Pharmacol.*, **26**, 239–242.

Ahmad, R.A.S. and Watson, R.D.S. (1990) Treatment of postural hypotension. *Drugs*, **39**, 75–85.

Bannister, R., DaCosta, D.F., McIntosh, C. and Mathias, C.J. (1984) Oral glucose but not oral xylose substantially lowers blood pressure in autonomic failure. *J. Physiol. (Lond.)*, **358**, 123.

Barnett, A.J. and Wagner, G.R. (1958) Severe orthostatic hypotension: case report and description of response to sympathicomimetic drugs. *Am. Heart J.*, **56**, 412–424.

Belzberg, A.S., Ong, M., Onrot, J. and Rangno, R.E. (1993) Mechanism of sympathetic orthostatic hypotension in primary autonomic failure. The role of blood distribution and effects of therapy. Manuscript in preparation.

Benedict, C.R. and Robertson, D. (1979) Angina pectoris and sudden death in the absence of atherosclerosis following ergotamine therapy for migraine. *Am. J. Med.*, **67**, 177–178.

Benowitz, N.L., Byrd, R., Schambela, M., Rosenberg, J. and Roizen, M.F. (1980) Dihydroergotamine treatment for orthostatic hypotension from Vacor rodenticide. *Ann. Int. Med.*, **92**, 387–388.

Bevegard, S., Castenfors, J. and Lindblad, L.E. (1974) Haemodynamic effects of dihydroergotamine in patients with postural hypotension. *Acta Med. Scand.*, **196**, 437–477.

Biaggioni, I. and Robertson, D. (1987) Endogenous restoration of norepinephrine by precursor therapy in dopamine-beta-hydroxylase deficiency. *Lancet*, **2**, 1170–1172.

Biaggioni, I., Robertson, R.M. and Robertson, D. (1993) Manipulation of norepinephrine metabolism with yohimbine in the treatment of autonomic failure. *J. Clin. Pharmacol.*, in press.

Biaggioni, I., Onrot, J., Parrish, C.K. and Robertson, D. (1987) The potent pressor effect of phenylpropanolamine in patients with autonomic impairment. *J. Am. Med. Assoc.*, **258**, 236–239.

Biaggioni, I., Zygmunt, D., Haile, V. and Robertson, D. (1990) Pressor effect of inhaled ergotamine in orthostatic hypotension. *Am. J. Cardiol.*, **65**, 89–92

Bickelmann, A.G., Lippschutz, E.J. and Brunjes, C.F. (1961) Hemodynamics of idiopathic orthostatic hypotension. *Am. J. Med.*, **30**, 26–38.

Bliddal, J. and Nielsen, I. (1970) Renin, aldosterone, and electrolytes in idiopathic orthostatic hypotension. *Dan. Med. Bull.*, **17**, 153–157.

Blomqvist, C.G. (1987) Orthostatic hypotension. *Hypertension*, **8**, 772–730.

Bobik, A., Jennings, G., Skews, H., Esler, M. and McLean, A. (1981) Low oral bioavailability of dihydroergotamine and first-pass extraction in patients with orthostatic hypotension. *Clin. Pharmacol. Ther.*, **30**, 673–679.

Brevetti, G., Chiariello, M., Gludice, P., De Michel, G., Mansi D. and Campanella, G. (1981) Effective treatment of orthostatic hypotension by propranolol in the Shy-Drager syndrome. *Am. Heart J.*, **102**, 938–941.

Brown, R.T., Polinsky, R.J. and Baucom, C.E. (1989) Euglycemic insulin-induced hypotension in autonomic failure. *Clin. Neuropharmacol.*, **12**, 227–231.

Campbell, I.W., Ewing, D.J. and Clarke, B.F. (1976) Therapeutic experience with fludrocortisone in diabetic postural hypotension. *Br. Med. J.*, **1**, 872–874.

Chobanian, A.V., Volicer, L., Liang, C.S., Kershaw, G. and Tifft, C. (1977) Use of propranolol in the treatment of idiopathic orthostatic hypotension. *Trans. Assoc. Am. Phys.*, **90**, 324–334.

Chobanian, A.V., Volicer, L., Tifft, C.P., Gavras, H., Liang, C.S. and Faxon, D. (1979) Mineralocorticoid-induced hypertension in patients with orthostatic hypotension. *N. Engl. J. Med.*, **301**, 68–73.

Chobanian, A.V., Tifft, C.P., Faxon, D.P., Creager, M.L.A. and Sackel, H. (1983) Treatment of orthostatic hypotension with ergotamine. *Circulation*, **67**, 602–609.

Crook, J.E., Robertson, D. and Whorton, A.R. (1981) Prostaglandin suppression: inability to correct severe idiopathic orthostatic hypotension. *South Med. J.*, **73**, 318–320.

Davidson, C., Smith, P. and Morgan, D.B. (1976) Diurnal pattern of water and electrolyte excretion and body weight in idiopathic orthostatic hypotension. The effect of three treatments. *Am. J. Med.*, **61**, 709–715.

Davies, B., Bannister, R. and Sever, P. (1978) Pressor amines and monoamine oxidase inhibitors for treatment of postural hypotension in autonomic failure: limitations and hazards. *Lancet*, **1**, 172–175.

Davies, B., Bannister, R., Sever, P. and Wilcox, E. (1979) The pressor actions of noradrenaline, angiotensin II and saralasin in chronic autonomic failure treated with fludrocortisone. *Br. J. Clin. Pharmacol.*, **8**, 253–260.

Davies, B., Bannister, R., Hensby, C. and Sever, P.S. (1980) The pressor actions of noradrenaline and angiotensin II in chronic autonomic failure treated with indomethacin. *Br. J. Clin. Pharmacol.*, **10**, 223–229.

Davies, B., Bannister, R., Mathias, C. and Sever, P. (1981) Pindolol in postural hypotension: the case for caution (letter). *Lancet*, **2**, 982–983.

De Mey, J. and Vanhoutte, P.M. (1981) Uneven distribution of postjunctional alpha-1 and alpha-2-like adrenoreceptors in canine arterial and venous smooth muscle. *Circ. Res.*, **48**, 875–884.

Diamond, M.A., Murray, R.H. and Schmid, P.G. (1970) Idiopathic postural hypotension; physiologic observations and report of a new mode of therapy. *J. Clin. Invest.*, **49**, 1341–1348.

Elliott, H.L. and Reid, J.L. (1983) Evidence for postjunctional vascular alpha-2 adrenoreceptors in peripheral vascular regulation in man. *J. Clin. Sci.*, **65**, 237–241.

Ghrist, D.G. and Brown, G.E. (1928) Postural hypotension with syncope: its successful treatment with ephedrin. *Am. J. Med. Sci.*, **175**, 336–349.

Gill, J.R. Jr. (1979) Neural control of renal tubular sodium reabsorption. *Nephron.*, **23**, 116–118.

Goldberg, M.R. and Robertson, D. (1983) Yohimbine: a pharmacological probe for study of the α_2-adrenoreptor. *Pharmacol. Rev.*, **35**, 143–180.

Goldberg, M.R., Hollister, A.S. and Robertson, D. (1983) Influence of yohimbine on blood pressure, autonomic reflexes and plasma catecholamines in humans. *Hypertension*, **5**, 772–778.

Goldberg, M.R., Robertson, D. and FitzGerald, G.A. (1985) Prostacyclin biosynthesis and platelet function in autonomic dysfunction. *Neurology*, **35**, 120–123.

Gordon, R.D., Kuchel, O., Liddle, G.W. and Island, D.P. (1967) Role of the sympathetic nervous system in regulating renin and aldosterone production in man. *J. Clin. Invest.*, **46**, 599–605.

Greenway, C.V. (1983) Role of splanchnic venous system in overall cardiovascular homeostasis. *Fed. Proc.*, **42**, 1678–1684.

Hickler, R.B., Thompson, G.R., Fox, L.M. and Hamlin, J.T. (1959) Successful treatment of orthostatic hypotension with 9-alpha-flurohydrocortisone. *N. Engl. J. Med.*, **261**, 788–791.

Hoeldtke, R.D. and Israel, B.C. (1989) Treatment of orthostatic hypotension with octreotide. *J. Clin. Endocrinol. Metab.*, **68**, 1051–1059.

Hoeldtke, R.D., Boden, G. and O'Dorisio, T.M. (1986) Treatment of postprandial hypotension with a somatostatin analogue. *Am. J. Med.*, **81**, 83–87.

Hoeldtke, R.D., Cavanaugh, S.T., Hughes, J.D. and Polansky, M. (1986) Treatment of orthostatic hypotension with dihydroergotamine and caffeine. *Ann. Int. Med.*, **105**, 168–173.

Hoeldtke, R.D., Dworkin, G.E., Gaspar, S.R., Israel, B.C. and Boden, G. (1989) Effect of the somatostatin analogue SMS-201-995 on the adrenergic response to glucose ingestion in patients with postprandial hypotension. *Am. J. Med.*, **86**, 673–677

Hui, K.K.P. and Conolly, M.E. (1981) Increased numbers of beta receptors in orthostatic hypotension due to autonomic dysfunction. *N. Engl. J. Med.*, **304**, 1473–1476.

Ibrahim, M.M., Tarazi, R.C., Dustan, H.P. and Bravo, E.L. (1974) Idiopathic orthostatic hypotension: circulatory dynamics in chronic autonomic insufficiency. *Am. J. Cardiol.*, **34**, 288–294.

Ibrahim, M.M., Tarazi, R.C., Shafer, W.H., Bravo, E.L. and Dustan, H.P. (1979) Unusual tyramine responsiveness in idiopathic orthostatic hypotension. *Med. J. Cairo. Univ.*, **47**, 49–55.

Jansen, R.W.M.M., Peeters, T.L., Lenders, J.W.M., van Lier H.J.J., V't Laar, A. and Hoefnagels, W.H.L. (1989) Somatostatin analog octreotide (SMS 201-995) prevents the decrease in blood pressure after oral glucose loading in the elderly. *J. Clin. Endocrinol. Metab.*, **68**, 752–756.

Jennings, G., Esler, M. and Holmes, R. (1979) Treatment of orthostatic hypotension with dihydroergotamine. *Br. Med. J.*, **2**, 307.

Johnson, J.R., Rowell, L.B., Niederberger, M. and Eisman, M.M. (1974) Human splanchnic and forearm vasoconstrictor responses to reduction of right atrial and aortic pressures. *Circ. Res.*, **34**, 515–524.

Jones, D. and Reid, J.L. (1980) Volume expansion and vasodilators in the treatment of idiopathic postural hypotension. *Postgrad. Med.*, **56**, 234–235.

Kafka, M.S., Polinsky, R.J., Williams, A., Kopin, I.J., Lake, C.R., Ebert, M.H. and Tokola, N.S. (1984) Alpha-adrenergic receptors in orthostatic hypotension syndromes. *Neurology*, **34**, 1121.

Kaufmann, H., Brannan, T., Krakoff, L., Yahr, M.D. and Mandeli, J. (1988) Treatment of orthostatic hypotension due to autonomic failure with a peripheral alpha-adrenergic agonist (midodrine). *Neurology*, **38**, 951–956.

Kochar, M.S., Itskowitz, H.D. and Albers, J.W. (1979) Treatment of orthostatic hypotension with indomethacin. *Am. Heart J.*, **98**, 271–280.

Korns, H.M. and Randall, W.L. (1939) Benzedrine and Paredrine in the treatment of orthostatic hypotension with supplementary case report. *Ann. Intern. Med.*, **12**, 253–255.

Kronenberg, M.W., Forman, M.B., Onrot, J. and Robertson, D. (1990) Enhanced left ventricular contractility in autonomic failure: assessment using pressure-volume relations. *J. Am. Coll. Cardiol.*, **15**, 1334–1342.

Kuchel, O., Buu, N.T., Gutkowska, J. and Genest, J. (1980) Treatment of severe orthostatic hypotension by metoclopramide. *Ann. Int. Med.*, **93**, 841–843.

Kuchel, O., Buu, N.T., Hamet, P., LaRochelle, P., Gutkowska, J., Schiffrin, E.L., *et al.* (1985) Orthostatic hypotension: a posture-induced hyperdopaminergic state. *Am. J. Med. Sci.*, **289**, 3–11.

LeCrubier, Y., Puech, A.J. and Des Lauriers, A. (1981) Favourable effects of yohimbine on clomipramine-induced orthostatic hypotension: a double-blind study. *Br J. Clin. Pharmacol.*, **12**, 90–93.

Lenders, J.W.M., Morre, H.L.C., Smits, P. and Thien, T. (1988) The effects of caffeine on the postprandial fall of blood pressure in the elderly. *Age and Ageing*, **17**, 236–240.

Lewis, R.K., Hazelrig, C.G., Fricke, F.J. and Russell, R.O. (1972) Therapy of idiopathic postural hypotension. *Arch. Intern. Med.*, **129**, 943–949.

Lipsitz, L.A. (1989) Orthostatic hypotension in the elderly. *N. Engl. J. Med.*, **321**, 952–957.

Lopes de Faria, S.R.G.F., Zanella, M.T., Amoriolo, A., Ribiero, A.B. and Chacra, A.R. (1988) Peripheral dopaminergic blockade for the treatment of diabetic orthostatic hypotension. *Clin. Pharmacol. Ther.*, **44**, 670–674.

Low, P.A., Thomas, J.E. and Dyck, P.J. (1978) The splanchnic autonomic outflow in Shy-Drager syndrome and idiopathic orthostatic hypotension. *Ann. Neurol.*, **4**, 511–514.

MacLean, A.R. and Allen, B.V. (1940) Orthostatic hypotension and orthostatic tachycardia. Treatment with the "head-up" bed. *JAMA*, **115**, 2162–2167.

MacLean, A.R., Allen, E.V. and Magath, T.B. (1944) Orthostatic tachycardia and orthostatic hypotension: defects in the return of venous blood to the heart. *Am. Heart J.*, **24**, 145–163.

Magrini, F., Ibrahim, M.M. and Tarazi, R.C. (1976) Abnormalities of supine hemodynamics in idiopathic orthostatic hypotension. *Cardiology*, **61**(Suppl I), 125–135.

Man in't Veld, A.J. and Schalekamp, M.A.D.H. (1981) Pindolol acts as a beta-adrenoceptor agonist in orthostatic hyptension: therapeutic implications. *Brit. Med. J.*, **282**, 929–931.

Man in't Veld, A.J., Boomsma, F.V.D., Meiracker, A.H. and Schalekamp, M.A.D.H. (1987) Effect of an unnatural noradrenaline precursor on sympathetic control and orthostatic hypotension in dopamine-beta-hydroyxlase deficiency. *Lancet.*, **2**, 1172–1175.

Mann, S., Altman, D.G., Raftery, E.B. and Bannister, R. (1983) Circadian variation of blood pressure in autonomic failure. *Circulation*, **68**, 477–483.

Mason, D.T., Kopin, I.J. and Braunwald, E. (1966) Abnormalities in reflex control of the circulation in familial dysautonomia. Effects of changes in posture on venous and arterial constriction in normal subjects and in patients with dysautonomia. *Am. J. Med.*, **41**, 898–909.

Mathias, C.J., Bannieser, R., Ardell, L. and Fenten, P. (1969) An assessment of various methods of treatment of idiopathic orthostatic hypotension. *Q. J. Med.*, **38**, 277–395.

Mathias, C.J., Fosbraey, P., daCosta, D.F., Thomely, A. and Bannister, R. (1986) Desmopressin reduces nocturnal polyuria, reverses overnight weight loss and improves morning postural hypotension in autonomic failure. *Br. Med. J.*, **293**, 353–354.

Mahias, C.J., Raimbach, S.J., Cortelli, P., Kooner, J.S. and Bannister, R. (1988) The somatostatin analogue SMS 201-995 inhibits peptide release and prevents glucose-induced hypotension in autonomic failure. *J. Neurol.*, **235**, S74–S75.

Mathias, C.J., daCosta, D.F., Fosbraey, P., Bannister, R., Wood, S.M., Bloom, S.R., *et al.* (1989) Cardiovascular, biochemical and hormonal changes during food-induced hypotension in chronic autonomic failure. *J. Neurol. Sci.*, **94**, 255–269.

Mellander, S. and Nordenfelt, I. (1970) Comparative effects of dihydroergotamine and noradrenaline on resistance, exchange and capacitance functions in the peripheral circulation. *Clin. Sci.*, **39**, 183.

Montastruc, J.L., Puech, A.J., Clanet, M., Guiraud-Chaumeil, B. and Rascol, A. (1981) La yohimbine dans le traitment de la maladie de Parkinson: resultats preliminaires. *Nov Presse Medicale*, **10**, 1331–1332.

Morales, A., Surridge, D.H.C., Marshall, P.G. and Fenemore, J. (1983) Nonhumoral pharmacological treatment of organic impotence. *J. Urol.*, **128**, 45–47.

Nanda, R.N., Johnson, R.H. and Keogh, H.J. (1976) Treatment of neurogenic orthostatic hypotension with a monoamine oxidase inhibitor and tyramine. *Lancet*, **2**, 1164–1167.

Niarchos, A.P., Magrini, F., Tarazi, R.C. and Bravo, E.L. (1966) Mechanism of spontaneous supine blood pressure variations in chronic autonomic insufficiency. *Am. J. Med.*, **65**, 547–552

Nordenfelt, I. and Mellander, S. (1972) Central haemodynamic effects of dihydroergotamine in patients with orthostatic hypotension. *Acta Med. Scand.*, **191**, 115–120.

Onrot, J., Hollister, A.S., Biaggioni, I. and Robertson, D. (1985a) Upright plasma catecholamine and renin responses in orthostatic hypotension. *Clin. Invest. Med.*, **8**, A129.

Onrot, J., Foreman, M., James, J., Robertson, D. and Kronenberg, M. (1985b) Hemodynamic responses to tilt in autonomic failure (abst). *Clin. Invest. Med.*, **8**, A64.

Onrot, J., Goldberg, M.R., Biaggioni, I., Hollister, A.S., Kincaid, D. and Robertson, D. (1985c) Hemodynamic and humural effects of caffeine in human autonomic failure: therapeutic implications for postprandial hypotension. *New Engl. J. Med.*, **313**, 549–554.

Onrot, J., Goldberg, M.R., Hollister, A.S., Biaggioni, I., Robertson, R.M. and Robertson, D. (1986) Management of chronic orthostatic hypotension. *Am. J. Med.*, **80**, 454–464.

Onrot, J., Goldberg, M.R., Biaggioni, I., Kincaid, D., Hollister, A.S. and Robertson, D. (1987a) Oral yohimbine in autonomic failure. *Neurology*, **37**, 215–220.

Onrot, J., Goldberg, M.R., Biaggioni, I., Hollister, A.S., Kincaid, D. and Robertson, D. (1987b) Postjunctional vascular smooth muscle alpha-2 adrenoreceptors in human autonomic failure. *Clin. Invest. Med.*, **10**, 26–31.

Page, E.B., Hickam, J.B., Sieker, H.O., McIntosh, H.D. and Pryor, W.W. (1955) Reflex venomotor activity in normal persons and in patients with postural hypotension. *Circulation*, **11**, 262–270.

Parks, J., Sandison, A.G., Skinner, S.L. and Wheland, R.F. (1961) Sympathomimetic drugs in orthostatic hypotension. *Lancet*, **1**, 1133–1136.

Qamar, M.I. and Read, A.E. (1986) The effect of feeding and sham feeding on the superior mesenteric artery blood flow in man. *J. Physiol. (Lond.)*, **377**, 59P.

Raab, W., Humphreys, R.J. and Lepeschkin, E. (1950) Potentiation of pressor effects of norepinephrine and epinephrine in man by desoxycorticosterone acetate. *J. Clin. Invest.*, **29**, 1397–1404.

Robertson, D., Frolich, J.C., Carr. R.K., Shand, D.G. and Oates, J.A. (1978) Effect of caffeine on plasma renin activity, catecholamines, and blood pressure. *N. Engl. J. Med.*, **298**, 181–186.

Robertson, D. (1979) Contraindication to the use of ocular phenylephrine in idiopathic orthostatic hypotension. *Am. J. Ophthal.*, **87**, 819–822.

Robertson, D., Wade, D. and Robertson, R.M. (1981) Postprandial alterations in cardiovascular hemodynamics in autonomic dysfunctional states. *Am. J. Cardiol.*, **48**, 1048–1052.

Robertson, D., Goldberg, M.R., Hollister, A.S., Wade, D. and Robertson, R.M. (1983) Clonidine raises blood pressure in idiopathic orthostatic hypotension. *Am. J. Med.*, **74**, 193–199.

Robertson, D., Hollister, A.S., Carey, E.L., Tung, C.S., Goldberg, M.R. and Robertson, R.M. (1984) Increased vascular beta2 adrenoreceptor responsiveness in autonomic dysfunction. *J. Am. Coll. Cardiol.*, **3**, 850–856.

Rohan-Chabot P. de, Said, G. and Ziza, J.M. (1980) Hypotension orthostatique idiopathique essential syndrome de Shy-Drager. *Ann. Med. Interne.*, **131**, 483–488.

Samueloff, S.L., Browse, N.L. and Shepherd, J.T. (1966) Response of capacity vessels in human limbs to head-up tilt and suction on lower body. *J. Appl. Physiol.*, **21**, 47–54.

Schatz, I.J., Miller, M.J. and Frame, B. (1976) Corticosteroids in the management of orthostatic hypotension. *Cardiology*, **61**(Suppl I), 280–289.

Schirger, A., Sheps, S.G., Thomas, J.E. and Fealy, R.D. (1981) Midodrine. A new agent in the management of idiopathic orthostatic hypotension and Shy-Drager syndrome. *Mayo Clin. Proc.*, **56**, 429–433.

Schmid, P.G., Eckstein, J.W. and Abboud, F.M. (1966) Effect of 9-alpha-fluorohydrocortisone forearm vascular responses to norepinephrine. *Circulation*, **34**, 620–626.

Seibyl, J.P., Krystal, J.H., Price, L.H. and Charney, D.S. (1989) Use of yohimbine to counteract nortriptyline-induced orthostatic hypotension. *J. Clin. Psychopharmacol.*, **9**, 67–68.

Sellers, R.H. (1969) Idiopathic orthostatic hypotension: report of successful treatment with a new form of therapy. *Am. J. Cardiol.*, **23**, 838–844.

Skagen, K. (1983) Augmented vasoconstrictor response to head up tilt in peripiheral tissues during beta-receptor blockade. *Eur. J. Clin. Pharmacol.*, **25**, 3–7.

Slaton, P.E. and Biglieri, E.G. (1967) Reduced aldosterone excretion in patients with autonomic insufficiency. *J. Clin. Endo.*, **27**, 37–45.

Smith, A.D. (1973) Mechanisms involved in the release of noradrenaline from sympathetic nerves. *Br. Med. Bull.*, **29**, 123–129.

Stead, E.A. Jr. and Ebert, R.V. (1941) Postural hypotension. A disease of the sympathetic nervous system. *Arch. Intern. Med.*, **67**, 546–562.

Thomas, J.E., Schirger, A., Fealey, R.D. and Sheps, S.G. (1981) Orthostatic hypotension. *Mayo Clin. Proc.*, **56**, 117–125.

Tikholov, K. (1976) Lechenie na ortostatichnata khipotoniia s dikhidrirani ergotaminovi proizvodni. *Vatr. Bol.*, **15**, 82–86.

Trendelenburg, V., Muskus, A., Fleming, W.W., *et al.* (1962) Effects of cocaine, denervation, and decentralization on the responses of the nictitating membrane to various sympathomimetic amines. *J. Pharmacol. Exp. Ther.*, **183**, 181–193.

Wilcox, C.S., Aminoff, J. and Slater, J.D.H. (1977) Sodium homeostasis in patients with autonomic failure. *Clin. Sci. Mol. Med.*, **52**, 321–328.

Wilcox, C.S., Puritz, R., Lightman, S.L., Bannister. R. and Aminoff, M.J. (1984) Plasma volume regulation in patients with progressive autonomic failure during changes in salt intake and posture. *J. Lab. Clin. Med.*, **104**, 331–339

Index

437

Printed and bound by CPI Group (UK) Ltd, Croydon, CR0 4YY

24/10/2024

01778291-0006